BOUNDARY LAYER STRUCTURES

Boundary Layer Structure

Modeling and Application to Air Pollution and Wind Energy

Edited by

HADASSAH KAPLAN and NATHAN DINAR

Israel Institute for Biological Research, Ness-Ziona, Israel

Reprinted from

Boundary-Layer Meteorology, Vol. 30, Nos. 1–4

D. Reidel Publishing Company

Dordrecht / Boston

ISBN-13: 978-94-009-6516-4 e-ISBN-13: 978-94-009-6514-0
DOI: 10.1007/978-94-009-6514-0

TABLE OF CONTENTS

PART III / WIND DISTRIBUTION IN COMPLEX TERRAIN

PART IV / DISPERSION OF POLLUTANTS IN THE ATMOSPHERE AND ITS DEPOSITION ON THE SURFACE

PART V / ROUND-TABLE DISCUSSION

ACKNOWLEDGEMENTS

The Organizing Committee of the 29th OHOLO Conference gratefully acknowledge the generous support of the following organizations:

The British Council
Tel-Aviv, Israel

Ministry of Science & Development
National Council for Research & Development
Jerusalem, Israel

Weizmann Institute of Science
Rehovot, Israel

The committee also wishes to thank:

Ms. Ilana Turner for her tireless assistance in the organization of the conference,

Dr. Steven R. Hanna for his skillful chairing of the Round Table Discussion and for the excellent editing of the corresponding text,

Dr. Yehuda Alexander and Ms. Judith Lempert for their invaluable help in the preparation of the questions and comments section of each lecture and of the Round Table Discussion,

And Ms. Nita Ben-David for her careful typing.

We would especially like to thank Dr. R.E. Munn for his invaluable help in editing this book.

OHOLO BIOLOGICAL CONFERENCES

organized and sponsored by the

ISRAEL INSTITUTE FOR BIOLOGICAL RESEARCH

Permanent Committee:

S. Cohen, M. Feldman, N. Grossowicz, I. Hertman,
A. Keynan, A. Kohn, M. Sela, G.A. Simon

Organizing Committee:
(29th Conference)

H. Kaplan, N. Dinar (Chairpersons)
D. Skibin, Y. Mahrer, Y. Tokatly

Secretariat & Technical Organization:

I. Turner, N. Ben-David, R. Pniel,
M. Navon, M. Ben-Yeffet

LIST OF PARTICIPANTS

ALEXANDER, Y.,Israel Institute for Biological Research,Ness-Ziona,Israel
ALMOG, H., Israel Electric Corporation, Haifa, Israel
ALPERT, P., Tel-Aviv University, Ramat-Aviv, Israel
ARIEL, I.,Israel Institute for Biological Research,Ness-Ziona,Israel
ARONSON, E.,Israel Institute for Biological Research,Ness-Ziona,Israel
ARYA, S.P.S., North Carolina State University, Raleigh, NC, U.S.A.
ASCULAI, E., Nuclear Research Center - Negev, Beer Sheba, Israel
AXENFELD, F., Dornier System GmbH, Friedrichshafen, Germany
AZOR, M., Israel Institute for Biological Research,Ness-Ziona,Israel
BALMOR, Y., Israel Electric Corporation, Haifa, Israel
BARKAN, Y.,Israel Institute for Biological Research,Ness-Ziona,Israel
BAUMGARTEN, Y., Tel-Aviv University, Ramat-Aviv, Israel
BERKOFSKY, L., Ben-Gurion University, Beer Sheba, Israel
BERKOWICZ, R.,National Agency of Environmental Protection,Roskilde,Denmark
BREIMAN, Y., Tel-Aviv University, Ramat-Aviv, Israel
CERMAK, J.E., Colorado State University, Fort Collins, CO, U.S.A.
COVALIU, Z., Atomic Energy Commission, Tel-Aviv, Israel
DAR, T.,Israel Institute for Biological Research,Ness-Ziona, Israel
DAYAN, N.,Israel Institute for Biological Research,Ness-Ziona,Israel
DEVIR, S.E.,Israel Institute for Biological Research,Ness-Ziona,Israel
DICKERSON, M.H.,Lawrence Livermore National Lab.,Livermore, CA, U.S.A.
DINAR, N.,Israel Institute for Biological Research, Ness-Ziona,Israel
DORON, E., Nuclear Research Center - Negev, Beer Sheba, Israel
EGAN, B.A.,Environmental Research & Technology,Inc.,Concord, MA, U.S.A.
ENDLICH, R.M., SRI International, Menlo Park, CA, U.S.A.
ENGER, L., Uppsala University, Uppsala, Sweden
EPPEL, A., Tel-Aviv, Israel
EYAL, E.,Israel Institute for Biological Research,Ness-Ziona,Israel
GAVZE, E., Hebrew University, Jerusalem, Israel
GIFFORD, F.A., Oak Ridge, TN, U.S.A.
GIVATI, R.,Israel Institute for Biological Research,Ness-Ziona,Israel
GLASER, E.,Assn of Towns for Environmental Protection,Hadera,Israel
GLUCK, P., Bar-Ilan University, Ramat-Gan, Israel
GOLDENBERG, M., Hebrew University, Jerusalem, Israel
GOLOMBEK, A.,Israel Institute for Biological Research,Ness-Ziona,Israel
GRABER, M., Ministry of Interior, Jerusalem, Israel
GUTEMARK, E., Tel-Aviv University, Ramat-Aviv, Israel
GUTMAN, L., Ben-Gurion University, Beer Sheba, Israel
HANNA, S.R.,Environmental Research & Technology,Inc.,Concord, MA, U.S.A.
HASSID, S., Technion, Haifa, Israel
HAZAN, N., Ministry of Defence, Israel
HÖGSTRÖM, U., Uppsala University, Uppsala, Sweden
HORST, T.W., Pacific Northwest Laboratory, Richland, WA, U.S.A.
HUNT, J.C.R., University of Cambridge, Cambridge, U.K.
ISRAELI, E.,Israel Institute for Biological Research,Ness-Ziona,Israel
JENSEN, A.B., Danish Boiler Owners Assn, Søborg, Denmark

JURGRAU, M., Ben-Gurion University, Beer Sheba, Israel
KAPLAN, H.,Israel Institute for Biological Research,Ness-Ziona,Israel
KAUFMANN, R., Dornier System GmbH, Friedrichshafen, Germany
KEIDAR, M.,Israel Institute for Biological Research,Ness-Ziona,Israel
KOLB, H.,Institut fuer Meteorologie & Geophysik der Universitaet Wien, Vienna, Austria
KOVACS, S., Israel Electric Corporation, Haifa, Israel
KRUSHKAL, E., Hebrew University, Jerusalem, Israel
LAHAT, A.,Israel Institute for Biological Research,Ness-Ziona,Israel
LAKRETZ, Y., Assn of Towns for Environmental Protection, Hadera,Israel
LEMPERT, J.,Israel Institute for Biological Research,Ness-Ziona,Israel
LERER, A., Israel Defence Forces
LEVANON, I., Soreq Nuclear Research Center, Yavneh, Israel
LIFSCHITZ, B., Assn of Towns for Environmental Protection,Hadera,Israel
LUEHRING,P.G.,GEOMET-Umweltberatung & Regionalplannung GmbH,Berlin,Germany
LUNINSKY, Y., Tel-Aviv University, Ramat-Aviv, Israel
LUMLEY, J.L., Cornell University, Ithaca, NY, U.S.A.
MANES, A., Israel Meteorological Service, Beit Dagan, Israel
MEGED, I.,Israel Institute for Biological Research,Ness-Ziona,Israel
MOLOD, A., Ben-Gurion University, Beer Sheba, Israel
NIEUWSTADT, F.M.T., Royal Netherlands Meteorological Institute, AE de Bilt, The Netherlands
PECHINGER, U.,Zentralanstalt fuer Meteorologie & Geodynamik,Vienna,Austria
PERNICK, A., Soreq Nuclear Research Center, Yavneh, Israel
PICK, K., Tel-Aviv University, Ramat-Aviv, Israel
POREH, M., Technion, Haifa, Israel
RAICHMAN, R.,Israel Institute for Biological Research,Ness-Ziona,Israel
RASHTY, E., Ministry of Defence, Israel
RAVIV,R., Israel Institute for Biological Research,Ness-Ziona,Israel
RINDSBERGER, M., Israel Meteorological Service, Beit Dagan, Israel
SCHERER, B.,Institut fuer Geophysikalische Wissenschaften,Berlin,Germany
SETTER, I., Israel Meteorological Service, Beit Dagan, Israel
SHINE, R., Israel Defence Forces
SIMON, G.A.,Israel Institute for Biological Research,Ness-Ziona,Israel
SIRKES, Z., Weizmann Institute of Science, Rehovot, Israel
SIVAN, Y.,Israel Institute for Biological Research,Ness-Ziona,Israel
SIVERTSEN, B., Institute for Air Research, Lilleström, Norway
SMEDMAN, A.S., Uppsala University, Uppsala, Sweden
SMITH, F.B., Meteorological Office, Bracknell, Berks, U.K.
STARKAND, Y., Weizmann Institute of Science, Rehovot, Israel
STEIN, U., Israel Defence Forces
STEINHORN, I., Ramat-Gan, Israel
STERN, E., Atomic Energy Commission, Tel-Aviv, Israel
STERN, R.,Institut fuer Geophysikalische Weissenschaften,Berlin,Germany
STUBLEY, G.D., University of Waterloo, Waterloo, Canada
TADMOR, J., Soreq Nuclear Research Center, Yavneh, Israel
TENNEKES,H.,Royal Netherlands Meteorological Institute, AE de Bilt, The Netherlands
TERLIUC, B., Nuclear Research Center, Beer Sheba, Israel
TOKATLY, Y., Israel Meteorological Service, Beit Dagan, Israel
TSINOBER, A., Tel-Aviv University, Ramat-Aviv, Israel

TZENTZIFER, A., Israel Aircraft Industries, Israel
ULRICH, W., University of Munich, Munich, Germany
WYGNANSKI, I., Tel-Aviv University, Ramat-Aviv, Israel
YAHAV, G., Ministry of Defence, Israel

PREFACE

In this volume, we present the lectures given during the 1984 OHOLO Conference, held in Zichron Yaacov, Israel. The Conference was organized by the Israel Institute for Biological Research, Department of Mathematics, which is involved in Environmental Risk Evaluation, and in Projects Estimating the Potential of Wind Energy. The lectures cover a broad spectrum of mathematical models, ranging from those that deal with the solution of atmospheric conservation equations, and to those models that yield empirical estimates based on real time measurements and thus are unique to the locale where measured.

The goal of the Conference was to allow scientists from various countries to meet and discuss topics of mutual interest, including the following:

1. Structure of the boundary layer - primarily models dealing in the understanding of the various processes of atmospheric energy transfer, and their influence on the size and composition of the boundary layer.

2. Advanced mathematical techniques for describing flow and diffusion - lectures on approximations and techniques for solving the diffusion and transport equations.

3. Flow over complex terrain - research into various aspects of the problem - mathematical models, physical models, experimental results.

4. Models of pollution transport and deposition.

As at the Conference, each lecture here is followed by a "Question and Comments Section. At the conclusion of the Conference, A Round Table Discussion was held, during which various points of view were presented in an attempt to formulate recommendations on the applicability of various models in different circumstances.

We feel that the conference furthers cooperation among the participants - and although we did not succeed in solving all the problems, we have at least made a start.

WELCOMING ADDRESS

Ladies and Gentlemen,

It is my great pleasure and honor to welcome you to the 29th. OHOLO Conference. I do this on behalf of the permanent OHOLO Organizing Committee, the Organizing Committee of this year's conference and on behalf of Prof. I. Hertman, Director of our Institute, and myself.

The Israel Institute for Biological Research has been organizing OHOLO Conferences for almost 30 years, and the responses that we have received have encouraged us to continue this tradition - that is, sponsoring conferences dedicated to a well defined topic to be discussed by a relatively limited number of participants.

The main purpose of the OHOLO Conferences as defined at the very beginning in 1956, was to encourage the dialogue between Israeli scientists and their colleagues from abroad. This has developed into a tradition of meetings with a rather informal atmosphere and ample time for direct contact between participants. We call these annual scientific meetings the OHOLO Conference, after the name of a place Oholo, located on the shores of the Lake of Galilee, where about 20 of these conferences took place.

I would like to describe briefly our Institute, the Israel Institute for Biological Research. It is a government research institute, project oriented and funded by contracts and grants. We put much emphasis on applied research, while at the same time, we do not neglect basic research which, we believe, is a fundamental necessity for any academic institution. Many of the institute's scientific staff also hold teaching appointments at universities.

Among the three Divisions of our Institute, namely biology, chemistry and environmental studies, the latter is the youngest as well as a very active one. This Division was entrusted with the task of organizing this year's conference, and having been involved with the preparations for this meeting, I assure you that they worked really hard. You may wish to know that the entire organization of this meeting is carried out by members of the permanent staff of our Institute, whereas every year a different group of scientists is appointed, according to the subject chosen.

This year's conference is devoted to studies relating to processes occurring in the planetary boundary layer. A better understanding of these processes will help us estimate pollutant dispersion in the atmosphere and the energy potential of the wind. These estimations will have a crucial influence on making the decisions taken at all levels of management, which have great impact on the quality of our lives as well as on our future.

I wish to take this opportunity to thank the organizers of this conference, Drs. H. Kaplan and N. Dinar as scientific directors and the technical staff, Mss. I. Turner, N. Ben-David, N. Duvdevani and Messrs. M. Navon, R. Pniel and M. Ben-Yeffet who worked tirelessly in preparing this meeting.

On behalf of the Organizing Committee, I wish you all, distinguished guests and colleagues, a fruitful and enjoyable meeting.

Gad A. Simon
Deputy Director
Israel Institute for Biological Research

OPENING LECTURE

TRANSPORT AND DIFFUSION IN COMPLEX TERRAIN (REVIEW)

Bruce A. Egan, Sc.D.
Environmental Research & Technology, Inc.
696 Virginia Road
Concord, MA 01742

ABSTRACT

This paper reviews activities over the past nine years involving the evaluation, verification and development of atmospheric transport and diffusion models applied to air pollution assessments in complex/ mountainous terrain settings. Results from experiments performed at different complex terrain settings under stably stratified flow conditions have been emphasized. Comparisons of model predictions to observations are made. Physical modeling laboratory tests simulating flow conditions for full-scale experiments complement the data bases and provide information on systematic variations of modeling parameters and input conditions for mathematical model development purposes. Principal findings from these and other experiments on specific technical issues are summarized.

1. INTRODUCTION

The development of methods to reliably assess the effects of atmospheric transport and dispersion on air pollution concentrations expected in regions of mountainous, complex terrain continues to be a major research effort in the field of air pollution meteorology. Understanding dispersion processes in mountainous terrain areas is of special concern because (1) it is common for sources of air pollutants to be located in river valleys, along coast lines or otherwise at the base of mountains, and (2) the influence of topography on local meteorological conditions and flows can be adverse from an air pollution transport point of view; and (3) for a given chimney height, nearby high terrain can be exposed to significantly higher concentrations than would be expected in a region of flat terrain.

A previous review by the author (Egan 1975) discussed the state-of-art of dispersion modeling for complex terrain settings as it existed some nine years ago. Although there were a number of pressing applications which required more knowledge of dispersion processes in complex terrain, there were no large, highly focused research efforts being undertaken to resolve the many technical issues outstanding.

Boundary-Layer Meteorology **30** (1984) 3–28. 0006–8314/84/0301–0003$03.90

That 1975 review, in fact, involved drawing information from studies which, for the most part, had research goals other than that of understanding how to estimate ground level air quality concentrations in mountainous areas. In contrast, there have been a number of efforts over the past several years which have been more directed to air quality applications and relevant to these issues.

In 1976, the US Energy Research and Development Administration sponsored a Workshop on Research Needs for Atmospheric Transport and Diffusion in Complex Terrain (Barr, et al, 1976). This workshop recommended that a multi-year program be initiated to address the air quality assessment aspects of oil shale development in the western USA and other energy conversion developments in complex terrain. The workshop recommendations formed the basis for the ongoing Atmospheric Studies in Complex Terrain (ASCOT) program (Dickerson and Gudiksen, 1980) sponsored by the U.S. Department of Energy. I will leave further descriptions to Dr. Dickerson who will be discussing this major program separately at this conference. The U.S. EPA sponsored a workshop focusing on dispersion model needs in 1979 (Hovind, et al, 1979). Proceedings from this workshop described the technological basis for a series of measurements and studies needed to develop reliable dispersion modeling techniques for complex terrain. An outgrowth of this workshop is the ongoing EPA-funded Complex Terrain Model Development (CTMD) program (Schiermeier et al, 1983) which I will describe in more detail later in this review.

The Electric Power Research Institute (EPRI) has undertaken a large Plume Model Validation and Development, (PMV&D) effort (Bowne, et al, 1983, 1984) based on extensive field measurements. In addition to having completed two experiment at a power plants located in flat and in moderately complex terrain, EPRI plans to perform another experiment in a fully complex terrain setting (stack height less than nearby terrain height).

Most recently, the AMS/EPA Cooperative Agreement sponsored a Workshop on Dispersion in Complex Terrain (Egan, 1984) for purposes of sharing information gained from various field measurement, physical modeling and model development efforts.

In addition to the above, the interest in complex terrain in the U.S. is manifested by the fact that about one sixth of the papers at the AMS Turbulence and Diffusion and Air Pollution Meteorology Conferences over the past eight years have been about complex terrain related efforts. These were about equally divided between theoretical development and application oriented topics. I mention the above statistics to illustrate the overall interest in this as a research topic and also to call attention to the fact that there are a multitude of applications testing the present state-of-art. Because of the site specific nature of much of the research performed, a challenge for investigators is to systematically separate general conclusions from those findings which are very much site specific.

For purposes of providing a review for this conference, I will first address progress made in some recent development work I have been involved in and then I will discuss progress made more generally in understanding some of the unique phenomena associated with the presence of complex terrain which need to be incorporated into dispersion modeling efforts.

2. U.S. EPA CTMD PROGRAM

Overview

The Environmental Sciences Research Laboratory of the U.S. Environmental Protection Agency is sponsoring the Complex Terrain Model Development (CTMD) program. This is a multi-year measurement and model development program to evaluate, develop and refine dispersion models which can be used to reliably calculate ground-surface air pollutant concentrations resulting from emission sources in or near mountainous terrain. The objectives of the program have been described by Holzworth (1980). Lavery et al (1982, 1983a, 1983b), Strimaitis et al (1983), and Schiermeier et al (1983) document progress of the project through last year. The primary emphasis of this effort is to improve our understanding of the dynamics of the interaction of an elevated plume with high terrain features. For this reason, the study has focussed on predicting one-hour average concentrations as a basic building block for a model. It was felt that the prediction of longer term averages (e.g., daily averages) depends critically on site-specific meteorological data-especially regarding the persistence of specific wind conditions. The program is "an integrated and highly coordinated effort that involves:

(1) model development/evaluation/improvement;

(2) scaled physical modeling in a fluid modeling laboratory;

(3) field measurements/experiments centered on isolated simple terrain features; and

(4) field measurements/experiments centered on a full-scale plant . . . in terrain with opportunities for plume impaction and other types of plume terrain interactions" (Holzworth, 1980).

Field Experiments

The program is now in its fourth year and two field experiments on isolated "simple" terrain features have been completed. The first feature was Cinder Cone Butte (CCB) near Boise, Idaho. CCB is a roughly axisymmetric, 100-meter high hill located in a broad, relatively flat river valley. The field program consisted of 18 multi-hour tracer gas experiments conducted primarily during stable flow conditions. Extensive meteorological, lidar and photographic measurements were taken. A "modelers' data archive" has been developed for 153 experimental hours.

The second field experiment was of similar design and with similar measurements and was conducted at Hogback Ridge near Farmington, New Mexico. In contrast to CCB, this feature is a nearly two-dimensional ridge but is also approximately 100 meters high at the experiment location. Of special interest was the differences of flow behavior for a two-dimensional vs. more three-dimensional shape. This experiment produced approximately 179 tracer-hours of data.

The third, full-scale, field experiment is planned for the fall of 1984 at the Tracy Power Plant located in the Truckee River Valley near Reno, Nevada. A feasibility study conducted in the fall of 1983 at this site, provided information to develop the study design in detail.

Physical Modeling Experiments

A parallel, complementary series of scaled fluid modeling experiments are being carried out at EPA's Meteorological Wind Tunnel and Water Channel/Towing Tank (Snyder (1980), Snyder and Lawson (1981); Snyder et al (1982); Thomson, et al (1983)). The physical model experiments performed under conditions similar to some of those experienced in the field studies have provided considerable insight with respect to the role of stratification and wind speed in determining flow behavior. Systematic variations of the flow and emission parameters under the controlled conditions of the fluid modeling facilities are providing detailed information on expected flow patterns. Some results and data comparisons are described later in this paper.

Mathematical Model Development and Evaluation

The EPA CTMD program has the ultimate objective of developing practical applied plume dispersion models which can be used routinely in air pollution regulatory activities. For these reasons the program was focussed on developing relatively simple algorithms, based on theory and the experiment results.

Modeling Stably Stratified Flows

The presence of mountainous terrain acts to distort the wind field causing accelerations and decelerations of the flow and associated deflections, contractions and expansions of "stream tubes" of air. The dynamics of this behavior in stable conditions depends upon the wind speed relative to the ambient density stratification of the flow and can be characterized by the Froude number, Fr, defined as

$$Fr = U/Nh \tag{1}$$

where U is the characteristic wind speed, N is the Brunt-Väisälä frequency given by

$$N = [-\frac{g}{\rho}\frac{\partial\rho}{\partial Z}]^{1/2} \quad (2)$$

h is the hill height and ρ is fluid density. Neutral stability behavior occurs for Fr >> 1 where inertial effects dominate the dynamics. Strongly stable behavior occurs for 0 < Fr < 1. Sheppard (1956) postulated that for a given stable lapse rate, one could calculate the horizontal velocity upwind which would enable air to surmount a hill by equating the kinetic energy of a fluid parcel to the potential energy change associated with lifting the parcel to the hillcrest. Hunt et al (1978) demonstrated experimentally the usefulness of the concept of a critical height defined as

$$H_c = h(1-Fr) \quad (3)$$

which, for constant density gradient and uniform velocity profile upwind, separated streamlines which could pass over the crest of a hill from those which could not. Snyder et al (1982) extended the energy argument and formulae to arbitrary velocity and density profiles with the following formula:

$$1/2\ \rho U^2(H_c) = \int_{H_c}^{h} \rho N^2(h-Z)dZ \quad (4)$$

where H_c must be determined iteratively.

The H_c concept as a means of determining whether a plume will impinge onto a hill or pass around or over it was also found by Ryan and Lamb (1984) to be valid at a 335 m isolated hill in the State of Washington.

This notion of considering two extremes of flow behavior defined by whether streamlines are above or below H_c has been a key element in the CTMD model development efforts and is illustrated schematically in Figure 1.

In the most recently published description of the Complex Terrain Dispersion Model (CTDM) developed in this program, CTDM versions 11083 and 11083E, (Lavery et al, 1983), the flow is treated as if it takes place in two discrete layers as described below.

The Wrap Component

The 'Wrap' component of CTDM, for dispersion below H_c, restricts the flow to travel in horizontal planes toward or around the sides of a hill. Concentrations along the upwind face of the hill are given by the equation

$$C(d,Z_R) = \frac{P(\theta_d)}{d} C_o(d,Z_R) = \frac{Q}{\sqrt{2\pi}\,\sigma_z(d)U}\frac{P(\theta_d)}{d}(\exp(-.5((Z_R-Z_s)/\sigma_z)^2)$$

$$+ \exp(-.5((Z_R + Z_s)/\sigma_z)^2)) \quad (5)$$

where d is distance from the source to the hillside receptor, Z_R is the receptor elevation, Z_s is the effective source height, and $P(\theta)$ is the probability density function (PDF) of wind directions over an hour averaging time. $P(\theta_d)$ is the probability that the wind blows along the stagnation wind direction. Use of the PDF allows incorporation of the actual distribution of wind directions into the computations. For a gaussian distribution of wind directions, one would have

$$P(\theta) = \frac{1}{\sqrt{2\pi}} \exp\left(-.5\left((\theta-\theta_m)/\sigma_\theta\right)^2\right) \tag{6}$$

and θ_m is the mean wind direction over the time period of interest. The improvements associated with using the more general PDF vs a gaussian distribution were discussed by Lavery et al (1982) and Strimaitis et al (1983). Notice that Z_R and Z_s in Eq. 5 are relative to the hill base elevation (Z=0) and that 'reflection' from the image sources is relative to Z=0, not the hill surface.

At receptors around to the side of the hill, the distance to receptor, s, is $d+s_a$ (see Fig. 2). To simulate the effects of further dispersion and the presence of the hill in 'reflecting' the plume, concentrations are calculated by treating the vertical distribution of plume material below H_c as a source distribution and integrating the response function or Green's function from the hill base to H_c. The equation is

$$C(s,Z_R) = \frac{P(\theta_d)}{s} \int_0^{H_c} \frac{C_o(d,Z)}{\sqrt{2\pi}\,\sigma_Z'}$$

$$\left[\exp\left(-.5((Z_R-Z)/\sigma_Z')^2\right) + \exp\left(-.5((Z_R+Z)/\sigma_Z')^2\right)\right]dZ \tag{7}$$

where C_o (d,z) is evaluated from Eq. 5 and

$$(\sigma_Z')^2 = \sigma_Z^2(s) - \sigma_Z^2(d) \tag{8}$$

A similar equation for considering reflection effects over uniform surfaces has been developed by Overcamp (1983).

The Lift Component

For flow above H_c a different algorithm, Lift, is used. Plume material is assumed to travel up and over a terrain feature but the concentration distribution is modified by the presence of the hill to spread over a wider area. This is simulated by a receptor location adjustment to the receptor location on the basis of the effective radius of curvature of the portion of the hill above H_c (See Lavery et al, 1983 for specifics). As shown in Figure 1, plume material below H_c impingement and the remaining distribution of material above H_c continues to diffuse with full reflection from the

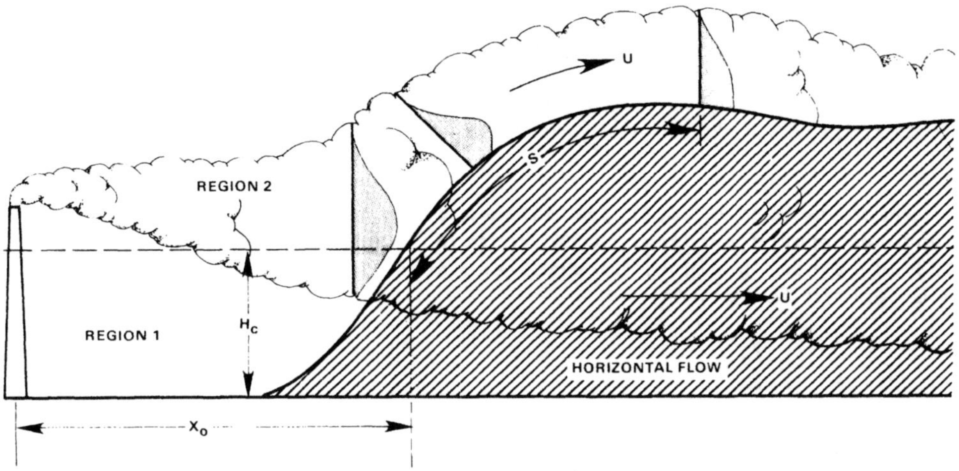

Figure 1 Dispersion and Flow Regions for Stratified Flow Around Hills

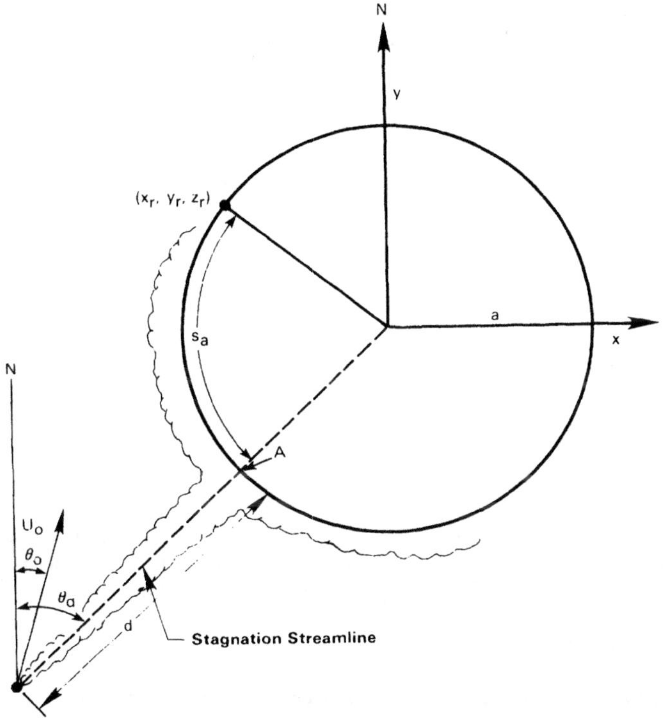

Figure 2 Plume Dispersion in the Region of Horizontal Flow

hill surface. Concentrations in this region are calculated by integrating the response function from H_c to infinity.

$$C(s,\theta,Z_p) = \frac{2\,P(\theta)}{s} \int_{H_c}^{\infty} \frac{C_o(s,Z)}{\sqrt{2\pi}\,\sigma_Z'} \exp\left(-.5((Z-H_c)/\sigma_Z')^2\right) dZ \tag{9}$$

where σ_Z' and $C_o(s,Z)$ are defined as in Eqs 8 and 5 and Z_p denotes ground surface. CTDM version 11083E differs from version 11083 as it has an algorithm which enhances the vertical dispersion via a terrain factor, T.

$$(\sigma_z')^2 = \frac{\sigma_z^{\,2}(s)}{T^2} - \sigma_Z^{\,2}(d) \tag{10}$$

where $T = PPC$ for $Z_R > Z_s$, $PPC = 0.5$,
and $T = 1-(1-PPC)(Z_R-H_c)/(Z_r-H_c)$ for $Z_R < Z_s$.

A comparison of model predictions of these versions of CTDM and the EPA Valley, Complex I and II models was made following model evaluation measures and methods of Fox (1981), Londergan (1980) and Venkatram (1982).

Defining model errors as the difference between observed and predicted

$$e_i = (C_o - C_p) \tag{11}$$

then the arithmetic mean, the noise and the variance of the residuals can be computed from

$$Ma = \frac{1}{N} \Sigma\, e_i$$

$$Sa = [\frac{1}{N} \Sigma\, (e_i - Ma)^2]^{1/2} \tag{12}$$

$$\overline{(e^2)} = [\frac{1}{N} \Sigma\, (e_i)^2]$$

Note that the total variance $\overline{(e^2)}$ is the sum of the squares of the bias, Ma, and the noise, S_a. Similarly for residuals e_i which are more nearly log-normally distributed, one can define

$$\zeta_i = \ln\, (C_o/C_p)_i$$

then the geometric mean, noise and variance are given by

$$Mg = \exp\, (\frac{1}{N} \Sigma\, \zeta i)$$

$$Sg = \exp \left(\frac{1}{N} \Sigma (\zeta_i - \ln Mg)^2\right)^{1/2} \quad (13)$$

$$\overline{(\zeta^2)} = \exp \left[\frac{1}{N} \Sigma \zeta_i^{\,2})\right]$$

Tables 1 and 2 summarize the arithmetic and geometric statistics for CCB and also for 17 hours of data available for testing from the Hogback Ridge experiment. For CCB, CTDM 11083-E ranks best by these measures than the others tested. For the Hogback Ridge, CTDM 11083 performed better.

TABLE 1
SUMMARY OF RESIDUAL STATISTICS BASED ON 153
DATA HOURS AT CCB. PEAK CONCENTRATIONS (UNPAIRED IN SPACE)

	Arithmetic Statistics ($\mu g/m^3$)			Geometric Statistics		
Model	Ma	Sa	$(\overline{e^2})^{1/2}$	Mg	Sg	$(\overline{\zeta^2})^{1/2}$
CTDM 11083E	3.3	27.4	27.6	1.37	3.2	3.34
CTDM 11083	2.8	33.7	33.8	2.08	6.9	7.89
Valley	-41	25.5	48.3	.24	3.0	6.06
COMPLEX I	-15	37.9	40.8	.65	3.8	4.07
COMPLEX II	-60.1	99.7	116.4	.38	5.0	6.54

TABLE 2
SUMMARY OF RESIDUAL STATISTICS BASED ON 17
DATA HOURS AT HBR. PEAK CONCENTRATIONS (UNPAIRED IN SPACE)

	Arithmetic Statistics ($\mu g/m^3$)			Geometric Statistics		
Model	Ma	Sa	$(\overline{e^2})^{1/2}$	Mg	Sg	$(\overline{\zeta^2})^{1/2}$
CTDM 11083E	27.5	71.1	76.2	2.00	5.7	6.51
CTDM 11083	28.4	70.6	76.1	2.05	4.3	5.08
Valley	-34.3	87.1	93.6	.24	3.7	6.93
COMPLEX I	-61.0	81.9	102.1	.26	4.1	7.03
COMPLEX II	-266.0	275.2	382.7	.11	12.4	28.5

The Wrap and Lift algorithms assume a level streamline at H_c, an obvious simplification. To explore the consequences of this simplification, Snyder and Lawson (1983) preformed towing tank tests producing streamline patterns and concentration distributions for a fully immersed (upside-down) hill in a stratified flow creating a 'natural' dividing streamline surface. These were then compared to patterns and concentrations on a partially immersed hill where the free surface was set at the former dividing streamline height. The results suggest that the assumption of a flat dividing streamline surface is reasonable for purposes of predicting the locations and values of maximum concentration on the windward hill side. When the maximum concentrations were expected on the lee side (from higher initial release heights), the agreement was not as good. This was apparently due to the presence of a hydraulic jump in the fully immersed case, not present in the partially immersed case.

Verification of Critical Steamline Concept

Because of the ability to perform experiments under controlled conditions using scale models and the ability to visualize flows better in a laboratory, the EPA FMF towing tank experiments (see previous references to Snyder et al) have been the primary mechanism used to explore many concepts associated with H_c. Significantly, Snyder et al (1982) and Castro et al (1983) have shown that the '(1-Fr)' dependence on dividing streamline height is largely independent of body shape (slopes, curvatures and aspect ratios).

The validity of the concept is also evident from the observed concentration patterns from the field experiments. Plumes heading toward the hill within the flow below H_c should produce a ground level concentration pattern where the peak values are on the upwind face of a hill. Plumes within the flow above H_c can flow over the top and produce peak concentrations on the lee side. Tables 3 and 4 summarize the number of hours for the CCB and HBR experiments, respectively, when the highest concentrations were observed on the windward or leeward sides classified by whether the plume elevation was above or below H_c. The associations of releases below H_c with windward peak observations and releases above H_c with leeward peaks may be seen for both data sets and are statistically significant for the CCB data set.

TABLE 3
NUMBER OF HOURS WHEN PEAK CONCENTRATIONS WERE OBSERVED ON THE WINDWARD VS. LEEWARD HILLSIDE QUADRANTS CATEGORIZED BY THE PLUME HEIGHT RELATIVE TO HC

	CCB			HBR		
	Windward	Leeward	Total	Windward	Leeward	Total
$Z_s<H_c$	27	11	38	56	0	56
$Z_s>H_c$	32	57	89	21	4	25
Total	59	68	127	77	4	81

An interesting difference in field study results at CCB vs HBR is the location of the maximum concentrations observed. At CCB, the peak concentrations were observed at an elevation less than the release height, under conditions when the release height was near H_c and the velocity gradient, dU/dz was large. At HBR, the peak concentration was on terrain above the release height but below H_c.

Recent Directions for CTDM

While the concept of dividing streamlines about H_c seems well founded, the specific treatment of trajectories and dispersion for the portion of the plume above H_c needs further refinement. We are now exploring alternative ways of forming terrain effect adjustment factors which can account separately for the effects of turbulence enhancement, plume centerline displacement, changes in the transport wind speed and horizontal streamline distortion.

Specifically, we are working with a modeling framework based on the following equations for the Lift component (valid for $H>H_c$):

$$C = \frac{Q e^{-(\ell/\sqrt{2}\, T_\ell \sigma_{ye})^2}}{2\pi u \sigma_{ye}\sigma_{ze}} \left\{ e^{-\left(\frac{H-H_c}{\sqrt{2}\sigma_{ze}}\right)^2} \left(1-\mathrm{erf}\left(\frac{H_c-H}{\sqrt{2}\,\sigma_{ze}} \; \frac{\sigma_z^{**}/T_z}{\sigma_z(s_o)}\right)\right) + e^{-\left(\frac{H+H_c}{\sqrt{2}\,\sigma_{ze}}\right)^2} \left(1-\mathrm{erf}\left(\frac{H_c+H}{\sqrt{2}\,\sigma_{ze}} \; \frac{\sigma_z^{**}/T_z}{\sigma_z(s_o)}\right)\right) \right\} \qquad (14)$$

where ℓ is the crosswind distance from the receptor to the centerline of the plume, s_o is the distance from the source to the point where the H_c surface intersects the hill, and

where $\sigma_{ze}^2 = \sigma_z^2(s_o) + \sigma_z^{**2}/T_z^2$; $\sigma_z^{**2} = \sigma_z^2(s) - \sigma_z^2(s_o)$

$\sigma_{ye}^2 = \sigma_y^2(s_o) + \sigma_y^{**2}/T_y^2$; $\sigma_y^{**2} = \sigma_y^2(s) - \sigma_y^2(s_o)$

and $T_z = T_h/T_{\sigma_z}$; $T_y = T_\ell/T_{\sigma_y}$

where T_h = net or average height change factor (plume centerline)
T_ℓ = net or average lateral deflection factor (plume centerline)

$T_{\sigma y}$, $T_{\sigma z}$ specify changes in the diffusivity in the lateral and vertical directions.

A similar approach is being taken to advance the Wrap component of CTDM. A second area for investigation is to simulate variations in

H_c over hourly averaging periods. The present equations assume a constant value for each hour and the model predictions seem more sensitive to changes in these values than the observed data indicates.

An important output of the program which should be useful also in noncomplex terrain settings is a model developed by Venkatram et al (1983, 1984) to estimate vertical dispersion rates under stable conditions from vertical wind fluctuation measurements. It will be of interest to test this algorithm at different experimental sites.

3. OTHER PROGRESS MADE ON SPECIFIC TOPICS

In this section, recent findings in several of the technical areas of special concern to simulating dispersion in complex terrain are reviewed.

Terrain Adjustment Factors

For plumes advecting over a hill, the basis for many applied diffusion models is the theory of turbulent plumes embedded within potential flow fields as introduced by Hunt and Mulhearn (1973) or the pseudo-potential flow approximations of Egan (1975). Using trajectory and deformation estimates from potential theory, one proceeds to estimate dispersion parameters normal to the trajectory centerline or to solve the diffusion equation to obtain expressions for the dispersion parameters and the ground surface concentration distribution. This method was applied by Hunt et al (1979), Isaacs et al (1979) and the underlying principles have been applied to buoyant plumes in flow about arbitrarily shaped three-dimensional hills by Weil et al (1981).

Further simplifications have been to utilize 'terrain correction factors' as coefficients or modifications of the argument within the exponential term for vertical dispersion in the gaussian equation. The approach being taken in the EPA CTMD program with respect to dividing streamline concepts and terrain factors was described earlier. Findings from other studies and of other investigator on this subject are noted below. Ellis and Liu (1979) systematically evaluated 288 alternative modifications to the EPA CRSTER model (U.S. EPA (1977)) by comparing predictions with the highest 3-hour and 24-hour measurements from seven monitors on moderately high terrain (monitors near or above stack tops elevation but probably below all final plume heights). The authors concluded that the best versions included an 'Intermediate Ground Displacement Procedure' whereby the effective plume height above stack base, h_e, was modified to an effective height above the terrain surface given by

$$h_e' = h_e - .5Z_R \tag{14a}$$

for stability classes A, B, C and D and

$$h_e' = h_e - Z_R \tag{14b}$$

for stability classes E and F. In a validation study involving occasions when a plume was lower than nearby terrain, Hanna (1983) used

$$h_e' = .4h_e \text{ for } h_e > H_c$$
and (15a,b)
$$h_e' = 0 \text{ for } h_e < H_c$$

H_c was defined as h(1-Fr) where h was the hill height. An alternative model which assumed direct impact for conditions when the plume was predicted to be below hill height overpredicted the observed concentrations significantly. Garrison and Basta (1983) came to the same conclusions independently, and with a different data base using formulations having a similar effect as those of (15a, b). Comparing concentrations predicted with the LAPPES models to observations on high terrain near several power plants, Slowik et al (1977) concluded that a minimum approach distance of .35 h_e during stable conditions yielded the best agreements. Burkhart et al (1983) compared more recent data to LAPPES and COMPLEX I model predictions demonstrating that systematic overpredictions of COMPLEX I were not found with LAPPES.

Rowe et al. (1980) investigated photographs of elevated plumes advecting over high terrain during stable conditions and concluded that potential flow estimates of plume trajectories calculated from a base level of H_c compare favorably with the observations. This supports the approaches taken by Hunt and Mulhearn (1973) and Hunt (1982) in applying potential flow theory as well as the 'cut-off-hill' approach tested in CTMD. One must, of course, include deformation effects in estimating the plume dimensions about plume trajectory centerlines computed.

Turbulence Intensities in Complex Terrain

New information is available on the effects of mountainous terrain on atmospheric turbulence intensities. Shearer and Minott (1977) and Minott et al (1977) analyzed several sets of dispersion data in rural, rolling terrain and in deep valleys and compared results with the standard Pasquill-Gifford (P-G) dispersion coefficients (Pasquill, 1961; Gifford, 1961). The comparisons showed that for near surface releases, and at any downwind distance, the vertical dispersion rates had a much narrower range of values as a function of stability class. The presence of the terrain apparently enhanced rates at the stable extreme and suppressed rates at the unstable extreme. Elevated releases (125 m) occurring during P-G Classes E and F produced dispersion rates closely following the E curve. Kao and Taylor (1977) reported vertical dispersion coefficients similar to those expected over flat terrain but found horizontal dispersion rates were significantly larger than those observed over flat terrain. Miller (1978) reported greater horizontal and vertical dispersion in complex terrain at Karlsrule and Jülich. Hanna (1980, 1981) and Hanna et al (1982) discuss the enhancement of turbulence due to wake effects of terrain feature results from experiments at the Widows Creek, Alabama and in the Geysers, Calif. geothermal area. Panofsky et al (1978) found that wind direction fluctuations downwind of a ridge were 2.5 times larger than for other flow orientations.

From the above it seems clear that upwind terrain has a more significant effect on enhancing crosswind dispersion than on vertical dispersion. From observations at CCB it was apparent that a hill can also affect crosswind dispersion rates to a limited extent upwind of the hill, especially for flow below H_c. This effect was also apparent at HBR where it could be more clearly associated with 'blocking' phenomena.

There is strong support for the use of on-site turbulence data in complex terrain settings as a means for improving model performance and reliability. This was demonstrated by Hanna et al (1984), Hanna (1983), Dittenhoefer (1983), and Strimaitis et al (1983).

Effects of Surface Reflection

One of the issues raised with respect to characterizing and parameterizing impaction during stable conditions on high terrain was the validity of the full doubling of surface concentrations due to "surface reflection" as is normally considered in flat terrain computations. Early methods for estimating impaction (e.g., Van der Hoven, 1972) had included this doubling and the factor is currently used in the USA EPA Valley and COMPLEX I and II models (Burt and Slater 1977) for direct impact conditions. Briggs (1973) postulated that for direct impaction the plume centerline concentration without reflection doubling should be used. Egan et al (1978, 1979) noted the doubling effect could imply a violation of the second law of thermodynamics and proposed a procedure on this basis to estimate the maximum effect that reflection would have on surface concentrations. The "reflection factor", was defined as the ratio of the ground level concentration divided by the elevated plume centerline concentration computed in the absence of a terrain object. It ranges between 1.0 (no reflection) and 2.0 (full reflection) and was shown to be a function of hill slope and the local rate of vertical spread of the plume in the region of direct impact on the hill surface. There is limited field and physical experiment data supporting the notion that full doubling does not occur during impaction. Cabe et al (1977) reported that a Gaussian model modified to ignore ground based reflection showed improved performance when applied to impaction on a steep escarpment. Williams et al (1982) drew from similar experiences to develop a model ignoring reflection effects for applications to the Four Corners region of the US Southwest. From physical modeling experiments, Hunt et al (1979) observed a reflection factor of 1.12 for a plume impinging on a three-dimensional hill and provided information on how the factor would vary with off-axis displacement. Peterson and Twombly (1982) in comparing data from stratified wind tunnel experiments to model predictions found best agreement when partial (vs. full) reflection was included.

In the CTDM modeling approach full reflection from the surface is always taking place but the integrated effect depends upon the time history of the plume as it approaches and passes along a surface. Thus "partial" reflection is an integrated effect.

The effect of surface reflection is difficult to verify directly with field experiments because it requires simultaneous measurements of surface concentrations and elevated plume centerline concentrations upwind of the region of impact where the concentrations are unaffected by the surface. The phenomenon is much more readily studied using physical models.

Predicting Concentrations on Lee-Sides

There has been increased attention to phenomena giving rise to high concentrations on the lee sides of hills. The "Good Engineering Practice" (GEP) regulations within the U.S. allow, within some legally set constraints, emission limit credit to stacks which must be built high enough to avoid downwash effects from upwind terrain. The regulations have inspired a number of physical modeling studies of smoke stacks placed in the lee of the mountains. A study which has produced some of the most general (vs. site specific) information on flow in the wake of two-dimensional ridges is that of Huber et al (1976). The study concluded that releases from a stack of 2 1/2 times as high as the ridge height would avoid the highly turbulent cavity region. Lott (1982) has analyzed data from the 14-station Widows Creek air quality network demonstrating the effects of wind speed, stability class, and stack exhaust parameters on concentration values. An ad-hoc reduction of estimated plume rise to simulate the effects of terrain-induced downwash was necessary to have mathematical model predictions compare favorably with the observed concentrations. Hunt and Snyder (1980) investigated the effects of an elevated source on concentrations on both the windward and leeward slopes of a bell-shaped hill in a stratified towing and in an unstratified wind tunnel. Detailed flow patterns were derived for conditions yielding a range of lee wave formation and flow separation effects. For moderately stratified conditions ($.4<Fr<1.5$) a plume originating upstream at a height above the dividing streamline height but not much higher than the hill (i.e. $h(1-Fr) <Z_r<h(1+Fr)$) could pass very close to the lee side surface-presumably resulting in large ground level concentrations on the lee side. Stratified wind tunnel tests simulating dispersion from an elevated release onto downwind terrain showed the tendency for even buoyant plumes to flow parallel to lee-side terrain (Weil, 1979).

Snyder (1983) provides an excellent summary of the effects of relative placement of an elevated release with respect to the windward and leeward slopes of terrain features. Defining an amplification factor, A, as the ratio of the maximum surface concentration found in the presence of the hill to the maximum concentration at ground level on level terrain in the absence of the hill, Snyder shows that a maximum amplifications occur for a source located downwind of two dimensional hills (see Table 4). The amplification factor was

TABLE 4
TERRAIN AMPLIFICATION FACTORS FOR TWO-DIMENSIONAL HILLS

		A		
		Source Location		
Hill AR	Z_r/h	Upwind	Top	Downwind
	1/4	1.5	0.9	3.4
8	1/2	1.1	0.6	3.0
	1	1.5	0.8	2.4
	1-1/2	1.2	0.8	1.7
	1/4	2.0	0.5	15.0
5	1/2	2.0	0.6	8.0
	1	1.7	0.9	5.6
	1-1/2	1.2	1.0	2.9
	1/4	2.8	0.3	7.5
3	1/2	2.5	0.7	6.4
	1	1.8	0.9	10.8
	1-1/2	1.9	0.9	7.8

greatest (15.0) for a relatively low release height ($Z_r/h=1/4$) in combination with a hill aspect ratio, AR, (streamwise half-length/ height) of 5. A three dimensional hill results in a smaller A on the lee side than for a ridge shape. For an upwind source, however, the amplification in greater for the three dimensional shape (See Table 5).

TABLE 5
SUMMARY OF TERRAIN AMPLIFICATION FACTORS
FOR NEUTRAL FLOW

Source Location	Hill Type	A
Downwind	Two-Dimensional	10-15
Downwind	Three-Dimensional	5-6
Upwind	Three-Dimensional	2-4
Upwind	Two-Dimensional	1-3
Top	Two-Dimensional	0.5-1

Slope and Mountain-Valley Flows

An improved understanding of nocturnal drainage flows in valleys is becoming available (Dickerson and Gudiksen, 1980). Manins and Sanford (1979) developed a two-dimensional model of Katabatic winds using a hydraulic flow approach and which incorporated entrainment assumptions. Rao and Snodgrass (1981) and Rao and Nappo (1983) have employed a turbulent energy closure equation to allow inclusion of ambient winds in two dimensional model for nocturnal slope flows showing good agreement with observations of wind and temperature profiles. Efforts to include the build up of cool air at the valley floor and to simulate along-valley axis drainage have been described by McNider et al (1981) and Yamada (1981). A thermodynamic model of temperature inversion breakup in deep valleys (Whiteman 1980) has led to a model (Whiteman and Allwine; 1983) of air pollution concentrations expected on the valley floor and sidewalls due to fumigation of elevated sources within valleys.

The issue of what characterizes cross wind conditions sufficient to ventilate deep valleys has been addressed by Bell and Thomson (1980) who found that Froude number based on Valley Side wall heights greater than 1.3 were necessary to avoid stagnation. The results found were independent of hill slopes.

4. SUMMARY AND CONCLUSIONS

This review has discussed some of the major accomplishments over the past eight or so years in developing capabilities to predict air quality concentrations in complex terrain. I would summarize these as follows:

(1) Large level-of-effort and comprehensive field experiments have provided a wealth of new information with which to test concepts. This is to be contrasted with information on only select aspects of dispersion in complex terrain a decade ago.

(2) The importance of the critical or dividing streamline height, H_c, as a concept to identify different flow regimes during stably-stratified conditions has been verified in both scale-model laboratory experiments and in full scale field studies. The estimation of the H_c from the integral form appears to be valid for a variety of hill slopes and is largely independent of hill aspect ratio, hill slope and shape.

(3) For neutrally stratified conditions or for flows above H_c, the notion of terrain adjustments for plumes initially below hill crests which modify the combined effect of reduced stand-off height and altered dispersion parameters provides a means for correcting the overpredictive tendency of models with level plume trajectories.

(4) Within the context of (2) and (3) above, relatively simple mathematical algorithms are emerging to estimate expected concentrations on hill slopes downwind of a source. The

concept of assuming a dead air region below H_c, for example, appears useful for a range of release heights above H_c.

(5) The site specific aspects of different complex terrain settings demand the use of appropriate meteorological input data in dispersion models. Meteorological data should include information on turbulence intensities in the region of interest and at elevations representative of plume height.

(6) Considerable new information and experience has emerged from extended use of physical modeling facilities over the past several years to investigate specific fluid dynamic phenomenon associated with complex terrain settings. Through verification with both theory and data from field experiments, the role of physical modeling in providing supportive information on large scale atmospheric flows is even more firmly established.

(7) The problem of estimating air quality concentrations in stagnant valley situations where the dispersion processes may be dominated by local, thermally driven flows is being approached with models attempting to simulate the flow conditions in some detail. Basic knowledge gained from field experiments should result in markedly improved dispersion models for these situations.

ACKNOWLEDGEMENT

I would express thanks to Mr. David Strimaitis for the many useful conversations about, and contributions to, this review.

REFERENCES

Barr, S., Luna, R.E., Clements, W.E., Church, H.W., and Sivinski, H.D., 1976: Workshop on Research Needs for Atmospheric Transport and Diffusion in Complex Terrain. Albuquerque, New Mexico. NTIS CONF-7609160.

Bell, R.C. and Thompson, R.O.R.Y., 1980: Valley Ventilation by Cross Winds. J. Fluid. Mech. 96, 4 p 757-767.

Bowne, N.E., Londergan, R.J., Murray, D.R. and Borenstein, H.S., 1983: Overview, Results, and Conclusions for the EPRI Model Validation and Development Project: Plains Site. EPRI EA-3074, Elect. Pow. Res. Inst., Palo Alto, Calif.

Bowne, N.E., Londergan, R.J., Murray, D.R., 1984: Summary and Conclusions for the EPRI Plume Model Validation and Development Project: Moderately Complex Terrain Site. EPRI, Palo Alto, Calif.

Briggs, G.A., 1973: "Diffusion Estimation for Small Emissions," ATDL Contribution File No. 79, Atmospheric Turbulence and Diffusion Laboratory, Oak Ridge, TN.

Burkhart, R.P., Londergan, R.J., Rothstein, R.A. and Borestein, H.S., 1983: Comparative Performance Evaluation of Two Complex Terrain Dispersion Models. 76th Annual Meeting of the Air. Pol. Centr. Assoc., Paper No. 83-47.4, Pittsburgh, Penna.

Burt, E.W. and Slater, H.H.: 1977, "Evaluation of the Valley Model." AMS-APCA Joint Conference on Applications of Air Pollution Meteorology, Salt Lake City, Utah.

Cabe, D.B., Eppright, B.R. and Stuart, J.D., 1977: "Application of the Gaussian Dispersion Model to Predicting Maximum Plume Concentrations in Abruptly Rising Terrain." 70th Annual Meeting of the Air Pollution Control Association, June 20-24, 1977, Ontario, Canada.

Castro, I.P., Snyder, W.H. and Marsh, G.L., 1983: Stratified flow over three-dimensional ridges. J. Fluid. Mech. 135, p 261-282.

Dickerson, M.H. and P.H. Gudiksen, 1980: The Department of Energy's Atmospheric Studies in Complex Terrain (ASCOT) program. Conference Papers Second Joint Conference on Applications of Air Pollution Meteorology. American Meteorlogical Society, 467-473.

Dittenhoefer, A.C., 1983: Measurements of Power Plant Dispersion in Hilly Terrain. 76th Annual Meeting of the Air Poll. Contr. Assoc., Paper No. 83-47.2, Pittsburgh, Penna.

Egan, B.A., 1984: "Workshop on Dispersion in Complex Terrain." (In preparation).

Egan, B.A., D'Errico, R. and Vaudo, C., 1979: "Estimating Air Quality Levels in Regions of High Terrain under Stable Atmospheric Conditions." Fourth Symposium on Atmospheric Turbulence, Diffusion and Air Quality, 15-18 January 1979. Reno, Nevada.

Egan, B.A., 1975: "Turbulent Diffusion in Complex Terrain." Lectures on Air Pollution and Environmental Impact Analyses, Workshop Proceedings, AMS, Boston, MA.

Egan, B.A., D'Errico, R. and Vaudo, C., 1978: "Assessing Air Quality Levels in Regions of Mountainous Terrain." World Meteorological Organization on Boundary-Layer Physics Applied to Specific Problems of Air Pollution, June 19-23, 1978, Norrkoping, Sweden.

Ellis, H.M. and Liu, P.C., 1979: Comparison of Predicted and Measured Concentrations for 54 Alternative Models of Plume Transport in Complex Terrain. 72nd Annual Meeting of the Air Pollution Control Association. June 24-29, 1979, Cincinnati, Ohio.

Fox, D.G., 1981: Judging Air Quality Model Performance: A Summary of the AMS Workshop on Dispersion Model Performance. Bull. Am. Meteor. Soc., 62, p 599-609.

Garrison, M.E. and Basta, M.F., 1983: Clinch River Revisited; A New Look at Dispersion Modeling in Complex Terrain. 76th Annual Meeting of the Air Poll. Cent. Assoc., Paper 83-47.1, Pittsburgh, Penna.

Gifford, F.A., 1961: Use of Routine Meteorological Observations for Estimating Atmospheric Dispersion. Nuclear Safety, 2(4), p 47-51.

Hanna, S.R., Briggs, G.A., and Hosker, R.P., Jr., 1982: Air-Pollution Meteorology in Complex Terrain. Handbook on Atmospheric Diffusion. Doc. No. DOE/TIC-11223. NTIS. U.S. Dept. of Commerce, Springfield, Virginia.

Hanna, S.R., 1983: Lateral Diffusion Due to Mesoscale Eddies with Period One to Two Hours at Cinder Cone Butte. Extended Abstracts, Sixth Symp. on Turb. and Diff. AMS, Boston, MA p 162-165.

Hanna, S.R., Egan, B.A., C.J. Vaudo and Curreri, A.J., 1984: A Complex Terrain Dispersion Model for Regulatory Applications at the Westvaco Luke Mill. Atmos. Envir. 18,4 pp 685-699

Hanna, S.R., 1981: Diurnal Variation of Horizontal Wind Direction Fluctuations in Complex Terrain at Geysers, Cal. Bound. Layer Meteor., 21, p 207-213.

Egan, B.A., 1984: "Workshop on Dispersion in Complex Terrain." (In preparation).

Egan, B.A., D'Errico, R. and Vaudo, C., 1979: "Estimating Air Quality Levels in Regions of High Terrain under Stable Atmospheric Conditions." Fourth Symposium on Atmospheric Turbulence, Diffusion and Air Quality, 15-18 January 1979. Reno, Nevada.

Egan, B.A., 1975: "Turbulent Diffusion in Complex Terrain." Lectures on Air Pollution and Environmental Impact Analyses, Workshop Proceedings, AMS, Boston, MA.

Egan, B.A., D'Errico, R. and Vaudo, C., 1978: "Assessing Air Quality Levels in Regions of Mountainous Terrain." World Meteorological Organization on Boundary-Layer Physics Applied to Specific Problems of Air Pollution, June 19-23, 1978, Norrkoping, Sweden.

Ellis, H.M. and Liu, P.C., 1979: Comparison of Predicted and Measured Concentrations for 54 Alternative Models of Plume Transport in Complex Terrain. 72nd Annual Meeting of the Air Pollution Control Association. June 24-29, 1979, Cincinnati, Ohio.

Fox, D.G., 1981: Judging Air Quality Model Performance: A Summary of the AMS Workshop on Dispersion Model Performance. Bull. Am. Meteor. Soc., 62, p 599-609.

Garrison, M.E. and Basta, M.F., 1983: Clinch River Revisited; A New Look at Dispersion Modeling in Complex Terrain. 76th Annual Meeting of the Air Poll. Cent. Assoc., Paper 83-47.1, Pittsburgh, Penna.

Gifford, F.A., 1961: Use of Routine Meteorological Observations for Estimating Atmospheric Dispersion. Nuclear Safety, 2(4), p 47-51.

Hanna, S.R., Briggs, G.A., and Hosker, R.P., Jr., 1982: Air-Pollution Meteorology in Complex Terrain. Handbook on Atmospheric Diffusion. Doc. No. DOE/TIC-11223. NTIS. U.S. Dept. of Commerce, Springfield, Virginia.

Hanna, S.R., 1983: Lateral Diffusion Due to Mesoscale Eddies with Period One to Two Hours at Cinder Cone Butte. Extended Abstracts, Sixth Symp. on Turb. and Diff. AMS, Boston, MA p 162-165.

Hanna, S.R., Egan, B.A., C.J. Vaudo and Curreri, A.J., 1984: A Complex Terrain Dispersion Model for Regulatory Applications at the Westvaco Luke Mill. Atmos. Envir. 18,4 pp 685-699

Hanna, S.R., 1981: Diurnal Variation of Horizontal Wind Direction Fluctuations in Complex Terrain at Geysers, Cal. Bound. Layer Meteor., 21, p 207-213.

Hanna, S.R., 1980: Measured σ_y and σ_θ in Complex Terrain near the TVA Widows Creek, Alabama Steam Plant, Atmos. Environ., 14: 401.

Holzworth, G.C., 1980: "The EPA Program for Dispersion Model Development for Sources in Complex Terrain." Second Joint AMS–APCA Conference on Applications of Air Pollution Meteorology, New Orleans, LA.

Hovind, E.L., Edelstein, M.W. and Sutherland, V.C., 1979: Workshop on Atmospheric Dispersion Models in Complex Terrain. EPA Doc. No. EPA-600/9-79-041.

Huber, A.H., Snyder, W.H., Thompson, R.S. and Lawson, R.E., Jr., 1976: Stack Placement in the Lee of a Mountain Ridge. A Wind Tunnel Study. U.S. EPA Doc. No. EPA-600/4-76-047. Research Triangle Park, NC, pp 45.

Hunt, J.C.R., and P.J. Mulhearn, 1973: "Turbulent dispersion from sources near two dimensional obstacles." J. Fluid Mech, 61 pp. 245–274.

Hunt, J.C.R., Puttock, J.S. and Snyder, W.H., 1979: "Turbulent Diffusion from a Point Source in Stratified and Neutral Flows around a Three-Dimensional Hill – Part I." Diffusion Equation Analysis. Atmospheric Environment, Vol. 13, p 1227–1239.

Hunt, J.C.R., Snyder, W.H. and Lawson, R.E., Jr., 1978: Flow Structure and Turbulent Diffusion Around a Three-Dimensional Hill: Fluid Modeling Study on Effects of Stratification: Pt. I: Flow Structure, U.S. EPA Report No. EPA 600/1-78-041, Research Triangle Park, NC.

Hunt, J.C.R. and Snyder, W.H., 1980: Experiments on stably and neutrally stratified flow over a model three-dimensional hill. J. Fluid Mech. 96, part 4, p 671–704.

Hunt, J.C.R., 1982: Recommended formulae for Gaussian Plume Modeling of Atmospheric Diffusion over Hills. Note to the U.K. Atmospheric Dispersion Working Group. Dep. of Appl. Math and Theo. Physics, U. of Cambridge, Cambridge, U.K.

Isaacs, R.G., Bass, A., Egan, B.A., 1979: Application of Potential Flow Theory to a Gaussian Point Source Diffusion Model in Complex Terrain, Fourth Symposium on Turbulence, Diffusion, and Air Pollution, AMS. Boston, MA.

Kao, S.K. and Taylor, G.H., 1977: Observations of Dispersion in Complex Terrain. Preprints, Joint Conf. on Appl. of Air Poll. Meteor. AMS, Boston, Mass. p 160–167.

Lavery, T.F., Greene, B.R., and Egan, B.A.: 1983a, "The EPA Complex Terrain Model Development Program." Extended Abstracts. Sixth Symposium on Turbulence and Diffusion. AMS, Boston, MA.

Lavery, T.F., Strimaitis, D.G., Venkatram, A., Greene, B.R., DiCristofaro, D.C. and Egan, B.A., 1983b, "EPA Complex Terrain Model Development." Third Milestone Report - 1983. U.S. EPA Doc. No. EPA-600-3-83-101, Research Triangle Park, NC.

Lavery, T.F., Bass, A., Strimaitis, D.G., Venkatram, A., Greene, B.R., Drivas, P.J. and Egan, B.A., 1982: "EPA Complex Terrain Model Development." First Milestone Report - 1981. U.S. EPA Doc. No. EPA-600/3-82-036, Research Triangle, Park, NC.

Londergan, R.J., 1980: Validation of Plume Models: Statistical Methods and Criteria. EA-1673-54. Electric Power Research Institute, Palo Alto, Calif.

Lott, R.A., 1982: Terrain-Induced Downwash Effects on Ground Level SO_2 Concentrations. Atmos. Envir. 16, 4 p 635-642.

Manins, P.C. and Sawford, B.L., 1979: A Model of Katabatic Winds. J. Atmos. Sci., 36 p 619-630.

McNider, R.T., Hanna, S.R., and Pielke, R.A., 1981: Numerical Simulation of the Transport and Dispersion of Pollutants in Slope and Mountain-Valley Flows. Fifth Symposium on Turbulence, Diffusion and Air Pollution. AMS, Boston, MA.

Miller, C.W., 1978: An Examination of Gaussian Plume Dispersion Parameters for Rough Terrain. Atmos. Enviro. 12, pp 1359-1364.

Minott, D.H., Shearer, D.L. and Marker, R.S., 1977: Development of Vertical Dispersion Coefficients for Deep-Valley Terrain. U.S. Energy Res. and Dev. Admin. Report to EY-76-C-02-4026. U.S. ERDA, Washington, D.C. 33 pp.

Overcamp, T.J., 1983: A Surface-Corrected Gaussian Model for Elevated Sources. J. Clim. and Appl. Meteor., 22, p 1111-1115.

Panofsky, H.A., Egolf, C.A., and Lipschutz, R., 1978: On Characteristics of Wind Direction Fluctuations in the Surface Layer Bound, Layer. Meteorol. 15, p439-446.

Pasquill, F., 1961: The Estimation of the Dispersion of Windborne Material. The Meteor. Magazine, 90(1063) p 33-49.

Peterson, R.L. and E.E. Twombly, 1982: "Site-Specific Concentration Estimates for Stable Plume Impact on Elevated Terrain." 75th Annual APCA meeting. June 20-25, New Orleans, Louisiana.

Rao, K.S. and Snodgrass, H.F, 1981: The Nocturnal Boundary Layer over a Sloping Surface. Fifth Symposium on Turbulence, Diffusion and Air Pollution. AMS, Boston, MA.

Rao, K.S. and Nappo, C.J., 1983: Comparison of Observations and Model Predictions of Nocturnal Drainage Flow. Sixth Symposium on Turbulence and Diffusion. AMS, Boston, MA.

Rowe, R.D., Lee, C.Z. and Havlena, J.J., 1980: The Elevation of Stable Atmospheric Plumes Above a Ridge. 73rd Annual Meeting of the Air Poll. Cont. Assoc., Paper 80-55.5. Pittsburgh, Penna.

Rowe, R.D., Benjamin, S.F., Chung, K.P., Havlena, J.J. and Lee, C.Z., 1981: Field Studies of Stable Air Flow over and around a Ridge. Atmos. Envir. 16, 4, p 643-653.

Ryan, W., and Lamb, B., 1984: Determination of Dividing Streamline Heights and Froude Numbers for Predicting Plume Transport in Complex Terrain. J. Air. Pol. Cont. Assoc. 34, 2 p 152-155.

Schiermeier, F.A., Lavery, T.F., Strimaitis, D.G., Venkatram, A., Greene, B.R. and Egan, B.A., 1983: EPA Model Development for Stable Plume Impingement on Elevated Terrain Obstacles. NATO/CCMS 14th Inter. Tech. Meeting on Air Pollution Modeling and Its Application, Copenhagen, Denmark, Sept. 27-30, 1983.

Shearer, D.L. and Minott, D.H., 1977: Development of Vertical Dispersion Coefficients for Rolling Terrain Environments. Preprints, Joint Conf. on Appl. of Air Poll. Meteor. AMS, Boston, Mass. p 160-167.

Slowik, A.A., Austin, J.M., and Pica, G.N., 1977: Plume Dispersion Modeling in Complex Terrain under Stable Atmospheric Conditions. 70th Annual Meeting of the Air Poll. Cont. Assoc. Paper 77-29.1, Pittsburgh, Penna.

Sheppard, P.A., 1956: "Airflow over Mountains," Quart. J. Roy. Meteorlogical Society, v. 82, p 528-9.

Snyder, W.H., 1980: "Towing Tank Studies in Support of Field Experiments at Cinder Cone Butte, Idaho. Phase III: Verification of Formula for Prediction of Dividing Streamline Height." U.S. EPA, Research Triangle Park, NC.

Snyder, W.H. and Lawson, R.E., Jr., 1981: "Laboratory Simulation of Stable Plume Dispersion over Cinder Cone Butte: Comparison with Field Data" Appendix to U.S. EPA Doc. No. EPA-600/3-82-036, Research Triangle Park, NC.

Snyder, W.H. and Lawson, R.E., Jr., 1983: Stable Plume Dispersion over an Isolated Hill; Releases above the Dividing-Streamline Height. Internal Report, U.S. Environmental Protection Agency, Research Triangle Park, NC. 72 pp.

Snyder, W.H., Thompson, R.S., Eskridge, R.E., Lawson, R.E., Jr., Castro, I.P., Lee, J.T., Hunt, J.C.R. and Ogawa, Y., 1982: The Structure of Strongly Stratified Flow over Hills, Dividing Streamline Concept. Appendix A, EPA 60013-83-015, EPA. Research Triangle Park, NC. 56 pp.

Snyder, W.H., 1983: Fluid Modeling of Terrain Aerodynamics and Plume Dispersion. A perspective view. Extended Abstracts. Sixth Symposium on Turbulence and Diffusion. AMS, Boston, MA p 317-320.

Strimaitis, D.G., Venkatram, A., Greene, B.R., Hanna, S., Heisler, S., Lavery, T.F., Bass, A. and Egan, B.A., 1983:, "EPA Complex Terrain Model Development." Second Milestone Report - 1982. U.S. EPA Doc. No. EPA-600/3-83-015, Research Triangle Park, NC.

Thompson, R.S., Snyder, W.H. and Lawson, R.E., Jr., 1983: "Laboratory Simulation of Neutral Plume Dispersion over Cinder Cone Butte: Comparison with Field Data" Appendix to U.S. EPA Doc. No. EPA-600/3-83-101, Research Triangle Park, NC.

U.S. EPA, 1977: User's Manual for Single-Source (CRSTER) Model, Office of Air Quality Planning and Standards. EPA-450/2-77-013. Research Triangle Park, NC.

U.S. EPA, 1980: Guideline on Air Quality Models, Proposed Revisions, EPA, Research Triangle Park, NC, 115 pp.

Van der Hoven, I. et al., 1972: "Southwest Energy Study," Report of the Meteorology Work Group. U.S. Department of Interior, Draft Report, March 1972. 113 p.

Van der Hoven, I., Ferber, G.J., Humphrey, P.A., Holzworth, G.C., Heffer, J.L., and Quiring, R.F., 1972: Report of the Meteorology Work Group: Appendix E, Southwest Energy Study. U.S. Dept. of the Interior, Washington, D.C. 180 pp.

Venkatram, A., Strimaitis, D. and DiCristofaro, D., 1984: A Semiempirical Model to Estimate Vertical Dispersion of Elevated Releases in the Stable Boundary Layer. (To be published in Atmos. Envir.)

Venkatram, A., Strimaitis, D., and Eberhard, W., 1983: Dispersion of Elevated Releases in the Stable Boundary Layer. Extended Abstracts Sixth Symposium on Turbulence and Diffusion. AMS, Boston, MA. p 297-299.

Venkatram, A., 1982: A Framework for Evaluating Air Quality Models. Boundary-Layer Meteor., 24, p 371-385.

Weil, J.C., 1979: Modeling of Buoyant Plume Dispersion in Complex Terrain. Report No. PPRP-35. Martin Marrietta Corp., Baltimore, Maryland.

Weil, J.C., S.C. Traugott, and Wong, D.K., 1981: Stack Plume Interaction and Flow Characteristics for a Notched Ridge. Report No. PPRP-61. Martin Marietta Corp. Baltimore, Maryland.

Whiteman, C.D., 1980: Breakup of Temperature Inversions in Colorado Mountain Valleys. Thesis. Dept. of Atmos. Science. Colorado State Univ., Ft. Collins, CO.

Whiteman, C.D. and Allwine, K.J., 1983: Valmet-A Valley Air Pollution Model. Doc. No. PNL-4728. Pacific Northwest Laboratory, Richland, Washington, 250 pp.

Williams, M.D., Mangeng, C.A., Barr, S. and Lewis, R., 1982: "Application and Development of the Four Corners Air Dispersion Model." 75th Annual Meeting of the Air Pollution Control Association, June 20-25, 1982, New Orleans, Louisiana.

Yamada, T., 1981: A Numerical Simulation of Nocturnal Drainage Flow. J. Meteor. Soc. Japan, 59. p 108-122.

PART I

BOUNDARY LAYER STRUCTURE

SOME ASPECTS OF THE TURBULENT STABLE BOUNDARY LAYER

F.T.M. Nieuwstadt
Royal Netherlands Meteorological Institute,
De Bilt, The Netherlands

ABSTRACT

We consider the structure of the stable boundary layer using the concept of local scaling. In this scaling approach turbulence variables, non-dimensionalized with measurements taken at the same height, can be expressed as a function of a single parameter z/Λ, where z is the height and Λ a local Obukhov length. One of the consequences is that locally scaled variables become constant above the surface layer. This behavior is illustrated with observations of the Richardson number. With local scaling as a closure hypothesis we then formulate a model of the stable boundary layer. Its solution for steady-state conditions is given. One result we obtain is the well-known Zilitinkevich equation for the boundary-layer height. A comparison of this equation with observations results in a reasonable agreement. Also we discuss some alternative expressions for the stable boundary-layer height and compare them with observations. Another result of our model is an explicit profile for the K-coefficient as a quadratic function of height. We discuss the consequences of this expression for the dispersion of a point source emission. We find that the time scale of diffusion in this case is about 5 hours.

1. INTRODUCTION

Turbulence in the stable, nocturnal boundary layer is less well understood than in its day-time counterpart the convective boundary layer. The main reason is that the stable boundary layer is usually influenced by other processes in addition to turbulence, i.e. gravity waves (e.g. Caughey and Readings, 1975; Einaudi and Finnigan, 1981 and Nai-Ping et al., 1983), long-wave radiation (Garratt and Brost, 1981 and André and Mahrt, 1982) and intermittency (Kondo et al., 1978). In this paper we restrict ourselves to a discussion of the vertical structure of a fully turbulent stable boundary layer. In particular we direct our attention to two subjects: the depth of the turbulent layer and the

Boundary-Layer Meteorology 30 (1984) 31–55. 0006–8314/84/0301–0031$03.75

vertical exchange coefficient.

We first consider the turbulent structure. Observations of the stable boundary layer have been discussed by Mahrt et al. (1979), Yamamoto et al. (1979), Caughey et al. (1979), Ueda et al. (1981), Hunt et al. (1983) and Nieuwstadt (1984a), while models have been developed by Brost and Wyngaard (1978), Rao and Snodgrass (1979), Yamada (1979), McPhee (1981) and Nieuwstadt (1984b). Neither observations nor models can be made to fit a single description of stable turbulence. However, the case of continuous turbulence found during relatively weak stratification seems more manageable. Its structure obeys a so-called local scaling hypothesis, which we will introduce here as an extension of Monin-Obukhov similarity theory valid in the surface layer. The results of local scaling are then applied in a boundary-layer model, which is able to produce realistic vertical profiles of turbulence variables.

Next we consider the height of the turbulent boundary layer, an important but also controversial parameter. To calculate this height two approaches are usually taken: diagnostic models, which give a direct relationship between the boundary-layer height and parameters such as turbulence or mean variables (Zilitinkevich, 1972; Yu, 1978; Mahrt and Heald, 1979; Nieuwstadt, 1981; Arya, 1981; Garratt, 1982b and Mahrt et al., 1982) and prognostic models which describe the boundary-layer height as a function of time by means of a rate equation (Zeman, 1979; Smeda, 1979; Mahrt, 1981; Nieuwstadt and Tennekes, 1981 and Garratt, 1982a). We discuss the merits of both type of models but in particular we will examine the diagnostic models by means of a comparison with observations. In addition we will consider the expressions for the boundary-layer height proposed by Stull (1983a, 1983b), which cannot be classified uniquely into the two groups introduced above.

Finally we turn to the dispersion of passive contaminants in a stable boundary layer. The few studies on this subject (Kerman, 1979; Hunt, 1982 and Pearson et al., 1983) have discussed several elements of this problem. A systematic approach does not yet exist. Our investigation here is based on local scaling, which produces an explicit profile of the vertical exchange coefficient. We will consider the solution of the diffusion equation with this profile for the case of an instantaneous point source emission.

2. VERTICAL STRUCTURE

The aim of this section is to give a description of the turbulent structure of the stable boundary layer as a function of height. Let us start with the surface layer which is defined as the bottom part of the boundary layer, where the fluxes of momentum and heat are approximately constant. The structure of the surface layer is usually described in terms of Monin-Obukhov (M-O) similarity theory, which employs these constant fluxes as scaling parameters (Monin and Yaglom, 1971). Let us give an example. Consider the mean velocity gradient, $\partial U/\partial z$. Its non-dimensional form is defined as

$$\phi_m = \frac{kz}{u_*}\frac{\partial U}{\partial z} , \tag{1}$$

where k is the Von Karman constant, z the height and u_* the friction velocity, which is defined as the square root of the kinematic surface stress. For the numerical value of the Von Karman constant we adopt here 0.4 ± 0.04 (Tennekes, 1982).

According to M-O similarity theory ϕ_m can be expressed as a function of only one parameter, z/L, where L is the Obukhov length. It is defined by

$$L = \frac{u_*^{\,2}}{k \frac{g}{T} T_*} . \tag{2}$$

Here, g/T is the buoyancy parameter and T_* the temperature scale which is related to the surface temperature flux $\overline{w\theta}_o$ by $T_* = - \overline{w\theta}_o/u_*$.

Experimental results (see e.g. Businger et al., 1971) agree closely with M-O scaling. For stable conditions (L > 0) the following relationship is found for ϕ_m,

$$\phi_m = 1 + \beta\, z/L , \tag{3}$$

where $\beta \sim 5$ (Dyer, 1974).

The surface layer covers about the lowest 10% of the boundary layer. Above the surface layer, fluxes can no longer be approximated as constant with height and consequently M-O similarity theory fails. However, M-O theory can be modified to describe the upper part of the stable boundary layer, if we adopt as scaling parameters the local kinematic stress, τ, and the local temperature flux, $\overline{w\theta}$, rather than the surface values u_* and T_*. This approach is called local scaling and its background is discussed in more detail by Nieuwstadt (1984a). As an example we consider the standard deviation of the vertical velocity fluctuations, σ_w. Local scaling asserts that $\sigma_w/\tau^{\frac{1}{2}} = f(z/\Lambda)$, where Λ is defined by

$$\Lambda = - \frac{\tau^{3/2}}{k \frac{g}{T} \overline{w\theta}} . \tag{4}$$

Comparison with (2) suggests that Λ can be interpreted as a 'local' Obukhov length. Observational evidence supporting local scaling is presented by Nieuwstadt (1984a). Laboratory data of Keisuke et al. (1983) seem also to agree with it.

As another example of local scaling we consider the gradient Richardson number

$$Ri = \frac{\frac{g}{T}\frac{\partial \Theta}{\partial z}}{\left\| \frac{\partial \tilde{U}}{\partial z} \right\|^{2}} , \tag{5}$$

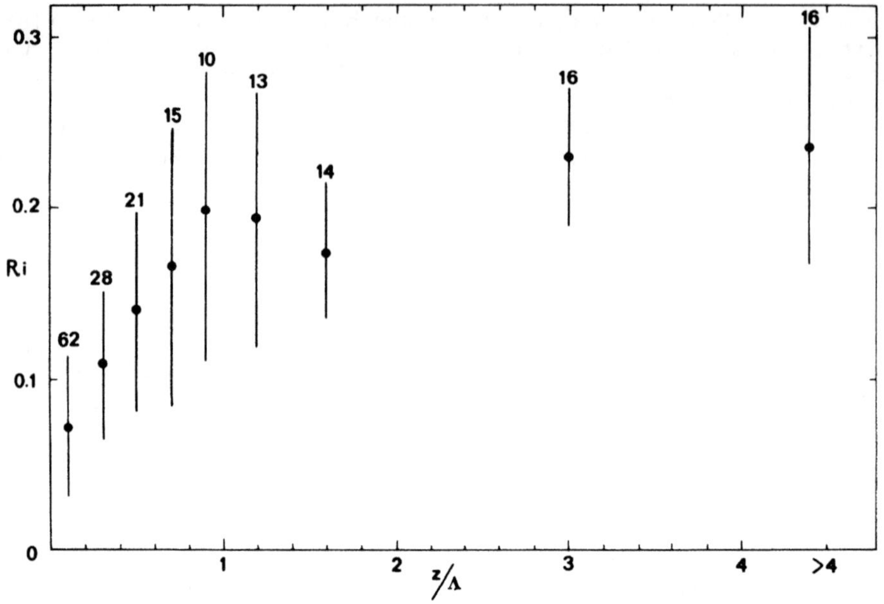

Figure 1. The Richardson number as a function of z/Λ. Each point is the average of a number of observations in a range of z/Λ values. The vertical bar gives the standard deviation. The number of observations is also indicated.

where $\partial \underset{\sim}{U}/\partial z$ is the gradient of the horizontal wind vector $\underset{\sim}{U} = (U,V)$ and where Θ is the mean temperature. Local scaling leads to a relationship of the form $Ri = Ri(z/\Lambda)$. The observations discussed by Nieuwstadt (1984a), which we summarize in Figure 1, indeed support such a relationship. This result implies that z/Λ and Ri can be uniquely related. Therefore, local scaling is in principle equivalent to the approach taken in other studies where turbulence parameters are expressed as a function of Ri (Kondo et al, 1978 and Ueda et al., 1981).

The data of Figure 1 give the impression that the gradient Richardson number approaches a limit for $z/\Lambda \to \infty$. This limit is usually denoted as the critical Richardson number and the data suggest a value in the neighbourhood of 0.2.

Another parameter closely related to Ri is the flux Richardson number, Ri_f. Its definition reads (Businger, 1982)

$$Ri_f = - \frac{\frac{g}{T}\,\overline{w\theta}}{\underset{\sim}{\tau} \cdot \frac{\partial \underset{\sim}{U}}{\partial z}}, \tag{6}$$

where $\underset{\sim}{\tau} = (\tau_x, \tau_y)$ is the kinematic stress vector. Local scaling predicts again a relationship between Ri_f and z/Λ, which turns out to exhibit a similar behavior as found for Ri. For instance, the flux-Richardson number approaches a limiting value, the critical flux-Richardson number for $z/\Lambda \to \infty$. According to the observations of Nieuwstadt (1984a) its value is ~ 0.2. Kondo et al. (1978) and Ueda et al. (1981) find for the critical Ri_f a value of 0.143 and 0.1 respectively. However, their studies do not suggest a value of the critical gradient Richardson number. The explanation of this result is that in their case the boundary layer is very stable and it becomes intermittently turbulent. In contrast, the observations of Nieuwstadt (1984a) are restricted to continuous turbulence. A critical value of the gradient Richardson number could perhaps be interpreted as the transition between these two states of turbulence.

Next, we concentrate on the limiting behavior of local scaling for large values of z/Λ. Stable stratification inhibits vertical motion and as a result reduces the length scale of turbulence. When this length scale becomes much smaller than the height above the surface, turbulence no longer feels the presence of the ground and as a consequence an explicit dependence on z disappears. The length scale of turbulence is proportional to Λ and in terms of local scaling this result means that dimensionless quantities approach a constant value for large z/Λ. This conclusion is for instance confirmed by the data shown in Figure 1. Further evidence is given by Nieuwstadt (1984a). A similar result is found for M-O theory when $z/L \to \infty$. For example, the velocity gradient ϕ_m (3) becomes in this limit equal to

$$\frac{L}{u_*}\frac{\partial U}{\partial z} = \frac{\beta}{k} \tag{7}$$

and we see that the explicit dependence on z has disappeared. Wyngaard (1973) has coined for this behaviour of the surface-layer profiles in stable circumstances the term local z-less stratification. Here, we extend it to the region above the surface layer for the case of locally scaled variables.

The results discussed above can be applied in a model for the horizontally homogeneous stable boundary layer. Its governing equations read

$$\frac{\partial \Theta}{\partial t} = - \frac{\partial \overline{w\theta}}{\partial z} , \tag{8}$$

$$\frac{\partial U}{\partial t} = f(V-V_g) + \frac{\partial \tau_x}{\partial z} , \tag{9}$$

$$\frac{\partial V}{\partial t} = - f(U-U_g) + \frac{\partial \tau_y}{\partial z} , \tag{10}$$

$$Ri = \text{const}, \; Ri_f = \text{const} , \tag{11}$$

where U_g, V_g are the components of the geostrophic wind and where the Coriolis parameter $f = 10^{-4}\ s^{-1}$. The equations (8), (9) and (10) express the energy and momentum balance of the mean flow. (Remember that τ_x and τ_y are kinematic stresses.) These equations are completed by the closure hypotheses (11), which, as we have seen above, follow from local z-less stratification.

It is expected that these equations give a reasonable description of the boundary layer for $h/L > 1$ except in a region close to the ground, where the Ri and Ri_f should explicitly depend on z according to M-O similarity theory.

A solution of the set of equations (8)-(11) can be found for the stationary case, which is defined as a boundary layer in which the turbulent fluxes are independent of time. The details of this solution are discussed by (Nieuwstadt, 1984b). Here, we will present only some results.

For the height of the boundary layer, h, we obtain

$$h = (\sqrt{3}\ k\ Ri_f)^{\frac{1}{2}} \left(\frac{u_* L}{f}\right)^{\frac{1}{2}} , \tag{12}$$

an expression derived originally by Zilitinkevich (1972). The constant $(\sqrt{3}\ k\ Ri_f)^{\frac{1}{2}} = 0.37$, which is close to the value 0.4 recommended by Brost and Wyngaard (1978) and Garratt (1982a). A further discussion of the stable boundary-layer height is postponed to section 3.

For the exchange coefficient K_M, which is defined by $\underset{\sim}{\tau} = K_M\ \partial \underset{\sim}{U}/\partial z$, the solution of (8)-(11) leads to

$$K_M = k\ Ri_f\ u_*\ L\ (1-z/h)^2 . \tag{13}$$

We will discuss the consequences of this profile for dispersion of pollutants in a stable boundary layer in section 4. Here, we note that in the limit $z/h \to 0$ the expression (13) is consistent with the velocity gradient (7) of M-O theory for $z/L \gg 1$, because in the surface layer $(L/u_*)\ \partial U/\partial z = u_*\ L/K_M$. As a result of this comparison we also obtain that $\beta = Ri_f^{-1}$.

The next results which follow from the solution of (8)-(11) are the profiles of the kinematic stress and the temperature flux. We find the expressions

$$\tau = u_*^2\ (1 - z/h)^{1.5} , \tag{14}$$

$$\overline{w\theta} = \overline{w\theta_o}\ (1 - z/h) . \tag{15}$$

These profiles agree very well with measurements taken at Cabauw (Nieuwstadt, 1984a). Additional evidence is presented in the studies of Mahrt et al. (1979), Caughey et al. (1979) and Yamamoto et al. (1979).

The vertical profiles (14) and (15) and the results of local scaling, in particular the limit condition of local z-less stratification, allow us to give a complete description of the vertical structure of turbulence in a stable boundary layer. As an example we consider σ_w. Local z-less stratification implies that $\sigma_w/\tau^{\frac{1}{2}}$ = constant. With (14) we then find

$$\sigma_w = 1.3\ u_* \ (1 - z/h)^{3/4} \tag{16}$$

where 1.3 u_* is the value of σ_w in the surface layer. Observations confirm this profile (Caughey et al. 1979 and Nieuwstadt, 1984a). In the latter study vertical profiles of other turbulent parameters are also discussed.

To conclude this section we note again that the derivation of the vertical profiles (14)-(16) from the solution of (8)-(11) is based on the assumption of steady-state conditions. However, in reality the nocturnal boundary layer is generally not stationary. Nevertheless, observations at Cabauw seem to agree well with these profiles. Our explanation is that because of the sufficiently small turbulent time scale the actual profiles of turbulence variables will not depart much from the steady-state curves in the rather weakly non-stationary or inhomogeneous conditions at Cabauw. Such a conclusion may not be drawn for the profiles of mean variables because as we will see later their time scales are much longer.

3. BOUNDARY-LAYER HEIGHT

This section deals with the height of a stable boundary layer. We will confine ourselves to the depth of the turbulent layer, which should not be confused with the inversion depth. The latter is generally larger (Mahrt et al., 1979; Nieuwstadt, 1981; Mahrt et al., 1982; André and Mahrt, 1982 and Wetzel, 1982).

In the previous section we have obtained Zilitinkevich's equation (12) for the height of the stationary boundary layer. Such an expression is called a diagnostic model because it calculates the boundary-layer height at each moment in terms of variables given for the same time. Expression (12) has received considerable attention in the literature (Yu, 1978; Mahrt and Heald,1979; Nieuwstadt, 1981; Arya, 1981; Garratt, 1982b and Mahrt et al., 1982). In general these studies indicate that (12) agrees reasonably well with observations. Usually one finds correlation coefficients of the order of 0.7.

Here, we present a new set of data of the stable boundary-layer height collected at the Cabauw experimental facility in the period of 1978 to 1979. The complete data set is given in Appendix A. First, we compare the observed heights with the expression (12), in which we take $(\sqrt{3}\ k\ Ri_f)^{\frac{1}{2}} \sim 0.4$. We exclude from the comparison the very weakly stable cases for which $h/L < 1$. A scatter plot is shown in Figure 2 and the statistical results of the comparison are given in Table 1. We find a reasonable agreement between Zilitinkevich's expression and the observations. The correlation coefficient is 0.77, which is of the same order of magnitude as the values reported by Arya (1981) and Mahrt et al. (1982).

The agreement is primarily due to the parameter u_* in (12), because the parameter T_* tends to become constant in a stable boundary layer (Venkatram, 1980 and Van Ulden and Holtslag, 1983). The latter authors find for clear nights at Cabauw that $T_* \simeq 0.09$. With this result we are

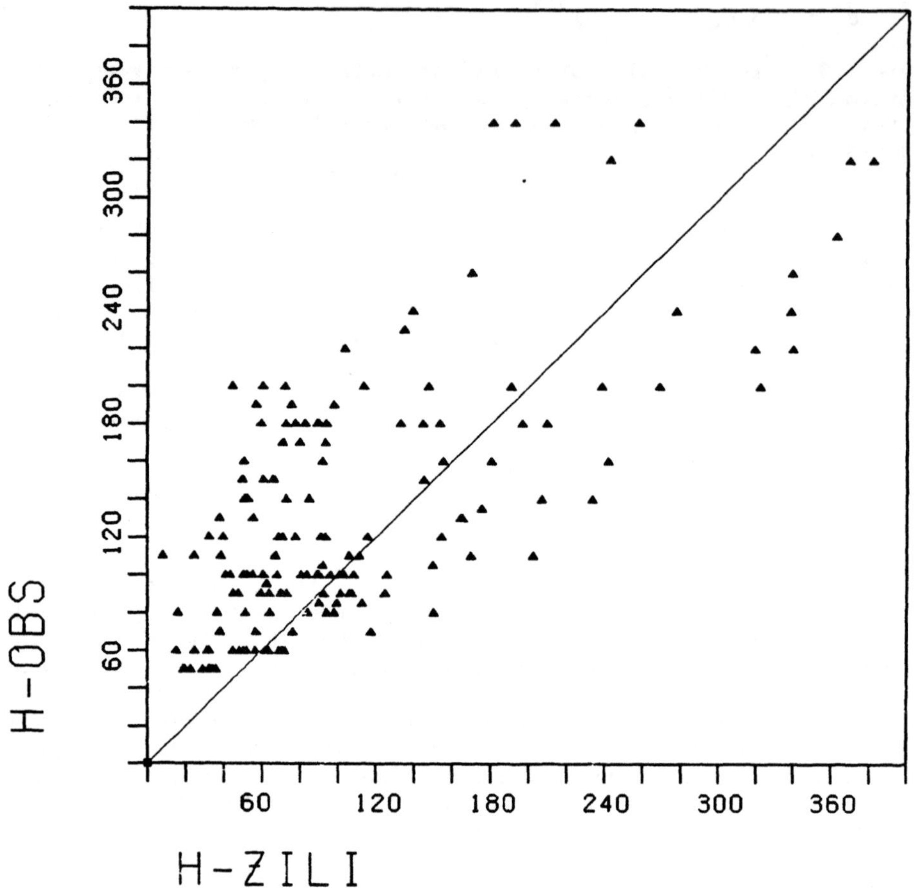

Figure 2. A comparison of Zilitinkevich's expression (12) with observed boundary-layer heights. The statistics are given in Table 1. The data are given in appendix A.

able to express the Zilitinkevich relationship solely in terms of the wind speed. With the aid of the wind profile which results by integrating the M-O expression (3) for ϕ_m we find

$$h = C_1 \, U_z^{3/2} \, \left\{\frac{1}{2} + \frac{1}{2} \left(1 - \frac{C_2}{U_z^2}\right)^{\frac{1}{2}}\right\}^{3/2} \tag{17}$$

with

$$C_1 = \left(\frac{\sqrt{3}\, Ri_f}{\frac{g}{T} \, f \, T_*}\right)^{\frac{1}{2}} \left(\frac{k}{\ln(z/z_o)}\right)^{3/2} \tag{18}$$

and

$$C_2 = \frac{4 \beta \frac{g}{T} T_* z \ln(z/z_o)}{k} . \quad (19)$$

Here, z_o is the roughness length, which appears as the integration constant in the integral of (3) and U_z is the windspeed at the observation height z = 10 m. For the conditions representative of Cabauw ($z_o \sim 0.2$ m) we find $C_1 \simeq 3.7$ and $C_2 \simeq 6.8$. For these values expression (17) is compared with the observations. The results are shown in Table 1. The agreement is poorer than for the Zilitinkevich's expression. The reason is that (17) allows only a solution for $U_{10} > C_2^{\frac{1}{2}} \sim 2.6$ m/s, so that low wind speed cases are eliminated from the comparison. Moreover, near this cut-off value of the wind speed the expression (17) underestimates the boundary-layer height considerably.

To eliminate this undesirable behaviour we propose a simplification of (17) by evaluating the term within the brackets with the average wind speed $\overline{U}_{10} \simeq 3.5$ m s^{-1}. The result reads

$$h = 28\, U_{10}^{3/2} . \quad (20)$$

In Table 1 we compare this expression with the observations. The agreement is not significantly different from that found with Zilitinkevich's expression (12). Thus (20) can be recommended as a simple diagnostic expression for estimation of the nocturnal boundary-layer height. The coefficient in this equation should be considered as representative only of the Cabauw region.

Another expression for the stable boundary-layer height has been discussed by Stull (1983a, 1983b). He introduces a heat-flux history

Table 1. Correlation coefficients and parameters in the linear regression equations. The tested regression equations are y = a x + b and y = c x where y are the observations and x are the model calculations. The correlation coefficient is r and the number of data is N.

model	Eq.(12)	Eq. (17)	Eq.(20)	Eq.(22)
N	157	128	212	220
$\overline{y}$	140	187	154	152
$\overline{x}$	118	370	149	416
$\{\overline{(y-x)^2}\}^{\frac{1}{2}}$	67	272	58	511
a	0.59 ± 0.04	0.20 ± 0.02	0.79 ± 0.05	0.09 ± 0.01
b	70 ± 6	109 ± 10	37 ± 9	112 ± 6
c	0.94 ± 0.03	0.42 ± 0.02	0.99 ± 0.02	0.21 ± 0.01
r	0.77	0.63	0.72	0.57

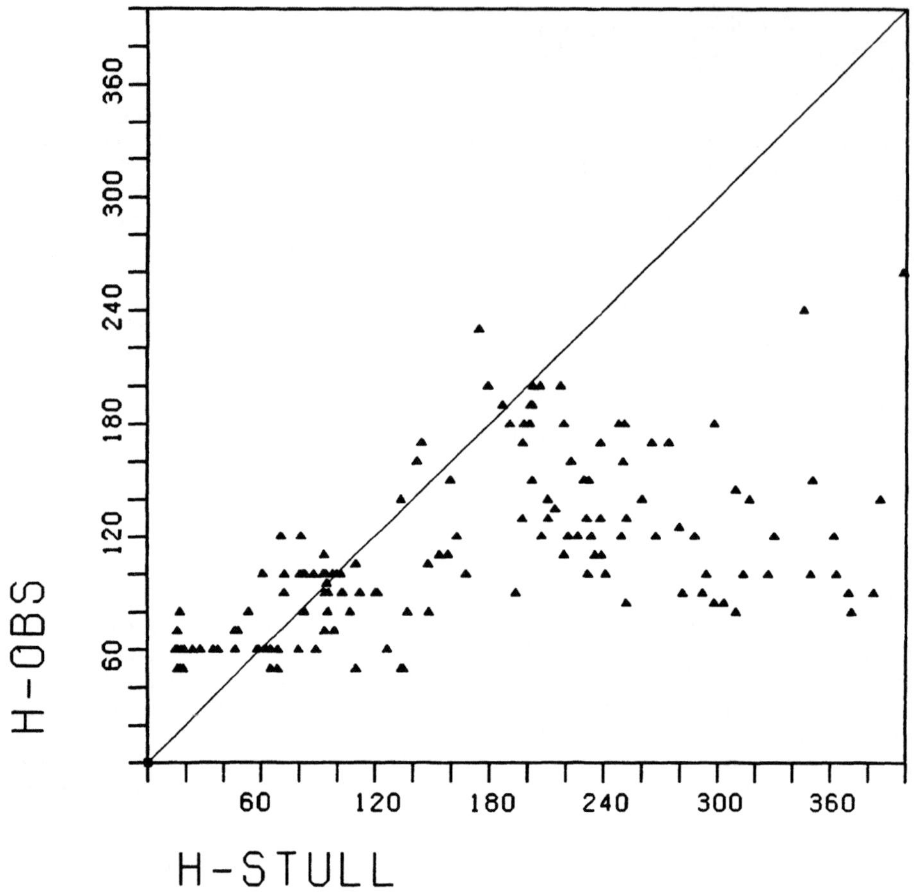

Figure 3. A comparison of Stull's expression for the boundary layer height (22) with observations. The statistics are given in Table 1. The data are given in appendix A.

length scale

$$H = \int_{t_o}^{t} \overline{w\theta}_o \, dt \;/\; \left((\Theta_s(t) - \Theta_s(t_o)\right), \tag{21}$$

where Θ_s is the mean potential temperature at the surface and where t_o is the time of transition when the temperature flux changes sign. For the depth of the turbulent boundary layer defined as the height where the temperature flux is equal to 5% of its surface value, Stull (1983b) proposes

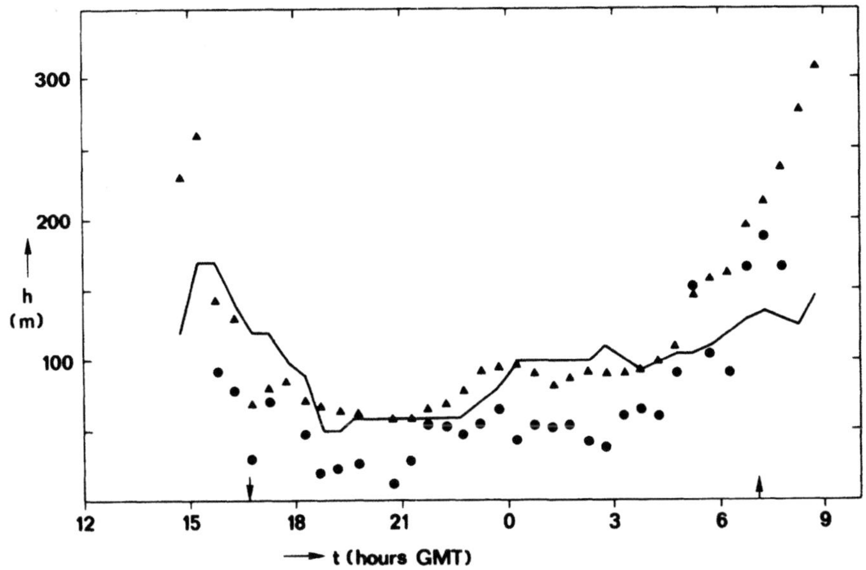

Figure 4. Time history of the observed boundary-layer height (solid line) for the night period of 8/9 February 1979. Times of sunset and sunrise have been indicated. The dots denote the boundary-layer height obtained with Zilitinkevich's equation (12). The triangles denote the values calculated with Stull's expression (22).

$$h = 5.35\ H\ . \tag{22}$$

We have compared this expression with our observations. A scatter diagram is shown in Figure 3 and the statistical results are given in Table 1. We find that the correlation is appreciably lower than for Zilitinkevich's expression. Moreover, (22) seems to overestimate the observations considerably, especially in cases with strong winds when the boundary-layer height is large. This suggests that (22) is more suited to estimate the inversion depth, which is the original interpretation put forward by Stull (1983a).

As an additional illustration we show in Figure 4 a typical example of the nocturnal boundary-layer height development. We have selected the observation period of February 8/9 1979. In that case the results of (22) are slightly superior to those of Zilitinkevich's expression. For the other periods given in Appendix A the comparison of (22) with observations is generally poorer, as should be also clear from the results shown in the scatter diagram of Figure 3.

In the previous section we have seen that a diagnostic model for the boundary-layer height, such as expression (12), is only valid for steady-state conditions. However, the nocturnal boundary layer, as we mentioned before, is in general not stationary. Therefore, several authors have developed prognostic equations for the stable boundary-layer height (Zeman, 1979; Smeda, 1979; Mahrt, 1981; Nieuwstadt and Tennekes, 1981 and Garratt, 1982a). These equations usually take the form of a relaxation process, in which the actual boundary-layer height approaches an equilibrium value. The equilibrium height is a diagnostic equation like (12). It turns out that the time scale of the relaxation proces is very large (Nieuwstadt and Tennekes, 1981). As a consequence, large deviations between the actual boundary-layer height and its equilibrium value might be expected. However, we have seen above that this is not the case. Our explanation is that the boundary-layer height at the end of the transition period around sunset is probably already close to its equilibrium. It remains near the equilibrium value because both the actual boundary-layer depth and the equilibrium height do not generally vary much during the course of a night period, as illustrated by our case study shown in Figure 4. This may explain the relative succes of diagnostic models. Garratt (1982b) goes as far as to attribute the discrepancy beween Zilitinkevich's expression (12) and observations primarily to statistical errors in the parameters which enter in the diagnostic equation.

Also we believe that the skill of prognostic models is primarily due to the use of an observed height as an initial value to start the calculation. Because of the large time scale of nocturnal boundary-layer development, this initial height is a good estimator for the boundary-layer deptht throughout the night. This effect has been stressed by Nieuwstadt and Tennekes (1981) and more recently by André (1983).

Thus we conclude that as long as no objective method exists to estimate the initial condition for the prognostic equations, diagnostic models for the nocturnal boundary-layer height should be preferred, despite the fact these models are theoretically not satisfactory in the stable boundary-layer. In this respect equations such as (20), which only depend on wind speed, seem to have practical value.

4. DIFFUSION

Let us now turn our attention to the dispersion of pollutants in a stable boundary layer. Turbulent diffusion of a passive contaminant is described by the equation of mass conservation. Here, we will restrict ourselves to a special case in which we consider the dispersion from an instantaneous surface source or, alternatively, the vertical dispersion of an area averaged concentration distribution. The diffusion equation then simplifies to

$$\frac{\partial C}{\partial t} = - \frac{\partial \overline{wc}}{\partial z} \tag{23}$$

where C is the mean concentration and $\overline{wc}$ the turbulent flux. We have seen in section 2 that for the stationary boundary layer our local scaling approach is equivalent to the gradient transfer hypothesis so that we may write

$$\overline{wc} = - K \frac{\partial C}{\partial z} .$$

For the K-coefficient we adopt the quadratic profile (13) and we will consider in this section its consequences for diffusion in a stable boundary layer.

Before we discuss the solution of (23) we rewrite this equation in the following dimensionless form

$$\frac{\partial C}{\partial \sigma} = \frac{\partial}{\partial \eta} \{(1-\eta)^2 \frac{\partial C}{\partial \eta}\} \tag{24}$$

where the coordinates σ and η are defined by

$$\eta = z/h, \quad \sigma = t/T . \tag{25}$$

We have introduced a characteristic time scale of diffusion

$$T = h^2/(k\ Ri_f\ u_* \ L). \tag{26}$$

With some realistic values for $h \sim 75$ m, $u_* \sim 0.15$ ms^{-1} and $L = 25$ m we find $T = \sim 5$ hours. An alternative estimate can be obtained by substitution of Zilitinkevich's expression (12) for h in (26) with as result $T = \sqrt{3}/f \sim 5$ hours. This result shows that diffusion in a stable boundary layer is a relatively slow process compared to neutral and unstable boundary layers where time scales are smaller. As a matter of fact we did already refer to such a large time scale in previous section when we discussed the non-stationarity of a stable boundary-layer height. Here, we find that the limited vertical mixing is responsible.

In the following we will discuss solutions of the dimensionless equation (24) for an instantaneous point source emission. These solutions should then satisfy the following initial and boundary conditions

$$C = \delta(\eta - \eta_s) \ , \ t = 0 \tag{27}$$

$$\overline{wc} = 0 \ , \qquad z = 0 \tag{28}$$

$$\overline{wc} = 0 \ , \qquad z = h \tag{29}$$

where $\eta_s = z_s/h$ and z_s is the emission height. We have normalized the amount of emitted material.

Let us first investigate this problem for an emission height $z_s = 0$ by means of a Laplace transform which is defined by

$$\chi = \int_0^\infty e^{-s\sigma}\ C\ d\sigma . \tag{30}$$

Substitution into (24) and application of the initial and boundary

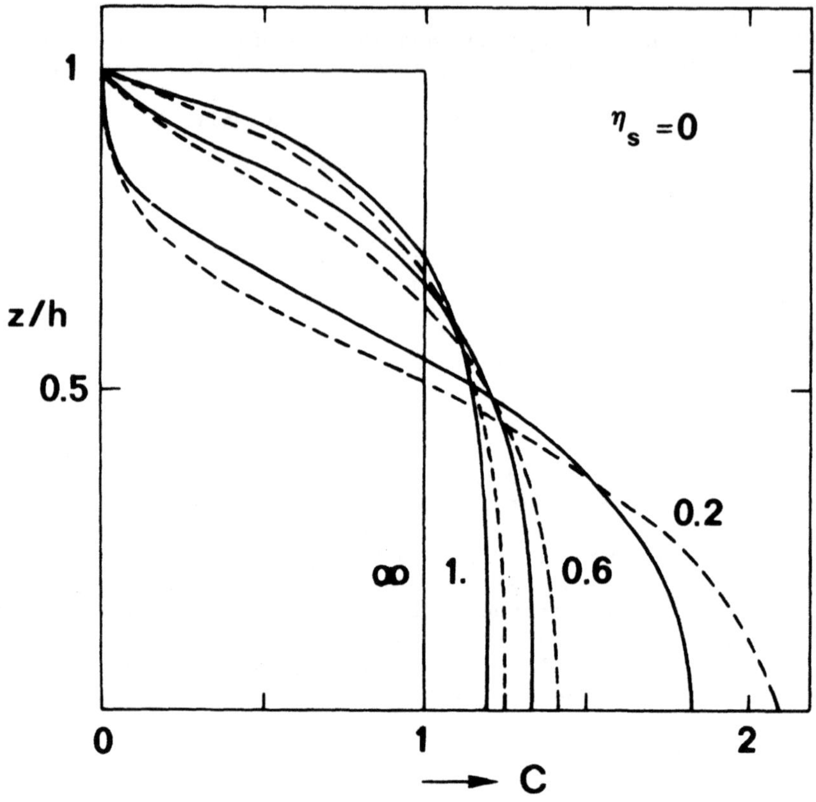

Figure 5. Vertical concentrations profiles for the emission from a ground level source for several dimensionless times. The solid line denotes the calculation with the quadratic K-profile (13). The dashed line gives the results calculated with the K-expression (36) for h/L = 2.

conditions lead to

$$\chi = \frac{2}{-1 + \sqrt{(1 + 4s)}} (1 - \eta)^{\frac{-1 + \sqrt{(1 + 4s)}}{2}} . \qquad (31)$$

The limit for $s \to 0$ can be interpreted in terms of the asymptotic behavior of the concentration distribution C for $t \to \infty$. We find the

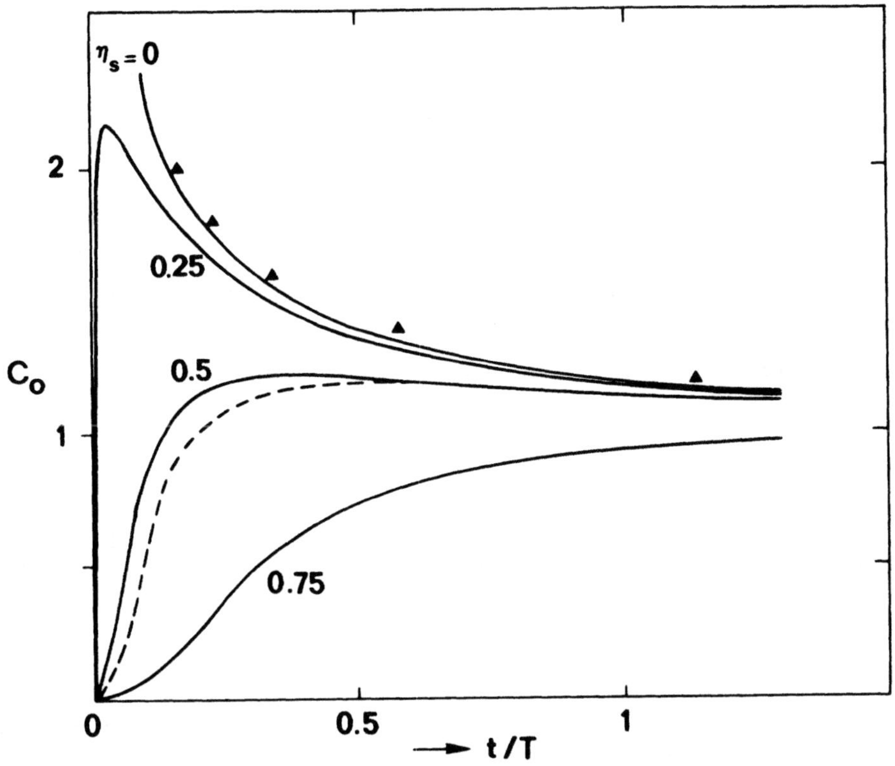

Figure 6. Ground-level concentrations as a function of dimensionless time for the emission from several source heights. The solid line denotes the calculation with the quadratic K-profile (13). The dashed line give the results calculated with the K-expression (36) for h/L = 2 and $z_s/h = 0.5$. The triangles are calculated with the aid of the expression $C_o = 0.53 / \overline{\eta}$, where $\overline{\eta}$ follows from (35).

following result

$$C = 1 \quad z/h < 1 - e^{-t/T} \tag{32}$$

$$C = 0 \quad 1 - e^{-t/T} < z/h < 1 \ .$$

We thus find that for $t \to \infty$ the concentration becomes uniformily distributed across the boundary layer, which is of course a familiar result. However, we also see that in this case the concentration is not reflected from the top of the boundary. The boundary layer fills up from below. Such a result should be interpreted in the light of the gradual decrease of turbulence to zero near the top of the boundary layer as demonstrated by e.g. the σ_w- profile (16).

The complete solution of our point source problem can only be found numerically. In Figure 5 we show the vertical concentration profile at several non-dimensional times for the case of a ground-level emission. The filling up of the boundary layer, as mentioned above, is clearly illustrated. Furthermore, in Figure 6 the ground-level concentration, C_o, is given as function of time for several source heights. We find that the emission from sources in the lower half of the boundary layer is mixed downward to ground level relatively quick. This effect is due to the shape of the K-profile (13), which has high K-values near the ground in comparison with the values in the upper part of the boundary layer.

For the case of an emission at ground level we may take a somewhat different approach. Let us introduce a mean height

$$\bar{\eta} = \int_o^1 \eta\, C\, d\eta. \tag{33}$$

The evolution of $\bar{\eta}$ as a function of time can be calculated by substitution of (33) in the diffusion equation (24). The result reads

$$\frac{d\bar{\eta}}{d\sigma} = C_o - 2 + 2\bar{\eta}\ , \tag{34}$$

where $C_o = C(\eta=0)$. The concentration C_o can be expressed in terms $\bar{\eta}$ by means of $C_o = \alpha/\bar{\eta}$, where the coefficient α varies from $2/\pi$ for $\sigma \to 0$ (Hunt, 1982) to 0.5 for $\sigma \to \infty$. The latter value folows by substitution of a uniform concentration distribution into (33). For the case of a constant value of α, equation (34) can be integrated with respect to σ and the result is:

$$\sigma = \ln\left(\frac{2}{\alpha}\bar{\eta}^2 - \frac{2}{\alpha}\bar{\eta} + 1\right)^{\frac{1}{4}} + \frac{1}{2\sqrt{(2\alpha-1)}}\left\{\tan^{-1}\left(\frac{2\bar{\eta}-1}{\sqrt{(2\alpha-1)}}\right) + \tan^{-1}\left(\frac{1}{\sqrt{(2\alpha-1)}}\right)\right\} \tag{35}$$

This expression, evaluated with $\alpha = 0.53$, is shown in Figure 6. We find that it agrees closely with the numerical results.

To conclude this section we note that the K-profile (13) cannot be correct close to the surface because it does not agree with the M-O similarity result (3) for ϕ_m. An interpolation equation which satisfies the correct limit for $z \to 0$ reads

$$\frac{K}{k\,Ri_f\,u_*\,L} = \frac{\beta\,\eta\,(1-\eta)^2}{1 + \beta\,\frac{h}{L}\,\eta} \tag{36}$$

where we have used the result found in section 2 that $\beta = Ri_f^{-1}$.

In Figure 5 and 6 we show the concentration distribution calculated, with this profile for the case h/L = 2. We find that the results are in general close to those found with the original K-profile (13).

5. CONCLUSIONS

We have examined several aspects of the turbulent atmospheric boundary layer in a stable circumstances. The conclusions can be summarized as follows

1. Turbulence in stable conditions can be described in terms of local scaling. In particular locally scaled variables become constant above the surface layer.

2. With the results of local scaling we are able to calculate vertical profiles of turbulence variables from a simple boundary-layer model.

3. Diagnostic equations for the boundary-layer height agree reasonably well with observations. For practical applications we provide an equation which expresses the boundary-layer height solely in terms of the wind speed. Prognostic models are theoretically more satisfactory but they need an initial value for the boundary-layer height.

4. Local scaling is consistent with a description of diffusion in terms of K-theory. For the case of a stationary boundary layer we find an explicit K-profile as a quadratic function of height. The vertically inhomogeneous structure of the stable boundary layer influences diffusion of passive contaminants strongly.

ACKNOWLEDGEMENT

Comments by R.A. Brost, A.A.M. Holtslag and J.R. Reiff on an early version of the manuscript are gratefully acknowledged. Furthermore, I thank T.L. van Stijn and A.A.M. Holtslag for making their computer programs available.

Appendix A.

Observations in the Stable Boundary Layer.

The observations have been collected at the Cabauw experimental facility (52° N) during clear nights. The following variables have been given with an averaging period of a half hour.

t_s = time of sunset (GMT)
t_r = time of sunrise (GMT)
t = starting time of a half hour measurement period (GMT)
$T_{0.6}$ = mean temperature (°C) at 0.6 m
U_{10} = mean wind speed (ms^{-1}) at 10 m
$\overline{w\theta}_o$ = temperature flux (°C ms^{-1}) measured at 20 m
u_* = friction velocity (ms^{-1}) measured at 20 m
h = boundary-layer height (m) measured with an acoustic sounder.

17 Febr. 1978

t_s = 16.58
t_r = 6.50

t	$T_{0.6}$	U_{10}	$-\overline{w\theta}_o$	u_*	h
1600	-1.8	4.1	0.007	0.29	100
1630	-2.5	4.2	0.013	0.29	120
1700	-3.6	2.8	0.008	0.17	110
1730	-4.7	1.7	0.003	0.09	90
1800	-5.5	1.9	0.001	0.03	80
1830	-5.5	3.0	0.008	0.14	90
1900	-5.6	2.9	0.007	0.16	90
1930	-5.5	2.7	0.009	0.17	80
2000	-6.0	2.3	0.003	0.07	50
2030	-7.0	2.1	0.001	0.04	50
2100	-7.3	1.9	0.003	0.15	50
2130	-6.9	2.4	0.004	0.09	60
2200	-6.8	1.7	0.003	0.08	50

20 Febr. 1978

t_s = 17.03
t_r = 6.43

t	$T_{0.6}$	U_{10}	$-\overline{w\theta}_o$	u_*	h
1600	-0.8	4.9	0.010	0.43	-
1630	-1.6	4.1	0.017	0.35	100
1700	-2.0	4.8	0.023	0.43	140
1730	-2.3	4.7	0.023	0.42	160
1800	-2.7	4.5	0.023	0.37	150
1830	-3.4	4.1	0.017	0.34	150

t	$T_{0.6}$	U_{10}	$-\overline{w\theta}_o$	u_*	h
1900	-3.8	4.2	0.021	0.40	280
1930	-3.9	4.7	0.023	0.48	260
2000	-4.0	4.8	0.028	0.56	240
2030	-4.4	4.6	0.020	0.44	230
2100	-4.7	4.5	0.023	0.47	210
2130	-4.9	4.3	0.021	0.44	200
2200	-5.1	3.9	0.021	0.43	200
2230	-5.3	3.5	0.016	0.37	200
2300	-5.4	3.5	0.016	0.37	200
2330	-5.4	3.5	0.017	0.39	190
0	-5.3	3.5	0.015	0.39	200
30	-4.8	3.5	0.016	0.38	220
100	-4.8	4.1	0.015	0.40	220
130	-5.0	3.8	0.020	0.38	260
200	-5.2	4.3	0.018	0.44	260
230	-5.4	4.2	-	-	240
300	-6.0	3.3	0.021	0.38	220
330	-6.1	3.9	0.022	0.37	220
400	-6.3	3.7	0.022	0.34	200
430	-6.3	4.1	0.025	0.39	200
500	-6.4	3.8	0.020	0.40	210
530	-6.4	4.2	0.018	0.42	220
600	-6.1	3.9	0.016	0.40	230
630	-6.0	3.9	0.016	0.42	220
700	-5.7	4.1	0.010	0.43	300

18 May 1978

t_s = 19.32
t_r = 3.40

t	$T_{0.6}$	U_{10}	$-\overline{w\theta}_o$	u_*	h
1700	15.6	5.2	0.004	0.45	-
1730	14.9	4.7	0.009	0.43	-
1800	14.2	4.2	0.012	0.36	-
1830	13.4	4.4	0.019	0.42	-
1900	12.8	4.8	0.024	0.44	-
1930	12.4	5.1	0.026	0.45	-
2000	11.6	4.2	0.021	0.39	-
2030	11.1	3.9	0.021	0.37	-
2100	11.0	3.8	0.016	0.35	-
2130	10.8	4.0	0.020	0.36	-
2200	10.9	4.4	0.034	0.43	-
2230	10.8	4.4	0.024	0.39	-
2300	10.6	4.1	0.020	0.36	-
2330	10.4	4.0	0.016	0.32	-
0	10.0	3.9	0.016	0.32	240
30	8.9	3.1	0.014	0.22	240
100	8.3	3.0	0.011	0.18	220
130	8.4	3.1	0.011	0.22	200
200	8.2	3.2	0.015	0.24	180
230	7.9	3.4	0.015	0.22	180
300	7.7	3.3	0.016	0.24	160
330	7.5	3.4	0.017	0.24	150

29 May 1978

t_s = 19.47
t_r = 3.27

t	$T_{0.6}$	U_{10}	$-\overline{w\theta}_o$	u_*	h
1730	19.6	-	0.002	0.32	-
1800	18.8	-	0.009	0.27	-
1830	18.0	-	0.012	0.27	-
1900	17.1	-	0.015	0.28	-
1930	16.2	-	0.013	0.25	-
2000	15.5	-	0.011	0.20	-
2030	15.1	-	0.013	0.22	-
2100	14.8	-	0.012	0.22	-
2130	14.1	-	0.006	0.14	-
2200	13.9	3.3	0.011	0.20	230
2230	13.5	3.0	0.010	0.19	200
2300	13.1	2.8	0.012	0.18	190
2330	12.9	2.5	0.008	0.16	180
t	$T_{0.6}$	U_{10}	$-\overline{w\theta}_o$	u_*	h
0	12.7	2.5	0.008	0.15	180
30	12.5	2.6	0.003	0.11	170
100	12.4	2.3	0.006	0.13	190
130	12.3	2.4	0.003	0.10	180
200	12.1	2.1	0.005	0.14	180
230	12.1	2.2	0.004	0.12	200
300	11.8	2.1	0.004	0.09	200
330	11.6	2.4	0.002	0.09	190
400	12.0	2.3	0.005	0.13	180
430	12.8	1.8	0.003	0.12	180
500	13.9	2.1	0.002	0.16	180
530	14.7	2.3	0.000	0.20	150

30 May 1978

t_s = 19.49
t_r = 3.26

t	$T_{0.6}$	U_{10}	$-\overline{w\theta}_o$	u_*	h
1630	23.5	3.4	0.015	0.32	-
1700	23.1	2.7	0.022	0.28	-
1730	22.9	2.8	0.020	0.32	-
1800	22.3	2.4	0.014	0.24	-
1830	21.3	2.5	0.013	0.20	120
1900	20.3	2.8	0.021	0.23	100
1930	19.8	3.0	0.024	0.27	120
2000	18.6	2.8	0.009	0.14	120
2030	17.9	2.4	0.002	0.03	110
2100	17.3	2.6	0.001	0.04	110
2130	16.8	2.9	0.004	0.08	120
2200	16.8	3.1	0.003	0.08	130
2230	16.3	3.0	0.005	0.10	150
2300	15.7	2.9	0.008	0.14	180
2330	16.3	3.5	0.029	0.27	180
0	15.5	2.8	0.020	0.20	160
30	14.4	2.6	0.008	0.13	150
100	13.6	2.5	0.006	0.12	200
130	14.0	2.9	0.012	0.16	170
200	13.5	2.8	0.007	0.13	150
230	13.1	2.9	0.007	0.12	130
300	12.6	2.8	-	-	120
330	13.0	3.4	0.016	0.70	120
400	13.0	2.8	0.013	0.13	140
430	14.1	3.1	0.024	0.20	140
500	15.2	3.3	0.024	0.33	140
530	16.0	3.3	0.009	0.36	100

1 June 1978

t_s = 19.50
t_r = 3.25

t	$T_{0.6}$	U_{10}	$-\overline{w\theta}_o$	u_*	h
1700	23.1	2.1	0.007	0.16	120
1730	23.0	2.2	0.008	0.14	120
1800	23.1	2.2	0.010	0.16	100
1830	22.7	2.5	0.012	0.19	100
1900	21.9	2.6	0.010	0.15	100
1930	21.1	2.8	0.014	0.19	100
2000	20.4	2.8	0.020	0.21	90
2030	19.5	2.6	0.010	0.14	100
2100	18.9	2.5	0.006	0.11	160
2130	18.4	2.4	0.006	0.11	140
2200	18.0	2.5	0.014	0.18	120
2230	17.8	2.5	0.011	0.14	110
2300	17.4	2.4	0.009	0.13	100
2330	17.2	2.8	0.016	0.18	85
0	17.6	2.8	0.014	0.19	90
30	17.2	2.7	0.018	0.21	85
100	16.6	2.6	0.016	0.19	85
130	16.3	2.7	0.018	0.19	80
200	16.5	3.2	0.035	0.29	80
230	16.0	3.0	0.018	0.22	90
300	15.5	2.6	0.013	0.17	100
330	15.5	2.9	0.019	0.19	90
400	15.8	3.2	–	–	110
430	16.2	3.1	0.028	0.29	110
500	16.7	3.1	0.012	0.25	110
530	17.6	3.0	0.009	0.26	80
600	18.6	2.8	0.000	0.25	100

12 Sept 1978

t_s = 18.04
t_r = 5.13

t	$T_{0.6}$	U_{10}	$-\overline{w\theta}_o$	u_*	h
1700	13.9	5.0	0.008	0.45	–
1730	12.9	3.9	0.017	0.34	–
1800	11.1	3.2	0.010	0.24	–
1830	10.9	3.2	0.007	0.19	–
1900	10.4	2.9	0.006	0.18	–
1930	8.7	3.0	0.002	0.15	–
2000	8.5	3.2	0.005	0.16	70
2030	8.9	3.4	0.000	0.06	80
2100	9.4	2.6	0.001	0.08	90
2130	8.9	2.3	0.001	0.08	90
2200	9.6	2.4	0.003	0.12	80
2230	10.3	3.0	0.012	0.25	90
2300	10.8	3.4	0.009	0.27	130
2330	10.6	3.7	0.009	0.31	170
0	11.0	4.1	0.014	0.35	240
30	11.1	3.6	0.009	0.33	260
100	11.4	4.0	0.008	0.36	280
130	11.7	4.3	0.007	0.41	280
200	11.9	4.4	0.006	0.42	270
230	12.0	5.3	0.003	0.37	270
300	12.1	5.1	0.005	0.34	270
330	12.2	4.6	0.005	0.30	270

25 Sept. 1978

t_s = 17.31
t_r = 5.32

t	$T_{0.6}$	U_{10}	$-\overline{w\theta}_o$	u_*	h
1600	14.4	3.7	0.003	0.25	–
1630	13.4	2.6	0.006	0.18	60
1700	12.3	2.1	0.001	0.09	60
1730	9.1	1.5	0.000	0.03	50
1800	8.6	1.3	0.000	0.03	50
1830	7.7	1.7	0.001	0.05	50
1900	7.3	1.5	0.000	0.04	60
1930	7.1	1.9	0.001	0.06	60
2000	7.2	2.1	0.000	0.04	80
2030	6.1	1.7	0.000	0.03	70
2100	6.2	1.8	0.001	0.06	–
2130	6.3	1.6	0.000	0.06	60
2200	6.6	1.7	0.001	0.07	60
2230	6.6	1.9	–	–	60
2300	6.5	2.0	0.004	0.11	60
2330	6.3	2.3	0.004	0.11	60
0	6.1	2.3	0.007	0.13	60
30	5.9	2.0	0.003	0.10	60
100	6.1	2.7	0.007	0.10	70
130	5.4	2.1	0.005	0.13	70
200	5.8	2.2	0.003	0.09	80
230	6.1	2.1	0.004	0.14	100
300	6.8	2.7	0.005	0.14	100
330	7.2	2.7	0.003	0.12	100
400	7.5	2.7	0.007	0.18	90
430	7.3	2.4	0.003	0.10	90
500	7.1	2.5	0.007	0.16	100

19 December 1978

t_s = 15.30
t_r = 7.47

t	$T_{0.6}$	U_{10}	$-\overline{w\theta}_o$	u_*	h
1400	-1.1	6.0	0.016	0.52	270
1430	-0.9	5.7	0.016	0.55	300
1500	-1.0	5.6	0.026	0.52	300
1530	-1.5	5.1	0.029	0.50	300
1600	-1.8	5.8	0.034	0.52	300
1630	-2.3	5.2	0.023	0.42	320
1700	-2.4	5.9	0.032	0.56	340
1730	-2.9	6.2	-	-	340
1800	-3.2	5.8	0.031	0.55	-
1830	-3.2	5.8	0.029	0.56	-
1900	-3.3	5.4	0.024	0.52	-
1930	-3.5	5.1	0.018	0.46	380
2000	-3.8	4.4	0.016	0.39	340
2030	-4.0	4.4	0.012	0.36	340
2100	-4.1	4.2	0.011	0.37	350
2130	-4.3	4.5	0.010	0.38	330
2200	-4.6	4.1	0.013	0.35	320
2230	-4.6	4.1	0.010	0.36	-
2300	-5.0	3.5	0.008	0.27	-
2330	-5.2	3.6	0.008	0.28	-
0	-5.5	2.9	0.005	0.20	340
30	-5.6	3.1	0.006	0.21	340
100	-5.6	3.3	0.007	0.25	340
130	-5.6	3.0	0.006	0.22	340
200	-5.6	2.8	0.008	0.25	320
230	-5.5	3.3	0.007	0.21	260
300	-5.5	3.1	0.008	0.23	180
330	-5.3	3.1	0.008	0.25	200
400	-5.3	3.0	0.005	0.20	200
430	-5.3	2.9	0.006	0.21	140
500	-5.5	2.6	0.006	0.21	160
530	-5.5	3.2	0.007	0.31	-
600	-5.6	3.9	0.004	0.32	-
630	-5.7	3.6	0.004	0.27	-
700	-5.9	3.3	0.004	0.22	-
730	-5.9	3.2	0.004	0.27	230
800	-5.8	2.9	0.004	0.30	260

8 Febr. 1979

t_s = 16.40
t_r = 7.04

t	$T_{0.6}$	U_{10}	$-\overline{w\theta}_o$	u_*	h
1400	3.3	3.4	0.005	0.22	80
1430	2.8	3.6	0.008	0.30	120
1500	2.6	2.7	0.007	0.27	170
1530	1.8	1.6	0.004	0.13	170
1600	1.3	1.7	0.005	0.12	140
1630	-0.6	1.2	0.001	0.06	120
1700	-0.5	2.0	0.003	0.11	120
1730	-0.9	2.0	-	-	100
1800	-2.1	1.2	0.005	0.10	90
1830	-2.5	1.2	0.000	0.03	50
1900	-2.9	1.6	0.001	0.04	50
1930	-3.3	1.5	0.001	0.04	60
2000	-3.5	0.9	0.000	0.04	-
2030	-3.8	1.0	0.000	0.03	60
2100	-3.9	1.9	0.000	0.04	60
2130	-3.4	1.8	0.002	0.08	60
2200	-3.2	1.9	0.001	0.07	60
2230	-2.5	2.4	0.002	0.08	60
2300	-1.9	2.1	0.002	0.09	70
2330	-2.1	2.2	0.003	0.10	80
0	-2.1	2.1	0.001	0.07	100
30	-2.5	1.5	0.001	0.07	100
100	-3.3	2.0	0.001	0.08	100
130	-3.3	2.0	0.002	0.09	100
200	-2.9	2.6	0.001	0.06	100
230	-3.1	2.2	0.001	0.06	110
300	-3.4	2.2	0.003	0.10	100
330	-3.5	2.5	0.002	0.09	95
400	-3.4	2.4	0.003	0.10	100
430	-3.2	2.6	0.004	0.13	105
500	-2.3	2.9	0.012	0.22	105
530	-2.5	2.3	0.010	0.18	110
600	-2.8	2.2	0.007	0.15	120
630	-2.3	2.6	0.012	0.23	130
700	-2.5	2.6	0.013	0.24	135
730	-2.5	2.7	0.014	0.24	130
800	-1.9	3.2	0.008	0.28	125
830	-1.5	3.5	0.003	0.33	145

REFERENCES

André, J.C.: 1983, 'On the variability of the nocturnal boundary-layer depth', J. Atmos. Sci. 40, 2309-2311.

André, J.C. and Mahrt, L.: 1982, 'The nocturnal surface inversion and influence of clear-air radiative cooling', J. Atmos. Sci. 39, 864-878.

Arya, S.P.S.: 1981: 'Parameterizing the height of the stable atmospheric boundary layer', J. Appl. Meteorol. 20, 1192-1202.

Brost, R.A. and Wyngaard, J.C.: 1978, 'A model study of the stably stratified planetary boundary layer', J. Atmos. Sci. 35, 1427-1440.

Businger, J.A., Wyngaard, J.C., Izumi, Y., and Bradley, F.F.: 1971, 'Flux profile relationships in the atmospheric surface layer', J. Atmos. Sci. 28, 181-189.

Businger, J.A.: 1982, 'Equations and concepts', Atmospheric Turbulence and Air Pollution Modelling, F.T.M. Nieuwstadt and H. van Dop (eds.), D. Reidel Publishing Company, Dordrecht, Holland.

Caughey, S.J. and Readings, C.J.: 1975, 'An observation of waves and turbulence in the earth's boundary layer', Boundary-Layer Meteorol. 9, 279-296.

Caughey, S.J., Wyngaard, J.C. and Kaimal, J.C.: 1979, 'Turbulence in the evolving stable boundary layer', J. Atmos. Sci., 36, 1041-1052.

Dyer, A.J.: 1974, 'A Review of flux-profile relationships', Boundary-Layer Meteorol. 7, 363-372.

Einaudi, F. and Finnigan, J.J.: 1981, 'The interaction between an internal gravity wave and the planetary boundary layer. Part I: The linear analysis', Quart. J. Roy. Meteorol. Soc. 107, 793-806.

Garratt, J.R. and Brost, R.A.: 1981, 'Radiative cooling effects within and above the nocturnal boundary layer', J. Atmos. Sci., 2730-2746.

Garratt, J.R.: 1982a, 'Observations in the nocturnal boundary layer', Boundary-Layer Meteorol. 22, 21-48.

Garratt, J.R.: 1982b, 'Surface fluxes and the nocturnal boundary-layer height', J. Appl. Meteorol. 21, 725-729.

Hunt, J.C.R.: 1982, 'Diffusion in the stable boundary layer', Atmospheric Turbulence and Air Pollution Modelling, F.T.M. Nieuwstadt and H. van Dop (eds.), D. Reidel Publishing Company, Dordrecht Holland.

Hunt, J.C.R., Kaimal, J.C., Gaynor, J.E. and Korrell, A.: 1983, 'Observations of turbulence structure in stable layers at the Boulder Atmospheric Observatory'. Studies of Nocturnal Stable Layers at BAO, J.C. Kaimal (ed.), NOAA/ERL, Boulder.

Keisuke, F., Masamomoto, N., and Ueda, H.: 1983, 'A laboratory experiment on momentum and heat transfer in the stratified surface layer', Quart. J. R. Meteorol. Soc. 109, 661-676.

Kerman, B.R.: 1979, 'A similarity model for maximum ground-level concentration in a height-invariant, stably stratified atmospheric boundary layer', Boundary-Layer Meteorol. 17, 297-313.

Kondo, J., Kanechika, O. and Yasuda, N.: 1978, 'Heat and momentum transfer under strong stability in the atmospheric surface layer', J. Atmos. Sci. 35, 1012-1021.

Mahrt, L.: 1981, 'Modelling the depth of the stable boundary-layer', Boundary-Layer Meteorol. 21, 3-19.
Mahrt, L. and Heald, R.C.: 1979, 'Comments on determining height of the nocturnal boundary layer', J. Appl. Meteorol. 36, 383.
Mahrt, L., Heald, R.C., Lenschow, D.H. and Stankov, B.B.: 1979, 'An observational study of the structure of the nocturnal boundary layer', Boundary-Layer Meteorol. 17, 247-264.
Mahrt, L., André, J.C. and Heald, R.C.: 1982, 'On the depth of the nocturnal boundary layer', J. Appl. Meteorol. 21, 90-92.
McPhee, M.: 1981, 'An analytic similarity theory for the planetary boundary layer stabilized by surface buoyancy', Boundary-Layer Meteorol. 21, 325-339.
Monin, A.S. and Yaglom, A.M.: 1971, 'Statistical Fluid Mechanics. Vol. I', M.I.T. Press, Cambridge, Mass.
Nai-Ping, L., Neff, W.D. and Kaimal, J.C.: 1983, 'Wave and turbulence structure in a disturbed nocturnal inversion', Boundary-Layer Meteorol. 26, 141-155.
Nieuwstadt, F.T.M.: 1981, 'The steady-state height and resistance laws of the nocturnal boundary layer: Theory compared with Cabauw observations', Boundary-Layer Meteorol. 20, 3-17.
Nieuwstadt, F.T.M.: 1984a, ' The turbulent structure of the stable nocturnal boundary layer', submitted to J. Atmos. Sci.
Nieuwstadt, F.T.M.: 1984b, 'A model for the stationary, stable boundary layer', Proceedings of the conference on Models of Turbulence and Diffusion in Stably Stratified Regions of the Natural Environment, March 1983, Cambridge.
Nieuwstadt, F.T.M. and Tennekes, H.: 1981, 'A rate equation for the nocturnal boundary-layer height', J. Atmos. Sci. 38, 1418-1428.
Pearson, H.J., Putlock, J.S. and Hunt, J.C.R.: 1983, 'A statistical model of fluid-element motions and vertical diffusion in a homogeneous statified turbulent flow', J. Fluid Mech. 129, 219-249.
Rao, K.S. and Snodgrass, H.F.: 1979, 'Some parameterizations of the nocturnal boundary layer', Boundary-Layer Meteorol. 17, 15-28.
Smeda, M.: 1979, 'Incorporation of planetary boundary-layer processes into numerical forecasting models', Boundary-Layer Meteorol. 16, 115-129.
Stull, R.B.: 1983a, 'A heat-flux history length scale for the nocturnal boundary layer', Tellus 35A, 219-230.
Stull, R.B.: 1983b, 'Integral scales for the nocturnal boundary layer part 2: heat budget, transport, and energy implications', J. Climate Appl. Meteor. 22, 1932-1941.
Tennekes, H.: 1982, 'Similarity relations, scaling laws and spectral dynamics', Atmospheric Turbulence and Air Pollution Modelling, F.T.M. Nieuwstadt and H. van Dop (eds.), D. Reidel Publishing Company, Dordrecht, Holland.
Ueda, H., Mitsumoto, S., and Komori, S.: 1981, 'Buoyancy effects on the turbulent transport processes in the lower atmosphere', Quart. J. R. Meteorol. Soc. 107, 561-578.

Van Ulden, A.P. and Holtslag, A.A.M.: 1983, 'The stability of the atmospheric surface layer', Preprint volume 6th symposium on turbulence and diffusion, March 1983, Boston, American Meteorological Society.
Venkatram, A.: 1980, 'Estimating the Monin-Obukhov length in the stable boundary layer for dispersion calculations', Boundary-Layer Meteorol. 19, 481-485.
Wetzel, P.: 1982, 'Toward parameterization of the stably boundary layer', J. Appl. Meteorol. 21, 7-13.
Wyngaard, J.C.: 1973, 'On surface layer turbulence', Workshop on Micrometeorology. D.A. Haugen (ed.), American Meteorological Society, Boston.
Yamada, T.: 1979, 'PBL similarity profiles determined from a level-2 turbulence closure model', Boundary-Layer Meteorol. 17, 333-351.
Yamamoto, S., Yokoyama, O. and Gamo, M.: 1979, 'Observational study on the turbulent structure of the atmospheric boundary layer under stable conditions', J. Meteorol. Soc. Japan 57, 423-431.
Yu, T.: 1978, 'Determining height of the nocturnal boundary layer', J. Appl. Meteorol. 17, 28-33.
Zeman, O.: 1979, 'Parameterization of the dynamics of stable boundary layers and nocturnal jets', J. Atmos. Sci. 36, 792-804.
Zilitinkevich, S.S.: 1972, 'On the determination of the height of the Ekman boundary layer', Boundary-Layer Meteorol. 3, 141-145.

Questions, answers and remarks

J. TADMOR: In your lecture, evidence was shown of the constancy of the Richardson number up to a height of 300-400m. What is the behavior of the Richardson number at higher heights?

F.T.M.NIEUWSTADT: Only in the region where turbulence predominates is the Richardson number approximately constant with height. Above it, the Richardson number usually increases strongly due to the influence of radiation and subsidence, which produce a positive temperature gradient above the turbulent boundary layer.

J.C.R. HUNT: Calculations and measurements of the stable boundary layer by Brost and Wyngaard show that even very small slopes (of the order of 1/1000) can significantly change the mean velocity profile and turning of the wind. Surely it is important to emphasize that the universal structure obtained by Dr. Nieuwstadt is only valid for very flat terrain.

F.T.M.NIEUWSTADT Indeed the stable boundary layer is very sensitive to slope effects. In particular the mean profile seems susceptible as it also is to the influence of non-stationarity. The turbulence profiles might not be as sensitive in the presence of weak slopes. The results of local scaling should still be valid in such circumstances.

J.C.R. HUNT: Could you comment on the value of Richardson number at the top of the stable boundary layer (SBL)? Does the Richardson number not change from zero to infinity over the night and does this affect the structure of the SBL?

F.T.M.NIEUWSTADT: It is true that the Richardson number above the boundary layer varies from a low value at the beginning of the night to large values later on. This depends mainly on the temperature and the wind profile as a function of time, e.g. due to the process of radiation. It does not influence the structure within the boundary layer, except possibly in the region of the boundary layer top.

PARAMETRIC RELATIONS FOR THE ATMOSPHERIC BOUNDARY LAYER

S.P.S. Arya
Department of Marine, Earth and Atmospheric Sciences
North Carolina State University, Raleigh, NC 27695, USA

Abstract: Some parametric relations for the atmospheric planetary boundary layer (PBL) are suggested for possible use in the various atmospheric circulation and air quality models, as well as in other applications. These are for parameterizing the mean wind and temperature profiles, the vertical fluxes of momentum, heat and moisture, the variances of velocity fluctuations and length and time scales in the PBL. The parametric relations for the PBL height, the vertical velocity at the top of the PBL and the total energy dissipation in the PBL are also discussed. Experimental and/or theoretical bases for the various parametric relation are given. Some of the suggested parameterizations should be considered as tentative, until they are properly validated.

1. INTRODUCTION

In various applications, ranging from the local air quality modeling to the global climate simulations, some aspects or properties of the atmospheric planetary boundary layer (PBL) must be incorporated through simple parametric relations which may have been developed earlier on the basis of observations, theory, or numerical modeling of the PBL. For example, in a sophisticated dispersion or air quality model, one may have to prescribe the vertical distributions of mean wind speed and direction, variances of velocity components, scales of turbulence and the boundary layer depth, as functions of stability, surface roughness, geostrophic winds, etc. In most of the global general circulation models, the surface fluxes of momentum, heat and moisture must be parameterized. Other large scale models of the atmosphere may require the parameterization of the mean vertical motion (Ekman pumping) at the top of the PBL, or the total energy dissipation in the PBL. The mesoscale models of the atmosphere, in general, must accurately parameterize most aspects of the PBL, because mesoscale circulations are intimately related to and usually have their roots in the PBL which may not be adequately resolved in such models. Such parameterizations are reviewed here, which should be quite relevant to atmospheric modeling over the world's oceans and flatter parts of the continents.

Boundary-Layer Meteorology **30** (1984) 57–73. 0006–8314/84/0301–0057$04.15

Parameteric relations for the PBL may also be used for inferring certain turbulent quantities, such as variances and fluxes, from more easily measured mean winds, temperatures, etc. These may be used for extrapolating information at other levels from measurements at one or two levels in the PBL.

In view of the above applications and needs, we have conducted a thorough search of the parametric relations that have been proposed for the atmospheric boundary layer. Based on this, our recommended relations or schemes for parameterizing the various properties of the PBL over a homogeneous and flat terrain are presented in the following sections. These parametric relations should be considered as tentative, and may be replaced by better and more proven relations as they become available.

2. MEAN WIND AND TEMPERATURE PROFILES

For this, the PBL of depth h may be divided into a surface layer of depth $h_s = 0.1\ h$ and an outer layer. In unstable and convective conditions, the outer layer may be further sub-divided into a mixed layer, which extends up to the base of capping inversion, and the stable transition layer at the top.

2.1 Surface layer (SL)

Mean wind ($\bar{u}$) and temperature ($\bar{\theta}$) profiles in the fully developed region of the surface layer, well above the tops of roughness elements, in which the vertical fluxes remain nearly constant with height, are considered to be well established from micrometeorological observations at homogeneous sites. These are best represented in the framework of the Monin-Obukhov similarity theory as

$$\frac{\bar{u}}{u_*} = \frac{1}{\kappa}\left[\ln \frac{z}{z_o} - \psi_m\left(\frac{z}{L}\right)\right]$$

$$\frac{\bar{\theta}-\bar{\theta}_o}{\theta_*} = \frac{\alpha}{\kappa}\left[\ln \frac{z}{z_o} - \psi_h\left(\frac{z}{L}\right)\right] \quad (1)$$

for $1.5\ h_o < z \leq h_s$. Here, h_o is the representative height of roughness elements, z_o the roughness parameter, $\bar{\theta}_o$ the mean surface temperature, u_* the friction velocity, $\theta_* = -Q_o/u_*$ the temperature scale defined from the kinematic surface heat flux Q_o, $L = -u_*^3/\kappa\beta Q_{vo}$ the Obukhov length, κ von Karman constant, β the buoyancy parameter, Q_{vo} the virtual heat flux, α an empirical constant, and ψ_m (z/L) and ψ_h (z/L) are the stability-dependent similarity functions related to the dimensionless wind shear $(\kappa z/u_*)\partial\bar{u}/\partial z = \phi_m$ (z/L) and the dimensionless temperature gradient $(\kappa z/\theta_*)\partial\bar{\theta}/\partial z = \phi_h(z/L)$.

The Monin-Obukhov (M-O) similarity functions have been empirically determined from micrometeorological flux-profile measurements by many investigators (Brutsaert, 1982). Particularly simple and adequate for many practical applications are the so-called Businger-Dyer formulas (Businger, 1973) with $\alpha=1$

$$\phi_h = \phi_m^2 = (1-15\,\frac{z}{L})^{-\frac{1}{2}}\ ,\ \text{for}\ \frac{z}{L} < 0$$
$$\phi_h = \phi_m = 1 + 5\,\frac{z}{L}\ ,\ \text{for}\ \frac{z}{L} \geq 0 \qquad (2)$$

which, after integration, yield

$$\psi_m = \ln\left(\frac{1+x^2}{2}\right) + 2\ln\left(\frac{1+x}{2}\right) - 2\tan^{-1}(x) + \frac{\pi}{2}$$
$$\psi_h = 2\ln\left(\frac{1+x^2}{2}\right)\ ,\ \text{for}\ \frac{z}{L} < 0 \qquad (3)$$
$$\psi_m = \psi_h = -5\,\frac{z}{L}\ ,\ \text{for}\ \frac{z}{L} \geq 0$$

where $x = (1-15\,\frac{z}{L})^{\frac{1}{4}}$.

The M-O similarity parameter z/L is simply related to and can be determined from the gradient Richardson number Ri as (Arya,1982)

$$z/L = Ri,\ \text{for}\ Ri \leqslant 0$$
$$z/L = Ri/\,(1-5Ri),\ \text{for}\ Ri > 0 \qquad (4)$$

2.2 Outer layer

The wind and temperature profiles in the outer layer are highly dependent on stability. Under stable condition, these are characterized by strong gradients with substantial turning of wind with height. Due to essential decoupling of various layers and the sensitivity of profiles to even small surface slope, changes in the net radiation with height and other factors, there is not a strong basis for generalizing the mean profile in the stably stratified nocturnal PBL (Brost and Wyngaard, 1978; Caughey et al., 1979; Nieuwstadt, 1983). There is some observational evidence (Wetzel, 1982; Wesely and Coulter, 1983), however, for considering the mean wind speed ($\bar{V}$) and potential temperature profiles as approximately linear up to the level of the low-level jet, which often coincides with the top of the PBL. Therefore, we can tentatively recommend the use of parametric relations.

$$\bar{V} = \bar{V}_s + \frac{z-h_s}{h-h_s}\,(\bar{V}_h - \bar{V}_s)$$
$$\bar{\theta} = \bar{\theta}_s + \frac{z-h_s}{h-h_s}\,(\bar{\theta}_h - \bar{\theta}_s) \qquad (5)$$

for $h/L > 0$, where the subscripts s and h refer to the top of the surface layer and the PBL, respectively. $\bar{V}_s$ and $\bar{\theta}_s$ can be obtained from the surface layer profile relations.

Under stable and convective conditions, the mean velocity and temperature distributions can be considered more or less uniform, independent of height, in the mixed layer. Following Garratt et al. (1982), we recommend the relations

$$\frac{\bar{u}_m}{u_*} = \frac{1}{\kappa}\left[\ln\frac{h}{z_o} - \frac{1}{2}\ln\left|\frac{h}{L}\right| - 2.3\right]$$

$$\frac{\bar{v}_m}{u_*} \simeq 0 \qquad (6)$$

where $\bar{u}_m$ and $\bar{v}_m$ are the layer-averaged velocity components in the mixed layer, with x-axis oriented in the direction of the surface wind or shear. In the free-convective limit ($-h/L>1000$), Wyngaard et al. (1974) and Arya (1978) obtained slightly different relations

$$\frac{\bar{u}_m}{u_*} = \frac{1}{\kappa}\left[\ln\left(-\frac{L}{z_o}\right)-a\right]$$

$$\frac{\bar{v}_m}{u_*} = 0 \qquad (7)$$

in which the constant $a \simeq 0$ (it actually is weakly dependent on h_s/L).

The mixed-layer potential temperature $\bar{\theta}_m$ is an evolving function of time and can be predicted from the rate equations for $\bar{\theta}_m$ and h in the context of an appropriate slab model (Driedonks, 1982; Arya and Byun, 1983). The simplest encroachment model gives

$$\bar{\theta}_m(t) = \bar{\theta}_m(t_o) + \gamma_\theta\,[h(t) - h(t_o)] \qquad (8)$$

where $\gamma_\theta \equiv \partial\bar{\theta}_+/\partial z$ is the potential temperature gradient in the free atmosphere. Eq. (8) should be adequate for most practical applications.

The transition layer above the mixed layer is generally characterized by strong velocity and temperature gradients. The mean profiles in this layer are often assumed to be linear between the mixed layer values at the bottom and the free-atmospheric values (e.g. geostrophic winds) at the top of the layer (Garratt et. al, 1982). The thickness of the transition layer Δh is assumed to be infinitesimal in certain jump models of the convective PBL, but in reality, Δh is in the range 0.1h to 0.3h.

3. SURFACE FLUXES AND RESISTANCE LAWS

The surface fluxes of momentum, heat and moisture (τ_o, H_o, E_o) are usually parameterized in terms of mean wind speed ($\bar{V}_r$) potential temperature ($\bar{\theta}_r$) and specific humidity ($\bar{q}_r$) at some reference level (z_r) in the PBL and the 'representative' surface temperature $\bar{\theta}_o$ and specific humidity $\bar{q}_o$, through the following bulk transfer relations:

$$\tau_o = \rho C_D \bar{V}_r^2$$
$$H_o = -\rho c_p C_H \bar{V}_r (\bar{\theta}_r - \bar{\theta}_o) \qquad (9)$$
$$E_o = -\rho C_E \bar{V}_r (\bar{q}_r - \bar{q}_o)$$

Here, ρ is the density, c_p is the specific heat and C_D, C_H and C_E are the bulk transfer coefficients which in general depend on the reference height, PBL height, surface roughness, bulk stability and possibly other parameters. In crude parameterizations of the PBL in certain GCM's, C_D, C_H and C_E are assumed equal and assigned fixed numerical values. In some cases, different values (typically, 2×10^{-3}) are assigned for land (typically, (3×10^{-3}) and ocean (typically, 1.5×10^{-3}) surfaces and some allowance made for different stability conditions. In more accurate parameterizations of the surface fluxes, the bulk transfer coefficients should be specified on the basis of our improved understanding of the PBL which has resulted from systematic experiments and sophisticated numerical models.

3.1 Parameterizations based on the surface-layer similarity

The best known region of the atmospheric PBL is the surface layer for which the observations represented in the framework of M-O similarity theory have yielded some universal flux-profile relations. The parameterizations based on these relations will also be expected to be most accurate and reliable. For these to be applicable, the reference level z_r, which would normally be the first interior grid level in a model, should fall within the surface layer. Very few models would have such a fine resolution, however. This method of parameterization can still be used in models with coarser resolution, if mean variables at a fixed level in the SL are calculated, for example, from the dynamic forecast equations for this level (Benoit, 1976).

The parametric relations based on the M-O similarity theory can be expressed in the form of Eqs. (9) (Arya, 1977), with the bulk transfer coefficients given as

$$C_D = C_D\left(\frac{z_r}{z_o}, Ri_b\right)$$
$$C_H = C_H\left(\frac{z_r}{z_o}, Ri_b\right) \qquad (10)$$
$$C_E = C_E\left(\frac{z_r}{z_o}, Ri_b\right)$$

where

$$Ri_b = \beta z_r (\bar{\theta}_{vr} - \bar{\theta}_{vo}) / \bar{V}_r^2 \qquad (11)$$

is a bulk Richardson number. The virtual potential temperatures θ_{vr} and θ_{vo} are used to account for the buoyancy effect of water vapor.

The bulk transfer coefficients C_D, C_H, etc., are unique functions of z_r/z_o and Ri_b whose forms depend on the particular flux-profile relations adopted for the SL. The coefficients based on the flux-profile relations of Businger et al (1971) are given by Arya (1977). These show a monotonic decrease of C_D and C_H with increasing stability, reducing to zero as $Ri_b \to 0.21$. Benoit (1976) has proposed a more elaborate and refined scheme in which extreme conditions of free convection on the one side and very stable stratification approaching the critical Richardson number on the other, are specially taken into account. In the former, the transfer coefficients are held constant at their maximum values, which are presumably reached at the transition between forced and free convection and are given as functions of the ratios z_r/z_o and z_r/h. The values of C_D, C_H etc., in very stable conditions are also recommended to be held constant for $Ri_b \geq 0.15$.

The above parametric relations can also be used for estimating surface fluxes from observations of mean wind, temperature, etc. at one height in the SL and the appropriate surface properties z_o, $\bar{\theta}_o$, etc.

3.2 Parameterizations based on similarity matching

In most large scale atmospheric models with coarse vertical resolution, no information is available on flow properties in the SL. The first interior grid level may lie near the top of the PBL or even outside the PBL. The information available is at best representative of the PBL as a whole. In such cases, the following drag and other transfer relations based on the matching of the surface and outer-layer similarity profiles are found to be useful (Deardorff, 1972; Arya, 1977, 1978).

$$\begin{aligned} \kappa\bar{u}_m/u_* &= -(\ln \hat{z}_o + A_m) \\ \kappa\bar{v}_m/u_* &= -B_m \text{ sign } f \\ \kappa(\bar{\theta}_m-\bar{\theta}_o)/\theta_* &= -(\ln\hat{z}_o + C_m) \\ \kappa(\bar{q}_m-\bar{q}_o)/q_* &= -(\ln\hat{z}_o + D_m) \end{aligned} \qquad (12)$$

Here, f is the Coriolis parameter, $\hat{z}_o$ is the roughness parameter normalized by the scale height h or u_*/f, A_m, B_m, etc. are some universal functions of stability parameter (h/L or u_*/fL) and, possibly, some other (baroclinity, entrainment, etc.) parameters (Arya, 1977, 1978). Detailed comparisons with observations in unstable and convective PBLs (Garratt et al., 1982) have indicated that

$$\begin{aligned} A_m &\simeq \frac{1}{2} \ln(-h/L) + 2.3 \\ B_m &\simeq 0, \text{ for } h/L < -2 \end{aligned} \qquad (13)$$

and that these are little affected by baroclinity, the scale-height ratio, advection and entrainment. One would expect C_m and D_m to be similar to A_m and not be overly sensitive to these other processes.

The dependence of A_m, B_m, etc., on the stability parameter h/L, is much stronger in stably stratified PBL (Arya, 1977; Yamada, 1976). One can justifiably ignore any weak dependence of these similarity parameters, on baroclinity, advection, etc. Furthermore, numerical models of the stationary, stable PBL predict the forms of the above functions to be linear for h/L>2 (Arya, 1977), i.e.,

$$A_m = a_0 - a_1(h/L)$$
$$B_m = b_0 + b_1(h/L) \qquad (14)$$
$$C_m = c_0 - c_1(h/L)$$

which are also consistent with observations (Yamada, 1976). The values of the various constants a_0, a_1, etc., are far from being well established, however (Arya, 1977).

For near-neutral conditions ($-2 \lesssim h/L \lesssim 2$), suitable interpolations may be used between the values of A_m, B_m, etc., at $h/L = \pm 2$.

The implicit relations (12) can be easily inverted and expressed into more convenient explicit forms of bulk transfer relations

$$\tan^{-1}(\bar{v}_m/\bar{u}_m) = \alpha_m(h/z_o, Ri_m)$$
$$\tau_o = \rho C_D(h/z_o, Ri_m)\bar{V}_m^2$$
$$H_o = \rho c_p C_H(h/z_o, Ri_m)\bar{V}_m(\bar{\theta}_o - \bar{\theta}_m) \qquad (15)$$
$$E_o = \rho C_E(h/z_o, Ri_m)\bar{V}_m(\bar{q}_o - \bar{q}_m)$$

where the angle (α_m) between the surface stress and the layer-averaged velocity, and the transfer coefficients C_D, C_H, etc. are functions of h/z_o and the layer averaged bulk Richardson number

$$Ri_m = \beta h(\bar{\theta}_{vm} - \bar{\theta}_{vo})/\bar{V}_m^2 \qquad (16)$$

which can be evaluated and represented in the forms of nomograms (Arya, 1977), knowing the forms of the similarity functions A_m, B_m, etc.

Note that over the expected ranges of h/z_o and Ri_m, the drag and other transfer coefficients can vary over more than a hundred-fold range and are quite sensitive to the stability parameter Ri_m. Therefore, constant assigned values of these for all the terrain and stability conditions would hardly seem justifiable.

The alternative forms of flux parameterizations, involving surface geostrophic winds, or mean variables at the top of the PBL, have also been proposed (Clarke and Hess, 1974); Meljarejo and Deardorff, 1974; Arya, 1975; Zilitinkevich, 1975). But these are not recommended here, because of the similarity functions involved in them are much more sensitive to baroclinity, the scale-height ratio, entrainment and advection, as compared to the similarity functions in Eqs. (12) (Arya and Wyngaard, 1975; Arya, 1978; Garratt, et al., 1982). However, the geostrophic drag relations may be used for specific applications in which the information on geostrophic rather than actual winds is more readily available. For the barotrophic PBL, in particular, the simplest geostrophic drag law, which can be derived from the equations of motion (Deardorff, 1972; Brost and Wyngaard, 1978) is

$$\tau_o = c\rho f h G_o \sin \alpha_o \qquad (17)$$

in which G_o is the geostrophic wind, α_o is the angle between the surface wind and geostrophic wind and c is a weak function of stability; $c = 1$ for the moderately unstable and convective conditions ($h/L<-2$) and $c \simeq 0.6$ for neutral and stable conditions.

The parametric relations based on mean variables at $z = h$ might be more appropriate for large scale atmospheric models with the first interior grid level z_1 above the PBL in which $\bar{u}_h$, $\bar{\theta}_h$, etc. could be extrapolated from the computed variables at z_1 and z_2 at each time step.

4. TURBULENT FLUXES AND VARIANCES IN THE PBL

In fine resolution models, there may be several grid points lying within the PBL. In these, the vertical fluxes at all the interior grid levels will have to be parameterized. The parameterization of flux profiles in the PBL should also be useful in other applications. Ideally, it should be based on observations taken under various stability conditions.

In unstable and convective PBLs, the fluxes at the top of the mixed layer can be determined in terms of the entrainment velocity W_e and the jumps ($\Delta\bar{u}$, $\Delta\bar{v}$, etc.) in the mean variable across the transition layer, using the well-known entrainment relations (Driedonks, 1982):

$$\begin{aligned}
(\overline{u'w'})_h &= -W_e \Delta\bar{u} \\
(\overline{v'w'})_h &= -W_e \Delta\bar{v} \\
(\overline{\theta'w'})_h &= -W_e \Delta\bar{\theta} \\
(\overline{q'w'})_h &= -W_e \Delta\bar{q}
\end{aligned} \qquad (18)$$

Furthermore, the heat and moisture fluxes within the mixed layer are known to have approximately linear variation between the values at the surface and the top of the mixed layer. For the momentum fluxes, the linear profiles are only expected under barotropic conditions; in the presence of baroclinity the profiles may develop significant curvatures leading to large flux values in the middle of the PBL (Wyngaard et al.,

1974; Kaimal *et al*., 1976).

There is limited observational evidence also suggesting linear variation of fluxes in the stably stratified PBL (Caughey *et al*., 1979). But more recently, Nieuwstadt (1983) has demonstrated, from comparisons between Minnesota and Cabauw observations, that due to the effects of nonstationarity, gravity waves and radiation, no generalizable similarity flux profiles may be expected in the nocturnal PBL.

A more satisfactory scheme for parameterizing the fluxes in the stably stratified PBL might be the use of the gradient transport relations

$$\begin{aligned} \overline{u'w'} &= -K_M\, \partial\bar{u}/\partial z \\ \overline{v'w'} &= -K_M\, \partial\bar{v}/\partial z \\ \overline{\theta'w'} &= -K_H\, \partial\bar{\theta}/\partial z \\ \overline{q'w'} &= -K_E\, \partial\bar{q}/\partial z \end{aligned} \qquad (19)$$

in which one may prescribe $K_M = K_H = K_E \simeq 0.15\, \lambda_{mw}\, \sigma_w$, where λ_{mw} and σ_w are the length and velocity scales based on vertical velocity fluctuations whose parameterization will be discussed later.

The variances or standard deviations of velocity fluctuations in PBL must be known for realistic modeling of turbulent transport and diffusion in the PBL. These are normally scaled by friction velocity (u_*) or the convective velocity (w_*), depending on the stability regime of the PBL.

For unstable and convective PBLs, the following parametric relations may be suggested on the basis of observations by Panofsky *et al*., (1977), Kaimal, *et al*., (1976) and Caughey (1982).

$$\begin{aligned} \frac{\sigma_u}{u_*} &= \frac{\sigma_v}{u_*} = (12 - 0.5\,\frac{h}{L})^{1/3}, \quad \text{for } \frac{z}{h} \leq 0.1 \\ \frac{\sigma_w}{u_*} &= 1.3(1-3\,\frac{z}{L})^{1/3}, \quad \text{for } \frac{z}{h} \leq 0.1 \\ \frac{\sigma_u}{w_*} &= \frac{\sigma_v}{w_*} = \frac{\sigma_w}{w_*} \simeq 0.6, \quad \text{for } 0.1 < \frac{z}{h} < 0.8 \end{aligned} \qquad (20)$$

In the upper part of the mixed layer and the interfacial transition layer, the values of σ_u, σ_v and σ_w are expected to depend on the entrainment parameter W_e/w_*.

For the stable PBL, the limited Minnesota data presented by Caughey *et al*. (1979), in the framework of the PBL similarity theory, suggests the linear relations

$$\frac{\sigma_u}{u_*} \simeq 2.4(1 - \frac{z}{h})$$
$$\frac{\sigma_w}{u_*} \simeq 1.6(1 - \frac{z}{h}). \qquad (21)$$

σ_v/u_* may be assumed to lie between σ_u/u_* and σ_w/u_*.

More recently, Nieuwstadt (1983) has suggested on the basis of Cabauw data that σ_u, σ_v and σ_w bear constant ratios with the local friction velocity $u_{*L} = [(\overline{u'w'})^2 + (\overline{v'w'})^2]^{1/4}$. However, the parameterizations based on height-dependent velocity and temperature scales require the knowledge of momentum and heat fluxes in the PBL, which are more difficult to get than the variances.

The neutral stability being a rare condition in the atmosphere, there are virtually no observations of turbulence in a neutral PBL. On the basis of observations in near-neutral surface layer and the results of the large eddy simulation model of Deardorff (1970), we suggest the following parametric relations:

$$\sigma_u/u_* = 2.5 \exp |-3fz/u_*|$$
$$\sigma_v/u_* = 1.8 \exp |-3fz/u_*| \qquad (22)$$
$$\sigma_w/u_* = 1.3 \exp |-3fz/u_*|$$

5. LENGTH AND TIME SCALES

In statistical models of diffusion, the integral time or length scales of motion are required to be specified. Related scales are the frequencies n_m or the wavelengths $\lambda_m = \overline{V}/n_m$ at which the variance spectra of fluctuating velocity components have their peaks.

Spectral measurements in the convective PBL (Kaimal et al., 1976; Caughey and Palmer, 1979) have given the peak wave lengths λ_{mu}, λ_{mv} and λ_{mw}, normalized by the PBL height h, as functions of z/h, which can be approximated as

$$\lambda_{mu}/h \simeq \lambda_{mv}/h \simeq 1.5$$
$$\lambda_{mw}/h = 5.9z/h, \text{for } z/h \leq 0.1 \qquad (23)$$
$$\lambda_{mw}/h = 1.8[1-\exp(-4z/h)-0.0003\exp(8z/h)], \text{for } z/h > 0.1$$

For the nocturnal stable PBL, Caughey et al., (1979) have represented the peak wave lengths in the various spectra as function of z/h. Hanna (1982) has suggested that the following simple power laws can be fitted to these data from the Minnesota experiment:

$$\lambda_{mu}/h = 1.5\ (z/h)^{0.5}$$

$$\lambda_{mv}/h = 0.7\ (z/h)^{0.5} \qquad (24)$$

$$\lambda_{mw}/h = 1.0\ (z/h)^{0.8}$$

These should be considered as highly tentative, since the validity of similarity scaling in the nocturnal PBL has not been established (Nieuwstadt, 1983).

The Eulerian integral time scale T_E, being inversely proportional to the frequency n_m, is related to λ_m as (Hanna, 1981)

$$T_E = 0.17\ \lambda_m/\bar{V}. \qquad (25)$$

Most needed in dispersion calculations is the Lagrangian integral time scale T_L, which is approximately related to T_E as (Pasquill, 1974)

$$T_L/T_E = \beta \simeq b/i \qquad (26)$$

where $i = \sigma_{u,v,w}/\bar{V}$ is the appropriate turbulence intensity and b is a constant. From the simultaneous Eulerian and Lagrangian measurements of turbulence in the day time convective PBL, Hanna (1981, 1982) has estimated $\beta \simeq 1.7$ and $b \simeq 0.6$ (this is substantially higher than the recommended value of $b = 0.4$ by Pasquill (1974) on the basis of older and less accurate observations).

Substituting from Eq. (25) into (26).

$$T_L = 0.10\ \lambda_m/\sigma \qquad (27)$$

which in conjunction with Eqs. (23) and (24) would lead to the parameteric relations for T_L, as given by Hanna (1981, 1982).

In most of the parametric relations discussed earlier. the PBL height h is an important length scale which must be given or parameterized. In the daytime unstable and convective conditions, the PBL height over a land surface rapidly grows with time in response to the integrated sensible heat flux from the surface, mean vertical motion at the top of the PBL, mechanical mixing and other factors governing the entrainment rate at the top of the PBL.

A variety of theoretical models have been proposed in the literature for predicting the depth of the unstable PBL over a flat and homogeneous terrain. These are based on the rate equations for h and other associated parameters, such as the potential temperature in the mixed layer $\bar{\theta}_m$, temperature jump at the inversion base $\Delta\bar{\theta}$, etc. The models differ mainly in their choice of closure assumptions and representation of the transition layer above the mixed layer. It has been shown (Driedonks, 1982, Arya and Byun, 1983) that there are not very significant differences in the predicted h by various models, except for the crudest encroachment type models. The constant heat-flux-ratio model proposed by Tennekes (1973) and others might be adequate for parameterizing the daytime mixed layer height. The rate equations to be solved are:

$$\frac{\partial h}{\partial t} = C\,\frac{Q_o}{\Delta\theta} + \bar{w}_h$$

$$\frac{\partial\Delta\bar{\theta}}{\partial t} = Q_o\left(\frac{C}{\Delta\bar{\theta}}\,\frac{\partial\theta_+}{\partial z} - \frac{1+C}{h}\right) \qquad (27)$$

where $C \simeq 0.2$.

The evolution of the stable nocturnal PBL is rather slow and it is not clear whether a simple diagnostic relation for h might be adequate or one must carry an appropriate rate equation for the same. Arya (1981) and Garratt (1982) have recommended and provided considerable observational support for the similarity relation

$$h = d(u_* L/|f|)^{1/2} \qquad (29)$$

in which $d \simeq 0.4$. On the other hand, Nieuwstadt (1981), Wetzel (1982) and Nieuwstadt and Tennekes (1981) have argued against the use of a diagnostic equation, such as (29), and instead have proposed a rate equation of the form

$$\frac{\partial h}{\partial t} = -\frac{h-h_e}{T_D} \qquad (30)$$

in which h_e is the equilibrium PBL height and T_D is a time scale of PBL evolution which is shown to increase monotonically with time after the evening transition. Accordingly, the PBL height evolves at a decreasing rate and may never quite attain its equilibrium height which is given by Eq. (29) or a similar diagnostic relation. A rate equation for h may be preferrable during the initial hour or two, but after that, since h may not deviate from h_e by more than 10-15%, one may use Eq. (29) as well. More recently, Nieuwstadt (1984) has also recommended the use of Eq. (29) on the basis of its satisfactory agreement with Cabauw observations.

6. VERTICAL VELOCITY AT THE TOP OF THE PBL

Ignoring the local and advective accelerations, and assuming that the momentum fluxes vanish at z = h, the integration of the equations of mean motion and continuity with respect to z from 0 to h yields

$$\frac{\tau_o}{\rho f} = h(\bar{v}_m - \bar{v}_{gm})$$

$$0 = h(\bar{u}_m - \bar{u}_{gm}) \qquad (31)$$

$$\bar{w}_h = -\frac{\partial}{\partial x}(h\bar{u}_m) - \frac{\partial}{\partial y}(h\bar{v}_m) + \bar{u}_h \frac{\partial h}{\partial x} + \bar{v}_h \frac{\partial h}{\partial y} .$$

Differentiating the first of Eqs. (31) w.r.t.y and the second w.r.t.x., and substituting in the third, we have

$$\bar{w}_h = -\frac{\partial}{\partial y}\left(\frac{\tau_o}{\rho f}\right) + (\bar{u}_h - u_{gm})\frac{\partial h}{\partial x} + (\bar{v}_h - v_{gm})\frac{\partial h}{\partial y}$$

$$- h\left(\frac{\partial u_{gm}}{\partial x} + \frac{\partial v_{gm}}{\partial y}\right) \qquad (32)$$

where u_{gm} and v_{gm} are the layer-averaged components of geostrophic wind.

In the above expression for $\bar{w}_h$ the first term represents the curl of $\tau_o/\rho f$ which is expected to be the primary term under most circumstances. The second and third terms represent the contribution to $\bar{w}_h$ due to spatial changes in the PBL height; these are zero for a horizontally homogeneous PBL. The last term in (32) involves the divergence of the layer-averaged geostrophic winds and is probably negligible for all practical purposes.

For PBL over a complex terrain, one must also consider the contribution to $\bar{w}_h$ due to the terrain slope. For a flat and homogeneous terrain, however, eq. (32) reduces to the well know expression

$$\bar{w}_h = \frac{1}{\rho} \nabla \times (\tau_o/f). \qquad (33)$$

7. ENERGY DISSIPATION IN THE PBL

Following Lettau (1962), one can derive an expression for the total energy dissipation E_D in terms of the surface drag and the layer-averaged winds

$$E_D = \int_o^h \rho\varepsilon dz = \int_o^h \tau_o \cdot \frac{\partial \bar{V}}{\partial z} dz \qquad (34)$$

where ε is the local energy dissipation . Using the relations

$$\int_o^h \frac{\partial}{\partial z}(\tau_o \cdot \bar{V})\, dz = 0 \qquad (35)$$

$$\bar{V} \cdot \frac{\partial \tau}{\partial z} = V_g \cdot \frac{\partial \tau}{\partial z} , \qquad (36)$$

one gets from Eq. (34)

$$E_D = V_{go} \cdot \tau_o + \int_o^h \tau \cdot \frac{\partial V_g}{\partial z} dz \qquad (37)$$

Here the primary term is the dot product of surface geostrophic wind and stress vectors and the second term represents the contribution of baroclinity in the PBL.

The above expression for the total energy dissipation in the PBL is expected to be valid only far away from the equator where advective accelerations can be neglected; Eq. (34) is the more generally valid expression for E_D.

Acknowledgement

This material is based upon the research work supported by the National Science Foundation under grants ATM 80-24048 and INT82-19710 to North Carolina State University. The author wishes to thank Professor M.P. Singh for inviting him to visit the Center for Atmospheric and Fluids Sciences, Indian Institute of Technology, New Delhi, during the course of this study.

References

Arya, S.P.S.: 1975, 'Geostrophic Drag and Heat Transfer Relations for the Atmospheric Boundary Layer', Quart. J. Roy. Meteorol. Soc. 101, 147-161.

Arya, S.P.S: 1977, 'Suggested Revisions to Certain Boundary Layer Parameterization Schemes Used in Atmospheric Circulation Models', Mon. Weath. Rev. 105, 215-227.

Arya, S.P.S.: 1978, 'Comparative Effects of Stability, Baroclinity, and the Scale-Height Ratio on Drag Laws for the Atmospheric Boundary-Layer', J. Atmos. Sci. 35, 40-46.

Arya, S.P.S.: 1981, 'Parameterizing the Height of the Stable Atmospheric Boundary Layer', J. Appl. Meteorol. 20, 1192-1202.

Arya, S.P.S.: 1982, 'Atmospheric Boundary Layer Over Homogeneous Terrain', Chapter 6, Engineering Meteorology, E.J. Plate, editor, Elsevier Scientific Publishing Co., Amsterdam, the Netherlands.

Arya, S.P.S. and Byun, D.W.: 1983, 'Rate Equations for the PBL Depth (Urban vs. Rural)', presented at the AMS Speciality Conference on Air Quality Modeling of the Nonhomogeneous, Nonstationary Urban Boundary Layer, October 31-November 4, 1983, Baltimore, U.S.A.

Arya, S.P.S. and Wyngaard, J.C.: 1975, 'Effect of Baroclinity on Wind Profiles and the Geostrophic Drag Law for the Convective Planetary Boundary Layer', J. Atmos. Sci. 32, 767-778.

Benoit, R.: 1976, 'A Comprehensive Parameterization of Atmospheric Boundary Layer for General Circulation Models', Ph.D. Dissertation, McGill University and National Center for Atmospheric Research, 278 pp.

Brost, R.A. and Wyngaard, J.C.: 1978, 'A Model Study of the Stably Stratified Planetary Boundary Layer', J. Atmos. Sci. 35, 1427-1440.

Brutsaert, W.H.: 1982, 'Evaporation into the Atmosphere', D. Reidel Publishing Co., Dordrecht, Holland, 299 p.

Businger, J.A.: 1973, 'Turbulent Transfer in the Atmospheric Surface Layer', Workshop on Micrometeorology, D.A. Haugen, editor, American Meteorological Society, 67-100.

Businger, J.A., Wyngaard, J.C., Izumi, Y. and Bradley, E.F.: 1971, 'Flux-Profile Relationships in the Atmospheric Surface Layer', J. Atmos. Sci. 28, 181-189.

Caughey, S.J.: 1982, 'Observed Characteristics of the Atmospheric Boundary Layer', Atmospheric Turbulence and Air Pollution Modeling, F.T.M. Nieuwstadt and H. van Dop, editors, D. Reidel Pub. Co., 107-158.

Caughey, S.J. and Palmer, S.G.: 1979, 'Some Aspects of Turbulence Structure through the Depth of the Convective Boundary Layer', Quart. J. Roy. Meteorol. Soc. 105, 811-827.

Caughey, S.J., Wyngaard, J.C. and Kaimal, J.C.: 1979, 'Turbulence in the Evolving Stable Boundary Layer', J. Atmos. Sci. 36, 1041-1052.

Clarke, R.H. and Hess, G.D.: 1974, 'Geostrophic Departure and the Functions A and B of Rossby-Number Similarity Theory', Boundary-Layer Meteorol. 7, 267-287.

Deardorff, J.W.: 1970, 'A Three-dimensional Numerical Investigation of the Idealized Planetary Boundary Layer', Geophys. Fluid Dyn. 1, 377-410.

Deardorff, J.W.: 1972, 'Parameterization of the Planetary Boundary Layer for use in General Circulation Models', Mon. Weath. Rev. 100, 93-106.

Driedonks, A.G.M.: 1982, 'Models and Observations of the Growth of the Atmospheric Boundary Layer', Boundary-Layer Meteorol. 23, 283-306.

Garratt, J.R.: 1982, 'Observations in the Nocturnal Boundary Layer', Boundary-Layer Meteorol. 22, 21-48.

Garratt, J.R., Wyngaard, J.C. and Francey, R.J.: 1982, 'Winds in the Atmospheric Boundary Layer-Prediction and Observations', J. Atmos. Sci. 39, 1307-1316.

Hanna, S.R.: 1981, 'Lagrangian and Eulerian Time-scale Relations in the Day-time Boundary Layer', J. Appl. Meteorol. 20, 242-249.

Hanna, S.R.: 1982, 'Applications in Air Pollution Modeling', Atmospheric Turbulence and Air Polluction Modelling, F.T.M. Nieuwstadt and H. van Dop, editors, D. Reidel Pub. Co., 275-310.

Kaimal, J.C., Wyngaard, J.C., Haugen, D.A. Coté, O.R. Izumi, Y., Caughey, S.J. and Readings, C.J.: 1976, 'Turbulence Structure in the Convective Boundary Layer', J. Atmos. Sci. 33, 2152-2169.

Lettau, H.H.: 1962, 'Theoretical Wind Spirals in the Boundary Layer of a Barotropic Atmosphere', Beitr, zur Phys. der Atmos. 35, 195-212.

Meljarejo, J.W. and Deardorff, J.W.: 1974, 'Stability Functions for the Boundary Layer Resistance Laws based upon Observed Boundary-Layer Heights', J. Atmos. Sci. 31, 1324-1333.

Nieuwstadt, F.T.M: 1981, 'The Steady State Height and Resistance Laws of the Nocturnal Boundary Layer, Theory Compared with Cabauw Observations', Boundary-Layer Meteorol. 20, 3-17.

Nieuwstadt, F.T.M.: 1983, 'The Turbulent Structure of the Stable Nocturnal Boundary Layer', Presented at the AMS Speciality Conference on Air Quality Modeling of the Nonhomogeneous, Nonstationary Urban Boundary Layer, Oct. 31 - Nov. 4, Baltimore, U.S.A.

Nieuwstadt, F.T.M.: 1984, 'Some Aspects of the Turbulent Stable Boundary Layer', Presented at the 29th OHOLO Conference on Boundary Layer Structure-Modelling and Application to Air Pollution and Wind

Energy, March 25-28, Zichron Ya'acov, Israel.
Nieuwstadt, F.T.M. and Tennekes, H.: 1981, 'A Rate Equation for the Nocturnal Boundary-layer Height', J. Atmos. Sci. 38, 1418-1428.
Panofsky, H.A., Tennekes, H., Lenschow, D.H., Wyngaard, J.C.: 1977, 'The Characteristics of Turbulent Velocity Components in the Surface Layer under Convective Conditions', Boundary-Layer Meteorol. 11, 355-361.
Pasquill, F.: 1974, 'Atmospheric Diffusion', 2nd Ed., John Wiley and Sons, New York, 429 pp.
Tennekes, H.: 1973, 'A Model for the Dynamics of Inversion above a Convective Boundary Layer', J. Atmos. Sci. 30, 558-567.
Wesely, M.L. and Coulter, R.L.: 1983, 'Meteorological Pollutant Profiles under very Stable Conditions', Extended Abstracts, Sixth Symposium on Turbulence and Diffusion, March 22-25, 1983, Boston, Mass., American Meteorological Society, 294-296.
Wetzel, P.J.: 1982, 'Toward Parameterization of the Stable Boundary Layer', J. Appl. Meteorol. 21, 7-13.
Wyngaard, J.C., Arya, S.P.S. and Coté, O.R.: 1974, 'Some aspects of the Structure of Convective Planetary Boundary Layers', J. Atmos. Sci. 31, 747-754.
Yamada, T.; 1976, 'On the Similarity Functions A, B and C of the Planetary Boundary Layer', J. Atmos. Sci. 33, 781-793.
Zilitinkevich, S.S.: 1975, 'Resistance Laws and Prediction Equations for the Depth of the Planetary Boundary Layer', J. Atmos. Sci. 32, 741-752.

Questions, answers and comments

L.N. GUTMAN: Can you show us any parametric relations concerning a planetary boundary layer developing over uneven terrain?

S.P.S. ARYA: No generalisable parametric relations can be given for the PBL over complex terrain, since the terrain can vary in a variety of ways, introducing a host of new length, velocity and temperature scales. For certain classes of complex terrain features, however, it is possible to obtain the parametric relations on the basis of systematic experiments and model studies. For example, in the case of gently and uniformly sloping terrain, the suggested wind profiles and resistance laws can be generalised to include the effect of terrain slope, using the model results of Gutman and Meljarejo (1982).

L. BERKOFSKY: In your $\bar{w}_h$ equations you have terms $\bar{u}_h$, $\bar{v}_h$. How do you determine these wind components?

S.P.S. ARYA: If we assume that at the top of the PBL, winds are in geostropic balance, then $\bar{u}_h$ and $\bar{v}_h$ are determined from the geostropic wind field, i.e. , $\bar{u}_h=u_{gh}$ $\bar{v}_h=v_{gh}$.

L. BERKOFSKY: It is probable that the parametrization of $\bar{w}_h$ is not simple.

S.P.S. ARYA: The parametrization of the vertical velocity at the top of the PBL is indeed difficult and requires a good estimate of the horizontal divergence of the wind field in the PBL or the spatial variation (curl) of the surface stress vector over a flat terrain. The addition of complex topography makes such a parameterization extremely difficult.

ENTRAINMENT EFFECTS IN THE WELL-MIXED ATMOSPHERIC BOUNDARY LAYER

A.G.M. Driedonks and H. Tennekes
Royal Netherlands Meteorological Institute,
de Bilt, Netherlands

ABSTRACT

We discuss the structure and evolution of a cloud-free atmospheric boundary layer (ABL) during daytime over land, starting from a shallow ABL at sunrise and developing into a deep ABL with strong convection in the afternoon. The structure of the turbulence in the lower half of a convective ABL capped by an inversion is reasonably well understood. Less is known about the details of the turbulence in higher regions affected by entrainment, because of the difficulty in taking turbulence measurements there. For the evolution in time of the height of the ABL and its mean potential temperature mixed-layer models have been developed that give satisfactory agreement with observations. It has been shown that for many practical applications accurate knowledge of forcing functions and boundary conditions is more important than a refinement of the entrainment hypothesis. Observations show that the assumption of well-mixedness of first-order moments of conservative variables is not valid for all quantities. A simple similarity relation for the inclusion of the effect of entrainment on the shape of the vertical profiles is given.

1. INTRODUCTION

In this study we consider the structure and dynamics of a cloud-free atmospheric boundary layer (ABL) during daytime over land, starting from a shallow ABL at sunrise and developing into a deep ABL with strong convection in the afternoon. The growth of the ABL is associated with entrainment, bringing air with different properties from above down into the turbulent boundary layer. The way in which this entrainment proceeds depends to a large extent on the forcing mechanisms that maintain the

Boundary-Layer Meteorology 30 (1984) 75–105. 0006-8314/84/0301-0075$04.65

turbulence in the ABL (e.g. a surface heat flux), and the counteracting influences from the stable stratification in the air aloft.

Entrainment and surface fluxes together will change the properties of the ABL. E.g. the surface heat flux together with the negative heat flux at the top will change the mean ABL-temperature, and when air with higher wind speed is entrained at the top this will, together with the surface friction, change the mean ABL wind. When air pollution is confined to the ABL, entrainment of non-polluted air will decrease the mean concentration. Thus the calculation of the height of the ABL and of its mean properties as functions of time is of great practical importance.

In order to understand the way entrainment works it is necessary to have insight in the way the turbulence in the ABL behaves, how it is maintained and what its structure is in the vertical direction. Knowledge about this structure is also necessary if we want to calculate the statistics of the distribution of air pollution in the ABL as a function of height or distance from the source. For an elaborate example of this type, see Lamb (1982). In the past twenty years considerable progress has been made in our understanding of the ABL through a number of experiments. Especially the well-mixed boundary layer during daytime over land has been the subject of several investigations. Knowledge has been gained from field experiments (Caughey, 1982), from laboratory experiments (Willis and Deardorff, 1974; Deardorff et al., 1980), and from detailed numerical simulations of the large-eddy structure of turbulence (Deardorff, 1974, 1980). This knowledge is now reasonably comprehensive as far as the turbulence up to a few hundred meters height in the convective ABL is concerned. Often it may be assumed that the turbulence is in local equilibrium with the external conditions, i.e. that the turbulent time scale of the large eddies is much smaller than the time scale of the change in e.g. the surface heat flux. Thus the structure of the ABL is not explicitly dependent on time. The structure of the turbulence higher up in the ABL and in the vicinity of a capping inversion is affected by the entrainment itself. Especially the higher-order statistics of the turbulence, such as variances and fluxes, are influenced, and sometimes entrainment effects are seen very deep down into the ABL. Comparatively little knowledge exists on the details of the turbulence in the higher regions in the ABL and on the "diffusivity" of the conditions at the top downward (Wyngaard and Brost, 1984).

The problem of the growth of the boundary layer during daytime has been given much attention (see Tennekes and Driedonks, 1981). Most widely used are slab models that predict the evolution of the ABL-height h(t) and of the vertically averaged values of first-order moments like potential temperature Θ. Essential in these models is the derivation of a rate equation for h(t), taken from a parameterized form of the turbulent kinetic energy budget. This parameterization has given rise to a huge amount of literature, particularly because our knowledge of the details of the processes in the entrainment zone is rather poor. Especially the details of the evolution of an ABL in which not only convection but also wind shear at the surface and at the top of the ABL is generating turbulence, are not yet completely solved for all limiting cases (Deardorff, 1983). However, it seems that for most practical conditions in the atmosphere satisfactory agreement with observations has been obtained (Driedonks, 1982b). It is very difficult to obtain observations in the real atmosphere against which more refined entrainment hypotheses can be tested. Besides, it has been demonstrated that the "external" factors that control the growth of the ABL (such as surface heat flux, mean subsidence, stability of the air aloft) do influence in most cases the evolution of the ABL even more than the particular entrainment relation used.

Slab models for the mixed layer do usually assume the vertical profiles of all first-order moments of conservative variables in the ABL to be well-mixed up to the inversion height h. Turbulence is assumed to be strong enough to maintain these flat profiles. However, observations suggest that this might be a reasonable assumption for potential temperature, but not for all other scalar quantities. Water vapour often shows a distinct gradient over the mixed layer, which can also be the case for passive contaminants. In order to account for this we allow the shape of these vertical gradients to be influenced by the entrainment rate itself. This can be done in a tentative way by some simple similarity arguments which, however, have not yet been tested satisfactorily against observations.

2. STRUCTURE OF THE WELL-MIXED BOUNDARY LAYER

The turbulent structure in a convective boundary layer has been investigated both through experiments (Caughey, 1982) and through large-eddy numerical simulation (Deardorff, 1974). It is recognized that already at relatively small values of $-h/L \sim 5$ (h being the ABL-height, L the Monin-Obukhov length) the ABL is driven into a convective state with strong effects of convection on the turbulence. Once the ABL is convective its structure does not depend on the value of $-h/L$ any more. A well-known typical picture of the convective ABL is given in Figure 1, where different regions are distinguished: the surface layer, the mixed layer, and the entrainment zone. A rather extensive description of the observational characteristics can be found in Caughey (1982). We will stress here some important points.

The surface layer

In the surface layer (~ the lowest 10% of the ABL) numerous experiments have confirmed that Monin-Obukhov similarity generally holds. According to this, the surface layer structure is completely determined by the friction velocity u_*, the surface heat flux $\overline{\theta w}_s$, the

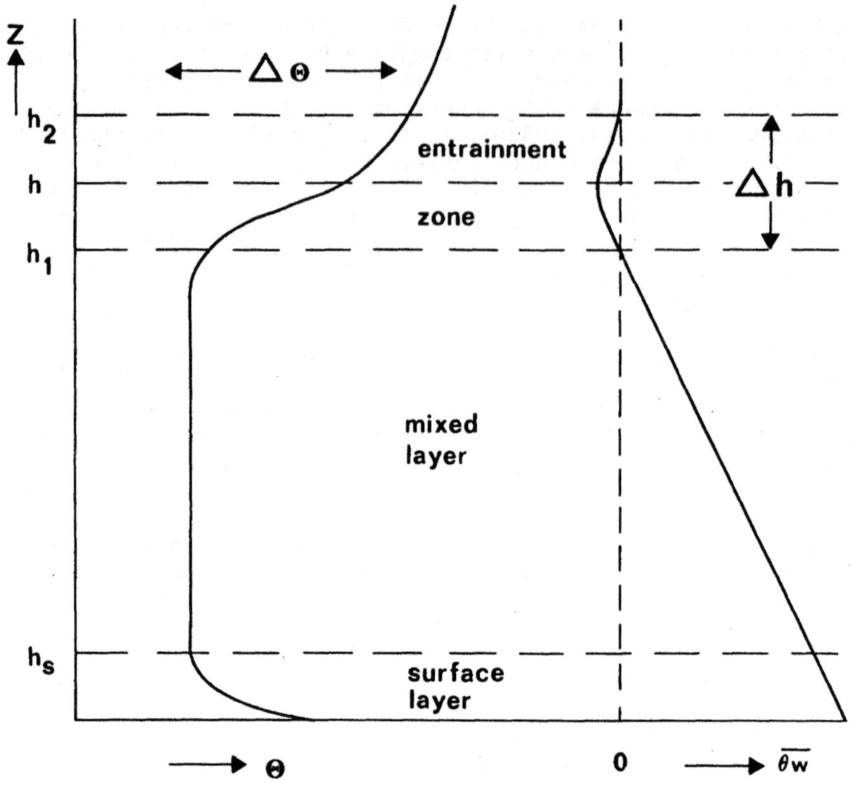

Figure 1. Typical structure of the vertical profiles of potential temperature and heat flux in a convective boundary layer.

buoyancy parameter g/T, and the height z. When we also consider some passive scalar we add its surface flux $\overline{cw}_s$. Thus, surface layer variables, if properly scaled, are universal functions of z/L, with $L = -u_*^3\ T/kg\ \overline{\theta w}_s$. This Monin-Obukhov similarity holds for many surface-layer quantities, but not for the statistics of the fluctuating horizontal velocity components u' and v'. The latter are influenced by the large eddies in the ABL and thus should depend on h (Panofsky et al., 1977).

Local free convection

In the upper part of the surface layer one might expect under certain conditions ($z > -L$ and $z \ll h$) that local free convection scaling holds, in which the friction velocity u_* stops being an important scaling parameter while h has not yet begun. Local free convection scaling leads to explicit predictions of the behavior of non-dimensionalized quantities that are to a large extent supported by observations (see e.g. Wyngaard, 1971; Caughey, 1982). The condition $z \ll h$ is to be interpreted as $z < 0.1$ h, thus a layer with local free convection scaling should be present for $-h/L > 10$. In measurements the height up to which this scaling holds is seen to vary quite a lot. In Caughey (1982) it is shown that σ_w^2 exhibits local free convective scaling up to $z \sim 0.1$ h above which it starts to deviate, while in the same experiment the structure parameters for temperature and velocity (C_T^2 and C_v^2) follow local free convection scaling as high as $z \sim 0.5$ h. However, Wyngaard and LeMone (1980) present results from other experiments in which local free convection scaling for C_T^2 breaks down already at $z \sim 0.1$ h. In their results the deviations from this scaling are not to be attributed to the necessity to include h as an important parameter, but rather to the effect of entrainment being different for different experiments. This is one of the complicating factors in finding scaling relations above the surface layer that we will discuss now.

The mixed layer

Above the surface layer ($z \sim 0.1$ h) observations indicate that in a convective ABL the profiles of potential temperature (and also of specific humidity, wind speed, etc.) are almost flat, independent of z. The existence of this mixed layer is typical for an ABL in convective conditions. The parameter measuring the amount of "convectiveness" is $-h/L$, and it has been shown that the existence of this mixed layer and its characteristic structure becomes independent of the value of $-h/L$ once it is larger than about 5 (Deardorff, 1974). Thus the transition to a convective ABL occurs rather abruptly and its structure remains essentially unchanged (in terms of properly non-dimensionalized variables) when the amount of convection increases. This has led to the idea of a simple convective scaling or mixed-layer scaling. The scaling parameters then are the surface heat flux $\overline{\theta w}_s$, the buoyancy parameter g/T, the ABL-height h, the height z, and if relevant the surface fluxes of passive scalars $\overline{cw}_s$. According to convective scaling all variables of the mixed layer become universal functions of z/h, when they are properly non-dimensionalized with the convective velocity scale $w_* = ((g/T)\ \overline{\theta w}_s\ h)^{1/3}$, temperature scale $\theta_* = \overline{\theta w}_s/w_*$, length scale h, and "passive scalar scale" $c_* = \overline{cw}_s/w_*$. The universal functions can be determined experimentally and should apply to <u>all</u> convective boundary layers. However, in a real ABL there are several complications that can cause this simple mixed-layer scaling to break down. Wyngaard (1983) discusses some of these problems. Here we will consider the effect of entrainment. At the top of the ABL entrainment brings air with different properties down into the mixed layer. In general the structure of this mixed layer will not only be influenced by processes at the surface (e.g. $\overline{\theta w}_s$), but also by the processes at the top of the ABL. In simple mixed-layer scaling these entrainment effects are not considered and so for this scaling to apply, the entrainment-induced contribution has to be small. It appears that even in one experiment the importance of this contribution can be different for different turbulence statistics and that, for the same turbulence statistics, it can vary from one experiment to the other.

The vertical profile of Θ seems least influenced by entrainment effects. It shows up to be almost independent of height in the mixed layer regardless (to a great extent) of the conditions at the top. This does not hold for all first-order moments, e.g. humidity or wind speed may exhibit non-zero vertical gradients within the mixed layer that probably depend on the sign and magnitude of the fluxes of these quantities at the top in relation to the corresponding surface fluxes and the intensity of the turbulence (w_*). We will return to this later. Nevertheless it can be generally stated that entrainment effects on mean profiles of first-order moments are usually not very large.

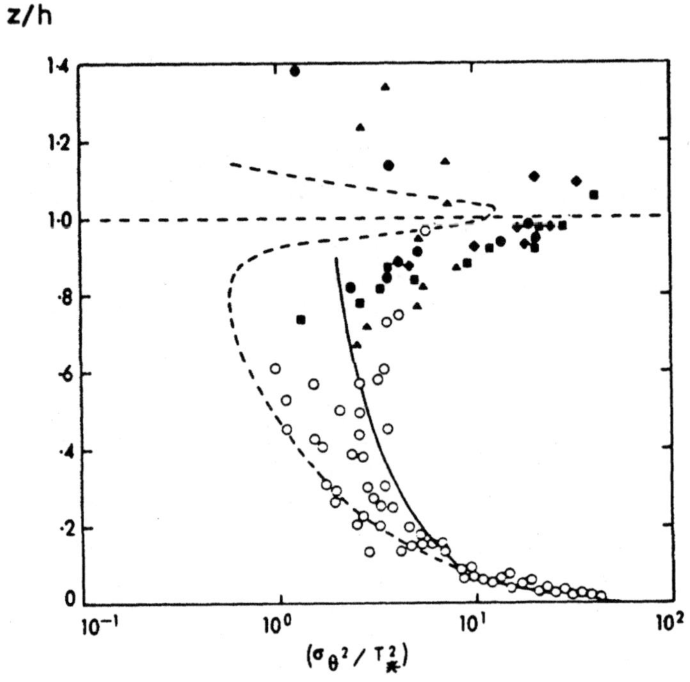

Figure 2. The normalized temperature variance in a convective boundary layer, from Caughey (1982). Open symbols refer to the Minnesota experiment, closed symbols to the Ashchurch experiment. The dashed line is from laboratory experiments (Willis and Deardorff, 1974), the solid line is the local free convection prediction. The symbol T_* in the figure is the same as θ_* in the text.

Higher-order turbulence statistics often do show entrainment effects of considerable magnitude making simple mixed-layer scaling questionable in many cases. One example is the variance of temperature fluctuations in Figure 2 (taken from Caughey, 1982). The normalized σ_θ^2 is shown as a function of z/h. In the lower 10% of the mixed layer local free convection similarity holds, collapsing the data on a single curve. Above this height the data do not seem to follow any universal behavior with z/h at all. At least part of the scatter must be attributed to entrainment effects violating simple mixed-layer scaling down to heights as low as ~ 0.1 h. In the upper part of the ABL ($z > 0.7$ h) the variance increases, associated with the capping inversion. In this region entrainment effects are very large and simple mixed-layer scaling obviously does not apply.

The large scatter in the σ_θ^2-data in Figure 2 in the mixed layer is probably due to entrainment effects on the low-frequency contribution to the variance. The high-frequency contributions are contained in the structure parameter for temperature C_T^2. For the same experiments as in Figure 2 the non-dimensionalized C_T^2 does follow mixed-layer similarity as high as $z \sim 0.5$ h quite well (Figure 3a, from Wyngaard (1983)). It is even seen that for these data the local free convection behavior (which in fact does not include h as a scaling parameter) holds even in the mixed layer. However, C_T^2-data from other experiments (Figure 3b) show strong entrainment effects as low as $z \sim 0.1$ h.

Other examples are the fluxes of heat ($\overline{\theta w}$) and passive scalars ($\overline{cw}$). Since the mean profiles of Θ and C are essentially flat, the fluxes have to be linear in the mixed layer, from their values near the surface to a value $\overline{\theta w}_h$ and $\overline{cw}_h$ near the top. For simple mixed-layer similarity theory to apply these fluxes at z = h have to be small. In a convective boundary layer usually $\overline{\theta w}_h \sim -0.2\ \overline{\theta w}_s$. This is not a particularly small ratio; however, since the ratio is constant, the inclusion of entrainment effects in simple mixed-layer similarity for the heat flux is quite straightforward: instead of $\overline{\theta w}(z) = \overline{\theta w}_s(1-z/h)$, we can state $\overline{\theta w}(z) = \overline{\theta w}_s(1- 1.2\ z/h)$. For other scalar fluxes $\overline{cw}$ such a simple generalization is not possible, since the jump at the top ΔC is involved which is an independent parameter. Thus the effect of entrainment on the profile of $\overline{cw}$ can be different in the mixed layer, dependent on the conditions at the top.

In general one could state that local free convection scaling usually holds up to about $z \sim 0.1$ h. Above this height vertical profiles of first-order moments in the mixed layer are least influenced by entrainment up to $z \sim 0.8$ h. Higher-order statistics of turbulence can be drastically influenced by entrainment effects, even as low as $z \sim 0.1$ h. Thus for any general scaling of the mixed layer we have to include variables from the top as scaling parameters. Such a generalization has not yet been achieved.

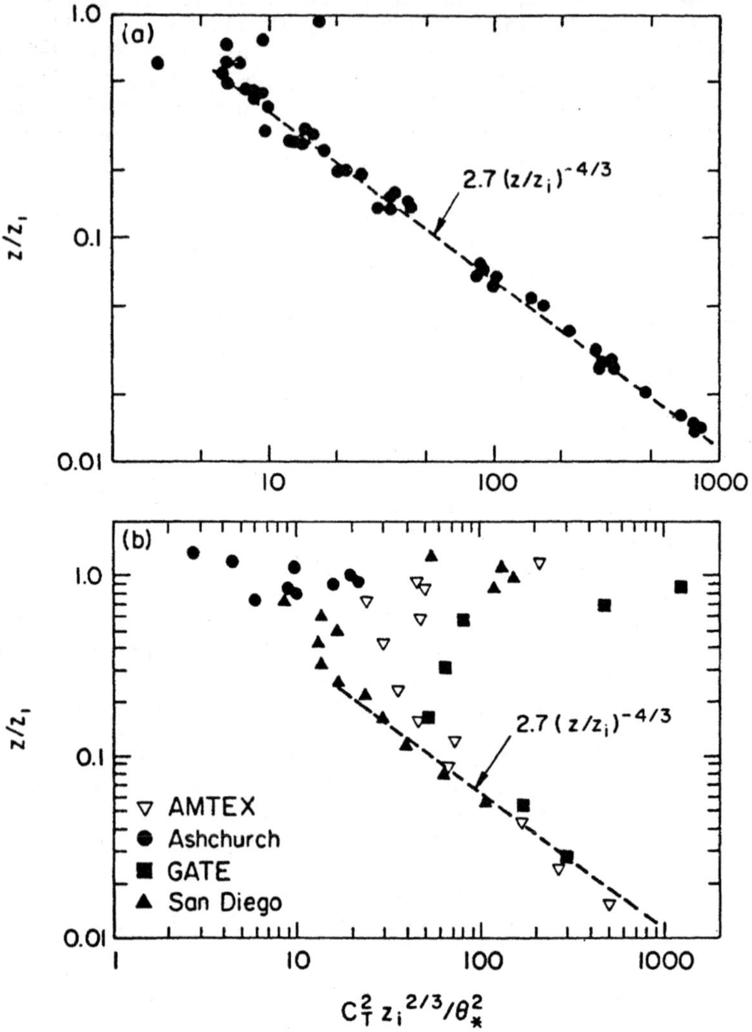

Figure 3. Behavior of the temperature structure parameter in the convective boundary layer in different experiments (from Wyngaard, 1983). Upper panel: from the Minnesota experiment where entrainment effects were small. Dashed line is the local free convection prediction. Lower panel: data from other experiments where entrainment effects were large ($z_i = h$).

The entrainment zone

At the top of the mixed layer we find the entrainment zone (Figure 1) in which the structure is dominated by the way in which the entrainment process actually proceeds. There is appreciably less knowledge about the details of the entrainment zone than about the lower part of the mixed layer. This is due to the great difficulty in taking accurate measurements in this region where the turbulence is highly intermittent. Also, due to the fact that entrainment is essentially a small-scale process, detailed large-eddy numerical simulation has not provided results for the entrainment zone (Deardorff, 1974, 1980). Yet some general insight in the entrainment process has been obtained. Measurements by Rayment and Readings (1974) in the atmosphere show that convective bubbles impinge on the stable air aloft and can penetrate a certain distance. Wave-like disturbances form on the edges of these domes, induced by the enhanced shear. These small-scale waves break, mixing non-turbulent air into the turbulent region. Thus the local boundary between turbulent and non-turbulent fluid is fairly sharp, convective bubbles mainly acting to force undulations on the interface and increasing the local shear on which the small-scale entrainment feeds. However, although the local interface may be sharp, after ensemble-averaging (or horizontal averaging) the vertical profiles will show an entrainment zone of finite depth, related to the amplitude of the undulations caused by the thermals. This amplitude depends on the vertical velocity of the impinging bubbles ($\sim w_*$) and on the stability of the air aloft. A conservative estimate for the amplitude is w_*/N with N the Brunt-Vaisala frequency ($N^2 = (g/T)\ \partial\Theta/\partial z$) in the air aloft. For $w_* = 1$ m/s, $\partial\Theta/\partial z = 3$ K/km, this gives an entrainment zone of ~ 100 m depth. A radiosonde ascent, giving an instantaneous sample of the ABL-height, must be considered to have this type of inaccuracy in estimating the ensemble-averaged ABL height (Driedonks, 1982b). When the stability in the air aloft is low, the entrainment zone can be very thick.

Several attempts have been made to formulate scaling laws for the entrainment zone. Wyngaard and LeMone (1980) derive under certain conditions an appealingly simple relation for the values of C_T^2, averaged over the entrainment zone, that agrees fairly well with the observations of Figure 3 in the vicinity of z = h. It remains a problem how to match this entrainment-zone scaling with the results in the lower part of the mixed layer in order to scale entrainment effects there. Probably the most extensive studies on the interfacial layer have been done by Deardorff et al. (1980) and Deardorff (1983). They formulate relations for $\Delta h/h$ (for definition of Δh see Figure 1) and for the Δh-averaged values of terms in the turbulent kinetic energy budget, e.g. the dissipation. An example is the expression from Deardorff et al. (1980) relating $\Delta h/h$ in a convective ABL to an overall Richardson-number

$$Ri_* = \frac{g}{T}\ \Delta\Theta\ h/w_*^2,$$

in the form

$$\frac{\Delta h}{h} = 0.21 + 1.31 \, Ri_*^{-1},$$

which however has an unexpected non-zero limit when $Ri_* \to \infty$. In general a number of empirical constants and closure functions is involved and "present experimental data are not nearly comprehensive enough to pin down these functions", especially in cases where there is not only convection but also wind shear at the top of the mixed layer generating turbulence.

Some interesting consequences of entrainment-zone dynamics for air pollution dispersion were reported by Deardorff and Willis (1982). They investigated the transient behavior of a plume of contaminants, initially lying in the stable air aloft, and incorporated into the convective boundary layer as it deepens (fumigation). They found that the line-averaged concentration at the ground reached its maximum value a considerable time after the fumigation process starts. Measured in dimensionless units of time, $T_+ = t\, w_*/h$, this can be $T_+ = 7\text{–}10$, depending on the entrainment rate. For typical atmospheric values h = 500, $w_* = 1$ this means a time delay of one or two hours. Deardorff and Willis attribute this time delay to the undulations of the interface in the entrainment zone, causing the fumigant to be intercepted spottily by the mixed layer and stretching the total period of time during which this interception takes place.

The overall conclusion for a convective boundary layer must be that to a large extent its structure is reasonably well understood. There are some remaining problems, related to the details of the structure of the entrainment zone and to the downward extension into the mixed layer of these entrainment effects on higher-order turbulence statistics such as variances or fluxes.

In this section we did not consider the evolution of the ABL as a function of time. Yet this behavior of h(t) and the mean profiles as a function of external forcing plays an essential role in the application of ABL-dynamics. We consider this in the next section.

3. THE ENTRAINMENT RATE

The upper edge of a turbulent boundary layer forms the interface between turbulent air within and non-turbulent air above the boundary layer. When the ABL is well-stirred it will tend to grow into the overlying quiet air. The rate at which this occurs, relative to the mean flow, is the entrainment rate. We will consider horizontally homogeneous conditions but allow for a mean vertical velocity $\overline{w}_h$ at the ABL top. The entrainment rate is there defined as

$$w_e = \frac{\partial h}{\partial t} - \overline{w}_h. \tag{1}$$

The entrainment rate determines the evolution of the ABL-height and the amount of air with different properties brought down from above into the ABL. It will depend on the structure of the turbulence in the ABL and on the opposing influences at the top.

We will consider an ABL that is well-stirred, i.e. in which the turbulence is strong enough to maintain a mixed layer with almost flat vertical profiles of mean potential temperature, wind speeds, etc., from the top of the surface layer up to the base of the entrainment zone as in Figure 1. This means that we exclude the stable nighttime boundary layer. When shortly after sunrise the ABL starts to grow, the surface heat flux is still small and convection not yet well developed. Yet we will assume that very quickly a situation exists in which the profiles in the bulk of the ABL can be clearly distinguished from the entrainment zone. This is justified by observations.

In the ABL that we consider, not only convection from the surface is producing turbulent kinetic energy (TKE). We will also have to take into account the TKE generated by wind shear. However, we exclude production of TKE by some other distinct internal source as in the case when radiational processes are important, e.g. when a cloud deck is present in the ABL (Deardorff, 1980).

In such a well-stirred ABL there are in general three distinct mechanisms that produce turbulence and possibly entrainment: a) convection from the surface, associated with the velocity scale w_*, b) friction at the surface, associated with the friction velocity u_*, and c) shear-generation of turbulence by wind shear at the top, associated with the velocity jump $\Delta \underset{\sim}{U}$ over the entrainment zone.

Opposed to these entrainment-producing mechanisms in the ABL are the counteracting influences at the top. There, relatively warmer air has to be brought down into the ABL at the expense of turbulent kinetic energy. So the main counteracting influence on w_e is the jump in potential temperature $\Delta\Theta$ (or rather virtual potential temperature $\Delta\Theta_v$).

An enormous amount of models has been developed to relate the entrainment rate w_e to the afore-mentioned production mechanisms and the opposing influences at the top. For some cases consensus has been reached, but up till now no general entrainment relation has been formulated that is valid for all possible combinations of the three production mechanisms mentioned and that is fully tested against observations. The most recent effort is by Deardorff (1983). The main reasons for the disagreement are the difficulties encountered in the parameterization of the amount of TKE produced in the ABL that is actually available for entrainment. Since the dissipation in the ABL is large, this available energy is only a small fraction of the total production rendering parameterizations sensitive to small inaccuracies. Another problem is the lack of observations from either the laboratory or the real atmosphere that are comprehensive enough to test these parameterizations, especially in the case of shear-generated turbulence. E.g. it has been recently shown by Scranton and Lindberg (1983) and confirmed by Deardorff (1983) that the results of annulus experiments, in which a rotating screen generates u_* and $\Delta\underline{U}$ values, are not to be trusted since they are all contaminated by wall effects and depend on the particular annulus geometry used.

In this paper we will take a rather practical point of view and bias our approach toward relatively simple models for the ABL that have been tested to give satisfactory results for the evolution of h(t) and $\Theta_m(t)$ in comparison with atmospheric observations (Θ_m is the vertically averaged mean temperature in the ABL).

We start with a simplification of the structure of the mixed layer in that we use a zero-order jump model in which the thickness of the entrainment zone is negligibly thin and in which the thickness of the surface layer is very small compared to h so that its effects on the mean budgets for the ABL also can be neglected. Thus we simplify the structure of Figure 1 to the one given in Figure 4, in which Θ can be read as Θ_v when humidity is included. The equations associated with this type of model are extensively discussed by Tennekes and Driedonks (1981) and Driedonks (1982 a, b), so we will not give the details here.

First, equations for the vertically averaged mean variables in the ABL are needed, e.g. for Θ_m. These equations are easy to derive. In the context of well-mixed profiles for all conserved first-order moments they read for any scalar quantity Φ:

$$\frac{\partial \Phi_m}{\partial t} = (\overline{\phi w}_s - \overline{\phi w}_h)/h, \qquad (2)$$

and for the wind-components we find

$$\frac{\partial U_m}{\partial t} = + f(V_m - V_{gm}) + (\overline{uw}_s - \overline{uw}_h)/h, \qquad (3)$$

$$\frac{\partial V_m}{\partial t} = - f(U_m - U_{gm}) + (\overline{vw}_s - \overline{vw}_h)/h. \qquad (4)$$

In these equations an index s denotes the value of the flux at the surface and an index h the value of the flux at z = h, caused by entrainment. Furthermore, U_{gm} and V_{gm} are the vertically averaged values of the geostrophic wind in the ABL and f is the Coriolis parameter.

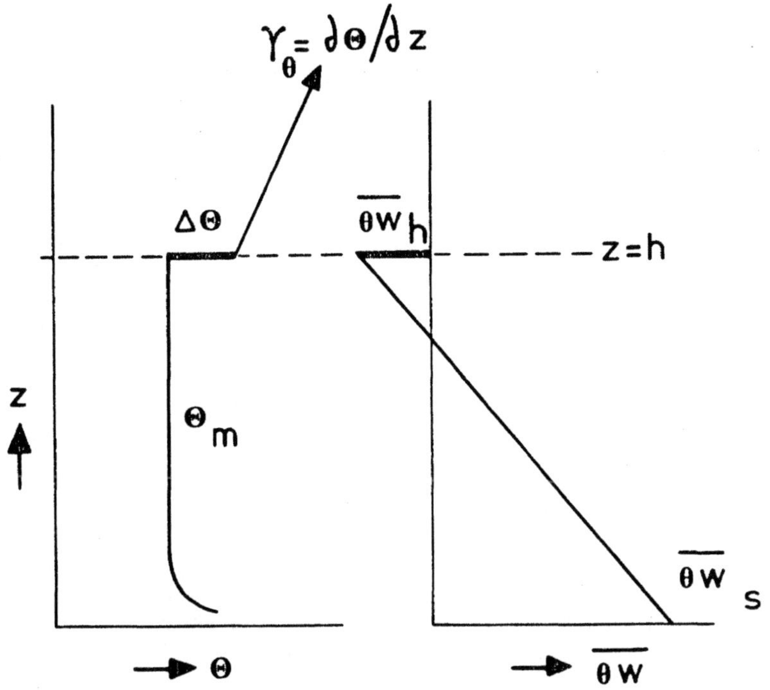

Figure 4. Schematic profiles of Θ and $\overline{\theta w}$ in a slab model.

Equations relating the fluxes at z = h to the jumps in the corresponding quantities are given by

$$-\overline{\phi w}_h = w_e \, \Delta\Phi, \tag{5}$$

where ϕ denotes any of the variables under consideration (e.g. Θ).

Additional equations for the jumps at z = h are required (although they also could be diagnostically determined at each time from the profiles in the overlying air and the mean values in the ABL). They read for scalar quantities Φ (e.g. ΔΘ):

$$\frac{\partial}{\partial t} \Delta\Phi = \gamma_\phi \, w_e - \frac{\partial \Phi_m}{\partial t}, \tag{6}$$

where $\gamma_\phi = \partial\Phi/\partial z$ refers to the lapse rate of Φ just above the ABL. For the wind components they read

$$\frac{\partial}{\partial t} \Delta U = \gamma_u \, w_e - (\overline{uw}_s - \overline{uw}_h)/h + f \, \Delta V, \tag{7}$$

$$\frac{\partial}{\partial t} \Delta V = \gamma_v w_e - (\overline{vw}_s - \overline{vw}_h)/h - f \Delta U. \tag{8}$$

The set of equations obtained is not yet closed and an additional closure equation has to be provided (semi)-empirically. This equation is usually derived from consideration of the budget of TKE since entrainment goes at the expense of TKE and it is the buoyancy flux $\overline{\theta w}_h$ (or $\overline{\theta_v w}_h$) that appears in this budget. However, closure can only be achieved after parameterization of all the terms (except of course the buoyancy flux) in the TKE-budget. It is here that entrainment models start to diverge.

Following Tennekes and Driedonks (1981) we consider the TKE-budget at z = h. This reads

$$\underset{T_h}{\left(\frac{\partial \bar{e}}{\partial t}\right)_h} = \underset{P_h}{- \overline{uw}_h \left(\frac{\partial U}{\partial z}\right)_h - \overline{vw}_h \left(\frac{\partial V}{\partial z}\right)_h} + \underset{B_h}{\frac{g}{T} \overline{\theta w}_h} - \underset{F_h}{\left(\frac{\partial}{\partial z} \left(\overline{ew} + \frac{\overline{pw}}{\rho}\right)\right)_h} - \underset{D_h}{\varepsilon_h}, \tag{9}$$

where $\bar{e}$ and e are the mean and fluctuating turbulent kinetic energy and ε the dissipation rate. Other symbols are according to the usual notation. The symbolic notation refers to the temporal rate of change (T_h) of $\bar{e}$, the shear production (P_h), the buoyancy term (B_h), the flux convergence (F_h) and the dissipation (D_h).

In parameterizing the TKE-budget we have to provide scaling laws for each of the terms. Here we simply assume that all length scales are proportional to h, all turbulent velocities scale on w_m, where w_m is a representative velocity scale for the bulk of the ABL and that $(\partial U/\partial z)_h$, $(\partial V/\partial z)_h$ do scale on $\Delta U/h$, $\Delta V/h$. In this way the term T_h becomes in parameterized form: $T_h = C_t \frac{w_m^2}{h} w_e$, and the dissipation as well as the flux convergence do both scale on w_m^3/h. Thus we take both terms together. In this simple parameterization the TKE-budget reads (after use of $-\overline{uw}_h = w_e \Delta U$, etc.).

$$C_t \frac{w_m^2}{h} w_e = C_M\{(\Delta U)^2 + (\Delta V)^2\} \frac{1}{h} w_e + \frac{g}{T} \overline{\theta w}_h + C_F \frac{w_m^3}{h}, \tag{10}$$

and after some rearranging we can write:

$$\frac{w_e}{w_m} = \frac{C_F}{C_t - C_M\{(\Delta U)^2 + (\Delta V)^2\}/w_m^2 + Ri}, \tag{11}$$

with Ri a Richardson number defined as

$$Ri = \frac{g}{T} \frac{h\,\Delta\Theta}{w_m^2} \quad . \tag{12}$$

For the velocity scale w_m, Tennekes and Driedonks (1981) take an interpolation between the convective velocity scale w_* and the friction velocity u_* in the form

$$w_m^3 = w_*^3 + (A'/C_F)u_*^3 \quad . \tag{13}$$

In eq. (10)-(13), C_t, C_M, C_F, and A' are all empirical constants, still quite a lot for such a simple model.

A few properties of this parameterization must be remarked. First, if ΔU, $\Delta V \to 0$, $u_* \to 0$, and if $w_e \ll w_m$ (usually the case), then this parameterization reduces to the well-known constant heat flux ratio $\overline{\theta w}_h = - C_F\, \overline{\theta w}_s$. Further, when ΔU, $\Delta V \to 0$, and $Ri \to 0$, then the entrainment rate remains finite with $w_e/w_m = C_F/C_t$.

In the parameterization (10) we refrained from introducing a separate length scale Δh associated with the entrainment zone, and from estimating the local dissipation and velocity gradients with this length scale. A few comments are in order here. First, if $\Delta h/h$ would be a constant, then it could be simply incorporated in the constants in eq. 10. However, this is not the case. Second, in Zeman and Tennekes (1977) $\Delta h/h$ is made a function of the stability aloft, γ_θ. However, this seems too arbitrary since it neglects the effect of all other quantities, such as $\Delta\theta$, ΔU, on $\Delta h/h$. Also, their approach has an unattractive consequence in leading to negative entrainment rates under certain conditions. Therefore we did not include their dissipation formulation. An extensive elaboration on this issue was given by Deardorff (1983). He makes $\Delta h/h$ a function of three different Richardson numbers:

$$Ri_* = \frac{g}{T} \frac{h\,\Delta\Theta}{w_*^2} ,$$

$$Ri_\tau = \frac{g}{T} \frac{h\,\Delta\Theta}{u_*^2} , \tag{14}$$

$$Ri_v = \frac{g}{T} \frac{h\,\Delta\Theta}{(\Delta U)^2 + (\Delta V)^2} \quad .$$

His approach is roughly equivalent to making the constants C_t, C_M, and C_F in (10) functions of these same three Richardson numbers. Deardorff derives these functions from a limited number of laboratory experiments and some suitable assumptions. Although his results are applicable to the full range of values of Ri_*, Ri_τ and Ri_v, it is also clear that the very limited amount of experimental data on which they are based still leaves a lot of uncertainties. At the moment we doubt that it will lead to better predictions in the atmospheric problem we are considering.

The entrainment equation (10) has an unattractive consequence that was pointed out by Driedonks (1982b) and Manins (1982). When we neglect for a moment the C_t-term (usually small) then it appears that the entrainment rate w_e will become negative when $Ri_v < C_M$. This, of course, is unrealistic. Manins (1982) capitalized on this issue by stating that entrainment takes place only at the critical value $Ri_v = C_M$ and that $w_e = 0$ when $Ri_v > C_M$. Thus entrainment proceeds in jumps; whenever Ri_v becomes subcritical the ABL-height grows suddenly such that Ri_v becomes critical again. We do not consider this approach very attractive since it may lead to entrainment rates that differ from timestep to timestep easily by an order of magnitude. Driedonks (1982b) also noted this singular behavior of eq. (10), but instead assumed tentatively that the effect of ΔU on the entrainment could be incorporated in the other shear generation term associated with the surface friction. He omitted the C_M-term completely from (10) and instead changed the value of A' in (13) to another empirical constant A. Thus he used as entrainment relations:

$$C_t \frac{w_m^2}{h} w_e = \frac{g}{T} \overline{\theta w}_h + C_F \frac{w_m^3}{h} ,$$

$$w_m^3 = w_*^3 + (A/C_F) u_*^3 . \qquad (15)$$

This is the form we use from now on.

Comparison with observations

It is clear that the theoretical issues on the entrainment formulation are not yet fully resolved. However, as said, our approach is practical, and we will consider how well these entrainment relations perform in the actual prediction of the behavior of h(t), $\Theta_m(t)$, etc. in the ABL from sunrise till the afternoon. Therefore we consider the observations of Driedonks (1982b) on ten days with varying conditions, taken near the 200 m meteorological mast at Cabauw in the Netherlands. During these ten days, spread over 2 years, values of h(t), $\Theta_m(t)$, and wind profiles were taken during the day from sodar observations, mast measurements, and radiosoundings. The gradients in the stable air aloft were also estimated this way. The surface fluxes of heat and moisture, and the friction velocity u_* were obtained from eddy-correlation measurements. The subsidence $\overline{w}_h$ was estimated from a large-scale analysis, and found to be very small compared with w_e. Thus it was possible to integrate the equations for the mixed layer in time, starting at sunrise, with proper measurements of all the conditions at the boundaries, and use various entrainment formulations for this integration. We will give here some results for two different versions of eq. (15), one without any friction included (A = 0), and one with A = 5. The value of C_F in both cases was 0.2. In the integrations presented here the C_t-term was omitted because it did not change the results for the cases considered. However, for practical application we recommend the use of this term with $C_t = 1.5$, in order to keep the entrainment rate always finite. The calculated values of h(t) are plotted against observations in Figure 5 and 6. It is clear that the results with A = 0 (no effect of friction on entrainment) are in general far too low, especially when h is small, which usually means values in the morning hours. After inclusion of the frictional influence, with A = 5, the results are quite good and the remaining scatter was estimated to be

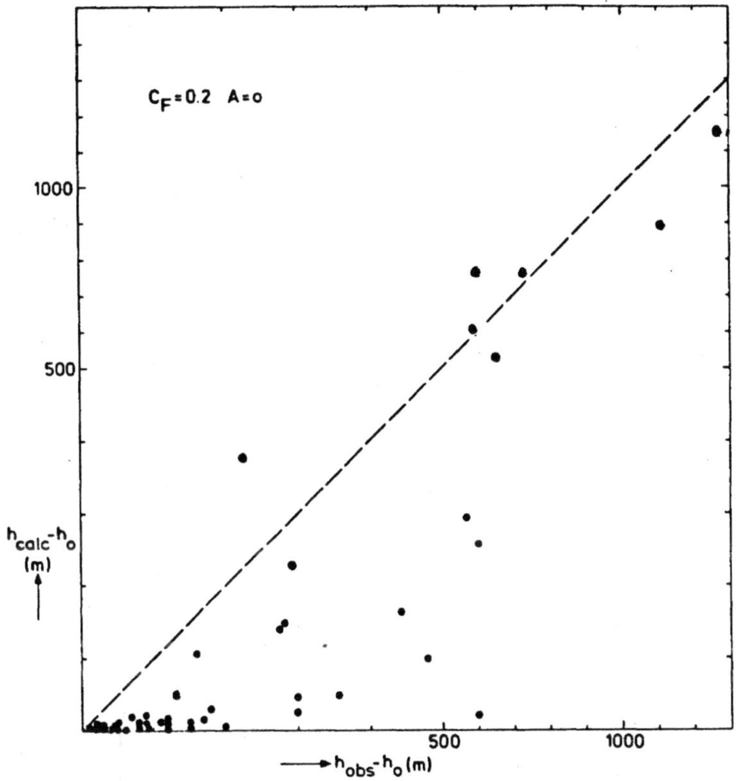

Figure 5. Observed values of ABL-height $h_{obs}-h_o$ (h_o is initial value) compared with calculations with entrainment eq. (15) where frictional effects are excluded ($C_F = 0.2$, A = 0).

within the accuracy of the measurements. In Figure 7a–c, the results from calculations with eq. (15), with $C_F = 0.2$, $A = 5$, are compared with the observations during the course of each of the ten days of observations. It appears that a rather simple entrainment formulation gives good results for the behavior of the ABL-height during the course of a day, provided frictional effects are included. More complicated entrainment formulations, also tested by Driedonks (1982b), did not lead to any improvements. The formulation used here has been applied by Reiff et al. (1983) in an operational forecast model, giving good results over a wide range of conditions during a long period of time.

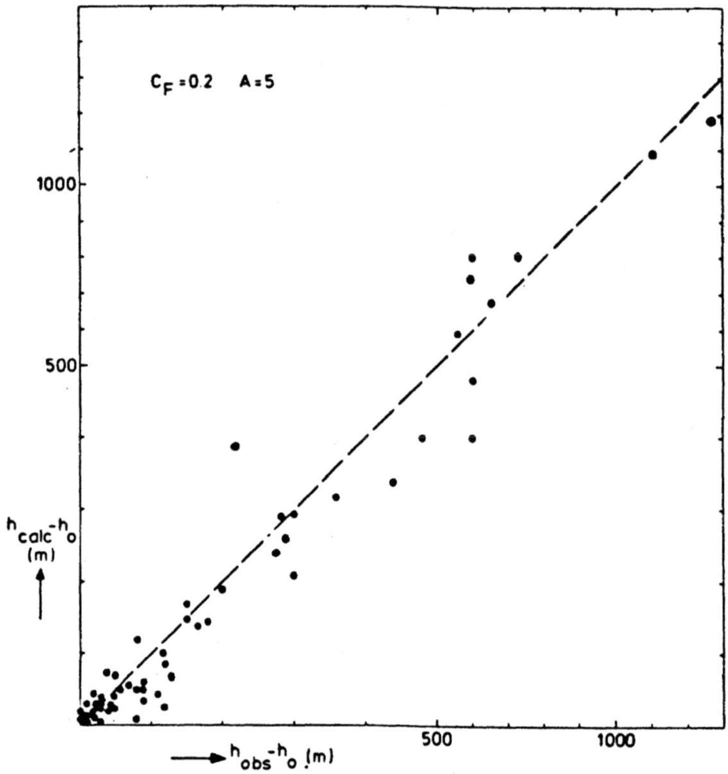

Figure 6. As Figure 5 but now with friction included ($C_F = 0.2$, $A = 5$).

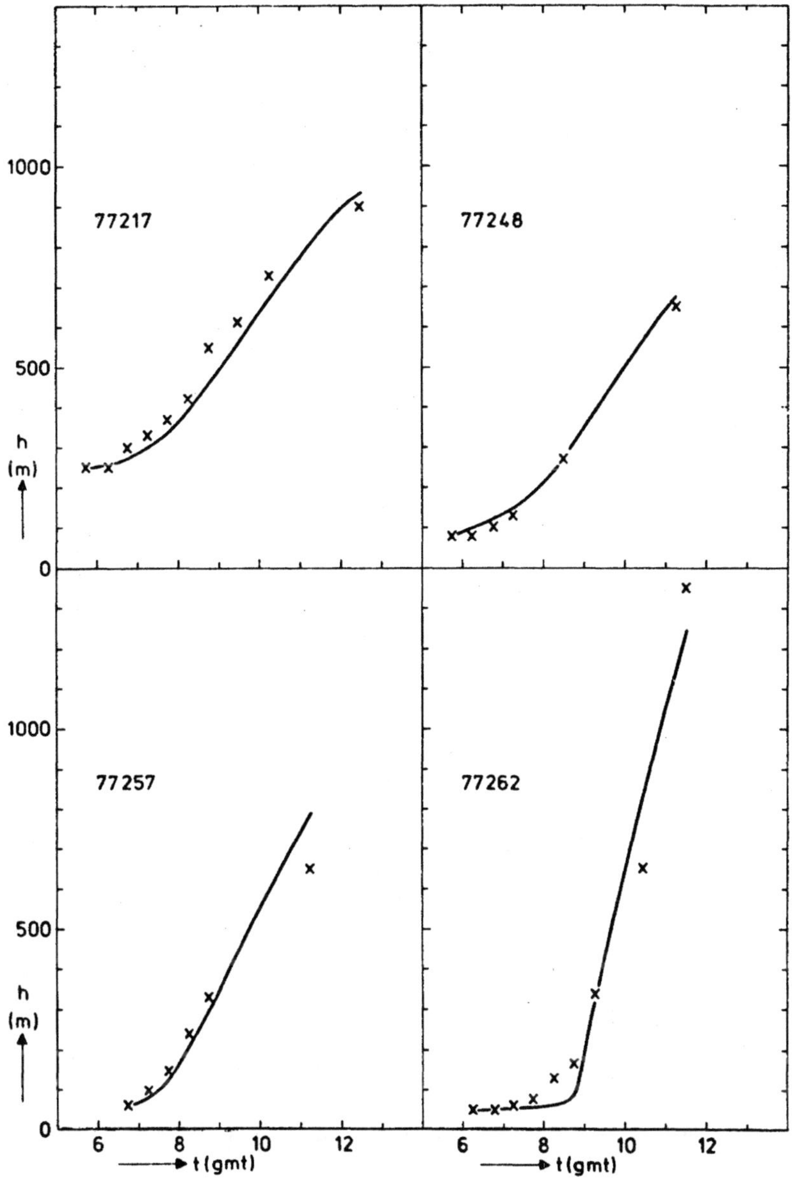

Figure 7a-c. Observed (x) and calculated (——) values of ABL-height h during the course of ten days. The date is indicated by year and day number (e.g. 77217). Calculations with eq. (15), $C_F = 0.2$, $A = 5$.

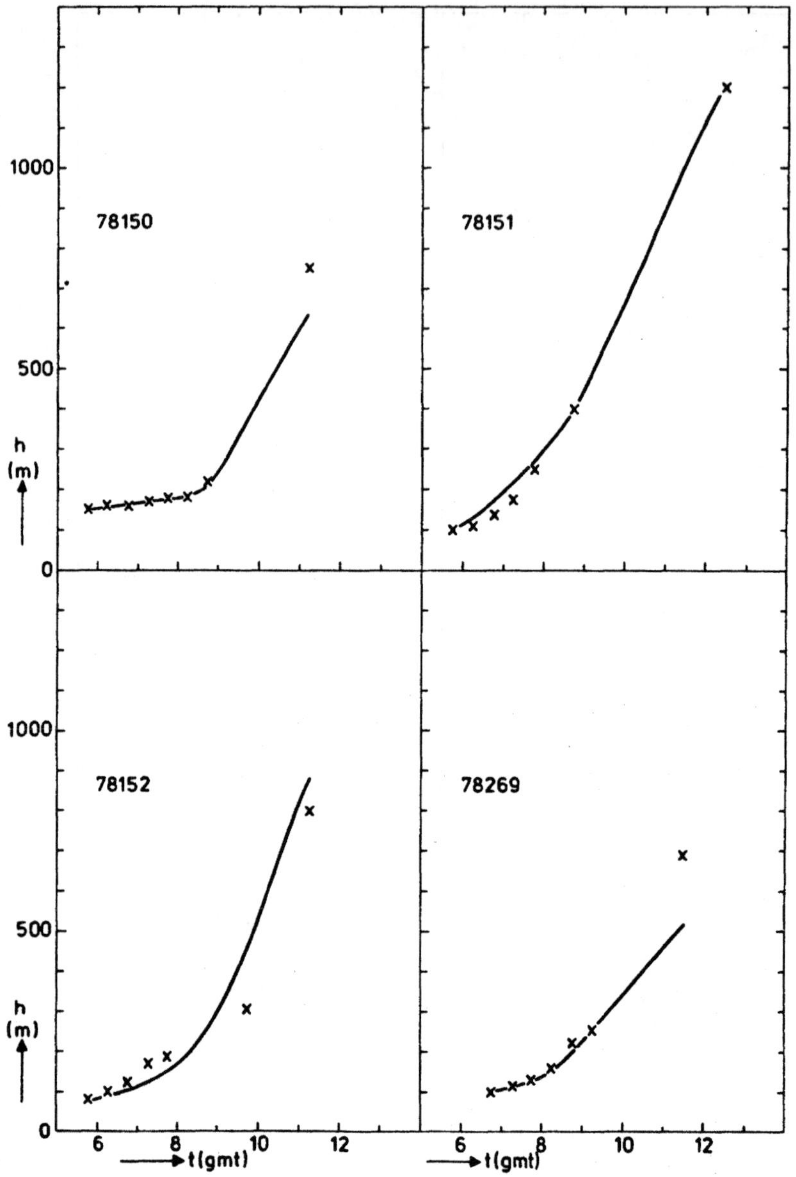

Figure 7b. For explanation see Figure 7a.

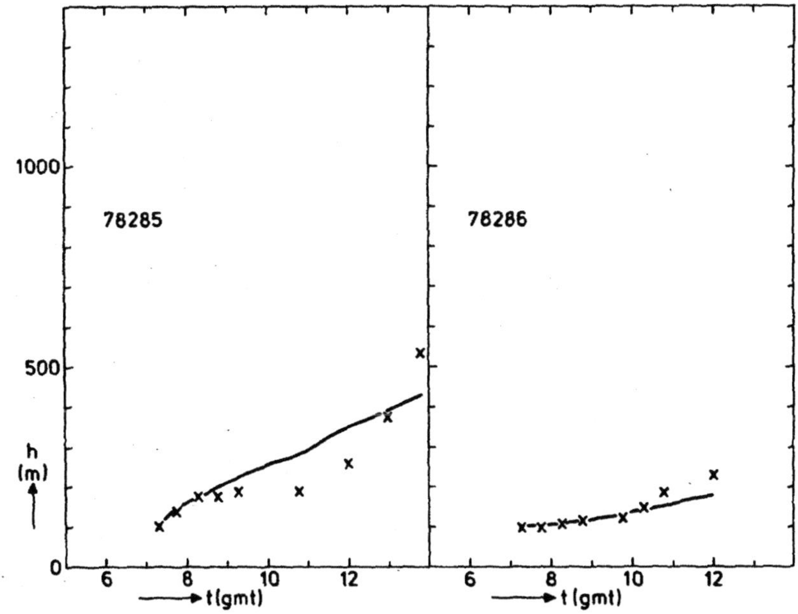

Figure 7c. For explanation see Figure 7a.

Sensitivity to conditions at the boundaries

Considerable attention has been given to the formulation of entrainment equations for the ABL. The issues are mainly related to the effect of the turbulence in the ABL on this entrainment, i.e. to the internal dynamics of the ABL. The conditions at the bounding edges, such as the heat flux at the surface, stability aloft, subsidence, etc., are taken for granted. These conditions may be controllable in a laboratory environment, but in the real ABL it is hard to prescribe these conditions with an acceptable accuracy. Driedonks (1982a) analysed the sensitivity of the solutions for h(t) and $\Theta_m(t)$ in a simple convective model to variations in the entrainment parameters as well as in the surface heat input and the stability aloft. He concluded that relatively small errors in the latter have about the same effect as large changes in the entrainment parameter. For practical applications therefore accurate estimates of the surface fluxes are essential. Holtslag and Van Ulden (1983) give a procedure for this which is applied in an operational environment by Reiff et al. (1983).

4. ENTRAINMENT EFFECTS ON THE MEAN PROFILES

In the preceding section we used the assumption that all first-order moments of ABL-variables, such as Θ, q and a concentration of a passive scalar C are completely mixed up to z = h, so that their values become independent of z. We assumed the turbulence in the ABL to be strong enough to maintain these profiles and the conditions neither at the top nor at the bottom to influence the shape. This is usually a good approximation for the Θ or Θ_v-profile in a convective ABL. However, it is often observed that other quantities, especially the specific humidity q, exhibit a distinct gradient over the ABL (André et al., 1979). This might be explained from the schematic picture in Figure 8. The air above the ABL is usually warmer but also dryer than the air within the ABL. Therefore the flux of heat at z = h will be downward, while the moisture flux is upward. Thus $\overline{\theta w}_s$ and $\overline{\theta w}_h$ have opposite sign (acting to increase the ABL-temperature from both sides) while $\overline{qw}_s$ and $\overline{qw}_h$ have the same sign, entrainment acting to decrease the ABL-humidity. Obviously this different flux behavior might have effects on the structure of the mean profiles induced by entrainment.

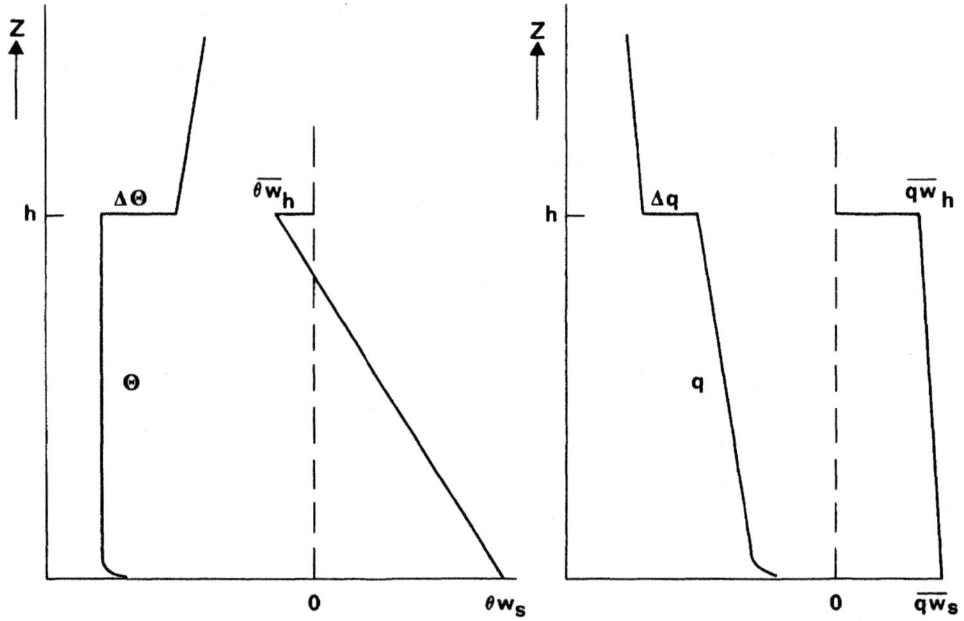

Figure 8. Schematic profiles of Θ and q and their fluxes as observed in a convective boundary layer, where q is not constant with height.

We therefore relax the assumption of well-mixedness a bit and allow the mean profiles to depend on z. We consider a passive scalar C with fluxes $\overline{cw}_s$ and $\overline{cw}_h$ at the ground and at z = h respectively. We assume that the shape of the profile of C(z,t) behaves in a self-similar way, i.e. that explicit time dependence occurs only in the vertically-averaged value $C_m(t)$. The existence of such a self-similar profile requires that the turbulent time scale in the ABL, $\tau_{turb} \sim h/w_*$, is small compared with external time scales imposed by the boundary conditions, i.e. by the changing surface flux $\overline{cw}_s(t)$ and the flux at the top $\overline{cw}_h(t)$. We will denote these time scales by τ_s and τ_h respectively. The time scale τ_s is usually of the order of a few hours, except when a sharp transition occurs. The turbulent time scale τ_{turb} is of the order of 10 minutes, thus the requirement $\tau_{turb} \ll \tau_s$ is usually fulfilled. We remark, however, that this may not be the case during the early morning hours, shortly after sunrise, when the surface heat flux changes quickly or in case where $\overline{cw}_s$ changes suddenly e.g. due to changing emission rates. We cannot expect self-similar solutions to hold during these transient periods with rapidly changing surface fluxes. However, we will assume that this is the case only during a short time and that usually $\tau_{turb} \ll \tau_s$. The other requirement is $\tau_{turb} \ll \tau_h$. We define a suitable scale for $\overline{cw}_h(t)$ as

$$\tau_h = \frac{\overline{cw}_h}{\partial\, \overline{cw}_h/\partial t} = \frac{w_e\, \Delta C}{w_e\, \partial \Delta C/\partial t + \Delta C\, \partial w_e/\partial t}, \tag{16}$$

from which it follows that the requirement $\tau_{turb} \ll \tau_h$ is fulfilled when $\tau_{turb} \ll \tau_{\Delta C}$ and $\tau_{turb} \ll \tau_{w_e}$, with $\tau_{\Delta C}$ and τ_{w_e} defined as $\Delta C/(\partial \Delta C/\partial t)$ and $w_e/(\partial w_e/\partial t)$ respectively. Now τ_{w_e} is associated with the time scale of the surface heat flux (assumed to be large compared to τ_{turb}) and the time scale for $\Delta\Theta$. Thus it is sufficient if $\tau_{turb} \ll \tau_{\Delta C}$ and $\tau_{turb} \ll \tau_{\Delta\Theta}$. We can further estimate (if the gradients aloft are small)

$$\tau_{\Delta C}^{-1} \sim \frac{\partial \Delta C/\partial t}{\Delta C} \sim \frac{\overline{cw}_s}{h\, \Delta C} + \frac{w_e}{h}, \tag{17}$$

and similarly for $\Delta\Theta$. Thus we fulfill the requirement that $\tau_{turb} \ll \tau_h$ if

$$\frac{\overline{cw}_s}{w_*\, \Delta C} = \frac{\overline{cw}_s}{\overline{cw}_h} \cdot \frac{w_e}{w_*} \ll 1,$$

$$\frac{\overline{\Theta w}_s}{\overline{\Theta w}_h} \cdot \frac{w_e}{w_*} \ll 1, \qquad \text{and} \tag{18}$$

$$\frac{w_e}{w_*} << 1 \; .$$

Since for very small ΔC, $\Delta\Theta$ their time scales are not very significant to our problem, we will consider the self-similarity condition at the top to be fulfilled when

$$\frac{w_e}{w_*} << 1 \; . \tag{19}$$

In a self-similar shape of the profile of C in the ABL the time does not explicity occur. In non-dimensionalized form this shape will depend on z/h and on the non-dimensional boundary conditions, i.e. on $\overline{cw}_h/\overline{cw}_s$. It will not depend explicitly on w_e/w_*, since we assumed this ratio to be asymptotically small. After scaling C itself with ΔC, we can then write:

$$C(z,t) = C_m(t) + \Delta C . F(z/h, \; \overline{cw}_h/\overline{cw}_s), \tag{20}$$

where F is some universal function. How does this assumption influence the budgets for the vertically averaged quantities? To look at this we integrate the local budget for C,

$$\frac{\partial C}{\partial t} = - \frac{\partial}{\partial z} \overline{cw} \; , \tag{21}$$

over the ABL up to $z = h_-$ (just below $z = h$) and find

$$h \cdot \frac{\partial C_m}{\partial t} = \overline{cw}_s - \overline{cw}_h + (C_{h-} - C_m)w_e , \tag{22}$$

or, using $-\overline{cw}_h = w_e \; \Delta C$ and defining C_{h+} to be the value of C just above $z = h$:

$$h \cdot \frac{\partial C_m}{\partial t} = \overline{cw}_s + (C_{h+} - C_m)w_e . \tag{23}$$

Thus we see that the equations are still very similar to eq. (2) for the flat profiles.

We are especially interested in the mean gradient over the ABL, which in self-similar form can be written as

$$(\partial C/\partial z)_m = \frac{\overline{cw}_s}{w_* \; h} \; G(\overline{cw}_h/\overline{cw}_s), \tag{24}$$

with G(x) some "universal" function of $x = \overline{cw}_h/\overline{cw}_s$. If C denotes potential temperature Θ (or virtual potential temperature Θ_v), observations indicate that $(\partial\Theta/\partial z)_m \simeq 0$ in a convective boundary layer and $\overline{\theta w}_h/\overline{\theta w}_s \simeq -0.2$ Thus we expect G(−0.2) = 0. Furthermore, if $\overline{cw}_s \to 0$ but $\overline{cw}_h \neq 0$, then $(\partial C/\partial z)_m$ should lose its dependence on $\overline{cw}_s$. This implies that $G(x) \to a_2 x$ for $x \to \pm\infty$, with a_2 some constant. We combine this and assume a simple linear interpolation between these two cases:

$$(\partial C/\partial z)_m = a_1 \frac{\overline{cw}_s}{w_* h} + a_2 \frac{\overline{cw}_h}{w_* h} , \qquad (25)$$

where a_2/a_1 should be of the order 5 (= 1/0.2). Note that both a_1 and a_2 will be negative. Thus we see that the entrainment-induced flux has an effect on the mean gradient in the ABL that is about 5 times larger than that of the surface flux. Wyngaard and Brost (1984) found from large-eddy simulations a ratio of the same order, but it was quite sensitive to the precise definition of z = h in their model. André et al. (1979) found about the same ratio from humidity observations, however, with an enormous scatter, as did André and Mahrt (1983) with laboratory data. In as yet unpublished results by Driedonks et al. this type of relation is tested with humidity data taken from our 200 m mast and from radiosoundings. They also found an enormous scatter and rather small correlations. The difficulties in testing a relation like (25) is partly caused by the indirect and perhaps inaccurate way to estimate $\overline{cw}_h$, but it can also be that effects that are as yet neglected are important. It is perfectly well possible that the coefficients in (25) are functions of a set of suitable Richardson numbers.

5. CONCLUSIONS

We considered some important effects of entrainment on the structure and the evolution of the ABL over land during a clear day. Entrainment occurs at the top of a well-stirred ABL, brings quiet air with different properties downward into the turbulent ABL, and causes the ABL-height to increase. Thus entrainment means essentially the incorporation of mass into the ABL, and this mass flux at the top is for a dry atmosphere equivalent with a (downward) heat flux, $\overline{\theta w}_h$. The heart of the entrainment problem is to give an estimate of this mass flux, its dependence on the structure of the turbulence in the ABL and the conditions in the quiet air above.

The structure of the turbulence in an entraining ABL depends on the conditions at the lower surface but also on the entrainment itself. Thus, in the absence of any distinct sources within the ABL, the structure of the turbulence has a surface-induced contribution and an entrainment induced contribution. Near the surface the turbulence does not sense large effects from the conditions that exist high up at the top of the ABL. At this very top, however, the turbulent structure will of course be dominated by entrainment effects. It is not yet clear how these two contributions add together in the bulk of the ABL. It appears that entrainment effects on the turbulent structure can sometimes, or for some statistics, be felt deep down into the ABL. As yet no general similarity framework has been developed that provides proper scaling laws for these effects throughout the whole ABL.

Because of these uncertainties about the detailed structure of the turbulence there is still a considerable lack of consensus on how to estimate $\overline{\theta w}_h$ or the entrainment rate for a wide range of conditions. Yet we have shown that for common conditions met in the atmosphere, the growth of the ABL-height could be calculated with a relatively simple model whose results compare favourably with a set of observations. Essential for obtaining good agreement throughout the whole day is to allow for both convectively generated turbulence (through the surface heat flux) and mechanically generated turbulence (through surface friction), the latter being important in the early morning hours. Good agreement between observations and model calculation was obtained with entrainment equations (15) with values of the constants $C_F = 0.2$, $A = 5$, $C_t = 1.5$.

Entrainment does not only depend on the internal dynamics but also on the conditions at the bounding edges, e.g. the surface heat flux, surface friction, the vertical profiles in the air aloft, and on the mean subsidence. Especially the ABL-height directly depends on the latter. For entrainment calculations in atmospheric problems, accurate knowledge of these external factors is at least as, if not more, important than the use of a particular entrainment formulation.

In simple models for the shape of the mean profiles in the convective ABL it is often assumed that they are all flat, independent of z, in the bulk of the ABL, and that this shape does not depend on the fluxes at the bounding edges. From observations it has been shown that this assumption is not always valid. From similarity arguments a simple scaling for the mean gradient over the mixed layer in terms of the surface flux and the flux at the top can be formulated. We estimated the entrainment-induced contribution to this gradient to be about five times larger than the contribution from the surface flux. However, this relation has as yet not been tested satisfactorily against observations.

6. SYMBOLS AND NOTATION

- vectors are indicated by a tilde under the variable, e.g. $\underset{\sim}{U}$ is the horizontal velocity vector.
- indexing:
  - m : vertically averaged value over the mixed layer
  - s : surface value
  - h : value at the top of the mixed layer.

- Symbols

A, A', a_1, a_2	constants
C	ensemble-averaged concentration of a scalar
$\overline{cw}$	vertical turbulent flux of a scalar
ΔC	jump in C at the top of the mixed layer
c_*	scale for concentration fluctuations, $(= \overline{cw}_s/w_*)$
C_t, C_M, C_F	constants in entrainment equation
C_T^2	structure parameter for temperature fluctuations, $C_T^2 = \overline{[T(\underset{\sim}{x},t)-T(\underset{\sim}{x}+\underset{\sim}{r},t)]^2}/r^{2/3}$
C_v^2	structure parameter for velocity fluctuations
$\overline{e}$	ensemble-averaged value of turbulent kinetic energy
$\overline{ew}$	vertical flux of turb. kin. energy
f	Coriolis parameter
g	gravitational acceleration
h	mixed-layer height
Δh	thickness of entrainment zone
k	von Karman constant
L	Obukhov length, $L = -u_*^3\, T/(gk\, \overline{\theta w}_s)$
N	Brunt-Vaisala frequency, $N^2 = (g/T)\partial\Theta/\partial z$
p	pressure
q	specific humidity
Ri	Richardson number, $Ri = (g/T)\ h\ \Delta\Theta/w_m^2$
Ri_*	$Ri_* = (g/T)\ h\ \Delta\Theta/w_*^2$
Ri_τ	$Ri_\tau = (g/T)\ h\ \Delta\Theta/u_*^2$
Ri_v	$Ri_v = (g/T)\ h\ \Delta\Theta/\lVert\Delta\underset{\sim}{U}\rVert^2$
T	representative boundary-layer temperature
T_+	dimensionless time, $T_+ = t\ w_*/h$
t	time
U, V	horizontal wind components (ensemble-averaged)
$\overline{uw}$, $\overline{vw}$	vertical turbulent fluxes of horizontal momentum
U_g, V_g	geostrophic wind
ΔU, ΔV	jump in wind components at the top of the mixed layer
u_*	surface friction velocity, $u_*^2 = (\overline{uw}_s^2 + \overline{vw}_s^2)^{\frac{1}{2}}$
w_*	convective velocity scale, $w_*^3 = (g/T)h\ \overline{\theta w}_s$
w_e	entrainment velocity
w_m	turbulent velocity scale for the mixed layer
$\overline{w}_h$	large-scale vertical velocity at $z = h$.
z	vertical coordinate
$\gamma_\theta, \gamma_u, \gamma_v, \gamma_\phi$	vertical gradient of indicated variable above the mixed layer

ε	dissipation rate of turbulent kinetic energy
Θ	ensemble-averaged potential temperature
$\Delta\Theta$	jump in Θ at the top of the mixed layer
$\overline{\theta w}$	vertical turbulent heat flux
θ_*	temperature scale (= $\overline{\theta w}_s/w_*$)
Θ_v	virtual potential temperature
ρ	representative boundary-layer density
σ_θ^2, σ_w^2	variances of temperature, vertical velocity
τ_{turb}	turbulent time scale, (= h/w_*)
τ_s	time scale associated with changes in surface fluxes
τ_h	time scale associated with changes in conditions at the top of the mixed layer
$\tau_{\Delta C}$, $\tau_{\Delta\Theta}$	time scale associated with changes in ΔC, $\Delta\Theta$
τ_{w_e}	time scale associated with the change in entrainment velocity
Φ, $\Delta\Phi$	ensemble-averaged value of variable Φ and jump at the top of the mixed layer
$\overline{\phi w}$	vertical turbulent flux

7. REFERENCES

André, J.C., Lacarrère, P., and L.J. Mahrt, 1979: "Sur la distribution verticale de l'humidité dans la couche limite convective". J. Rech. Atmos., 13, 135-146.

Caughey, S.J., 1982: "Observed characteristics of the atmospheric boundary layer". In: Atmospheric Turbulence and Air Pollution Modelling. Eds. F.T.M. Nieuwstadt and H. van Dop. Reidel Publ., Holland. pp. 107-158.

Deardorff, J.W., 1974: "Three-dimensional numerical study of the height and mean structure of a heated planetary boundary layer". Boundary-Layer Meteor., 7, 199-226.

Deardorff, J.W., 1980: "Stratocumulus-capped Mixed Layer derived from a three-dimensional model". Boundary-Layer Meteor., 18, 495-527.

Deardorff, J.W., 1983: "A multi-limit mixed-layer entrainment formulation". J. Phys. Oceanogr., 13, 988-1002.

Deardorff, J.W., Willis, G.E., and B.H. Stockton, 1980: "Laboratory studies of the entrainment zone of a convectively mixed layer". J. Fluid Mech., 100, 41-46.

Deardorff, J.W., and G.E. Willis, 1982: "Ground-level concentrations due to fumigation into an entraining mixed layer". Atmos. Env., 16, 1159-1170.

Driedonks, A.G.M., 1982a: "Sensitivity analysis of the equations for a convective mixed layer". Boundary-Layer Meteor., 22, 475-480.

Driedonks, A.G.M., 1982b: "Models and observations of the growth of the atmospheric boundary layer". Boundary-Layer Meteor., 23, 283-306.

Holtslag, A.A.M. and A.P. van Ulden, 1983: "A simple scheme for daytime estimates of the surface fluxes from routine weather data". J. Clim. and Appl. Meteor., 22, 517-529.

Lamb, R.G., 1982: "Diffusion in the Convective Boundary Layer". In: Atmospheric Turbulence and Air Pollution Modelling. Eds. F.T.M. Nieuwstadt and H. van Dop. Reidel Publ., Holland, pp. 159-230.

Mahrt, L.J. and J.C. André, 1983: "On the stratification of turbulent mixed layers". J. Geophys. Research, 88, 2662-2666.

Manins, P.C., 1982: "The daytime planetary boundary layer: a new interpretation of Wangara data". Quart. J. Roy. Met. Soc., 180, 689-705.

Panofsky, H.A., Tennekes, H., Lenschow, D.H., and J.C. Wyngaard, 1977: "The characteristics of turbulent velocity components in the surface layer under convective conditions". Boundary-Layer Meteor., 11, 355-361.

Rayment, R., and C.J. Readings, 1974: "A case study of the structure and energetics of an inversion". Quart. J. Roy. Met. Soc., 100, 221-233.

Reiff, J., Blaauboer, D., de Bruin, H.A.R., van Ulden, A.P., and G. Cats, 1983: "An air mass transformation model for short-range weather forecasting". Mon. Wea. Rev., in press.

Scranton, D.R., and W.R. Lindber, 1983: "An experimental study of entraining, stress-driven, stratified flow in an annulus". Physics of Fluids, 26, 1198-1205.

Tennekes, H. and A.G.M. Driedonks, 1981: "Basic entrainment equations for the atmospheric boundary layer". Boundary-Layer Meteor., 20, 515-531.

Wyngaard, J.C., Coté, O.R., and Y. Izumi, 1971: "Local free convection similarity and the budgets of shear stress and heat flux", J. Atmos. Sci., 28, 1171-1182.

Wyngaard, J.C., and M.A. LeMone, 1980: "Behavior of the refractive index structure parameter in the entraining convective boundary layer". J. Atmos. Sci., 37, 1573-1585.

Wyngaard, J.C., 1983: "Lectures on the planetary boundary layer". In: Mesoscale Meteorology - Theories, Observations, and Models. Eds. D.K. Lilly and Tzvi Gal-Chen. Reidel Publ., Holland, pp. 603-650.

Wyngaard, J.C., and R.A. Brost, 1984: "Top-down and bottom-up diffusion in the convective boundary layer". J. Atmos. Sci., in press.

Zeman, O., and H. Tennekes, 1977: "Parameterization of the turbulent energy budget at the top of the daytime atmospheric boundary layer". J. Atmos. Sci., 34, 111-123.

Questions, answers and comments

S.P.S. ARYA: In parameterizing the energy balance equations at the inversion base, the mechanical production terms are often neglected. Yet, according to the critical Richardson number for the transition layer, these terms may be as large as four times the buoyancy term one is trying to evaluate. What is the justification of neglecting the shear production terms? How would the magnitudes (or constants) of other terms be affected if one were to include these terms?

H. TENNEKES: In our parameterization, the mechanical production terms are neglected because they are more trouble than they are worth. A possible rationale for this is that shear production in the entrainment zone creates small-scale turbulent energy, which is dissipated rapidly, implying that it cannot contribute efficiently to the entrainment process. A critical Richardson number does not imply anything on the net efficiency of energy conversion as far as entrainment is concerned. In Driedonks' model, mechanical production terms are represented by a relatively large coefficient in front of u_*^3 in the interpolation formula for $\sigma_w^3 (= w_m^3)$. The core of my message is that encroachment, with $\Delta\theta \equiv 0$ and $\overline{\theta w}_h = 0$, handles 80% of the problem, that a crude parameterization of $\overline{\theta w}_h$ adds about 10%, and that further requirements tend to get lost in the unavoidable inaccuracies of most experiments.

J.C.R. HUNT: I think that Prof. Tennekes would agree that unlike the stable boundary layer, the convective boundary layer structure is not particularly sensitive to small undulation and slopes of the ground.

H. TENNEKES: Yes, I agree.

J.C.R. HUNT: In your discussion of the inversion layer you did not discuss the energy loss from the layer by internal waves which may affect the energy available for increasing the height of the mixed layer. Is this likely to have a significant effect on the convective boundary layer as a whole?

H. TENNEKES: I have not found a satisfactory way of parameterizing the energy loss to internal waves above the boundary layer. Otto Zeman and several others have made unsuccessful attempts. The growth of the convective boundary layer is, to my knowledge, not significantly affected by this mechanism. In the parameterization, this effect would show up as a separate term only if the scaling of the radiative loss were distinctly different from that of the effects that are accounted for explicitly. In any case, entrainment parameterization is subject to the law of diminishing returns: slight improvements in the model are swamped by the lack of accuracy inherent in most experiments. At the 5% level, it does not make much sense to invest heavily in sophisticated refinement of the theory.

L.N. GUTMAN: What is the difference between a "well mixed layer" and a "mixed layer"? Why did you use the term "well mixed layer" in your lecture?

H. TENNEKES: The words "mixed layer" and "well mixed" layer are used interchangeably. When I say "well mixed", I emphasize the uniform distribution of potential temperature. Humidity and momentum are not mixed that well.

L.N. GUTMAN: What can you say about horizontally non-homogeneous mixed layer models?

H. TENNEKES: Entrainment is a local process, which can be incorporated easily into models that take advection and other non-homogenous effects into account. In a stationary situation with uniform wind, all that needs to be done is to replace t by x/u.

PART II

THEORY AND ADVANCED MATHEMATICAL TECHNIQUES IN MODELING AIR FLOW AND ATMOSPHERIC DIFFUSION

SECOND ORDER MODELING OF TURBULENT TRANSPORT IN THE SURFACE MIXED LAYER*

J. L. Lumley, P. Mansfield
Sibley School of Mechanical and Aerospace Engineering
Upson and Grumman Halls, Cornell University
Ithaca, NY 14853 USA

A model for the simultaneous transport of heat and water vapor is presented. In an effort to resolve the structure of the entrainment region at the inversion base, models are constructed so as to satisfy realizability as far as possible. Density anomaly and water vapor mixture fraction (specific humidity) are taken as the basic variables. Algebraic expressions for the third moments are derived from first principles, and contain no adjustable constants. Separate equations are carried for the dissipation of each variance, constructed to give rational behavior of all time scale ratios. New forms for relaxation and cross-dissipation terms are constructed in such a way as to guarantee realizability. We describe how realizability was used as a tool to construct these models. We present preliminary results without mean velocity gradients for a dry surface mixed layer leaving the land and starting over water, producing a stable internal humidity boundary layer, but with large fluxes of sensible heat and water vapor (local advection).

1. INTRODUCTION

Ten years ago Rao et al (1974) published calculations based on a second order model for local advection of momentum, heat and moisture. The authors remark in section 5. Conclusions: "We emphasize that our model uses relatively simple closure expressions that are in the nature of first approximations. All models of this type are in an evolutionary stage, and closure refinements are to be expected as new flows are modeled." We are attempting to construct a model for the same physical process as that described by Rao et al (1974), but incorporating the most advanced closure refinements that have been developed during the last ten years.

---

*Prepared for presentation at the 29th OHOLO Conference on Boundary Layer Structure - Modelling and Application to Air Pollution and Wind Energy, Zichron Ya'acov, Israel, March 25-28, 1984. Supported in part by the U.S. Office of Naval Research under the following programs: Physical Oceanography (Code 422PO), Power (Code 473); in part by the U.S. National Science Foundation under grant no. ATM 79-22006; and in part by the U.S. Air Force Geophysics Laboratory.

Boundary-Layer Meteorology 30 (1984) 109–142. 0006-8314/84/0301-0109$05.40

Specifically, the model used by Rao et al (1974):

1) did not incorporate "rapid" terms in the various pressure correlations (i.e. - terms proportional to the mean velocity gradient and the buoyancy);

2) did not carry separate equations for the dissipation of temperature and humidity variance;

3) used simple gradient transport forms for the third moments, with empirically determined constants;

4) used constant relaxation coefficients for the various pressure correlations.

During the past five years considerable progress has been made in the (one-dimensional) modeling of turbulent transport in the dry atmospheric surface mixed layer. We summarize here primarily the work of our own group during this period. In particular, the role of buoyancy in modifying the vertical turbulent transport is now well understood (Lumley et al, 1978), and a comprehensive model for the dry atmospheric surface mixed layer exists, including passive contaminant transport (Zeman & Lumley, 1976a, b, c; 1979). To date this model has been applied only in one-dimensional simulations. The model, however, is in principle three-dimensional, and it is our intention to apply the model to turbulent transport over complex terrain.

The additional complications which arise as a result of the presence of water vapor have, to a certain extent, been confronted. The basic problem here is the parameterization of the cross-diffusivity (between heat and contaminant). A first attempt to deal with this question was made by Zeman & Lumley (1976a) but with only partial success. A successful model was subsequently produced by Lumley (1978b). The extension to the case of an active contaminant (one that contributes to the buoyancy) is relatively simple. This has been done very recently in another situation (Lumley, 1980a; Zeman & Lumley, 1982, 83); specifically, the convective model was extended to include the case of thermohaline convection in sea water, producing a model which changed smoothly from turbulent buoyant convection (driven by the net density difference resulting both from heat and salt) to a salt-fingering regime (driven by the difference in diffusivities).

Extensive studies have been carried out on the behavior of scalar fluctuations, alone, or in the presence of fluctuations of a different scalar. For example, Warhaft & Lumley (1978a, b), and Newman et al (1977, 81) described in detail the way in which a single passive scalar decays, giving particular attention to the behavior of the thermal/mechanical time scale ratio which, surprisingly, is a function of the ratio of the peak wavenumbers of the thermal and mechanical spectra. The decay of the heat flux has been studied extensively (Warhaft & Lumley, 1978a, b) and more recently (Sirivat & Warhaft, 1981b; Shih & Lumley, 1981) it has become clear that there is a necessary dependence of the relaxation constant on the time scale ratio, which was predicted by Lumley (1978a). Warhaft (1980a,b) and Sirivat & Warhaft (1980, 1981a) have studied the behavior of temperature fluctuations under strain, the behavior and interaction of temperature fluctuations from two diferent sources, and the behavior and interaction of simultaneous fluctuations of temperature and helium. All these recent developments have not yet been incorporated into a comprehensive atmospheric dispersion model.

In Lumley (1978a) a rational technique for obtaining equations for the third moments (the fluxes of the variances and fluxes) was developed. This technique produces forms very similar to those used by Zeman & Lumley (1976a, b, c; 1979), but with certain refinements: in particular, the values of the constants (which were previously determined by comparison with experiment) are determined by the technique, and a rational form is produced for the flux of dissipation. These transport forms have been applied (Taulbee & Lumley, 1981; Janicka & Lumley, 1981) to isothermal wake and jet situations with excellent results, essentially predicting the third moments within experimental error without adjustable constants.

It is perhaps worthwhile to underscore that the model of Zeman & Lumley (1976a, b, c; 1979) is the only second-order model currently available that can handle purely convective situations, without mean wind shear, because the influence of buoyancy on the turbulent transport is handled exactly. In other models, the influence of buoyancy on the transport is neglected, but is included in the second order equations. This means that, in these models, buoyancy is a small perturbation on the shear.

These recently discovered advanced turbulence models have yet to be applied to multidimensional flows. In the presence of complex topography, major distortions of the mean flow can dominate the dispersion process. Little is known about turbulent transport in such situations even with the most elementary Gaussian-plume dispersion models.

Our ultimate goal is the development of a modeling capability for simultaneous transport of heat and water vapor over non-uniform terrain with non-uniform surface distribution of heat and moisture flux and of roughness. This goal must be approached in stages. Although we present here fairly general arguments, we present only one-dimensional calculations without mean velocity gradients. More complexity will be added later. Our emphasis here is on developing (in so far as possible) fully realizable models which can resolve the complex structure in the entrainment region at the inversion base, where all quantities vanish at varying rates.

2. CHOICE OF FUNDAMENTAL VARIABLES

We choose to work in density anomaly and water vapor mixture fraction as our fundamental variables. This is equivalent to working in virtual potential temperature and specific humidity. The pair of variables potential temperature and mixture fraction would also have been satisfactory, but would have made the transport much more difficult to calculate; since both potential temperature and mixture fraction appear in the momentum equation, it would be necessary to calculate simultaneously all third moments involving either potential temperature or water vapor mixture fraction in order to obtain the velocity field. By working in terms of density anomaly and mixture fraction, since only density anomaly appears in the momentum equation, we need calculate only the third moments that contain the density anomaly in order to obtain the velocity field. The third moments that contain the water vapor mixture fraction can be calculated separately afterward. We use the Boussinesq approximation, and we take the diffusivities for heat and water vapor to be equal (their ratio is 0.837 at 20°C and 1 atmosphere). We designate by ρ and ρ' the instantaneous density anomaly and its fluctuating part, and by ρ_0 the density of an adiabatic atmosphere. We designate by F and f the mean and fluctuating part of the water vapor mixture fraction (mass of water vapor per mass of mixture). $\beta_i = g_i/\rho_0$, where $x_3 = z$ increases upward, and $g_i = (0, 0, -g)$ (so that, if $\rho > 0$, $\partial_t u_3 = \partial_t w < 0$). The instantaneous equations for the fluctuating quantities are

$$\partial_t u_i + u_{i,j}U_j + U_{i,j}u_j + u_{i,j}u_j - \langle u_{i,j}u_j\rangle$$
$$= - p_{,i}/\rho_0 + \beta_i\rho' + \nu u_{i,jj} \tag{1}$$

$$\partial_t\rho' + \rho'_{,i}U_i + \langle\rho\rangle_{,i}u_i + \rho'_{,i}u_i - \langle\rho'_{,i}u_i\rangle$$
$$= \gamma\rho'_{,ii} \tag{2}$$

$$\partial_t f + f_{,i}U_i + F_{,i}u_i + f_{,i}u_i - \langle f_{,i}u_i\rangle$$
$$= \gamma f_{,ii} \tag{3}$$

For compactness, we use Cartesian tensor notation, so that $\partial_i A = A_{,i}$, and repeated indices are summed. ν and γ are molecular diffusivities for momentum and mass.

We take as our fundamental variables the mean fields U_i (and the mean pressure P), F and $\langle\rho\rangle$, and the second moments of u_i, ρ' and f: $\langle u_iu_j\rangle$, $\langle u_i\rho'\rangle$, $\langle u_if\rangle$ - the fluxes of momentum, density anomaly and water vapor mixture fraction. We will need in addition equations for length or time scales for the three fields. We use equations for the mechanical dissipation $\langle\varepsilon\rangle$, the dissipation of density anomaly variance $\langle\varepsilon_{\rho'}\rangle$, and the dissipation of water vapor mixture fraction variance $\langle\varepsilon_f\rangle$. We model the cross dissipation $\langle\varepsilon_{f\rho'}\rangle$ directly. The basic equations are (where we assume large Reynolds, Schmidt and Péclet numbers):

$$\partial_t U_i + U_{i,j}U_j + \langle u_iu_j\rangle_{,j} = -P_{,i}/\rho_0 + \beta_i\langle\rho\rangle \tag{4}$$

$$\partial_t\langle\rho\rangle + \langle\rho\rangle_{,i}U_i + \langle\rho'u_i\rangle_{,i} = 0$$

$$\partial_t F + F_{,i}U_i + \langle fu_i\rangle_{,i} = 0 \tag{6}$$

$$\begin{aligned}&\partial_t\langle u_iu_j\rangle + \langle u_iu_j\rangle_{,k}U_k + U_{i,k}\langle u_ku_j\rangle \\ &+ U_{j,k}\langle u_ku_i\rangle + \big(\langle u_iu_ju_k\rangle + \langle pu_j\rangle\delta_{ik}/\rho_0 \\ &+ \langle pu_i\rangle\delta_{jk}/\rho_0\big)_{,k} = \langle p(u_{i,j} + u_{j,i})\rangle/\rho_0 \\ &+ \beta_i\langle\rho'u_j\rangle + \beta_j\langle\rho'u_i\rangle - 2\nu\langle u_{i,k}u_{j,k}\rangle\end{aligned} \tag{7}$$

$$\begin{aligned}&\partial_t\langle\rho'u_i\rangle + \langle\rho'u_i\rangle_{,j}U_j + U_{i,j}\langle\rho'u_j\rangle \\ &+ \langle\rho\rangle_{,j}\langle u_iu_j\rangle + \big(\langle\rho'u_iu_j\rangle + \langle p\rho'\rangle\delta_{ij}/\rho_0\big)_{,j} \\ &= \langle p\rho'_{,i}\rangle/\rho_0 + \beta_i\langle\rho'^2\rangle - (\gamma + \nu)\langle\rho'_{,j}u_{i,j}\rangle\end{aligned} \tag{8}$$

$$\begin{aligned}&\partial_t\langle fu_i\rangle + \langle fu_i\rangle_{,j}U_j + U_{i,j}\langle fu_j\rangle \\ &+ F_{,j}\langle u_iu_j\rangle + \big(\langle fu_iu_j\rangle + \langle pf\rangle\delta_{ij}/\rho_0\big)_{,j} \\ &= \langle pf_{,i}\rangle/\rho_0 + \beta_i\langle f\rho'\rangle - (\gamma + \nu)\langle f_{,j}u_{i,j}\rangle\end{aligned} \tag{9}$$

$$\begin{aligned}&\partial_t\langle f\rho'\rangle + \langle f\rho'\rangle_{,j}U_j + \langle\rho\rangle_{,i}\langle u_if\rangle + F_{,i}\langle u_if\rangle \\ &+ \langle f\rho'u_i\rangle_{,i} = -\,2\gamma\langle f_{,i}\rho'_{,i}\rangle = -2\langle\varepsilon_{f\rho'}\rangle\end{aligned} \tag{10}$$

$$\begin{aligned}&\partial_t\langle\rho'^2\rangle + \langle\rho'^2\rangle_{,i}U_i + 2\langle\rho\rangle_{,i}\langle\rho'u_i\rangle \\ &+ \langle\rho'^2u_i\rangle_{,i} = -2\gamma\langle\rho'_{,i}\rho'_{,i}\rangle = -2\langle\varepsilon_{\rho'}\rangle\end{aligned} \tag{11}$$

$$\begin{aligned}&\partial_t\langle f^2\rangle + \langle f^2\rangle_{,i}U_i + 2F_{,i}\langle fu_i\rangle \\ &+ \langle f^2u_i\rangle_{,i} = -2\gamma\langle f_{,i}f_{,i}\rangle = -2\langle\varepsilon_f\rangle\end{aligned} \tag{12}$$

$$\partial_t\langle\varepsilon\rangle + \langle\varepsilon\rangle_{,i}U_i + \langle\varepsilon u_i\rangle_{,i} = -\,\big(\langle\varepsilon\rangle^2/\langle q^2\rangle\big)\psi \tag{13}$$

$$\partial_t\langle\varepsilon_f\rangle + \langle\varepsilon_f\rangle_{,i}U_i + \langle\varepsilon_f u_i\rangle_{,i} = -\,\big(\langle\varepsilon_f\rangle^2/\langle f^2\rangle\big)\psi^f \tag{14}$$

$$\begin{aligned}&\partial_t\langle\varepsilon_{\rho'}\rangle + \langle\varepsilon_{\rho'}\rangle_{,i}U_i + \langle\varepsilon_{\rho'}u_i\rangle_{,i} \\ &= -\,\big(\langle\varepsilon_{\rho'}\rangle^2/\langle\rho'^2\rangle\big)\psi^{\rho'}\end{aligned} \tag{15}$$

Various terms must be modeled: the pressure correlations, the transport terms, the cross-dissipations [equations (7) and (10)] and the terms ψ, $\psi^{\rho'}$, ψ^{f} appearing in the dissipation equations, representing production and destruction of the corresponding dissipation. The general philosophy of this modeling process has been discussed elsewhere (see, e.g.- Lumley, 1978a). We will discuss the modeling process in detail only to the extent to which it differs from that presented in Lumley (1978a) or in Zeman & Lumley (1976a, b, c; 1979).

3. MODELING THE PRESSURE TERMS

As usual, we split the pressure strain correlation in equation (7) into two parts, a rapid part $p^{(1)}$ and a return-to-isotropy part $p^{(2)}$. We combine the $p^{(2)}$ part with the anisotropic part of the dissipation:

$$\langle p^{(2)}(u_{i,j} + u_{j,i})\rangle/\rho_0 - 2\nu\langle u_{i,k}u_{j,k}\rangle + 2\langle\varepsilon\rangle\delta_{ij}/3$$
$$= -\langle\varepsilon\rangle C_1 b_{ij} \tag{16}$$

where $b_{ij} = \langle u_i u_j\rangle/\langle q^2\rangle - \delta_{ij}/3$. The quantity C_1 was taken as constant by Zeman & Lumley (1976a, b, c; 1979), but in Lumley (1978a) it was shown in detail that realizability requires that C_1 vary with the anisotropy. Here we use a form based on an inequality involving the relaxation coefficient in the density anomaly flux equation. We write

$$\langle p^{(2)}\rho'_{,i}\rangle/\rho_0 - (\gamma + \nu)\langle\rho'_{,j}u_{i,j}\rangle = -C^{\rho'}\langle\rho' u_i\rangle\langle\varepsilon\rangle/\langle q^2\rangle \tag{17}$$

and

$$\langle p^{(2)}f_{,i}\rangle/\rho_0 - (\gamma + \nu)\langle f_{,j}u_{i,j}\rangle = -C^{f}\langle f u_i\rangle\langle\varepsilon\rangle/\langle q^2\rangle \tag{18}$$

Again, $C^{\rho'}$ and C^{f} (or the corresponding quantities) were taken constant by Zeman & Lumley (1976a, b, c; 1979). However, it was shown by Shih & Lumley (1981) that they must also be functions of anisotropy and other invariants in order to satisfy realizability. Specifically, we require

$$C^{\rho'} = r + C_1/2 - (C_1/2 - 1)Q/M + LF_D^{1/2} \tag{19}$$

and a similar relation for C^{f}, where r is the time scale ratio, $r = (\langle\varepsilon_{\rho'}\rangle/\langle\rho'^2\rangle)/(\langle\varepsilon\rangle/\langle q^2\rangle)$, L is an as yet undetermined function,

$$M = D_{ij}b_{jk}D_{ki} - b_{ij}D_{ij} + Q \tag{20}$$

$$Q = (1 - D_{ij}D_{ji})/6 \tag{21}$$

$$F_D = 9D_{ij}D_{jk}D_{ki} - (27/2)D_{ij}D_{ji} + 9/2 \tag{22}$$

$$D_{ij} = \left(\langle\rho'^2\rangle\langle u_i u_j\rangle - \langle\rho' u_i\rangle\langle\rho' u_j\rangle\right)$$
$$/\left(\langle\rho'^2\rangle\langle q^2\rangle - \langle\rho' u_p\rangle\langle\rho' u_p\rangle\right) \tag{23}$$

Now, in order to maintain Schwartz's inequality on all the density anomaly fluxes, it is necessary and sufficient (Shih & Lumley, 1981) to require that $0 \leq F_D \leq 1$, and expression (19) assures this. M has zeros only when the D_{ij} field is isotropic or one-dimensional, and the numerator of this term in (19) has the same zeros, so that the term is bounded. However, depending on the choice of C_1, the upper bound for the third term in (19) may be large enough to drive $C^{\rho'}$ negative. This is not only unphysical, but extremely unstable computationally. Hence, C_1 must satisfy a certain bound. The worst case occurs when both F_D and F_b are close to zero, where $F_b = 1 + 27III + 9II$, where $II = -(1/2)b_{ij}b_{ji}$, $III = (1/3)b_{ij}b_{jk}b_{ki}$; that is, one component of the density anomaly flux and of the velocity field are vanishing, which usually occurs just under the inversion base. In a vertically axisymmetric turbulence with $F_D = 0$, the expression for $C^{\rho'}$ takes on the form

$$C^{\rho'} = r + C_1/2 - (C_1/2 - 1)/3(b_{33} + 1/3) \tag{24}$$

The numerator $C_1/2 - 1$ will also go to zero when $b_{33} + 1/3$ vanishes.

If $F_D \to 0$, then $\rho' \propto u_3$, and $\langle\varepsilon_{\rho'}\rangle/\langle\rho'^2\rangle = \gamma\langle u_{3,j}u_{3,j}\rangle/\langle u_3^2\rangle$. In order to make use of this expression, we must relate the anisotropy of the dissipation to that of the energy containing range.

If we define $d_{ij} = \nu\langle u_{i,k}u_{j,k}\rangle/\langle\varepsilon\rangle - \delta_{ij}/3$, by analogy with b_{ij}, describing the anisotropy of the dissipation scales, then we expect that, as $F_b \to 0$ and arbitrary Re (i.e.- one component of the energy vanishes in some coordinate system) that the corresponding component of the dissipation will vanish also. In a general situation, we would expect the other two components of b_{ij} in principal axes to be unequal, and correspondingly the other two components of d_{ij} to be unequal also, but closer to each other than those of b_{ij}. In an axisymmetric situation such as we have here, if the axial componment of the energy vanishes, the other two components will be equal, and hence also the other two components of d_{ij} will be equal; hence, here we can say that if $F_b \to 0$ for arbitrary Reynolds number, $d_{ij} = b_{ij}$. In general, for $F_b = O(1)$ and $Re >> 1$, we expect $d_{ij} \approx b_{ij}/Re^{1/2}$. Hence, at least in an axisymmetric situation, we can write something like $d_{ij} = b_{ij}/(1 + aF^nRe^{1/2})$ for some a and n. For a non-axisymmetric situation, we will have to devise a more complex expression: it is probably acceptable that d_{ij} and b_{ij} have the same principal axes, so that $d_{ij} = A_1 b_{ij} + A_2(b_{ik}b_{kj} + 2II\delta_{ij}/3)$, where A_1 and A_2 are functions of II and III and Re. These will have to be determined from the conditions relating the behavior of d_{ij} to b_{ij} when $F_b \to 0$, and for large Re.

With our simplified form for d_{ij}, we obtain

$$r = [1 + 3b_{33}/(1 + aF_b{}^n Re^{1/2})]/[1 + 3b_{33}] \tag{25}$$

and substituting this in (24) we have

$$C_1 - 2 \leq 2\{2/(1 - 3\psi) - 1 - 1/(1 + aF_b{}^n Re^{1/2})\}, \ \psi < 1/3$$

$$C_1 - 2 \geq 0 \ , \ \psi > 1/3 \tag{26}$$

We have introduced for convenience $\psi = b_{33} + 1/3$, which is just $\langle w^2\rangle/\langle q^2\rangle$. This is rather specific to a particular atmospheric situation. It would be more useful if it were placed in terms of F_b, which is independent of the coordinate system. We can write $4F_b/27 = \psi - 2\psi^2 + \psi^3 = g$, say. This relation may be approximately inverted, giving

$$\psi_s = g + 2g^2 + 7g^3 + 30g^4 + 143g^5; \ \psi_s \leq 0.24 \ @ \ F_b = 1 \tag{27}$$

It is unfortunately necessary to carry such high powers to obtain an adequate model for larger f. Finally, we take

$$C_1 - 2 = 2\{2/(1 - 3\psi_s) - 1 - 1/(1 + 0.75F_b{}^3 Re^{1/2})\}\times$$
$$(1 - \exp[-(1 - F_b)/0.35]) \tag{28}$$

This matches very well the data of Choi & Lumley (1984), and is in general behavior very similar to the expression used by Zeman & Lumley (1982, 3), $(7F_b{}^2(1 - F_b)^{1/2})$, but the latter slightly violates the inequality for $\psi < 0.17$ roughly. One would think that this small difference would be irrelevant, but our program seemed always to find this region, and force $C^{\rho'}$ negative. The concave behavior of the expression used by Zeman & Lumley (1982, 3) was necessitated by stability considerations in modeling combined turbulent buoyant convection and thermo-haline convection; if the curve was not concave, the flow tended away from thermohaline convection rather than toward it in situations physically favoring such convection. Simpler, as well as more general, expressions are presented in Choi & Lumley (1984).

In expression (19), L must be determined. This was given by Shih & Lumley (1981) as $1.1r^2$, but this corresponded to a relatively narrow range of values of r. We found in calculation that this expression was not safe in regions near the inversion base, where r could attain large values as both dissipations and energies went to zero. In addition, it vanishes as $r \to 0$, whereas it is easy to convince oneself that mixing to decrease the ρ'–u_3 correlation coefficient should not cease at $r = 0$ (which Shih & Lumley's form would force). Finally, Shih & Lumley's (1981) form was for isotropic turbulence. The collection of terms in (19) (excluding the last one) are analogous to the 2 in the expression for C_1; that is, they correspond to a fixed value of the correlation coefficient. In anisotropic situations, one would expect to find in L the additional real mixing (which will decrease the correlation coefficient) which is represented by $C_1 - 2$.

Incorporating these ideas, we have taken

$$L = \left[1.1 + 0.55\mathrm{Tanh}[4(r - 1)]\right](C_1 - 1) \tag{29}$$

which is essentially equivalent to the expression of Shih & Lumley (1981) for r near unity and isotropic turbulence.

The rapid terms we use are the same as those used by Zeman & Lumley (1976a). That is,

$$\langle p^{(1)}(u_{i,j} + u_{j,i})\rangle/\rho_0 = (3/10)\beta_j\big(\langle\rho' u_i\rangle\delta_{jk} + \langle\rho' u_k\rangle\delta_{ji} - (2/3)\langle\rho' u_j\rangle\delta_{ik}\big)$$

$$\langle p^{(1)}\rho'_{,i}\rangle/\rho_0 = \beta_j\langle\rho'^2\rangle\delta_{ij}/3$$

$$\langle p^{(1)}f_{,i}\rangle/\rho_0 = \beta_j\langle f\rho'\rangle\delta_{ij}/3 \tag{30}$$

Another rapid term is required, that involving the mean velocity gradients. We have not yet carried out a calculation involving mean velocity gradients, so we do not reproduce here a suitable form. Shih (1984) has derived a number of possible forms, which we intend to try.

These rapid terms do not satisfy realizability; they are, in fact, the only part of the model left which does not. We have been wrestling with the problem of devising terms which do, but have so far not been successful. We will see in the discussion of results that intermittent failures of the model near the inversion base, requiring arbitrary fixes, can be attributed entirely to these terms.

4. TRANSPORT TERMS

In Lumley (1978a) a formalism was presented which leads to expressions for the third moments having no adjustable constants. These expressions do not constitute a model in the usual sense, but more nearly an exact solution to the equations of motion, subject to certain restrictions. Briefly, we begin with an equation for the moment generating function (equivalent to the probability density). We require that this produce the model equations (above) for the second moments in the homogeneous case, with the modeled pressure correlations, and that the equation produce a Gaussian density as a stable solution in a homogeneous flow (agreeing approximately with experiment), toward which all other solutions relax with a single time constant proportional to $\langle q^2\rangle/\langle\varepsilon\rangle$. Then a perturbation expansion is carried out about the homogeneous state, corresponding to a fine grained turbulence. Finally, the expressions for the third moments are obtained from the resulting expression for the moment generating function. These expressions contain the relaxation coefficients from the second moment equations (which are set by comparison with homogeneous experiments), but contain no new adjustable constants. The way in which the analysis is carried out is reminiscent of non-equilibrium statistical mechanics of a mixture, and the resulting expressions are similar, the second order variances and fluxes playing the role of species, and fluxes of one species being produced by gradients of all the species. This is similar to the way in which the flux of heat in salt water is produced not only by the temperature gradient, but also by the salinity gradient, and the salt flux is also produced by both gradients (the Soret and Dufour effects).

4.1 Flux and Variance Transport

In Lumley (1978a), however, explicit forms were presented only for the case of a passive scalar. If buoyancy is dynamically important, it is necessary to include terms corresponding to the "rapid" part of the pressure - the part that is linear in the buoyancy. The form of the rapid buoyant terms was worked out in Lumley et al (1978), and these can easily be combined with the development of Lumley (1978a). In the latter, we showed that the terms involving the mean velocity gradient were generally of lower order, and could be neglected. However, we will be examining situations in which this may not be true. For example, just at the inversion base, we may expect to find a strong shear, which may be quite significant. Since, due to the inclusion of the buoyancy, the equations for the third moments are quite cross-coupled, we will have to resort to a matrix inversion in any event; it will not make the equations more complicated to carry the terms in the mean velocity gradient. Hence, we must compute the rapid terms in the mean velocity gradient. This has been done at some length in Ettestad & Lumley (1983). The technique is perfectly standard: we will illustrate with one of the terms. Consider the spectral representation

$$-\langle p^{(1)}{}_{,k}\rho' u_\ell\rangle/\rho_0 = 2U_{i,j}\int \psi_{j\ell}\kappa_i\kappa_k/\kappa^2 d\underline{\kappa} \tag{31}$$

where the spectral density tensor must satisfy

$$\int \psi_{j\ell} d\underline{\kappa} = \langle \rho' u_j u_\ell\rangle \tag{32}$$

Equation (31) is obtained from the part of the pressure that is linear in the mean velocity gradient:

$$-p^{(1)}{}_{,ii}/\rho_0 = 2U_{i,j}u_{j,i} \tag{33}$$

by assuming homogeneity and using Fourier transforms. We write for the integral part

$$I_{ikj\ell} = \int \psi_{j\ell}\kappa_i\kappa_k/\kappa^2 d\underline{\kappa} \tag{34}$$

and we have the conditions arising from symmetry, normalization and incompressibility respectively:

$$I_{ikj\ell} = I_{kij\ell};\ I_{iij\ell} = \langle \rho' u_j u_\ell\rangle;\ I_{ikk\ell} = 0 \tag{35}$$

If we now assume the most general possible linear expression in

$\langle\rho' u_i u_j\rangle$, all the coefficients are determined, giving

$$I_{ikj\ell} = (2/5)\delta_{ik}\langle\rho' u_j u_\ell\rangle - (1/10)(\delta_{ij}\langle\rho' u_k u_\ell\rangle + \delta_{kj}\langle\rho' u_i u_\ell\rangle) \tag{36}$$

When this is combined with the transpose of (1) and the result combined with the production terms in the equation for $\langle\rho' u_k u_\ell\rangle$ we obtain

$$(2/5)S_{kp}\langle u_p\rho' u_\ell\rangle + (2/5)S_{\ell p}\langle u_p\rho' u_k\rangle \tag{37}$$

where S_{ij} is the strain rate of the mean velocity field. In fact, the effect of all the rapid velocity terms is the same, resulting simply in the replacement of $U_{i,j}$ by $(2/5)S_{ij}$ in all the production terms. Combining the buoyant rapid terms with the mechanical rapid terms, and using the same development as used in Lumley (1978a), resulting in the same relaxation coefficients as given there, we obtain:

$$\begin{aligned}&\langle\rho'^2\rangle_{,j}\langle u_i u_j\rangle + 2\langle\rho' u_i\rangle_{,j}\langle\rho' u_j\rangle\\ &= (2/3)\langle\rho'^3\rangle\beta_i - 2\langle\rho\rangle_{,j}\langle\rho' u_i u_j\rangle - (2/5)S_{ij}\langle\rho'^2 u_j\rangle\\ &-2(\langle\varepsilon_{\rho'}\rangle/\langle\rho'^2\rangle)[1 + C^{\rho'}((\langle\varepsilon\rangle/\langle q^2\rangle)(\langle\rho'^2\rangle/\langle\varepsilon_{\rho'}\rangle)]\langle\rho'^2 u_i\rangle\end{aligned} \tag{38}$$

$$\begin{aligned}&\langle\rho' u_i\rangle_{,k}\langle u_k u_j\rangle + \langle\rho' u_j\rangle_{,k}\langle u_k u_i\rangle\\ &+ \langle u_i u_j\rangle_{,k}\langle\rho' u_k\rangle\\ &= (1/5)\ \delta_{ij}\beta_\ell\langle\rho'^2 u_\ell\rangle + (7/10)\left(\beta_j\langle\rho'^2 u_i\rangle + \beta_i\langle\rho'^2 u_j\rangle\right)\\ &-\langle\rho\rangle_{,k}\langle u_i u_j u_k\rangle - (2/5)S_{ik}\langle\rho' u_j u_k\rangle\\ &-(2/5)S_{jk}\langle\rho' u_i u_k\rangle\\ &- \left[C_1((\langle\varepsilon\rangle/\langle q^2\rangle)\left(\langle\rho' u_i u_j\rangle - \langle q^2\rho'\rangle\delta_{ij}/3\right)\right.\\ &\left.+ (2/3)((\langle\varepsilon\rangle/\langle q^2\rangle)\langle q^2\rho'\rangle\delta_{ij} + 2C^{\rho'}((\langle\varepsilon\rangle/\langle q^2\rangle)\langle\rho' u_i u_j\rangle\right]\end{aligned} \tag{39}$$

$$\begin{aligned}&\langle u_i u_j\rangle_{,p}\langle u_k u_p\rangle + \langle u_i u_k\rangle_{,p}\langle u_j u_p\rangle\\ &+ \langle u_j u_k\rangle_{,p}\langle u_p u_i\rangle = (7/10)\left(\beta_i\langle\rho' u_k u_j\rangle\right.\\ &\left.+ \beta_j\langle\rho' u_i u_k\rangle + \beta_k\langle\rho' u_i u_j\rangle\right)\\ &+ (1/10)\left(\delta_{ik}\beta_p\langle\rho' u_p u_j\rangle + \delta_{ij}\beta_p\langle\rho' u_p u_k\rangle\right.\\ &\left.+ \delta_{kj}\beta_p\langle\rho' u_p u_i\rangle\right) - (2/5)S_{i\ell}\langle u_\ell u_j u_k\rangle\\ &- (2/5)S_{j\ell}\langle u_\ell u_i u_k\rangle - (2/5)S_{k\ell}\langle u_\ell u_i u_j\rangle\end{aligned}$$

$$- 3C_1(\langle\varepsilon\rangle/\langle q^2\rangle)[\langle u_iu_ju_k\rangle - (1/9)(\delta_{ij}\langle q^2u_k\rangle$$
$$+ \delta_{ik}\langle q^2u_j\rangle + \delta_{jk}\langle q^2u_i\rangle)] - (2/3)(\langle\varepsilon\rangle/\langle q^2\rangle)(\delta_{ij}\langle q^2u_k\rangle$$
$$+ \delta_{ik}\langle q^2u_j\rangle + \delta_{jk}\langle q^2u_i\rangle) \tag{40}$$

These equations are closed with the exception of $\langle\rho'^3\rangle$. A similar equation for this becomes

$$\langle\rho\rangle_{,j}\langle\rho'^2u_j\rangle + \langle\rho'^2\rangle_{,j}\langle\rho'u_j\rangle = -2\langle\rho'^3\rangle(\langle\varepsilon_{\rho'}\rangle/\langle\rho'^2\rangle) \tag{41}$$

This can be substituted directly in (8) to give

$$\langle\rho'^2\rangle_{,j}\langle u_iu_j\rangle + 2\langle\rho'u_i\rangle_{,j}\langle\rho'u_j\rangle$$
$$= -(1/3)\beta_1(\langle\rho'^2\rangle/\langle\varepsilon_{\rho'}\rangle)(\langle\rho\rangle_{,j}\langle\rho'^2u_j\rangle + \langle\rho'^2\rangle_{,j}\langle\rho'u_j\rangle)$$
$$-2\langle\rho\rangle_{,j}\langle\rho'u_iu_j\rangle - (2/5)S_{ij}\langle\rho'^2u_j\rangle$$
$$-2(\langle\varepsilon_{\rho'}\rangle/\langle\rho'^2\rangle)(1 + C_{\rho'}(\langle\varepsilon\rangle/\langle q^2\rangle)(\langle\rho'^2\rangle/\langle\varepsilon_{\rho'}\rangle))\langle\rho'^2u_i\rangle \tag{38'}$$

This will give in the general three-dimensional case a 19×19 matrix: ten components of $\langle u_iu_ju_k\rangle$, six components of $\langle\rho'u_iu_j\rangle$ and three components of $\langle\rho'^2u_i\rangle$. If there is no mean velocity, so that the turbulence is axisymmetric in the vertical, we have only five components: $\langle\rho'^2w\rangle$, $\langle\rho'u^2\rangle$, $\langle\rho'w^2\rangle$, $\langle u^2w\rangle$, and $\langle w^3\rangle$. If there is a plane shear, this gives a twelve by twelve matrix: $\langle\rho'^2u\rangle$, $\langle\rho'^2w\rangle$; $\langle\rho'v^2\rangle$, $\langle\rho'u^2\rangle$, $\langle\rho'uw\rangle$, $\langle\rho'w^2\rangle$; $\langle u^3\rangle$, $\langle uv^2\rangle$, $\langle uw^2\rangle$, $\langle u^2w\rangle$, $\langle v^2w\rangle$, $\langle w^3\rangle$. It is tempting to use a machine algebra program such as MAKCYMA to solve these equations; however, such programs are quite inefficient above 6×6. We have therefore decided to invert the matrices numerically at each step.

For the water vapor mixture fraction f, we must work out the rapid buoyant terms (the rapid terms in the mean velocity gradient are the same as for the mechanical equations; that is, they have the result of replacing each $U_{i,j}$ in the equations with $(2/5)S_{ij}$). For the buoyant terms, we proceed exactly as above, expressing each integral as a linear function of the appropriate moment, and applying all symmetry, normalization and incompressibility conditions. This serves to determine the coefficients in all cases but one. In attempting to determine the form of $\langle p^{(1)}{}_{,i}fu_k\rangle$, we have no incompressibility condition to apply, and hence there is an undetermnined coefficient. With no justification whatever, we require the same incompressibility condition as $\langle(p^{(1)}f)_{,i}u_k\rangle$.

The relaxation and dissipation terms are worked out exactly as in Lumley (1978a), forming the moment generating function for the combined velocity, water vapor mixture fraction and density anomaly, requiring this to produce at second order the second order equations with their relaxation and dissipation terms, and differentiating to obtain equations for the third moments. The resulting equations are:

$$(2/5)S_{ij}\langle f^2u_j\rangle + 2F_{,j}\langle fu_ju_i\rangle + \langle f^2\rangle_{,k}\langle u_iu_k\rangle$$
$$+ 2\langle fu_i\rangle_{,k}\langle fu_k\rangle = (2/3)\beta_i\langle f^2\rho'\rangle$$
$$- 2(\langle\varepsilon_f\rangle/\langle f^2\rangle)\left[1 + C_f(\langle\varepsilon\rangle/\langle q^2\rangle)(\langle f^2\rangle/\langle\varepsilon_f\rangle)\right]\langle f^2u_i\rangle \tag{42}$$

$$(2/5)S_{ik}\langle u_kfu_j\rangle + (2/5)S_{jk}\langle u_kfu_i\rangle + F_{,k}\langle u_iu_ju_k\rangle$$
$$+ \langle fu_i\rangle_{,k}\langle u_ku_j\rangle + \langle fu_j\rangle_{,k}\langle u_ku_i\rangle$$
$$+ (u_iu_j\rangle_{,k}\langle fu_k\rangle = (1/5)\delta_{ij}\beta_\ell\langle\rho'fu_\ell\rangle$$
$$+ (7/10)\left[\beta_j\langle\rho'fu_i\rangle + \beta_i\langle\rho'fu_j\rangle\right]$$
$$- (\langle\varepsilon\rangle/\langle q^2\rangle)\left[C_1(\langle fu_iu_j\rangle - \delta_{ij}\langle fq^2\rangle/3) + (2/3)\delta_{ij}\langle fq^2\rangle\right.$$
$$\left.+ 2C_f\langle fu_iu_j\rangle\right] \tag{43}$$

$$(2/5)S_{ij}\langle f\rho'u_j\rangle + \langle\rho\rangle_{,j}\langle fu_iu_j\rangle + F_{,j}\langle\rho'u_iu_j\rangle$$
$$+ \langle\rho'f\rangle_{,k}\langle u_iu_k\rangle + \langle\rho'u_i\rangle_{,k}\langle fu_k\rangle + \langle fu_i\rangle_{,k}\langle\rho'u_k\rangle$$
$$= (2/3)\beta_i\langle\rho'^2f\rangle - (\langle\varepsilon\rangle/\langle q^2\rangle)\left[C_{\rho'} + C_f\right.$$
$$\left.+ 2(\langle\varepsilon_{f\rho'}\rangle/\langle f\rho'\rangle)(\langle q^2\rangle/\langle\varepsilon\rangle)\right]\langle\rho'fu_i\rangle \tag{44}$$

$$2F_{,i}\langle u_if\rho'\rangle + \langle\rho\rangle_{,i}\langle u_if^2\rangle + 2\langle\rho'f\rangle_{,k}\langle fu_k\rangle$$
$$+ \langle f^2\rangle_{,k}\langle\rho'u_k\rangle$$
$$= - 2(\langle\varepsilon_f\rangle/\langle f^2\rangle)\left[1 + 2(\langle\varepsilon_{f\rho'}\rangle/\langle f\rho'\rangle)(\langle f^2\rangle/\langle\varepsilon_f\rangle\right]\langle f^2\rho'\rangle \tag{45}$$

$$2\langle\rho\rangle_{,i}\langle\rho'fu_i\rangle + F_{,i}\langle\rho'^2u_i\rangle + \langle\rho'^2\rangle_{,k}\langle fu_k\rangle$$
$$+ 2\langle\rho'f\rangle_{,k}\langle\rho'u_k\rangle$$
$$= - 2(\langle\varepsilon_{\rho'}\rangle/\langle\rho'^2\rangle)\left[1 + 2(\langle\varepsilon_{f\rho'}\rangle/\langle f\rho'\rangle)(\langle\rho'^2\rangle/\langle\varepsilon_{\rho'}\rangle)\right]\langle\rho'^2f\rangle \tag{46}$$

4.2 Dissipation Transport

In Lumley (1978a) transport terms for the equations of mechanical dissipation and dissipation of scalar variance were suggested, but these were not based on first principles; rather, they were derived from the soundly based transport terms for the velocity field by requiring that the dissipation and corresponding variance vanish together at an appropriate rate. In practice these transport terms have been of approximately the correct order of magnitude, but they have occasionally produced unphysical results. Consequently, we have derived transport terms for the dissipations that are as soundly based as those for the velocity field.

The procedure is the same as in Lumley (1978a). Begin by constructing an equation for the moment generating function, but include the velocity gradient also:

$$G = \ell n\langle \exp[ik_j u_j + im_{pq}u_{p,q}]\rangle \tag{47}$$

Now, if the equilibrium distribution is jointly Gaussian, and the field is homogeneous, and the Reynolds number high, so that the dissipation region is isotropic, the equilibrium F will have the form

$$G = -(1/2)\left(k_i k_j\langle u_i u_j\rangle + m_{rs}m_{rs}\langle\varepsilon\rangle/9\nu\right) \tag{48}$$

In this homogeneous, relaxing situation without disturbing mean fields, the equations of motion reduce to

$$\partial_t\langle u_i u_j\rangle = -C_1\langle\varepsilon\rangle b_{ij} - 2\langle\varepsilon\rangle\delta_{ij}/3$$

$$\partial_t\langle\varepsilon\rangle = -\psi_0\langle\varepsilon\rangle^2/\langle q^2\rangle \tag{49}$$

where ψ_0 is the value of ψ without production or anisotropy. Hence, if we differentiate equation (48), we obtain

$$\partial_t G = -(1/2)(\langle\varepsilon\rangle/\langle q^2\rangle)\left(k_i k_j[-C_1\langle q^2\rangle b_{ij} - 2\langle q^2\rangle\delta_{ij}/3]\right.$$

$$\left.+m_{rs}m_{rs}[-\langle\varepsilon\rangle\psi_0/9\nu]\right) \tag{50}$$

We now note that, for a Gaussian distribution

$$\partial^2 G/\partial ik_i\partial ik_j = G_{,ij} = \langle u_i u_j\rangle$$

$$\partial^2 G/\partial im_{pq}\partial im_{rs} = G_{;pq;rs} = \langle u_{p,q}u_{r,s}\rangle \tag{51}$$

where we have introduced a somewhat bizarre notation for convenience. If we replace these expressions in equation (50), we have finally a form which will relax to Gaussian with a single time constant:

$$\partial_t G = +(1/2)(\langle\varepsilon\rangle/\langle q^2\rangle)\left\{k_i k_j\left[C_1\left(G_{,ij} - \delta_{ij}G_{,pp}/3\right)\right.\right.$$

$$\left.\left.-2\delta_{ij}G_{,pp}/3\right] + m_{rs}m_{rs}\psi_0 G_{;pq;pq}/9\right\} \tag{52}$$

This form will produce the relaxation term in the ε transport equation. Differentiating equation (52), we obtain

$$\partial_t\langle\varepsilon u_i\rangle = -\psi_0\langle\varepsilon u_i\rangle\langle\varepsilon\rangle/\langle q^2\rangle \tag{53}$$

This describes the relaxation of the ε transport in a homogeneous flow without buoyancy or mean velocity. We assume that the relaxation term remains the same to first order when we add inhomogeneity and buoyancy. Now, we must work out the form for the inhomogeneous terms and for the buoyant terms.

The inhomogeneous terms which appear on the left hand side of the equation for $\langle\varepsilon u_i\rangle$ are [carrying out the same perturbation expansion as in Lumley (1978a), which results in the neglect of fourth order cumulants in the third moment equation]:

$$\langle\varepsilon\rangle_{,\ell}\langle u_j u_\ell\rangle + 2\nu\langle u_{i,k}u_j\rangle_{,\ell}\langle u_{i,k}u_\ell\rangle$$
$$+ 2\nu\langle u_j u_{i,\ell}\rangle\langle u_{\ell,k}u_{i,k}\rangle + 2\nu\langle u_j u_{\ell,k}\rangle\langle u_{i,k}u_{i,\ell}\rangle \tag{54}$$

We can show that most of these terms are negligible. For example, we can write

$$\nu\langle u_{\ell,k}u_{i,k}\rangle \simeq \langle\varepsilon\rangle[\delta_{\ell i}/3 + ab_{\ell i}/Re^{1/2}]$$
$$\nu\langle u_{i,k}u_{i,\ell}\rangle \approx \langle\varepsilon\rangle[\delta_{k\ell}/3 + cb_{k\ell}/Re^{1/2}] \tag{55}$$

where a and c are constants of order unity; this embodies the idea that the dissipative regions are more nearly isotropic than the energy containing regions. Of course, $b_{ij} = \langle u_i u_j\rangle/\langle q^2\rangle - \delta_{ij}/3$. See, for example, Tennekes & Lumley (1972) for similar forms. Using this, we find that the third and fourth terms in (54) are of order $Re^{-1/2}$ relative to the other terms, and thus may be neglected at large Reynolds numbers. $Re = \langle q^2\rangle^2/9\langle\varepsilon\rangle\nu$. So far as the second term is concerned, it is shown in Tennekes & Lumley (1972) that $\langle u_{i,j}u_k\rangle$ is of order u^2/ℓ, where u and ℓ are the scales of the energy containing range. The second term is thus of order $\nu u^4/\ell^3$, while the first term is of order u^5/ℓ^2; their ratio is Re^{-1}. Hence, only the first term in (54) remains.

Now, so far as the buoyant terms are concerned, write the equation for $\langle\varepsilon u_i\rangle$ (we suppress all but the relevant terms):

$$\partial_t\langle\varepsilon u_j\rangle + \ldots = -(2\nu/\rho_0)\langle p_{,ik}u_{i,k}u_j\rangle - \langle p_{,j}\varepsilon\rangle/\rho_0$$
$$+2\beta_i\nu\langle\rho'_{,k}u_{i,k}u_j\rangle + \beta_j\langle\rho'\varepsilon\rangle + \ldots \tag{56}$$

Now, using the usual expressions for the rapid part of the pressure, we may write

$$-\langle p_{,j}\varepsilon\rangle/\rho_0 = -\beta_i\int(\kappa_i\kappa_j/\kappa^2)\phi_{\varepsilon\rho'}d\underline{\kappa} \tag{57}$$

where

$$\int\phi_{\varepsilon\rho'}d\underline{\kappa} = \langle\varepsilon\rho'\rangle \tag{58}$$

Our experience has shown that a form for the integral in (57) which is responsive to anisotropy is necessary. We write

$$\int(\kappa_i\kappa_j/\kappa^2)\phi_{\varepsilon\rho'}d\underline{\kappa} = I_{ij}$$

$$I_{ij}/I_{pp} - \delta_{ij}/3 = b_{ij} \tag{59}$$

That is, we assume that the anisotropy induced in the structure of the scalar spectrum is the same as that induced in the energy containing eddies of the velocity field.

We can also write (in a homogeneous situation)

$$\langle p_{,ik}u_{i,k}u_j\rangle = -\langle p_{,k}u_{i,k}u_{j,i}\rangle \tag{60}$$

and at high Reynolds numbers the only non-zero contribution to the right hand side will be

$$-\langle p_{,j}\varepsilon\rangle/9\nu \tag{61}$$

since the other terms will be poorly correlated. Thus, the term in (60) can also be written in a simple form. Using the fact that $I_{pp} = \langle\varepsilon\rho'\rangle$, we have

$$I_{ij} = \langle\varepsilon\rho'\rangle(b_{ij} + \delta_{ij}/3) \tag{62}$$

We may also rewrite the third term on the right hand side of (56) as

$$\nu\langle\rho'_{,k}u_{i,k}u_j\rangle = -\nu\langle\rho' u_{i,k}u_{j,k}\rangle = -\langle\varepsilon\rho'\rangle\delta_{ij}/3 \tag{63}$$

where the first equality is justified by homogeneity and high Reynolds number [if estimates are made, as in Tennekes & Lumley 1972, of the relative magnitudes, the other terms are found to be at least $Re^{-1/4}$ compared to this], and the second by local isotropy at high Reynolds numbers. Thus, we may write the terms on the right hand side of (56) as

$$-\beta_i\langle\varepsilon\rho'\rangle(7b_{ij}/9 - 2\delta_{ij}/27) \tag{64}$$

Now we must obtain the equation for $\langle\varepsilon\rho'\rangle$:

$$\partial_t\langle\varepsilon\rho'\rangle + \ldots = -(2\nu/\rho_0)\langle p_{,ik}u_{i,k}\rho'\rangle + 2\beta_i\nu\langle\rho'_{,k}u_{i,k}\rho'\rangle \tag{65}$$

and we must carry out exactly similar manipulations here. We rewrite

$$\nu\langle\rho'_{,k}u_{i,k}\rho'\rangle = -\nu\langle\rho'_{,k}\rho'_{,k}u_i\rangle = -(\nu/\gamma)\langle\varepsilon_{\rho'}u_i\rangle$$

$$\langle p_{,ik}u_{i,k}\rho'\rangle = -\langle p_{,ik}u_i\rho'_{,k}\rangle - \langle p_{,ikk}u_i\rho'\rangle \tag{66}$$

justifying these in the same way. Again solving for the rapid part of the pressure term, we may write [with an assumption like (59)]

$$\nu\langle p_{,ik}u_i\rho'_{,k}\rangle/\rho_0 = +(\beta_j\nu/\gamma)\langle\varepsilon_{\rho'}u_i\rangle(b_{ij} + \delta_{ij}/3)$$

$$\nu\langle p_{,ikk}u_i\rho'\rangle/\rho_0 = -(\beta_j\nu/\gamma)\langle\varepsilon_{\rho'}u_j\rangle/3 \tag{67}$$

so that the terms on the right hand side of (65) become

$$-2(\beta_j\nu/\gamma)\langle\varepsilon_{\rho'}u_i\rangle(\delta_{ji} - b_{ji}) \tag{68}$$

Finally, the equation for $\langle\varepsilon_{\rho'}u_i\rangle$

$$\partial_t\langle\varepsilon_{\rho'}u_j\rangle + \ldots = -\langle p_{,j}\varepsilon_{\rho'}\rangle/\rho_0 + \beta_j\langle\varepsilon_{\rho'}\rho'\rangle \tag{69}$$

Using a form for $\langle p_{,j}\varepsilon_{\rho'}\rangle$ similar to that for $\langle p_{,j}\varepsilon\rangle$, we have

$$-\langle p_{,j}\varepsilon_{\rho'}\rangle/\rho_0 = -\beta_i\langle\varepsilon_{\rho'}\rho'\rangle(b_{ij} + \delta_{ij}/3) \tag{70}$$

so that the right hand side of (69) becomes

$$-\beta_i\langle\varepsilon_{\rho'}\rho'\rangle(b_{ij} - 2\delta_{ij}/3) \tag{71}$$

Finally the equation for $\langle\varepsilon_{\rho'}\rho'\rangle$ should be considered; it is found to have no buoyant terms on the right hand side.

We may now write the equations for these four quantities for the case without mean velocity shear:

$$\begin{aligned}
&\langle\varepsilon\rangle_{,\ell}\langle u_j u_\ell\rangle = -\psi_0\langle\varepsilon u_j\rangle\langle\varepsilon\rangle/\langle q^2\rangle \\
&- \beta_i\langle\varepsilon\rho'\rangle(7b_{ij}/9 - 2\delta_{ij}/27) \\
&\langle\varepsilon\rangle_{,\ell}\langle\rho' u_\ell\rangle = -\psi_0\langle\varepsilon\rho'\rangle\langle\varepsilon\rangle/\langle q^2\rangle \\
&- 2(\beta_j\nu/\gamma)\langle\varepsilon_{\rho'}u_i\rangle(\delta_{ji} - b_{ji}) \\
&\langle\varepsilon_{\rho'}\rangle_{,\ell}\langle u_\ell u_j\rangle = -\psi_0{}^{\rho'}\langle\varepsilon_{\rho'}u_j\rangle\langle\varepsilon_{\rho'}\rangle/\langle\rho'^2\rangle \\
&- \beta_i\langle\varepsilon_{\rho'}\rho'\rangle(b_{ij} - 2\delta_{ij}/3) \\
&\langle\varepsilon_{\rho'}\rangle_{,\ell}\langle\rho' u_\ell\rangle = -\psi_0{}^{\rho'}\langle\varepsilon_{\rho'}\rho'\rangle\langle\varepsilon_{\rho'}\rangle/\langle\rho'^2\rangle
\end{aligned} \tag{72}$$

These may readily be solved to give;

$$\begin{aligned}
&\langle\varepsilon_{\rho'}u_j\rangle = -(\langle\rho'^2\rangle/\langle\varepsilon_{\rho'}\rangle\psi_0{}^{\rho'})\{\langle u_\ell u_j\rangle \\
&- (\beta_i\langle\rho'^2\rangle/\langle\varepsilon_{\rho'}\rangle\psi_0{}^{\rho'})\langle\rho' u_\ell\rangle(b_{ij} - 2\delta_{ij}/3)\}\langle\varepsilon_{\rho'}\rangle_{,\ell} \\
&\langle\varepsilon u_j\rangle = -(\langle q^2\rangle/\langle\varepsilon\rangle\psi_0)\{[\langle u_j u_\ell\rangle \\
&- (\beta_i\langle q^2\rangle/\langle\varepsilon\rangle\psi_0)\langle\rho' u_\ell\rangle(7b_{ij}/9 - 2\delta_{ij}/27)]\langle\varepsilon\rangle_{,\ell} \\
&+ 2(\beta_p\beta_q\nu\langle q^2\rangle\langle\rho'^2\rangle/\langle\varepsilon\rangle\langle\varepsilon_{\rho'}\rangle\psi_0\psi_0{}^{\rho'}\gamma)(7b_{pj}/9 \\
&- 2\delta_{pj}/27)(\delta_{qi} - b_{qi})[\langle u_\ell u_i\rangle
\end{aligned}$$

$$- (\beta_j\langle\rho'^2\rangle/\langle\varepsilon_{\rho'}\rangle\psi_0{}^{\rho'})\langle\rho' u_\ell\rangle(b_{ji} - 2\delta_{ji}/3)]\langle\varepsilon_{\rho'}\rangle_{,\ell}\} \tag{73}$$

We see here the appearance of the forms obtained in Lumley et al (1978) which serve to increase or decrease the vertical transport according to the stability. The transport term for ε_f turns out to be exactly the same as that for $\varepsilon_{\rho'}$:

$$\langle\varepsilon_f u_j\rangle = -(\langle f^2\rangle/\langle\varepsilon_f\rangle\psi_0{}^f)\{\langle u_\ell u_j\rangle$$

$$- (\beta_i\langle f^2\rangle/\langle\varepsilon_f\rangle\psi_0{}^f)\langle\rho' u_\ell\rangle(b_{ij} - 2\delta_{ij}/3)\}\langle\varepsilon_f\rangle_{,\ell} \tag{74}$$

5. THE CROSS DISSIPATION

We do not carry an equation for $\langle\varepsilon_{\rho' f}\rangle$, the cross dissipation. However, it is easy to obtain an inequality based on realizability. If we take

$$\langle\rho'^2\rangle\langle f^2\rangle - \langle\rho' f\rangle^2 \geq 0 \tag{75}$$

and require that its time derivative vanish if it vanishes, and require that the dissipation terms separately satisfy this requirement (as discussed in Lumley, 1978a), we easily obtain

$$\langle\varepsilon_{\rho' f}\rangle = (\langle\rho' f\rangle/2)(\langle\rho'^2\rangle\langle f^2\rangle/\langle\rho' f\rangle^2)(\langle\varepsilon_{\rho'}\rangle/\langle\rho'^2\rangle + \langle\varepsilon_f\rangle/\langle f^2\rangle)$$

$$\text{if } \langle\rho'^2\rangle\langle f^2\rangle - \langle\rho' f\rangle^2 = 0 \tag{76}$$

We may satisfy this by setting

$$\langle\varepsilon_{\rho' f}\rangle = (\langle\rho' f\rangle/2)J(\langle\varepsilon_{\rho'}\rangle/\langle\rho'^2\rangle + \langle\varepsilon_f\rangle/\langle f^2\rangle) \tag{77}$$

where J is a function of $\langle\rho' f\rangle^2/\langle\rho'^2\rangle\langle f^2\rangle$, and $J \geq 1$, and $J = 1$ when $\langle\rho'^2\rangle\langle f^2\rangle - \langle\rho' f\rangle^2 = 0$. In the absence of other information, we have taken

$$J = 1 + 0.5(1 - \langle\rho' f\rangle^2/\langle\rho'^2\rangle\langle f^2\rangle) \tag{78}$$

There are other requirements that this expression does not satisfy, for example that

$$\langle\varepsilon_{\rho'}\rangle\langle\varepsilon_f\rangle - \langle\varepsilon_{\rho' f}\rangle^2 \geq 0 \tag{79}$$

We have made a considerable effort to obtain a simple interpolation form which will maintain both inequalities (75) and (79), without success. We use the form (77)-(78), and monitor the solution to check violations of (79), which occur rarely, and then only at the inversion base when all quantities are nearly zero.

6. THE EQUATION FOR THE SCALAR DISSIPATION

In atmospheric flows of the type that interest us here there is, and should be, always a region of negative production of density anomaly variance near the inversion base. This comes about because the density anomaly flux and the gradient of the mean density do not change sign in the same place. There is also, of course, a region of negative buoyant production of mechanical energy, since the sign of the density anomaly flux changes, but this does not concern us here. It is vitally important that the equation for the dissipation of density anomaly variance behave properly in this region. For example, suppose that $\langle\varepsilon_{\rho'}\rangle = 0$, but $\langle\varepsilon\rangle \neq 0$, $\langle q^2\rangle \neq 0$. If $\langle\rho'^2\rangle$ is being produced, then $\langle\varepsilon_{\rho'}\rangle$ should be produced also. Hence, the right hand side of the equation for $\langle\varepsilon_{\rho'}\rangle$ cannot be proportional to $\langle\varepsilon_{\rho'}\rangle$, for this would not permit production of $\langle\varepsilon_{\rho'}\rangle$ under these circumstances. On the other hand, if the production of density anomaly variance is negative, it should not be possible to reduce the level of $\langle\varepsilon_{\rho'}\rangle$ below zero. We can satisfy these requirements in a relatively simple way:

$$\partial_t\langle\varepsilon_{\rho'}\rangle = -\psi_0{}^{\rho'}\langle\varepsilon_{\rho'}\rangle^2/\langle\rho'^2\rangle + Z\Big[(\langle\varepsilon\rangle/\langle q^2\rangle)(P_{\rho'} + \big|P_{\rho'}\big|)/2 + (\langle\varepsilon_{\rho'}\rangle/\langle\rho'^2\rangle)(P_{\rho'} + \big|P_{\rho'}\big|)/2\Big] \tag{80}$$

Here we are using $P_{\rho'}$ to designate the production of density anomaly variance:

$$P_{\rho'} = -\langle\rho\rangle_{,i}\langle\rho' u_i\rangle \tag{81}$$

Hence, if $P_{\rho'} > 0$, we can produce $\langle\varepsilon_{\rho'}\rangle$ at a rate determined by the mechanical field, regardless of the value of $\langle\varepsilon_{\rho'}\rangle$, while if $P_{\rho'} < 0$, we can remove $\langle\varepsilon_{\rho'}\rangle$ at a rate determined by the density anomaly field; the removal will stop when the level of $\langle\varepsilon_{\rho'}\rangle$ reaches zero. Equation (80) can be rewritten as

$$\partial_t\langle\varepsilon_{\rho'}\rangle = -(\langle\varepsilon_{\rho'}\rangle^2/\langle\rho'^2\rangle)\Big\{\psi_0{}^{\rho'} - Z\Big[(P_{\rho'} + \big|P_{\rho'}\big|)/2r\langle\varepsilon_{\rho'}\rangle + (P_{\rho'} - \big|P_{\rho'}\big|)/2\langle\varepsilon_{\rho'}\rangle\Big]\Big\} \tag{82}$$

to place it in a more customary form. Note that, if $P_{\rho'} = \langle\varepsilon_{\rho'}\rangle$, and $P_m = \langle\varepsilon\rangle$, $P_b = 0$, $r = r_e$ [using the notation of Lumley (1978a) for the mechanical and buoyant production of mechanical energy, and for the time scale and its equilibrium value], we are reduced to the case described in Lumley (1978a), from which we obtain

$$Z = 2r_e + \psi_1 - 2 \tag{83}$$

where ψ_1 is the coefficient of the (negative) production term in the $\langle\varepsilon\rangle$ equation.

In the case of isotropic turbulence, we require that

$$\psi_0 \rho' = 2 - (2 - \psi_0)/r \tag{84}$$

For anisotropic turbulence, it is reasonable to expect that r will tend toward r_e, the equilibrium value [given by $3(\psi_0 - 2)/4$; see Lumley, 1978a]. This effect can be produced by a term of the form

$$\psi_0 \rho' = 2 - (2 - \psi_0)/r + g(II, r - r_e)$$

$$g(0, r - r_e) = 0; \; g(II, r - r_e)(r - r_e) \geq 0 \tag{85}$$

We used

$$g = \left[(2 - \psi_0)/r_e(r + F_D{}^3)\right] II4 \ln(r/r_e) \tag{86}$$

This vanishes when II vanishes as required. A term $r^{-1}\ln(r/r_e)$ would treat logarithmic excursions above and below r_e symmetrically, which seems desirable. Computationally, however, the behavior was too violent near r = 0. It is particularly important that r be kept near r_e when F_D is near zero. The term chosen has the behavior of becoming much stronger for small r when F_D is small, but of being computationally well-behaved as long as they are not both zero together, which does not happen.

7. INITIAL AND BOUNDARY CONDITIONS

In the problem we consider here, we have no mean velocity gradient. This will be added at the next level of complexity. We begin with no water vapor content, and develop a dry convectively driven mixed layer with an overlying inversion. At a time t_*, when the dry layer has attained an equilibrium structure and growth, we hold the net heat flux (sensible plus latent) constant, but change from all sensible to all latent. This corresponds roughly to the surface changing from a dry solid to a deep body of water at ambient air temperature, with the same incident radiation. Ignoring the effect of mean velocity gradients, and of streamwise turbulent transport (which is relatively small under most circumstances) this is a crude model of the passage of a dry boundary layer over a lake or rice paddy.

We take initial conditions essentially the same as those of Zeman & Lumley (1976a). That is, we take intially everywhere $F_{,3} = 0$, $\langle fu_3 \rangle = 0$, $\langle \rho' f \rangle = 0$, $\langle \varepsilon_f \rangle = 0$, $\langle f^2 \rangle = 0$, and

$$w_* = (g\langle \rho' u_3 \rangle_0 h_0/\rho_0)^{1/3} = 1 \text{ m/s}$$

$$\rho_* = \langle \rho' u3 \rangle 0/w_* = 6 \times 10^{-4} \text{ Kg/m}^3$$

$$h_0 = 2.11 \times 10^2 \text{ m}, \; \rho_0 = 1.25 \text{ Kg/m}^3, \; T_0 = 300 \text{ K} \tag{87}$$

where the subscript 0 corresponds to an initial value or a value at the lower boundary $z = z_b$ or both (e.g.- $\langle\rho' u_3\rangle_0$ is the initial value of $\langle\rho' u_3\rangle$ at z_b). Then we take throughout the field

$$\langle u_3^2\rangle/w_*^2 = 1.25(z/h)^{2/3}(\langle\rho' u_3\rangle/\langle\rho' u_3\rangle_0)^{2/3}$$

$$\langle\rho'^2\rangle/\rho_*^2 = 1.6(z/h)^{-2/3}(\langle\rho' u_3\rangle/\langle\rho' u_3\rangle_0)^{4/3}$$

$$h\partial_z\langle\rho\rangle/\rho_* = -0.55(z/h)^{-4/3}(\langle\rho' u_3\rangle/\langle\rho' u_3\rangle_0)^{2/3}$$

$$\langle\varepsilon\rangle h/w_*^3 = \langle\rho' u_3\rangle/\langle\rho' u_3\rangle_0 - 0.3(1 - 2z/h)\langle\rho' u_3\rangle/\langle\rho' u_3\rangle_0$$

$$\langle\varepsilon_{\rho'}\rangle h/\rho_*^2 w_* = (z/h)^{-4/3}(\langle\rho' u_3\rangle/\langle\rho' u_3\rangle_0)^{5/3}$$

$$\langle u_\perp^2\rangle = \langle u_3^2\rangle/2; \quad \langle\rho' u_3\rangle/\langle\rho' u_3\rangle_0 = 1 - z/h \tag{88}$$

The lower boundary is taken as $z/h = 0.1$. For computation, the equations are written in terms of variables non-dimensionalized by w_*, h_0, ρ_* etc.

The heat flux (sensible plus latent) is given by

$$H = \langle\rho\rangle c_p\langle\theta u_3\rangle + \lambda\langle\rho\rangle\langle f u_3\rangle \tag{89}$$

so that the density anomaly flux is given by

$$\langle\rho' u_3\rangle = -c\rho_0^2\langle f u_3\rangle - \rho_0\langle\theta u_3\rangle/T_0 \tag{90}$$

where $c_p = 10^3$ J/Kg, $\langle\rho\rangle = 1.25$ Kg/m^3, $\lambda = 2.4\times10^6$ J/Kg. Thus, before the transition we have

$$\langle\rho' u_3\rangle = -\rho_0\langle\theta u_3\rangle/T_0 = -\rho_0 H/T_0\langle\rho\rangle c_p = -HT_0/c_p \tag{91}$$

while after the transition we have

$$\langle\rho' u_3\rangle = -c\rho_0^2 H/\lambda\langle\rho\rangle = -c\rho_0 H/\lambda \tag{92}$$

where we are ignoring the difference between ρ_0 and $\langle\rho\rangle$. We have

$$c = 1/\rho_w - 1/\rho_a \tag{93}$$

where ρ_w is the density of water vapor and ρ_a is the density of dry air at mean ambient temperature and pressure. We take $\rho_w = 0.77$ Kg/m^3, so that $c = 4.86\times10^6$ m^3/Kg. We define $f_* = \langle f u_3\rangle_0/w_*$. At t_*, w_* drops to 0.423 m/s, ρ_* drops to 1.08×10^{-4} Kg/m^3 and f_* jumps from zero to 1.42×10^{-4}. We take at $t_* = 0$ $\langle f^2\rangle = 0$, $F = 0$, so that the water vapor content of the mixed layer is produced entirely by the sudden jump in surface flux. We take at the lower boundary $\langle\rho' f\rangle = -0.95\langle\rho'^2\rangle^{1/2}\langle f^2\rangle^{1/2}$ and zero elsewhere; i.e.- the density anomaly there is produced almost entirely by the water vapor.

So far as boundary conditions are concerned, we hold the value of the flux of sensible and latent heat fixed at the surface, but we allow the values of the other quantities to float, instead holding their power law behavior fixed. That is, we require them to begin from the surface with the same power law behavior that they would have if the solution obeyed similarity (as in the initial values). Thus, for example, if $A \propto z^n$, then (taking our lowest point at 0, and evaluating the derivative at 1/2, for reasons that will be made clear in the section on numerical method)

$$\partial_z A\big|_{1/2} = nA_{1/2}/z_{1/2} \tag{94}$$

Taking $A_{1/2} = (A_0 + A_1)/2$, and using a centered difference for the derivative at 1/2, for a second order accurate expression, we obtain

$$A_0 = A_1(1 - n\Delta z/2z_{1/2})/(1 + n\Delta z/2z_{1/2}) \tag{95}$$

where Δz is the mesh spacing. We take $z_1 = 0.1h_0$. The lowest point calculated from the equations is z_1.

8. NUMERICAL METHOD

Since the turbulent transport terms are needed on the grid points, and they are divergences of the third moments, the third moments are calculated on the half-grid points. All gradients needed are calculated using centered differences of grid-point values; coefficients at the half-grid points are constructed from averages of the grid point values of the primary variables. Since 1 is the lowest grid point calculated, the values of the third moments at 1/2 and 3/2 are needed to construct the flux divergence at 1. To construct the third moment at 1/2, the values of the primary variables at 0 and 1 are needed. The value at zero is obtained from the boundary condition (see previous section).

Recall that the flux, the divergence of which appears in each equation, depends in general on gradients of all the dependent variables. The differencing scheme used treats, in each equation for a particular dependent variable, the divergence of the terms proportional to gradients of that dependent variable implicitly, while the terms proportional to gradients of other dependent variables are treated explicitly. In a system with constant coefficients it is not difficult to show that such a system is unconditionally stable so long as the eigenvalues of the matrix of diffusion coefficients are all positive.

At every time step we check to see if all correlation coefficients are within the bounds imposed by Schwartz's inequality. Apart from the need for this to assure physical behavior, this is computationally necessary because a number of coefficients depend on the values of the correlation coefficients, and will be driven out of their bounds if the correlation coefficients are allowed to exceed their own bounds, usually producing instability. As we will see in the discussion section, violation of Schwartz's inequality generally occurs only at the inversion base, as the variables tend toward zero, usually at the first point different from the lower bound for the variables in question. It is necessary to clip the appropriate variables, in order to restore values within the bounds. This clipping occurs in practice about every ten or twenty steps. The system is very sensitive to the manner in which it is done, since it amounts to a restart with non-smooth initial conditions fairly frequently during a run. By trial and error we have evolved a system which is satisfactory. First, we clip when the correlation coefficient reaches 0.97 in absolute value. Suppose, for example that the $\langle\rho' u_3\rangle$ is in question. First, the variances $\langle\rho'^2\rangle$ and $\langle u_3^2\rangle$ are smoothed in the vicinity of the offending point, say n_0. This is done by replacing the variable by a linear combination of its value and the average of the neighboring values; i.e.- at the point n we replace f_n by

$$\beta_n f_n + (\beta_n - 1)(f_{n-1} + f_{n+1})/2 \tag{96}$$

where the coefficients are given by $\beta_n = \exp[-(n-n_0)^2/4]$, or in practice 0.78, 0.37, 0.11 and 0.0. The averaging thus extends three points on either side of the offending point. The effect is to introduce a little extra diffusion in this vicinity. Since the problem always occurs on the tail of the curve as it extends into the undisturbed inversion, the curvature is always positive, and the effect of the diffusion is to increase the values of the variances. If the bounds are still exceeded, a multiplicative correction is now calculated which will be applied to increase the standard deviation of ρ' and u_3 and reduce the magnitude of the correlation $\langle\rho' u_3\rangle$ by equal proportions, to reduce the correlation coefficient to 0.95 in absolute value. The correction is also applied three points on either side of the offending point, using the same Gaussian weighting factor. If the clipping procedure is not carried out with this care, abrupt changes in curvature will be produced which will make themselves felt through the transport, and which will ultimately cause instability. We clip $\langle\rho' u_3\rangle$ first; second, we clip $\langle f u_3\rangle$, and since u_3 has already been adjusted, we apply the smoothing only to $\langle f^2\rangle$, and the correction only to $\langle f^2\rangle$ and $\langle f u_3\rangle$; finally we clip $\langle\rho' f\rangle$, and since both ρ' and f have already been smoothed and corrected, we do no smoothing, and apply the correction only to $\langle\rho' f\rangle$.

The only equations treated differently are the equations for the mean density and mean water vapor mixture fraction. Since only the gradients of these quantities appear in the equations, we actually compute the equations for the gradients, which are subsequently integrated to obtain the quantities themselves. Since the gradients of these quantities appear in many places in the system, and since these gradients are proportional to the curvature of the density anomaly flux and the specific humidity flux respectively, it is important to avoid as much as possible the effect of clipping on these quantities. Accordingly, we calculate mean density anomaly gradient and mean specific humidity gradient from the new values before clipping. That is, the values of all other variables are updated, the mean gradients are calculated from these updated values, and then the variables are clipped if necessary. Hence, the fluxes used to compute the mean profiles have had at worst one time step to allow some smoothing of the results of clipping.

Since time scales and time scale ratios appear throughout the model, it is necessary to be sure that these scales and ratios have reasonable values. In addition, it is important to avoid dividing by zero. There is again no difficulty in the interior of the region; the problem only arises at the inversion base, where $\langle q^2 \rangle$, $\langle \varepsilon \rangle$ and the other variances and dissipations are vanishing; the relative rates at which they vanish are rather critical. We decided arbitrarily to restrict the time scales $\langle q^2 \rangle / \langle \varepsilon \rangle$, $\langle \rho'^2 \rangle / \langle \varepsilon_{\rho'} \rangle$, and $\langle f^2 \rangle / \langle \varepsilon_f \rangle$ to lie between 0.1 and 10 when non-dimensionalized by w_*, ρ_*, f_* and h_0. We placed a lower bound of $10^{-4}/3$ on each of the component energies (all quantities non-dimensionalized by w_*), so that $\langle q^2 \rangle$ has a lower bound of 10^{-4}. From the bound on the time scale, $\langle \varepsilon \rangle$ consequently has a lower bound of 10^{-5}. Other variances and dissipations are similarly bounded. As a result, the (very weak) turbulence in the inversion aloft is isotropic, and both density/mechanical and moisture/mechanical time scale ratios are unity there. Although we placed similar bounds on the time scale ratios to lie between 0.1 and 10, in fact these are never exceeded because the equations for the dissipations have been constructed to maintain these ratios within reasonable values. The time scale ratios generally fall to a few tenths just below the inversion base where the production of scalar variance is very large, and usually have a substantial peak just below this region. In most of the interior of the region, however, they are roughly two.

9. RESULTS

In figures 1-16 we show the results of the calculation. The captions (e.g.- Fig. 1, RPU3 at 1.4, 10.0, 42.0) require interpretation. The group of letters is an acronym, in this case for $\langle \rho' u_3 \rangle$. The numbers are eddy-turn-over times, based on w_* and h_0 initially. Note that after the water vapor is turned on (see below), w_* drops to about half its initial value; since t is still non-dimensionalized with the initial value, the number of physical eddy turn-over times after the transition is about half the number indicated. The vertical coordinate is non-dimensionalized by h_0. To reduce clutter, we have not labeled the individual curves; they can usually be identified because the inversion base continues to erode; later curves go to zero at a larger value of dimensionless altitude. All curves shown are for a dimensionless stability of 13.6, the same value for which Zeman & Lumley (1976a) gave most of their curves.

The dry boundary layer is allowed to develop for one eddy-turn-over time; at 1.0, the radiant flux is changed from sensible to latent heat, resulting in an abrupt drop in surface density anomaly flux. In Figure 1, the curve that is strongly convex to the left is $\langle \rho' u_3 \rangle$ at 1.4; the surface value has dropped, but the effect has not yet reached above about 0.35. In Figure 3, the value of $\langle \rho \rangle$ ("Rho Bar") is given, and it is seen that the surface layer in which the density anomaly flux is decreasing upward is stably stratified. This stable stratification lasts for about an eddy-turn-over time, as we will see when we discuss the water vapor mixture fraction. By a dimensionless time of 10.0, it is complete gone. During this recovery period from the change of surface conditions, the inversion base rises slowly, if at all. The slightly ragged appearance of the curves near the inversion base is partly a response to the clipping, and partly a result of poor resolution in that region. It is clear that a great deal is happening rapidly there, and we have only fifty points vertically for the whole computational domain. It would clearly be desirable to use either an adaptive or nested grid to better resolve the inversion base, and we are considering such a scheme. In addition, it is evident that at the inversion base, all variables are quite intermittent, and their probability densities certainly are not perturbations on a Gaussian form; a correction for intermittency in the transport terms here would certainly improve behavior.

We have not shown any plots of the third moments, for reasons of space. The only third moments for which data appear to be avaliable are $\langle q^2 w\rangle$ and $\langle w^3\rangle$; for these cases, our curves fall virtually on top of the Zeman & Lumley (1976b) curves for the same value of stability, and both lie in the middle of the data.

The values of $\langle \rho'^2\rangle$ and $\langle \varepsilon_{\rho'}\rangle$ (RPS and EPRP, Figs 2 and 4) are unexceptional, the large variances caused by entrainment at the inversion base.

We see from figure 5 that $\langle fu_3\rangle$ only reaches the inversion base at a dimensionless time of 4.0; before then, it is unaware of the existence of the overlying inversion. As the water vapor reaches the inversion and begins to fill the mixed layer uniformly, developing a larger and larger discontinuity in water vapor at the inversion base (see fig. 7, FB is F Bar, or $\bar{F}$), and larger and larger fluctuations in water vapor variance (Fig. 6), and dissipation of water vapor variance (Fig. 8, EPF is EPsilon F, or $\langle \varepsilon_f\rangle$), the layer oscillates somewhat, seen primarily in $\langle fu_3\rangle$; the region of the curve of $\langle fu_3\rangle$ that is convex up at 10 is part of a wave that makes several passes from surface to inversion base and back before finally settling down. This is not numerical, but is part of the model dynamics; whether it corresponds to reality is not known, but it seems likely.

In fig. 9, we see the correlation between density anomaly and water vapor. The correlation coefficient is set at 0.95 at the surface, on the grounds that water vapor is responsible for the entirety of the density amomaly. At the inversion base, an increase in density corresponds to the arrival of the water vapor in the (lower density) stably stratified region. The large value of the correlation here does not violate Schwarz's inequality, because the variance is also large here.

In figure 10, we have the values of the horizontal (U1S = $\langle u_1^2\rangle$) and vertical (U3S = $\langle u_3^2\rangle$) energies, the vertical being about 3.6 times the horizontal. The curve of $\langle \varepsilon\rangle$ in Fig. 12 is unremarkable. In Fig. 11 we show the values of the relaxation coefficients for momentum, water vapor and density anomaly. The momentum relaxation coefficient, C_1 (C01) rises from 3 to about 3.5 before it falls to 2 in the inversion. Recall that the classical value assumed for a constant value of this coefficient is 3.25. The relaxation coefficients for f and ρ' are between 6 and 7 in the core of the layer; recall that Zeman & Lumley (1876a) used a value of 7 for the corresponding constant values. The large value of $C_{\rho'}$ just below the inversion base is necessary to bring $\langle\rho' u_3\rangle$ down fast enough to keep pace with $\langle\rho'^2\rangle$ and $\langle u_3^2\rangle$ as they go to zero, and is brought about by the increased value of r (the density/mechanical time scale ratio) here.

In figure 13 we show the values of the ratio of the exchange coefficients for heat and water vapor, which is observed to be about 0.7. At a dimensionless time of 10, we have anomalous (including negative) values near the surface, and values near unity aloft; by a time of 42, however, when the layer has reached something like equilibrium again, the values are well behaved, and of the order of 0.7 through most of the layer. Near the inversion base, of course, the ratio becomes meaningless, and takes on large positive and negative values.

The values of the flux and gradient Bowen ratios $\beta_f = (c_p/\lambda)(\langle\theta w\rangle/\langle fw\rangle)$, and $\beta_g = (c_p/\lambda)(\partial_z\Theta/\partial_z F)$ are given in figures 14 and 15. The later times in both cases are the curves that are convex upward. These begin from zero (at least after equilibrium has been achieved) because the flux of sensible heat at the surface is set equal to zero in this calculation, since we convert all the incoming radiation to latent heat. In a more realistic calculation, there would probably still be some surface flux of sensible heat. In figure 16, we see the curve for potential temperature (Theta). This should be compared with figures 3 and 7. Note that the density anomaly is nearly constant through the layer, but consists more of water vapor near the surface, and more of heat aloft. As the water vapor accumulates at the inversion base, causing a larger and larger jump in density (in a destabilizing sense) a corresponding stabilizing jump in heat must develop, to restore the temperature to the undisturbed value in the inversion. This jump is seen in figure 16.

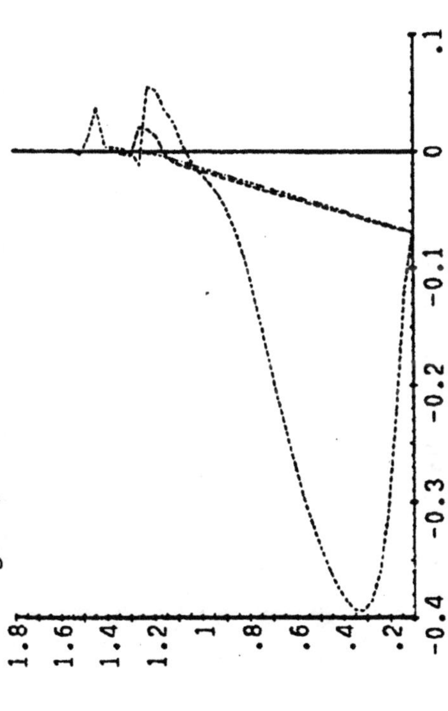

Fig. 1. RPU3 at 1.4, 10.0, 42.0

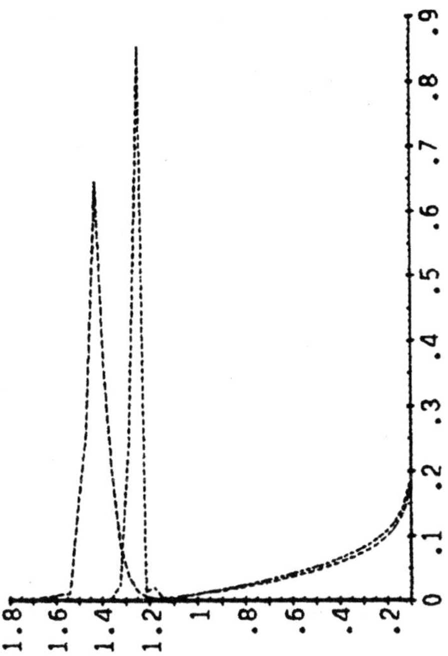

Fig. 2. RPS at 10.0, 42.0

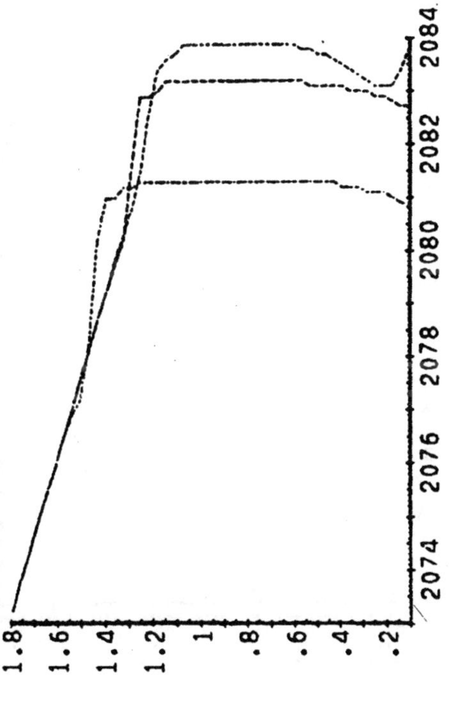

Fig. 3. RB at 1.4, 10.0, 42.0

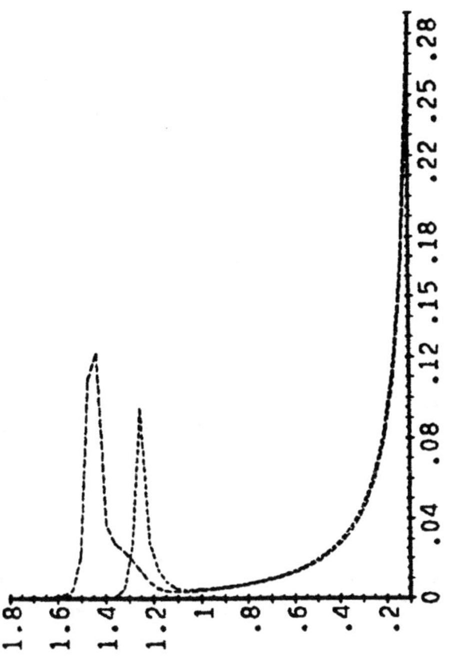

Fig. 4. EPRP at 10.0, 42.0

Fig. 5. FU3 at 1.2, 2, 4, 10, 42

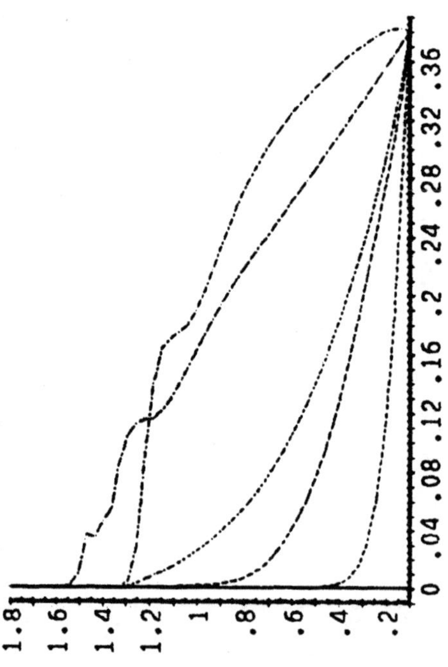

Fig. 6. FS at 1.2, 2, 4, 10, 42

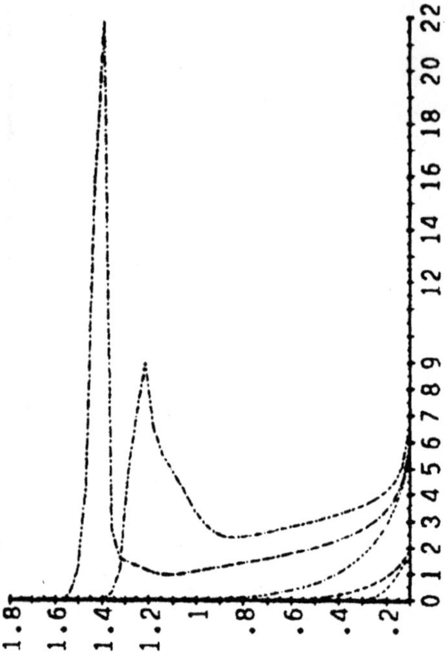

Fig. 7. FB at 2, 10, 42

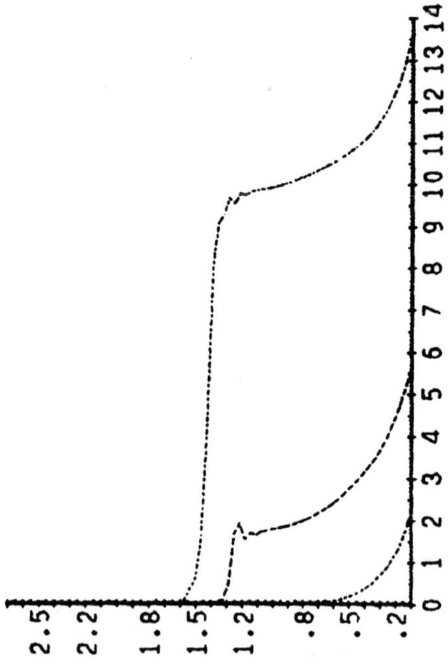

Fig. 8. EPF at 10, 42

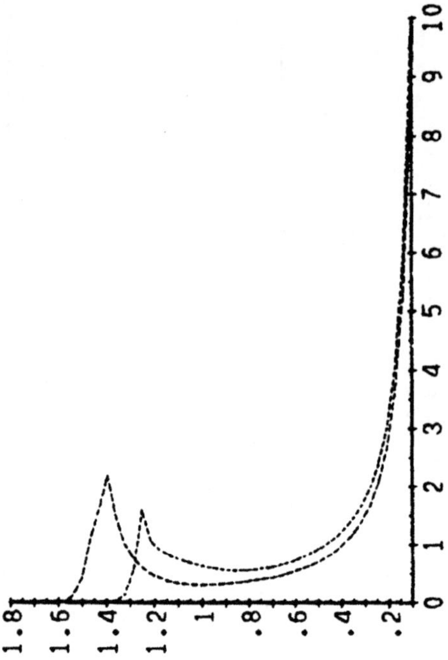

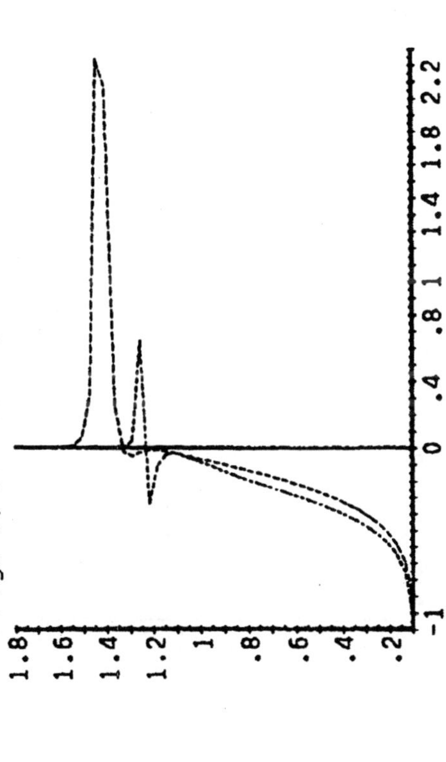

Fig. 9. RPF at 10. 42

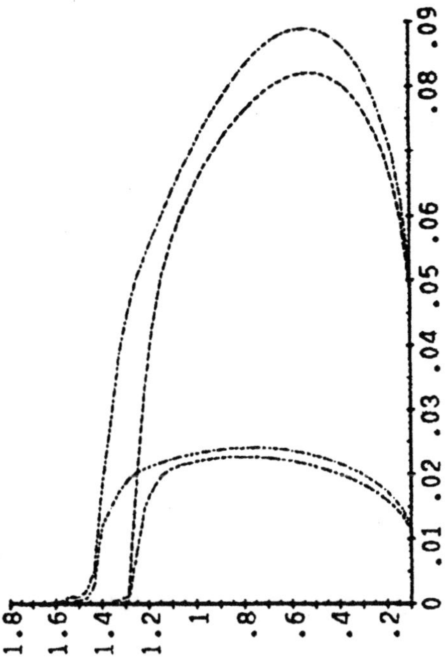

Fig. 10. U1S & U3S at 10. 42

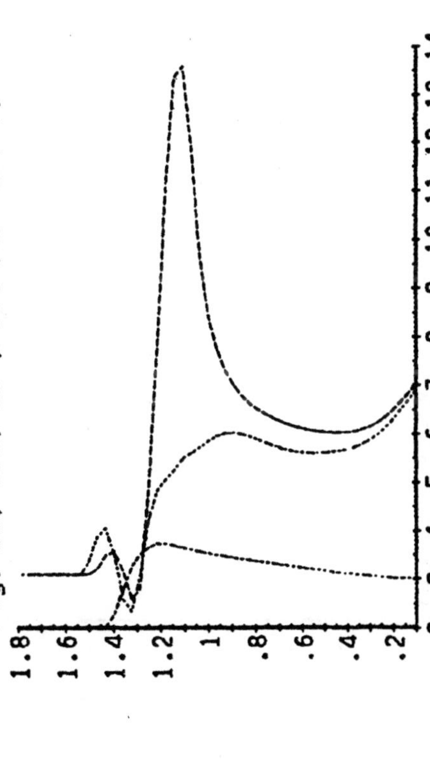

Fig. 11. CO1. COF. COR (max=13.6). at 10

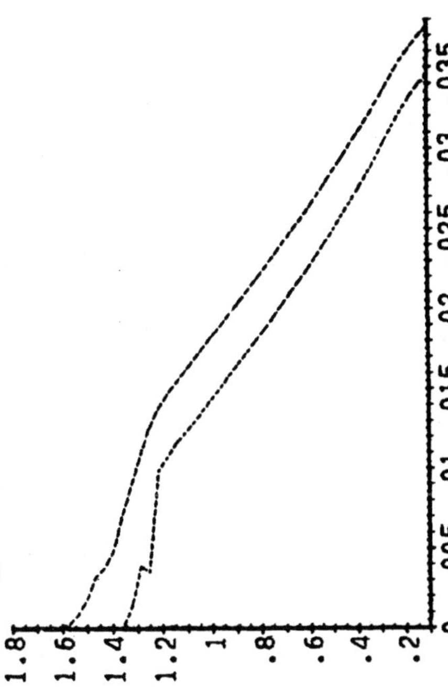

Fig. 12. EP at 10. 42

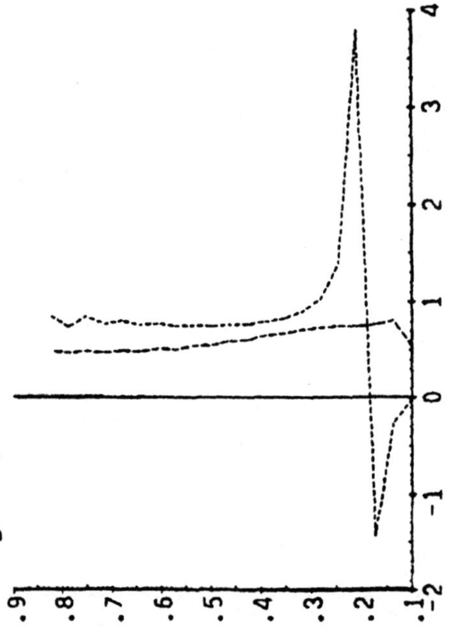

Fig. 13. Kh/Kw at 10, 42

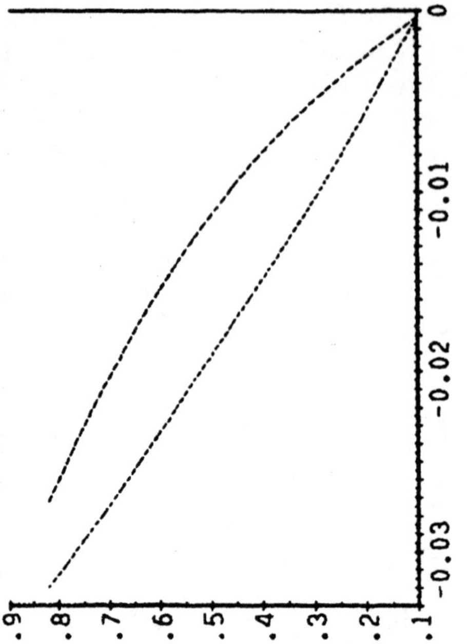

Fig. 14. BETAf at 10, 42

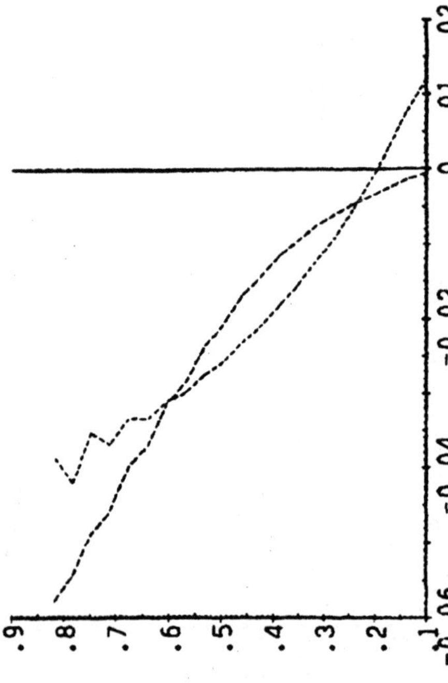

Fig. 15. BETAg at 10, 42

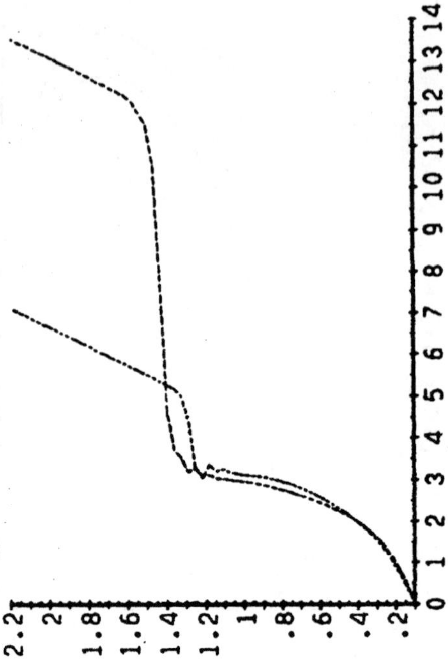

Fig. 16. Theta at 10, 42

10. BIBLIOGRAPHY

Choi, K.-S. & Lumley, J. L. 1984. The Return to Isotropy of Homogeneous Turbulence; Experiment and Modeling. Sibley School of Mechanical and Aerospace Engineering Report No. FDA 84-02. Ithaca, NY: Cornell. To be submitted to the Journal of Fluid Mechanics.

Ettestad, D. & Lumley, J. L. 1983. Parameterizing the effect of turbulent transport in swirling flows. I - Theoretical considerations. In Fourth Symposium of Turbulent Shear Flows, eds. L. J. S. Bradbury et al. Karlsruhe: The University.

Janicka, J. & Lumley, J. L. 1981. Second order modeling in non-constant density flows. Sibley School of Mechanical and Aerospace Engineering Report No. FDA 80-01. Ithaca, NY: Cornell.

Lumley, J. L. 1978a. Computational modeling of turbulent flows. In Advances in Applied Mechanics 18, ed. C.-S. Yih, pp. 123-176. New York: Academic Press.

Lumley, J. L. 1978b. Turbulent transport of passive contaminants and particles: fundamentals and advanced methods of numerical modeling. In Pollutant Dispersal, lecture series 1978-7. Rhode-St.-Genese, Belgium: Von Karman Institite for Fluid Dynamics.

Lumley, J. L. 1980a. Some fundamental aspects of turbulence with implications in geophysical flows. In Marine Turbulence, ed. J. C. J. Nihoul, pp. 175-192. Amsterdam, Netherlands: Elsevier.

Lumley, J. L., Zeman, O., & Siess, J. 1978. The influence of buoyancy on turbulent transport. J. Fluid Mech. 84: 581-597.

Newman, G. R., Launder, B. E. & Lumley, J. L. 1981. Modeling the behavior of homogeneous scalar turbulence. J. Fluid Mech. 111: 217-232.

Newman, G. R., Lumley, J. L. & Warhaft, Z. 1977. The decay of temperature fluctuations in isotropic turbulence. In 6th Australasian Hydraulics and Fluid Mechanics Conference, pp. 5-40. Australia: The Institution of Engineers.

Rao, K. S., Wyngaard, J. C. & Coté, O. R. 1974. Local advection of momentum, heat and moisture in micrometeorology. Boundary Layer Meteorology 7: 331-348.

Shih, T.-H. 1984. Second Order Modeling of Scalar Turbulent Flow. Ph. D. Thesis. Ithaca, NY: Cornell.

Shih, T.-H. & Lumley, J. L. 1981. Influence of time scale ratio on scalar flux relaxation: modeling Sirivat & Warhaft's homogeneous passive scalar fluctuations. Sibley School of Mechanical and Aerospace Engineering Report No. FDA-81-06. Ithaca NY: Cornell.

Sirivat, A. & Warhaft, Z. 1980. Preliminary measurements of Helium and temperature fluctuations in grid turbulence. Paper BA5 read at the American Physical Society, Division of Fluid Dynamics meeting at Cornell University, November 23-25, 1980. (Abstract p. 1074, Bulletin of the APS vol. 25, No. 9).

Sirivat, A. & Warhaft, Z. 1981a. The mixing of passive Helium and temperature fluctuations in grid turbulence. Submitted to J. Fluid Mech.

Sirivat, A. & Warhaft, Z. 1981b. Experiments on the evolution of temperature variance and flux in the presence of a linear thermal gradient in grid turbulence. In Preparation.

Taulbee, D. B. & Lumley, J. L. 1981. Implications regarding large eddy structures from attempts to model turbulent transport. In Applications of Fluid Mechanics and Heat Transfer to Energy and Environmental Problems, eds. D. D. Papailiou and D. K. Papailiou. Berlin/Heidelberg/New York: Springer. In Press.

Tennekes, H. & Lumley, J. L. 1972. A First Course in Turbulence. Cambridge, MA: MIT Press.

Warhaft, Z. 1980a. An experimental study of the effect of uniform strain on thermal fluctuations in grid generated turbulence. J. Fluid Mech. 99: 545-573.

Warhaft, Z. 1980b. The use of dual heat injection to infer scalar covariance decay in grid turbulence. J. Fluid Mech. In Press.

Warhaft, Z. & Lumley, J. L. 1978a. An experimental study of the decay of temperature fluctuations in grid-generated turbulence. J. Fluid Mech. 88: 659-684.

Warhaft, Z. & Lumley, J. L. 1978b. The decay of temperature fluctuations and heat flux in grid generated turbulence. In Lecture Notes in Physics, Vol. 76, Structure and Mechanisms of Turbulence II, ed H. Fiedler, pp. 113-123. Berlin/.Heidelberg/New York: Springer-Verlag.

Zeman, O. & Lumley, J. L. 1976a. Turbulence and diffusion modeling in buoyancy driven mixed layers. In Proceedings of Third Symposium on Atmospheric Turbulence, Diffusion and Air Quality, Raleigh, NC, pp. 38-45. Boston, MA: American Meteorological Society.

Zeman, O. & Lumley, J. L. 1976b. Modeling buoyancy-driven mixed layers. J. Atmos. Sci. 33, 10: 1974-88.

Zeman, O. & Lumley, J. L. 1976c. A Second-order model for buoyancy-driven mixed layers. In Proceedings, 9th ICHMT International Seminar "Turbulent Buoyant Convection", Dubrovnik, pp. 65-76. Washington, DC: Hemisphere Publishing.

Zeman, O. & Lumley, J. L. 1979. Buoyancy effects in entraining turbulent boundary layers: a second-order closure study. In Turbulent Shear Flows I, eds. F. Durst, B. E. Launder, F. W. Schmidt and J. H. Whitelaw, pp. 295-302. Berlin/Heidelberg: Springer-Verlag.

Zeman, O. & Lumley, J. L. 1982. Modeling salt-fingering structures. J. Marine Research 40,2: 315-330.

Zeman, O. & Lumley, J. L. 1983. Progress in the modeling of multi-layer salt-fingering structures. Mathematical Modeling 4: 73-85.

Questions, answers and comments

J.C.R. HUNT: In your development of third-order moments in the equation for second-order moments you assumed that the probability density function for the turbulent velocity components is approximately Gaussian. Yet, you are considering turbulent flows where the skewness is of the order of one. Is this consistent and does it matter?

J.L. LUMLEY: No, it is not completely consistent, but it does not appear to matter. There are many examples of perturbation expansions, the first term of which gives satisfactory answers when the expansion parameter is of order one, and this appears to be such a case. We have constructed a model to correct for the effects of intermittency (i.e. to handle the Dirac-spike and the continuous part of the distribution separately) in wakes and jet-like flows, and have found only a small correction. It is likely that the correction is larger for erosion of a very stable inversion (where the gradients of the intermittency would be larger), but I still do not expect a large correction based on our experience.

S.R. HANNA: How sensitive are your second-order closure models to the form of the probability density function to the turbulent velocity components?

J.L. LUMLEY: This equation is essentially the same as Hunt's. The basic model does not assume that the density is close to Gaussian, only that this is its equilibrium form. It is only when the fine grained assumption is introduced which allows neglect of fourth cumulants in the third cumulant equations that there is a suggestion that the form is not far from Gaussian. Even this is a little misleading; the assumption is really that the normalized fourth cumulant is small compared to three regardless of the shape of the distribution. As long as it is of order one, this is fairly safe. Worst case calculations for densities consisting of a Gaussian-plus-spike show that it is unlikely for the normalized fourth cumulant to be larger than one. The conclusion is: not very sensitive.

SOLUTIONS OF THE INTEGRAL EQUATION OF DIFFUSION AND THE RANDOM WALK MODEL FOR CONTINUOUS PLUMES AND INSTANTANEOUS PUFFS IN THE ATMOSPHERIC BOUNDARY LAYER

by F.B. Smith and D. Thomson
Meteorological Office (Met.O.14)
London Road, Bracknell,
Berkshire, England.

ABSTRACT

The integral equation method is related to the random walk modelling that has proved so effective and popular in recent years. The I.E. method, by using simple probability techniques, avoids the inefficient determination of thousands of trajectories in order to build up concentration profiles. In fact it is so simple and efficient it can be run on a conventional programmable calculator. The method is applied to passive material being released from an elevated source within a neutrally stable surface layer over a uniform surface, and also to an instantaneous release when the effect of wind shear is examined. The latter scenario is also studied using random walk techniques and a comparison of the solutions obtained. Agreement is very good, although downwind spread is shown to be quite sensitive to gridlength size in the I.E. method.

INTRODUCTION

Most, but not all, experimental studies of how material injected into the atmosphere is dispersed have, for very practical reasons, been within the surface layer (see Pasquill and Smith, 1983). Paralleling these have been the development of theoretical techniques for predicting vertical and crosswind plume growth. The concepts of an eddy diffusivity K(z) and the parabolic differential equation of eddy diffusion are very familiar. Except in neutral stability conditions the definition of K(z) requires a large measure of empiricism in which diffusion experiments have to be carried out in order to define K(z), which when generalised in terms of basic meteorological surface layer parameters can then be inserted back into the diffusion equation to yield plume behaviour in more general conditions. When the source is very close to the ground, the solutions of the diffusion equation are considered to yield a satisfactory description of vertical plume growth. But when the source is elevated or when considering crosswind dispersion the solutions are theoretically dubious.

Boundary-Layer Meteorology 30 (1984) 143–157. 0006–8314/84/0302–0143$02.25

A second very familiar technique, similarity theory, is based on dimensional considerations, but is strictly applicable only to ground-level sources, and yields only limited information on the distribution of concentration within the plume.

Higher-order closure techniques applied to the basic conservation equations of momentum, heat and injected material are in principle capable of giving considerable detail in a wide variety of conditions, but the method is complex, partially empirical, and laborious.

In recent years the relative simplicity of so-called random walk techniques has been widely exploited very successfully with the ready availability of computers. The basic ideas of the method are presented briefly in the next section. Above all it is a highly versatile method capable of yielding considerable detail of plume behaviour, given the spatial distribution of turbulence and details of the boundaries.

In the third section, we present a development of a rather new technique which is closely related in principle to the random walk method. It has been called the integral equation method (Smith, 1982), since it involves a step by step summing up (or integration) of all possible random motions of the particles in the plume through the definition of probabilities. It is almost as simple to formulate as the solutions have the virtue of being substantially easier and quicker to compute. In some simple situations the method can be programmed and solved using a hand-held programmable calculator.

In the fourth section the results of the integral equation method and the random walk method are applied to diffusion from an elevated crosswind line source in the neutral surface layer. The solutions using the two methods are in close agreement. Solutions are obtained for a continuous release, and for an instantaneous release in which the chief interest is in the particle distribution in the xz plane (i.e. alongwind and in the vertical) at given times. These distributions show the interaction of vertical diffusion and the shear of the horizontal wind.

2. THE RANDOM WALK MODEL.

The random walk method of modelling dispersion consists of simulating numerically the motion of many particles of the pollutant, in order to build up a picture of the concentration distribution. The method has been used previously by many authors, for example Reid (1979) and Ley (1982). The trajectory of each particle is simulated by modelling the evolution of the particle's vertical and downwind velocities over a succession of time-steps. The crosswind motion is not modelled here and so all concentrations from the model must be interpreted as crosswind integrated values. (x_i, z_i) will be used to denote the coordinates of the particle before the ith time step, and (u_i, w_i) will represent the particle's velocity components during the step.

The vertical motion of each particle is modelled by

$$w_{i+1} = w_i\left(1 - \Delta t/\tau(z_{i+1})\right) + \mu_{i+1} \tag{2.1}$$

where Δt is the length of the time-step, $\tau(z)$ is the Lagrangian time-scale for the particle's motion and μ_{i+1} is a Gaussian random number. The physical interpretation of this equation is that over a time Δt the particle loses a small fraction $\Delta t/\tau$ of its momentum to the surrounding air and in return receives a random impulse μ_{i+1}.

In homogeneous turbulence the mean and mean square vertical velocity of a particle of tracer over an ensemble of realisations is equal to zero and σ_w^2 respectively at all times, where σ_w^2 is the Eulerian mean square vertical velocity. If we force our model particles to satisfy this condition, ie if we force $\overline{w_i} = \overline{w_{i+1}} = 0, \overline{w_i^2} = \overline{w_{i+1}^2} = \sigma_w^2$, we obtain

$$\overline{\mu} = 0 \tag{2.2a}$$

$$\overline{\mu^2} = 2\frac{\Delta t}{\tau}\sigma_w^2 + O(\Delta t^2) \tag{2.2b}$$

for the mean and variance of the random variable μ. In inhomogeneous turbulence this derivation is no longer strictly valid; however if σ_w is constant, as in the neutral surface layer considered in this paper, the result is still correct (for a full explanation of this and of how to proceed when σ_w does vary, see Thomson 1983)

There are two assumptions in (2.1). The first is that w_{i+1} depends linearly on w_i as postulated by Smith (1968), and the second is that w_{i+1} depends only on w_i and is not influenced by μ_i, μ_{i-j}, w_{i-j} ($j \geqslant 1$) except through w_i. Hanna (1979) has presented some exerimental evidence in support of the first assumption. These assumptions connot be exactly true since (2.1) implies a discontinuous acceleration whose size $|-w_i\Delta t/\tau + \mu|/\Delta t$ tends to infinity as $\Delta t \rightarrow 0$. However atmospheric accelerations are large and are significantly correlated only over very short times. (of the order of the Kolmogorov time-scale - Monin and Yaglom 1975 p 370, pp 548-549).

The horizontal velocity of the particle is taken to be equal to the mean wind at the particle's height:

$$u_i = \overline{u}(z_i) \tag{2.3}$$

Horizontal turbulence could be easily included in this equation. However, Ley (1982) showed that in models of this type it has only a small effect on the results.

$\overline{u}(z)$, σ_w and τ are taken to be the standard similarity forms appropriate to a neutral surface layer:

$$u(z) = \frac{u_*}{k} \ln \frac{z}{z_o} \tag{2.4a}$$

$$\sigma_w = 1.3\, u_* \tag{2.4b}$$

$$\tau = \frac{k\, u_*\, z}{\sigma_w^2} \tag{2.4c}$$

k - is the von Karman constant (k=0.4)
where u_* is the friction velocity and z_o is the roughness length.

The time-step Δt must be chosen to be short in comparison to τ so that the particle motions are well resolved. Of course this is not possible at all heights if the profile of τ is as given by (2.4c). To solve this problem a modified profile is used in the model:

$$\tau = \begin{cases} \dfrac{ku_*z}{\sigma_w^2} & z > 0.1\text{m.} \\ \dfrac{ku_*0.1}{\sigma_w^2} & 0.1\text{m.} \geqslant z \geqslant 0 \end{cases}$$

By keeping τ constant below 0.1m, the difficulty is avoided. The particles are perfectly reflected at the ground. To avoid using a very small time-step when it is not necessary, Δt is allowed to vary as the particle moves. In the simulations presented in this paper $\frac{\Delta t}{\tau} = 0.05$ was used.

In a real cloud of pollution, the velocities of the fluid elements within the cloud are correlated. In the model however, each particle is assumed to move independently. Hence the concentrations obtained from the model must be regarded as ensemble average values.

The random walk approach has several advantages over the more classical diffusion equation approach. Firstly, and most importantly from a theoretical point of view, the velocities of the particles change smoothly in time, in line with Taylor's (1921) ideas about diffusion. In contrast the diffusion equation assumes very large velocities and very short time-scales and so is only strictly valid for molecular diffusion. Secondly, the method is very flexible and can be adapted to any turbulent flow provided the mean flow and turbulence statistics are known. Finally, because the method does not need an Eulerian grid, there are no problems with artificial diffusion when the cloud width is of the same order as the grid spacing.

A drawback of the technique is that the Lagrangian correlogram is fixed by the model (in homogeneous turbulence it is equal to $\exp(-t/\tau)$ where t is the time-lag between velocity measurements) and does not necessarily agree with the true correlogram. However Pasquill and Smith (1983 §3.5) have shown that dispersion characteristics depend mainly on σ_w and τ and are only weakly dependant on the shape of the correlogram.

THE INTEGRAL EQUATION OF DIFFUSION

The I.E. method is strongly related to the random walk method that has just been described. By introducing probability functions and summing up (or integrating) over all possibilities, considerable computational saving can be achieved, since it is no longer necessary to simulate dispersion by following thousands of distinct particle trajectories one after the other.

The general idea of the method can be best understood by considering its application to the simple case of dispersion in homogeneous turbulence. Later the technique will be applied to the main study of this paper, dispersion within the neutrally-stratified surface layer of the atmosphere.

The method, in one sense, is more primitive than the random walk method as it has been described above, which strictly could be referred to more accurately (but more pedantically) as a Markov chain simulation technique, since it describes turbulent motion correctly as a continuous process. The I.E. method is more akin to the original random walk, or drunkard's walk, concept. It describes diffusion as a collision process in which elements of marked fluid, originating from the source, undergo collisions with other elements, of unspecified origin, totally lose correlation with their previous motion and then travel with constant velocity until they collide again. In this sense the method is strictly analogous to Brownian diffusion of elemental particles. An element will experience collision at a point with a probability that in general depends on the location of the point. The probability is proportional to the local Lagrangian timescale τ . In stationary homogeneous turbulence τ is, of course, an absolute constant, whereas in other situations τ may vary in time or in space. The inherent difficulty of specifying τ is common to all theories of diffusion (even though the difficulty may be implicit since τ may not appear explicity in the theory), and represents the greatest barrier to specifying dispersion in terms of an Eulerian fixed-point description of turbulence, without recourse to empiricism, no matter what theory or technique is used. In this respect the I.E. method is no harder or easier than any other.

The method can recognise in a rational way the statistical properties of the turbulence at every location; in particular it can reflect the probability distribution of the velocity fluctuations, its variance, skewness and finite range.

Before describing the method, it should be noted that it has been simplified quite significantly from its earliest form which has been described by Smith (1984) and by Pasquill and Smith (1983).

The basic ideas of the method can be summarised as follows:

(i) calculations are performed on a grid of equal squares (or rectangles) of sides Δz and Δx.

(ii) collisions occur only at the mid-line of each square: at $\frac{\Delta x}{2}$ in from the side of the gridsquare.

(iii) some elements (or "particles") will collide whilst passing through a given square, others will not. Greater accuracy is achieved when the squares are sufficiently small that only a few percent undergo collision in a single square.

(iv) between collisions, elements travel with constant velocity.

(v) in the simpler forms described here, the turbulence is considered in the cross-flow z-direction only. In the along-flow x-direction, the velocity is given by the mean windspeed u.

(vi) the probability that an element will undergo a collision in a gridsquare is given by a =

(vii) if w is the turbulent velocity fluctuation in the z-direction, then the probability density function P(w) can be assumed zero for $|w| > w_m$, a specified maximum velocity. For $|w| < w_m$, P(w) is supposed given. In many cases it will be close to a Gaussian distribution and may be approximated by the analytical expression:

This satisfies the following:

In the homogeneous turbulence case, in which u and w_m are constant, the gridlengths Δz and Δx are chosen so that $u\Delta z \equiv w_m \Delta x$, and elements cannot diffuse further than one gridsquare either way in the z direction in the time taken to advance Δx in the x-direction.

(viii) Consider the gridsquare labelled (M,N) in Figure 1. M indicates that the gridsquare is M squares up in the z-direction from some arbitrarily selected origin. Similarly N indicates it is N squares along in the x direction from the same origin; which could be the source gridsquare, or could be the square adjacent to the surface immediately below the source in the case of diffusion in the atmospheric surface layer. In general let the source be at $(M_s,0)$, where M_s might be chosen to be zero.

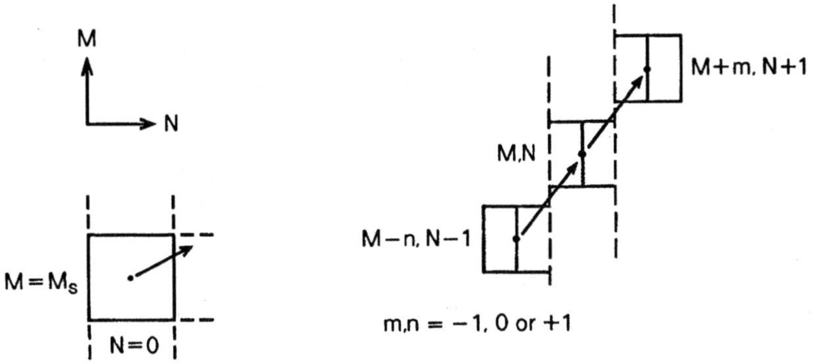

Figure 1. Definition of grid square notation

(ix) Define Q(M,m) as the probability that a "particle" undergoing collision within gridsquare (M,N) will subequently move to gridsquare (M+m,N+1), where m can take the value -1, 0 or 1.

Furthermore, define R(M,N,m) as the probability that a marked "particle" originating from the source, arriving at (M,N) will move to (M+m,N+1), regardless of whether or not it undergoes collision in (M,N).

If D(M,N) is the number of marked particles in (M,N), then the basic equations that govern this diffusion process are

$$D(M,N)R(M,N,m)=D(M,N).a.Q(M,m)+\sum_{n=-1,0,1} f(n,m)\ D(M-n,N-1).(1-a).R(M-n,N-1,n)$$

and $D(M,N+1)=\sum_{n=-1,0,1} R(M+n,N,-n).D(M+n,N)$

where in the first of these equations, the left-hand-side is the number of marked particles moving from (M,N) to (M+m,N+1), the first term on the right-hand-side is the number of particles which undergoing collision in (M,N) then move to (M+m,N+1) whereas the second term represents the sum of those marked particles in (M,N) that do not experience collision there and have come from (M-n,N-1) and proceed to (M+m,N+1). The factor f takes the following values:

$$f(n,m) = \begin{cases} 0.6 & \text{if } |n-m| = 0 \\ 0.4 & \text{if } |n-m| = 1 \text{ and } |n| = 1 \\ 0.2 & \text{if } |n-m| = 1 \text{ and } |n| = 0 \\ 0 & \text{if } |n-m| = 2 \text{ or if } |n| > 1 \text{ or } |m| > 1. \end{cases}$$

These values can be seen to preserve a uniform concentration distribution. The second equation sums up all the marked particles arriving in (M,N+1) from (M+1,N), (M,N) and (M-1,N).

In homogeneous turbulence, when P(w) given above is a good approximation,

$$Q(M,0) = \int_{-w_m}^{w_m} \left(1 - \frac{|w|}{w_m}\right)\frac{15}{16w_m}\left(1 - \frac{w^2}{w_m^2}\right)^2 dw = \frac{11}{16} = 0.6875$$

and $Q(M,\pm 1) = 0.15625$

At the source $D(M_s,0)$ = number of particles released, say 10^4.

and $R(M_s,0,m) \equiv Q(M_s,m), = Q(m)$ in homogeneous turbulence.

The choice of a, and hence Δz and Δx, is arbitrary, but to achieve reasonable smoothness it should be at most 0.1. Smaller values will give smoother solutions at the expense of more computing time.

Concentration is obtained by dividing D(M,N) by the wind speed u(M). The resulting concentration distribution for homogeneous turbulence is very quickly derived, and has moments in very close agreement with analytically - derived moments (see Smith 1984). Thus the second moment is in virtually complete agreement with Taylor's statistical theory result with an exponential Lagrangian correlogram with the same timescale :

$$\sigma^2(x) = \frac{2\tau\sigma_w^2}{u}\left(x + u\tau(\exp(-x/u\tau) - 1)\right)$$

The method gives more than just the moments of course: the ensemble average concentration distribution results, and in this case the distribution is very close indeed to being Gaussian.

Diffusion in the atmospheric neutral surface layer

(i) A continuous source

The formulation is very similar to that outlined above, the only basic difference is that both u and τ are now functions of z (or M).

A log-law is assumed for u: $u = \frac{u_*}{k}\ln\frac{z}{z_o}$

and a linear variation for τ : $\tau = ku_*z/\sigma_w^2$

Since $a = \Delta x/u\tau$ it follows that a also varies with z:

$$a = \Delta x\,\sigma_w^2/(u_*^2 z \ln z/z_o)$$

It is assumed that $\sigma_w = 1.3u_*$ following the evidence from surface layer studies in the field (Pasquill & Smith, 1983). Furthermore it will be assumed that a = 0.1 at $z = z_* \equiv 1000\,z_o$ (where z_o is the surface roughness).

Consequently $\Delta x = a u \tau = \left[\dfrac{1000\ z_0 \ln 1000 \times 0.1}{1.3^2}\right] = 408.743\ z_0$

and $\Delta z = \left[\dfrac{w_m \Delta x}{u(z_*)}\right] = a\ w_m \left[\dfrac{k\ u_*\ z_*}{\sigma_w^2}\right] = 81.4083\ z_0.$

The midpoint of the first gridsquare (M=0) above the surface is at a height $\frac{\Delta z}{2}$ and that of the gridsquare M is at a height $(M+\frac{1}{2})\Delta z$.

Thus $a = \left[(M+\frac{1}{2}) \times .11785056 \times \ln(81.4083(M+\frac{1}{2})) \right]^{-1}$.* In practice if a exceeds 1 in the very lowest layers, then a is put equal to 1.

$$Q(M,m) = \frac{15}{16}\left(c_m - Ac_m^3 + Bc_m^5\right) \text{ for } M \geqslant 1 \text{ and } m = -1,0,1$$

where $c_m = 0.144765 \ln (81.4083(M+\frac{1}{2}))$,

$B = \frac{1}{80}$ and $A = \frac{1}{6}$ if $m = 0$

and if $m = \pm 1$: $Q(M,m) = \frac{1}{2}(1-Q(M,0))$

For M = 0 put $Q(0,1) = 0.2603$
$Q(0,0) = 1-Q(0,1) = 0.7397$
and $Q(0,-1) = 0$, so that particles are fully reflected at the surface.

Note that except for $Q(0,m)$: $Q(M,-1) \equiv Q(M,1)$

The solution can now proceed using the two basic equations given in the previous section. Results are given later.

(ii) An instantaneous release (or puff)

The method can also be applied to an instantaneous release from a source near the ground in the neutral surface-stress layer. The aim is to learn something about the spatial distribution of particles in such a puff, as a function of time, resulting from the interaction of vertical turbulent spreading and the logarithmic wind shear.

It is suggested that because the diffusive spread is governed by small-scale eddies which are largely controlled and influenced by the proximity of the ground, there is relatively very little difference between the particle distribution for any single realisation of a release and that for the ensemble mean distribution. However if horizontal turbulence were to be taken into account, (which is no problem in principle), then the assumption would be less acceptable since variable larger-scale eddies would create significant differences from one realisation to the next.

*note that here we assume collision occurs at M+]/2

Let Δz remain as before. Since we are concerned with spread at a given time, rather than a given distance, Δx will be replaced by Δt. The probability of collision $a = \Delta t/\tau$ is again put equal to 0.1 at $z = z_*$.

Thus $$\Delta t = 81.4083 \frac{z_0}{w_m}$$

and $$a(M) = \frac{1.22837}{(M+\frac{1}{2})} \qquad M \geq 1, \qquad a(0) = 1.$$

Q(M,m) takes the same two values found for homogeneous turbulence

i.e. $Q(M,0) = 0.6875$ and $Q(M, \pm 1) = 0.15625$

Let the suffix k denote the kth package which contains particles that after a specified time $N\,\Delta t$ have travelled the same distance x_k. Suppose this package has moved from (M-m,N-1) to (M,N). Part of this package (ie a sub-package) now moves from (M,N) to (M+m,N+1). The laws governing this are:

$$D_k(M,N).R_k(M,N,m) = D_k(M,N).a(M).Q(m) + D_k(M-n,N-1).(1-a(M)).R(M-n,N-1,n).f(m,n$$

with meanings the same as those described earlier. f(m,n) values are also identical to their previous values.

Furthermore $D_k(M,N) = D_k(M-n,N-1).R(M-n,N-1,n)$

so that $$R_k(M,N,m) = a(M)Q(m) + (1-a(M))f(m,n) \qquad n=0,\pm 1$$

Now $$x_k(N+1) = x_k(N) + u((M+\tfrac{1}{2}(m+1))\Delta z) \times \Delta t.$$

Since $u = \frac{u_*}{k} \ln \frac{z}{z_o}$, if X is defined as x/z_o, then:

$$X_k(N+1) = X_k(N) + 59.1719 \ln \left[81.4083\,(M+\tfrac{1}{2}(m+1))\right]$$

At each new timestep the number of packages trebles and the number would soon become unmanageable if some sort of grouping was not imposed. Various options are available. The following was selected for its relative simplicity:

At N, find the maximum $X(N) = X_{max}$ and the minimum $X(N) = X_{min}$, over <u>all</u> packages regardless of M. Then define $\delta = (X_{max} - X_{min})/25$.

At each level (each M) group together all packages which have X_k lying between $(X_{min} + r\,\delta)$ and $(X_{min} + (r+1)\,\delta)$ where r runs from 0 to 24.

In this grouping together, the following rules are to be observed:

(i) if more than one package lies in an interval then $D(M,r) = \sum D_k(M,r)$, ie the new package conserves the number of particles from all the contributing packages, and

(ii) $D(M,r)\ R(M,N,m) = \sum D_k(M,r)\ R_k(M,N,m)$, and

(iii) $X(M,r) = \dfrac{\sum D_k X_k}{D(M,r)}$, averages the values of X_k.

If the packages are kept in order of X at each level, then in marching forward one timestep, the order will be preserved for those with m = 0, and those additional subpackages entering the level from levels immediately above and below (corresponding to m = -1 and +1) can be inserted quickly if their X's are tested against the X's (for m = o) with the same r. Results will now be presented.

Results

(a) The continuous source

Solutions of both the integral equation (I.E.) and random walk (R.W.) methods have been obtained for an elevated source S_1 at a height 2.5 Δz and for a near ground level source S_o at 0.5 Δz. The gridsizes, Δz and Δx, are as given in the previous section. Concentration profiles for a continuous release at S_1 are given in Figure 2 in non-dimensional form and in equivalent dimensional form for the following particular values of the parameters:

$z_o = 0.02\text{m}$, $\Delta z = 1.628\text{m}$, $\Delta x = 8.175\text{m}$, h_s = height of $S_1 = 4.07\text{m}$, $u_* = 1\text{ms}^{-1}$ and $u(h_s) = 13.29\text{ms}^{-1}$

The profiles are given at four downwind distances; N = 2 corresponds to x = 2Δx, N = 6 to x = 6 Δx, and so on. By N = 19 the profile is virtually identical to that from the near-groundlevel source S_o under the same conditions but the latter taken at a greater distance downwind (at approximately N = 30). The shape of the vertical profile is by then closely consistent with a simple form:

$$C(z) = C(o) \exp\left[-az^s\right]$$

where the value of s is found to be 5/4.

In general the solutions from the two methods are in very close agreement, although small but observable differences are often evident very close to the ground where the I.E. method, because it obtains its solutions on the basis of a grid, reflects the details of the rapid u(z) and τ (z) variations there rather too simplistically. The two

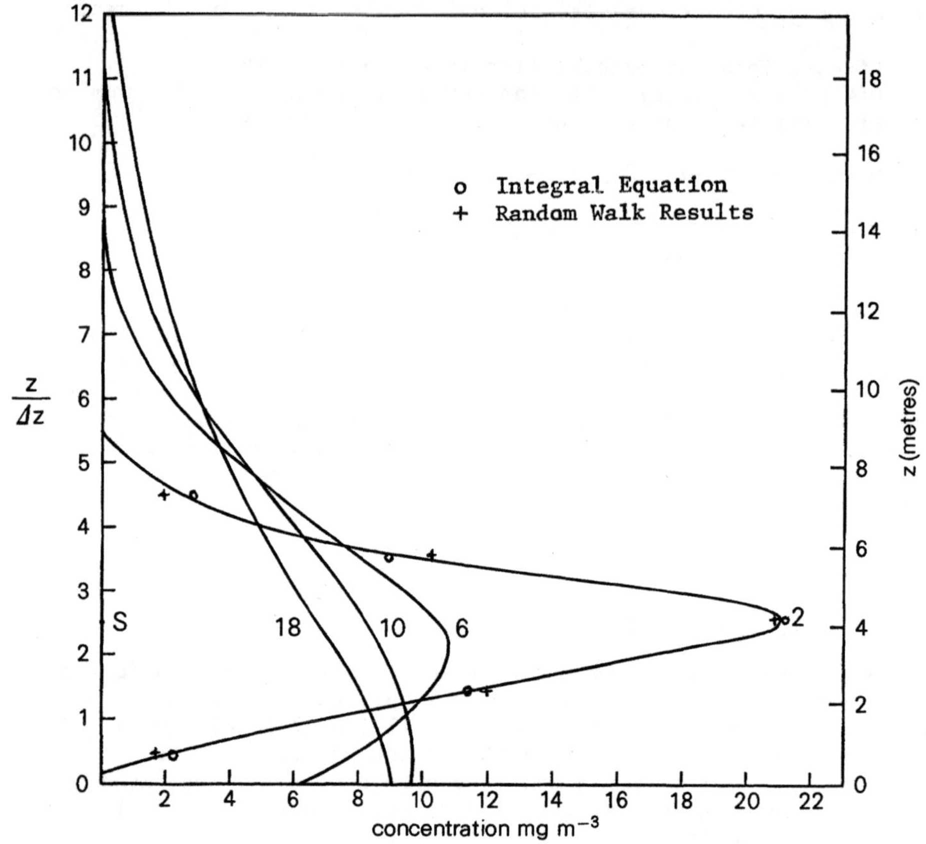

Figure 2. Concentration profiles at four values of N downwind from a source at $z = 2{\cdot}5\,\Delta z$ for a release rate of 1 gs^{-1} m^{-1}

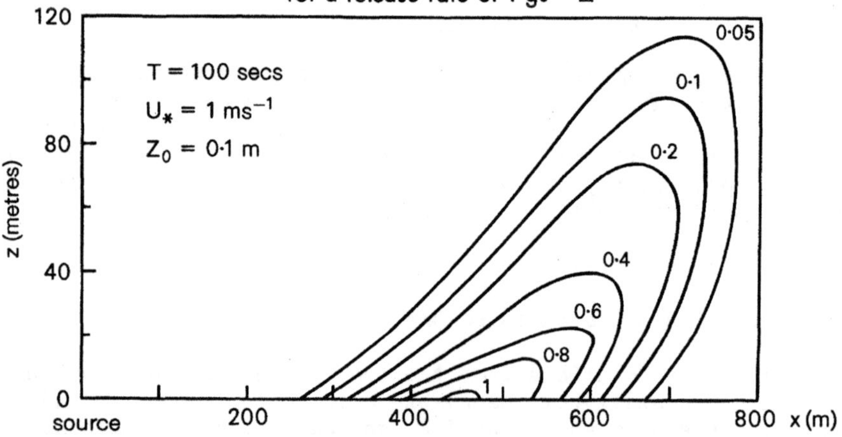

Figure 3. Contours of concentration c/c max within a cloud released instantaneously from a groundlevel source.

solutions are shown at N = 2 by a series of point values. Away from the lowest gridsquares the differences are very small, especially for larger values of N.

Table 1 gives values of σ_z/z_o for continuous releases from S_o and S_1 for different values of x/z_o, and it can be seen that σz grows almost parabolically for small x, where wind shear is dominant, but approaches a more linear variation at greater x when the linear form of τ predominates over the wind shear.

N	0	1	2	4	8	12	16	19
x/z_o	0	409	817	1635	3270	4905	6540	7766
σ_z/z_o (S_o)	40.7	67.1	85.0	115	168	216	259	289
σ_z/z_o (S_1)	203.5	207	216	238	281	320	357	384

Table 1. Plume growth from sources S_o at $\frac{1}{2}\,\Delta$ z and S_o at 2.5 Δz in a neutrally-stratified surface layer. σ_z is the root mean square particle elevation above ground.

(b) An instantaneous release

Solutions have been obtained using both methods for an instantaneous release from S_o , at a height h_s = 0.5 Δ z. They show the marked interaction of vertical diffusion and the shear of the horizontal wind u(z). An example is given in Figure 3 which shows contours of concentration after 100 seconds, obtained from the R.W. model. Table 2 compares the basic statistics at three different times from the two methods. The differences are small except for σx, the overall alongwind root-mean-square spread of all the particles in the cloud. As before the reason for the difference appears to be the relatively limited resolution in the lowest grid layer used in the I.E. method. A particle diffusing into this layer in the random walk model can closely approach the surface, move downwind very slowly (since u(z) is very small there) and and remain there for some time since τ is also very small. In contrast in the I.E. method the values of u and τ are defined by the values at z = 0.5Δz and therefore may under-estimate the alongwind shearing effect in the single layer, which the R.W. method captures.

The error is largely a matter of gridsize, of course, and reducing Δ z and Δx to a fifth of their previous values gives the value of σx in column 6 in Table 2, which is in much better agreement with the R.W. value, albeit at the expense of an increase in computational effort. These differences in σx are much less apparent if σx is calculated, not for the whole plume, but for a given value of z away from the lowest grid layer. For example at N = 19, M = 10, σx (R.W.) = 33 metres and σx (I.E.) = 32.4m with the original grid sizes, for z_o = 0.1, u_* = 1ms^{-1} and h_s = 4.07m

N	Method	$\bar{z}$	σ_z (taken about $\bar{z}$)	$\bar{x}$	σ_x	$\sigma_x/\bar{x}$	Plume tilt
	Units : metres: z_o=0.1, u_*=1.0ms^{-1}, h_s=4.07m						
3	R.W.	6.0	4.35	65.3	10.4	0.16	~1
	I.E.(orig.Δz)	7.7	5.56	70.3	6.7	0.095	
	I.E.($\frac{1}{5}$x origΔz)	6.0	4.68	-	10.7	-	
6	R.W.	8.5	7.16	134	25.7	0.19	~1
	I.E.(origΔz)	10.9	8.79	147	16.1	0.11	
19	R.W.	20	19	474	92	0.19	~1
	I.E.(origΔz)	22	19	510	64	0.13	

Table 2. Values of the main statistical parameters describing the cloud originating from an instantaneous release at N = 0, $z = \frac{1}{2}\Delta z$. Plume tilt is defined as the tangent of the best eye-fit linear regression line through the cloud.

The variation of $\bar{x}$ with time is rather faster than linear, as shown in Table 2, due to the increase in average windspeed as the cloud deepens. The variation of $\sigma_x/\bar{x}$, shown in the same Table, is almost constant beyond N = 6 at approximately 0.2, but below 6 increases with increasing N. Similarly the plume tilt above the very lowest grid-layers appears to be almost constant at 45° (tangent = 1) out to N = 19.

References

Hanna S.R. 1979 "Some statistics of Lagrangian and Eulerian Wind Fluctuations" J Appl Met 18, 518-525

Ley A.J. 1982 "A Random Walk Simulation of Two Dimensional Diffusion in the Neutral Surface Layer" Atmos Envir 16, 2799-2808

Monin A.S. and Yaglom A.M. 1975 "Statistical Fluid Mechanics" Vol 2 M.I.T. Press, Cambridge, Massachusetts.

Pasquill F. and Smith F.B. 1983 "Atmospheric Diffusion". 3rd Edition), Ellis Horwood Ltd, Chichester, England.

Reid J.D. 1979 "Markov Chain Simulations of Vertical Dispersion in the Neutral Surface Layer for Surface and Elevated Releases". Boundary Layer Met 16, 3-22.

Smith F.B. 1968 "Conditioned Particle Motion in an Homogeneous Turbulent Field" Atmos. Envir. 2, 491-508.

Smith F.B. 1984 "The Integral Equation of Diffusion", Air Pollution and Its Application III, Proceedings of the 13th International Technical Meeting, 1982, in France Ed. by C. De Wispelaere. Plenum Press, New York.

Taylor G.I. 1921 "Diffusion by Continuous Movements" Proc Lond Math Soc, Ser 2, 20, 196

Thomson D.J. 1983 "Random Walk Modelling of Diffusion in Inhomogeneous Turbulence". Internal Met Office Note T.D.N.144.

Questions, answers and comments

M. POREH: How universal is your method when applied to other stability classes?

F.B. SMITH: The scheme can be diversified by adding a bias velocity for stable or unstable layers, similar to Random-Walk-Method. P(w) is simply modified to include bias. Therefore, the scheme can be applied through the boundary layer, if we define τ , σ_w, etc. It is as versatile as the Random-Walk-Method.

THE RANDOM FORCE THEORY: APPLICATION TO MESO- AND LARGE-SCALE ATMOSPHERIC DIFFUSION

F. A. Gifford
109 Gorgas Lane
Oak Ridge, TN 37830

ABSTRACT: The physical basis for the random force theory of atmospheric diffusion is briefly reviewed. The random force equation has usually been introduced as an atmospheric analogy to Brownian motion. Here it is derived by assuming a type of Rayleigh friction in the Lagrangian equations of atmospheric motion. Consequences, including the Lagrangian correlation and spectrum, the particle-particle intercorrelation, and cluster and plume dispersion and meandering, are derived and compared with atmospheric observations. The characteristic time-scale of atmospheric diffusion is shown to be governed by the Coriolis effect, a result in good agreement with meso- and large-scale observations.

1. INTRODUCTION

Obukhov (1959) proposed that atmospheric diffusion could be represented by a continuous Markov process in the phase space consisting of an air particle's coordinates and velocity. The implications for turbulent diffusion theory of this proposition, which are far reaching, have been developed and summarized by Lin and Reid (1963), Monin and Yaglom (1967), and most recently in the detailed review by Yaglom (1977). These center around three important theoretical consequences. The first is that diffusing particles obey a differential equation of motion of the type of Langevin's equation for the velocity of Brownian motion,

$$dv/dt + Bv = a(t) \tag{1}$$

where v is (one component of) the particle's velocity, B^{-1} is the integral or outer time scale of the turbulence process, and a(t) is a random acceleration that is assumed to have a flat spectrum and zero mean. Application of this equation to the theory of atmospheric diffusion has been discussed by Chadam (1962), Krasnoff and Peskin (1971), and Levin (1971).

As is well known, Eq. (1) also implies the following velocity relationship, from which it can easily be derived (Gifford, 1982):

Boundary-Layer Meteorology 30 (1984) 159–175. 0006-8314/84/0302-0159$02.55

$$v(t+\Delta t) = Cv(t) + v'(t) \tag{2}$$

where v' is a random and Cv a correlated part of the velocity. This linear velocity relationship was first applied to the theory of atmospheric diffusion by Smith (1968), and Hanna (1979) demonstrated that it compares well to atmospheric turbulence data. A number of numerical, Monte-Carlo or random-walk type models of diffusion in the planetary boundary layer (PBL) have recently been proposed based on Eq. (2), which can be taken as the definition of a simple, continuous Markov process; these include the studies by Hanna (1979), Reid (1979), Durbin (1980), Sawford (1982b), Ley and Thompson (1983), and Thompson (1983).

The third consequence of Obukhov's assumption, that the diffusing particle's position and velocity probability density functions obey the Fokker-Planck equation, has not yet been exploited much in atmospheric applications [cf. however Levin (1971, 1973)] but is potentially quite important, since it defines the particle concentration distribution.

The present paper is concerned with the application of Eq. (1), the random-force, Langevin, or Brownian motion equation, to the problem of horizontal atmospheric diffusion to meso- and larger scales, as recently proposed by Gifford (1982). The atmospheric turbulent motions that diffuse particle clusters are driven by the large-scale, primarily horizontal, quasi-geostrophic motions of the atmosphere and, through these, ultimately by the pole-equator imbalance of solar radiation received by the earth's surface. Diffusive turbulent motions are observed to exist over a tremendous range of atmospheric scales, from the very small scale of dissipative motions to that of the great, planetary waves. Diffusive motions are particularly strong at the scale of the daytime, convective PBL (i.e. to about 1 km), and result in vigorous, highly variable and strongly stability-dependent diffusion there. At yet larger scales of motion, through that of the "spectral gap" (about 3 to 10 km) and beyond into the mesometeorological range (about 100 km), turbulence becomes increasingly two dimensional in the presence of the generally stable tropospheric lapse rate. A cascade of turbulent kinetic energy toward small scales, as described by Richardson (1926), which is accompanied by rapid diffusion at a rate increasing with scale (i.e. with cluster size, or plume width), can exist only up to a range of atmospheric motions that retains some 3-D turbulence character. At yet larger scales, up to the synoptic motions, there is a range of 2-D "geostrophic turbulence" (Charney, 1971) in which a down-scale cascade of enstrophy (m. s. vorticity), but not of eddy kinetic energy, can occur. One of the principal objects of the present paper is to show that there is a well-defined outer time scale of eddy diffusion which separates these large-scale, quasi-geostrophic motions from the 3-D, diffusive motions of smaller scale.

2. PROPERTIES OF THE RANDOM FORCE MODEL (RFM)

It seems most descriptive to refer to Eq. (1) as the "random-force" equation, a terminology introduced by Novikov (1963). This avoids any implication, present in the alternatives, that actual Brownian motions are being considered. The atmospheric motions to which Eq. (1) is to be applied are thus just atmospheric turbulence as ordinarily defined, in which the motion is assumed to be microscopically continuous. This equation is a (one-dimensional) linear, stochastic differential equation which can be solved for the horizontal (lateral) component of the particle velocity, v, by elementary methods, with the following result:

$$v(t) = v_0 \exp(-Bt) + \exp(-Bt)\int_0^t \exp(Br)a(r)dr \tag{3}$$

The quantity v_0 is the initial particle velocity, i.e. that of the air motion at the source at the initial time of the diffusion process. It can be thought of as the velocity that results from the operation of the assumed turbulence process up to time zero, i.e.

$$v_0 = \int_{-\infty}^{0} \exp(Br)a(r)dr \tag{3a}$$

2.1. Velocity Statistics

Many useful Lagrangian, or particle-attached turbulence and diffusion statistics follow directly from Eq. (3) if the usual assumptions of statistical homogeneity, stationarity, and zero mean ($\overline{v} = 0$) are applied. By squaring and forming the ensemble (overbar) average, the Lagrangian autocovariance is found to be

$$\overline{v(t)v(t+\Delta t)} = v_0^2 \exp(-Bt)\exp(-B(t+\Delta t)) + \\ +\exp(-B(2t+\Delta t))\int_0^t\int_0^{t+\Delta t} \exp[B(r+s)]\overline{a(r)a(s)}drds. \tag{4}$$

The double integral term can be evaluated by recognizing that the acceleration correlation $A = \overline{a(r)a(s)}$ drops to zero very quickly, in a time much less than 1/B. This assumption is a particularly good one in the atmosphere, where the time scale of A is small because it is governed by the atmosphere's molecular viscosity, ν, whereas the Lagrangian time-scale of the atmospheric turbulence, $t_L = 1/B$, is as we shall see many orders of magnitude larger, being governed by the eddy viscosity of the atmosphere. The result is that

$$\overline{v(t)v(t+\Delta t)} = \exp(-B\Delta t)[v_0^2\exp(-2Bt) + \overline{v^2}(1-\exp(-2Bt))]; \tag{5}$$

notice that it is time dependent. The velocity variance $\sigma_v(t)$ is formed by equating Δt to zero;

$$\sigma_v^2(t) = v_0^2\exp(-2Bt) + \overline{v^2}(1-\exp(-2Bt)) \tag{6}$$

Thus the velocity variance experienced by a diffusing particle is also a function of time, even in the assumed stationary flow, and becomes equal to the flow variance, $\overline{v^2}$, only after a sufficiently long diffusion time, t, on the order of several times $t_L = 1/B$.

By forming the quotient of Eqs. (5) and (6) it follows that the Lagrangian, particle-velocity autocorrelation function $R_L(\Delta t)$,

$$R_L(\Delta t) = \overline{v(t)v(t+\Delta t)}/\sigma_v^2(t) = \exp(-\Delta t/t_L) \tag{7}$$

is exponential. This result, which has usually been introduced as an assumption into models based on Eq. (2) (where $C = R_L$), is thus seen to be an analytical consequence of the RFM and moreover to be valid only in the sense defined by the time-dependent variance and covariance of Eqs. (6) and (5).

The kinetic-energy spectrum of the horizontal motion, $\overline{v^2}F(n)$, corresponding to Eq. (7) is given by

$$\overline{v^2}F(n) = 4\overline{v^2}\int_0^\infty R_L(t)\cos 2\pi nt\,dt = 4\overline{v^2}t_L[1+(2\pi t_L n)^2]^{-1}, \tag{8}$$

where n is frequency in Hz, and the customary meteorological normalization, $\int_0^\infty F(n)dn = 1$, has been applied. Thus the RFM has an energy spectrum that behaves as $(\overline{v^2}/\pi^2 t_L)n^{-2}$ over a broad range of frequencies, in agreement with the well-known equilibrium-theory result that $\overline{v^2}F(n) \propto n^{-2}$ [cf. Tennekes and Lumley (1972)]. Both the spectrum and the correlation are defined, in the RFM, explicitly in terms of the large-scale parameters of the flow, namely $\overline{v^2}$ and t_L.

2.2. Displacement Statistics

Since v = dy/dt, a second integration of Eq. (3) provides the lateral displacement of a particle, y, from which ensemble average displacement variances, defining the forms of clusters and plumes on an instantaneous basis, can be formed. These are

$$\sigma_{yT}^2(t) - \sigma_o^2 = 2\overline{v^2}t_L^2[T-(1-e^{-T}) - ((\overline{v^2}-v_o^2)/2\overline{v^2})(1-e^{-T})^2] \tag{9}$$

$$\sigma_{yR}^2(t) - \sigma_o^2 = 2\overline{v^2}t_L^2[T-(1-e^{-T}) - (1/2)(1-e^{-T})^2] \tag{10}$$

$$\sigma_{yC}^2(t) = 2\overline{v^2}t_L^2[(v_o^2/2\overline{v^2})(1-e^{-T})^2] \tag{11}$$

where $T=t/t_L$, σ_{yT} is the total (instantaneous) particle dispersion from the y=0 axis, σ_{yR} is the dispersion calculated relative to an ensemble-averaged cluster's centroid, and σ_{yC} is the dispersion of that centroid from the y=0 axis. Notice that the total m.s. dispersion equals the sum of the relative cluster dispersion and that of the centroid,

$$\sigma_{yT}^2(t) = \sigma_{yR}^2(t) + \sigma_{yC}^2(t) \tag{12}$$

as of course should be the case.

Equations (9)-(11) apply to small, finite sources, approximating points, over which the initial velocity values, v_o, can be taken to be equal. Lee and Stone (1983) extended these results to include uniformly distributed sources of arbitrary finite widths and release times. The displacement variances they derived have exactly the same forms as Eqs. (9)-(11), but the constant coefficients of each of the right-hand terms contain the space- and time-scales of Eulerian correlations in addition to $\overline{v^2}$ and v_o.

If Eq. (9) is further averaged over all possible conditions of the initial velocity, v_o, the result is

$$\sigma_{ya}^2(t) - \sigma_o^2 = 2\overline{v^2}t_L^2[T-(1-e^{-T})]. \qquad (13)$$

This has the same form as the equation for the averaged diffusion from a continuous point source that was first derived by Taylor (1921), apart from the initial σ_o. Figure 1 displays these various dispersions, Eqs. (9), (10) and (13), as functions of the total turbulence energy, $\overline{v^2}$, and the Lagrangian time scale, t_L. From these curves it can be seen that the RFM dispersion equations contain both the small and large time asymptotic limits of statistical diffusion theory, proportional to T and $T^{1/2}$, as well as an accelerating range of relative diffusion for which σ behaves like $T^{3/2}$.

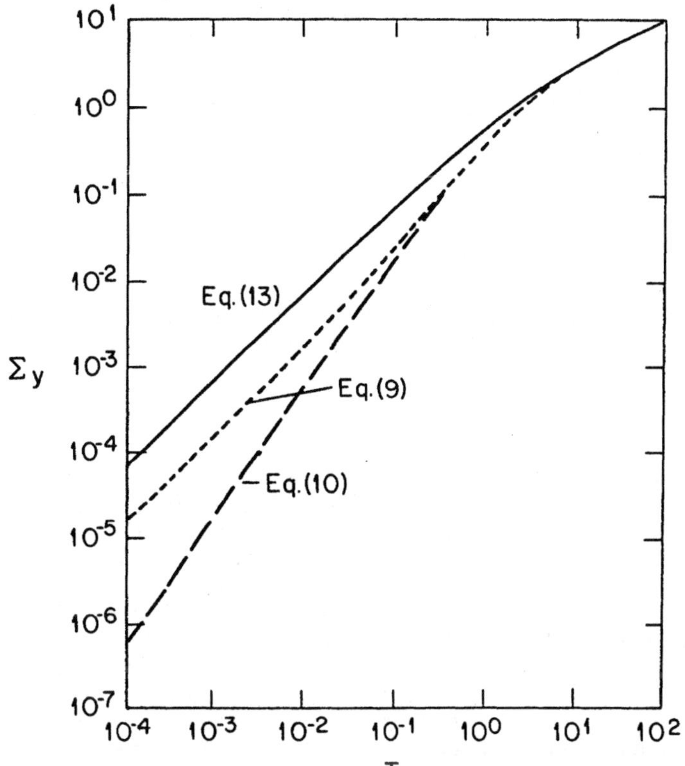

Figure 1. Non-dimensional horizontal dispersions, as functions of diffusion time, $T=t/t_L$, from a source; $\Sigma y^2 = (\sigma_y^2 - \sigma_o^2)/2\overline{v^2}t_L^2$. Curves are shown for: total instantaneous plume or cluster dispersion, σ_{yT}, from the axis y = 0, Eq. (9); dispersion relative to the instantaneous plume or cluster centroid, σ_{yR}, Eq. (10); and fully averaged, or Taylor dispersion, σ_{ya}, Eq. (13). Eq. (9) is illustrated for the value $(\overline{v^2}-\overline{v_o^2})/\overline{v^2} = 0.95$.

Any other particle displacement or velocity statistic can be derived from Eq. (3) in a similar way. The average particle displacement, $\bar{y}$, for instance is readily found to be given by

$$\overline{y}(t) = (v_0 t_L)(1-e^{-T}) \tag{14}$$

The single assumption made in deriving all these results from Eq. (1) is that the time scale of the acceleration correlation, which arises in forming the ensemble-averaged velocity squared, lagged product, and integral from Eq. (3) is very short compared with any diffusion time of interest. This is likely, as has been mentioned, to be an excellent assumption in the atmosphere for the purpose of determining such quantities as R_L, $F(n)$, and the total and averaged dispersions, σ_{yT} and σ_{ya}, which depend only on the large-scale turbulence parameters and on v_0. The relative dispersion, σ_{yR}, and the centroid dispersion, σ_{yC}, depend however on the intermediate spatial scale defined by the cluster's size at any diffusion time. Thus the ability of the RFM to describe relative diffusion needs further consideration, and we will return to this question in Section 5.

3. THE NATURE OF LARGE-SCALE ATMOSPHERIC DIFFUSION

The two large-scale parameters of the RFM, $\overline{v^2}$ and t_L, define an effective, or eddy viscosity of the atmosphere, $K = \overline{v^2} t_L$, and a characteristic length, L;

$$L \sim (\overline{v^2} t_L^2)^{1/2} \sim (K t_L)^{1/2} \sim \varepsilon^{-1/4} K^{3/4} \tag{15}$$

where ε is the average eddy-energy transfer rate of the atmosphere. The last of these relations is just a form of the Richardson-Obukhov "4/3-law" of diffusion. The length, L, characterizes the smallest size range of the large-scale, quasi-horizontal motions of the atmosphere that is capable of maintaining the energy-transfer rate to the small, 3-D PBL eddies, given the large-scale atmospheric "viscosity", K, as was pointed out by Monin (1972). By comparing the RFM cluster diffusion formula, Eq. (10), with a wide range of tropospheric diffusion measurements, Gifford (1982) found that, typically, $K \approx 5 \times 10^4 m^2 s^{-1}$ and $t_L \approx 10^4 s$. Similar values were found in a series of independent comparisons of Eq. (10) with long-range diffusion data by Barr (1983). From these values it follows that $L \sim 20$ km, which is near the scale of the so-called "spectral gap", or mesometeorological minimum, of the atmospheric energy spectrum.

Scales of motion greater than L, up to the synoptic, are quasi-horizontal and two-dimensional. In this range the largest eddies lose significant energy to the next smaller ones in a time $t_L \sim L^2/K \sim \varepsilon^{-1/3} L^{2/3}$ (see e.g. Monin, 1972, Golitsyn, 1973), which is the relaxation time of the diffusion process. Using the above values of K and t_L it is found that $\varepsilon \approx 5\ cm^2 s^{-3}$. This value of ε agrees well with other estimates of average tropospheric energy transfer that have been made by various authors using quite different lines of reasoning. Of course atmospheric motions are not characterized by a single, constant value of energy transfer, ε . On the contrary this quantity is known to vary widely in the troposphere, especially with height in the PBL, but also everywhere according to the local conditions of the turbulence. Nevertheless its average value, which is essentially determined by external properties such as astronomical quantities and the composition of air, can be expected to exert an influence that increases with the scale of atmospheric diffusion.

At scales smaller than L, in the range of typical 3-D PBL turbulence, energy is derived from the larger 2-D eddies both indirectly from synoptic-scale motions and directly, from the instability of the vertical shear layer created above the earth's surface by the frictional drag of the large-scale motions. The latter effect operates, however, at the comparatively slow rate of viscous decay, ν/t_L^2, which is much smaller than ε . Consequently the bulk of the energy that drives the small-scale, 3-D PBL turbulence is supplied by quasi-horizontal diffusive atmospheric motions whose scale is greater than L. These are maintained by the large, synoptic-scale motions from which they derive their energy. The exact nature and properties of this scale of motions, intermediate between the lower end of the synoptic scale and the 3-D PBL turbulence, as Golitsyn (1973) points out, have been very little studied. A number of interesting possibilities exist, including that of up-scale transfer of energy from large convective storms, as recently described by Lilly (1983). But whatever these may ultimately prove to be, they must necessarily involve a net transfer of turbulent kinetic energy to the boundary layer. The properties of diffusion are in this way always being directly influenced to some extent by the larger-scale (i.e. $>$ L) atmospheric motions.

The validity of applying the RFM to the atmosphere hinges on whether there is a wide range of scales of motion for which the diffusion is dominated by the outer time-scale of turbulence, t_L, defining an energy cascade in Richardson's sense. Instantaneous plume and cluster spreading in this range should occur at accelerating rates; i.e.

$$\sigma_{yR} \propto t^p, \quad 1 < p \leqq 3/2. \tag{16}$$

Kinetic energy, created at large scales, is transferred in this region to smaller and smaller scales and, finally, dissipated at molecular scales, to maintain the global energy balance.

Recent, detailed series of observations of the width of smelter plumes in Australia, to very large downwind distances (Carras and Williams, 1981), taken together with earlier diffusion data sets, suggest (Gifford, 1983) that accelerating diffusion exists for 10 to 15 hours or more of downwind travel from the source, i.e. to length scales of up to several hundred kilometers, and that the dominant time scale, t_L, is on the order of 10^4s. The Australian plume-width data are plotted in Figure 2, in the form of average values of the instantaneous plume dispersion, σ_{yR}, as a function of downwind travel time, or plume "age", t. The average for each point is over 15 contiguous σ_{yR}-values from the data set as ordered in time. Following a region extending to $t \approx 2$ hrs that is evidently dominated by source effects and active PBL convection ($p \sim 1$), a region of accelerating diffusion ($p > 1$) occurs, extending to

10-20 hrs, beyond which the diffusion rate decreases again, to $p<1$. The presence of an accelerating rate of plume spreading in a similar range was also noted independently in the earlier Australian smelter plume data studied by Bigg, Ayers, and Turvey (1978). The possibility of applying the RFM to horizontal diffusion to fairly large scales is encouraged by these comparisons, in that the necessary broad range of accelerating diffusion evidently exists.

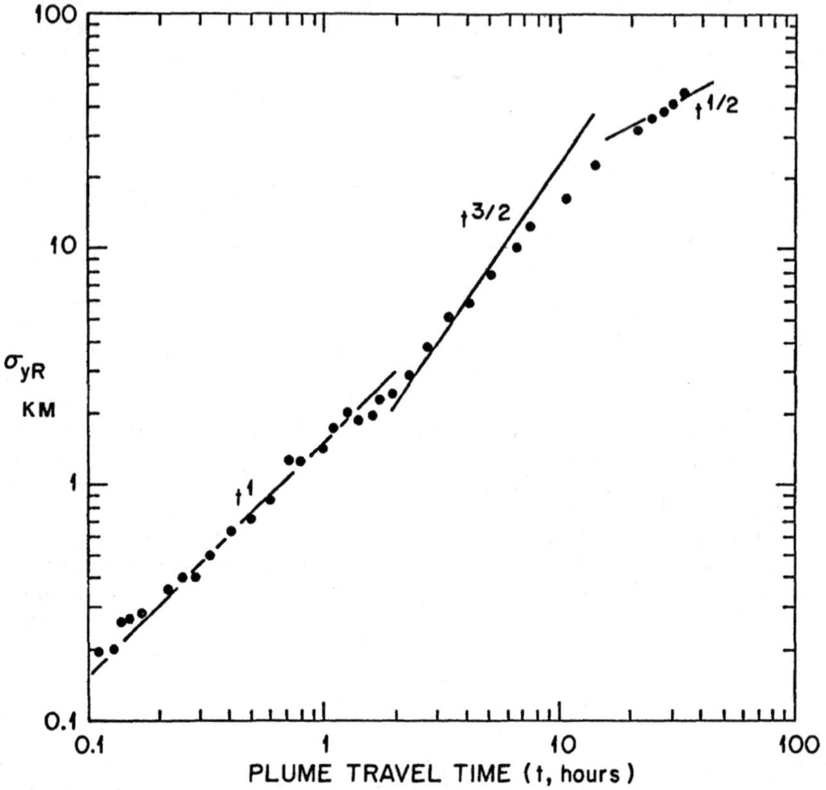

Figure 2. Experimental values of σ_{yR} based on Australian smelter plume-width measurements by Carras and Williams (1981). Each point is an average of 15 contiguous data points (cf. Gifford, 1983).

4. RFM BASED ON ATMOSPHERIC DYNAMICAL EQUATIONS

Lagrangian, (i.e. particle-attached) equations for the motion of an air particle in the atmosphere can be written in the following form (see e.g. Kao, 1962)

$$d^2x/dt^2 - f_z dy/dt = P_x + F_x \tag{17}$$

$$d^2y/dt^2 + f_z dx/dt = P_y + F_y \tag{18}$$

$$d^2z/dt^2 + f_y dx/dt = P_z + F_z - g \tag{19}$$

using standard notation. The particle displacements are x, y, z, toward the east, north, and vertical directions; $f_{y,z}$ are components of the Coriolis parameter, $F_{x,y,z}$ are components of the frictional force per unit mass, $P_{x,y,z}$ are those of the pressure force per unit mass, and g is the gravitational acceleration.

Consider a cluster consisting of a large number, N, of marked air particles, whose center of mass, or centroid, is given by

$$x_{m,j}(t) = N^{-1} \sum_{i=1}^{N} x_j(a_i,b_i,c_i,t) \tag{20}$$

where $x_j = x_j(a,b,c,t)$ and a,b,c are the initial values of x,y,z. The particle coordinates can be expressed in terms of the mean position as $x_j = x_{m,j} + x_{r,j}$, where $x_{r,j}$ are coordinates fixed to the cluster centroid. In addition, the frictional terms are assumed to be expressed in the following form;

$$F_x = -S(u-u_g) + \emptyset_x \tag{21}$$

$$F_y = -S(v-v_g) + \emptyset_y \tag{22}$$

$$F_z = -Sw + \emptyset_z \tag{23}$$

where: S is a frictional constant; the turbulent velocities u,v, and w, in the x,y, and z, directions, are expressed as departures from the geostrophic wind components u_g, v_g; and $\emptyset_x$, $\emptyset_y$, and $\emptyset_z$ are random components of the friction. This implies, as Kao (1962) pointed out, a kind of simple, Rayleigh friction, in which the (random) departure of the particle motion from the basic, quasi-horizontal geostrophic flow is balanced by a force proportional to the relative velocity. Rayleigh friction has been widely applied in atmospheric dynamics studies and found, in spite of its extremely simple form, to be useful in describing the role of friction in determining the details of many flows, including land-sea breezes, slope winds, and various upper atmospheric phenomena. In general linear friction occurs in solutions to dynamical equations of the viscous-flow type, and so it is consistent with the standard assumption that atmospheric turbulence can be described by means of a very large, "apparent" viscosity. This analogy to molecular friction is explicit, for instance, in the usual Ekman boundary layer solution.

If (a) Eqs. (20)-(23) are introduced into (17)-(19), (b) the horizontal components of the pressure force are expressed as geostrophic velocities, and (c) vertical motion of the cluster's centroid is considered negligible (i.e. the hydrostatic assumption is applied) and its horizontal motion unaffected by the random frictional components $\emptyset$, the result is two sets of equations, one for the cluster centroid motion and one for the particle displacements relative to that centroid (Kao, 1962, 1973). These important equations are

$$du_m/dt - f_z(v_m-v_g) + S(u_m-u_g) = 0 \tag{24}$$

$$dv_m/dt + f_z(u_m-u_g) + S(v_m-v_g) = 0 \tag{25}$$

for the mean cluster velocity; and

$$du_r/dt - f_z v_r + Su_r - \emptyset_x = 0 \tag{26}$$

$$dv_r/dt + f_z u_r + Sv_r - \emptyset_y = 0 \tag{27}$$

$$dw_r/dt + f_y u_r + Sw_r - \emptyset_z = 0 \tag{28}$$

for the turbulent relative velocity components. Kao, who was primarily interested in applying these equations to long-range tetroon trajectories in the troposphere, discussed the solution of Eqs. (24) and (25), for the mean cluster motion but did not solve Eqs. (26)-(28) for the relative, diffusive particle motions.

Since particle motions of most interest in the present study are quasi-horizontal, Eq. (28) can provisionally be neglected; or alternatively the motion can be regarded as a horizontal projection. Then if Eq. (27) is multiplied by $i = \sqrt{-1}$ and added to (26) the result is that

$$dV_r/dt + (S+if_z)V_r = \emptyset \tag{29}$$

where $V_r = u_r + iv_r$, and $\emptyset = \emptyset_x + i\emptyset_y$. This interesting result, derived here from the equations of motion assuming Rayleigh friction, is just a form of the random-force model, Eq. (1). The constant S was shown by Kao (1962), based on Angell's (1960) analysis of the 300 mb transosonde data, to equal 1.58×10^{-5} s^{-1}. Consequently the characteristic time of atmospheric diffusion, as described by the RFM Eq. (29), is essentially equal to $1/f_z$, and is completely dominated almost everywhere by the effect of the earth's rotation. The quantity S, which expresses the effect on relative diffusion of direct frictional drag, can influence the Lagrangian time scale only very near the equator.

Eq. (29) has several significant consequences. In the first place it supports the sizes of the Lagrangian time scales, t_L found by Gifford (1982) and Barr (1983) by comparing Eq. (10) to atmospheric diffusion data, since $t_L = 1/f_z \sim 10^4$ s. The implication is that there is a "natural" time scale that limits atmospheric diffusive motions, namely the period of inertial oscillations. Atmospheric values of t_L determined from short-range diffusion experiments, for instance those summarized by Li and Merony (1983), have typically been found to be highly variable, but usually between one and two orders of magnitude smaller than $1/f_z$. Indeed the point of view has been advanced that the scale of atmospheric diffusion is determined by that of the experiment. This is a most unsatisfactory state of affairs, but one that can easily be understood in the context of Eq. (29). If an "apparent" Lagrangian time scale for an experiment is defined as

$$t_{aL} = \int_0^{t_a} R_L(t)dt \tag{30}$$

as in effect is usually assumed in estimating t_L from diffusion data, then from Eq. (7) it follows that

$$t_{aL} = t_L[1-\exp(-t_a/t_L)], \tag{31}$$

where t_a is the averaging time, or duration, of an experiment. By expanding the exponential term in series it is found that, for small enough values of t_a, the apparent Lagrangian scale t_{aL} just equals t_a. For experimental periods t_a up to an hour, t_{aL} differs from t_a by less than 20%, for $t_L = 10^4$s. But of course if t_{aL} is assumed to equal t_L the diffusion will be incorrectly estimated because a range of diffusive motions of larger scale has been neglected.

In attempting to estimate or model atmospheric diffusion, even over the short times and distances of most experiments, it is essential to account for the influence on cluster or plume spreading, both instantaneous and time-averaged, of the total turbulence v^2 and the diffusion time scale t_L, a point Tennekes (1978) has also stressed. The RFM, Eq. (29), partitions atmospheric motion into a diffusive, turbulent component, the geostrophic departure, and a non-diffusive part, the geostrophic velocity. This, in principle, allows the total turbulence and its time scale to be determined without reference to an arbitrary or experimentally determined time scale.

A second important consequence is that all analytical results derived for Eq. (1) apply equally to Eq. (29), including the exponential autocorrelation, equilibrium (n^{-2}) frequency spectrum, and the various displacement equations. This means that the assumption of Rayleigh friction in the atmosphere, which is consistent with the often-used idea of a large, "apparent" atmospheric eddy-viscosity, is the atmospheric equivalent to the Markov-process assumption or Brownian-motion analogy. This includes in particular results related to the Fokker-Planck equation, such as Gaussian position and concentration distributions. Thus the above argument, leading from the Lagrangian equations of motion to the RFM Eq. (29), whose only significant assumption is that of Rayleigh friction, produces a large number of important consequences many of which can be checked experimentally. Indeed several of these, e.g. the exponential autocorrelation and Gaussian concentration distribution, have frequently been introduced as *a priori* assumptions into studies of atmospheric diffusion, because of their general reasonableness and agreement with experimental data.

5. TWO-PARTICLE INTERCORRELATION

It isn't self-evident that the assumption of independent particle motions, which is central to the above formulation of the RFM, correctly describes the instantaneous, relative diffusion of clusters at all diffusion times. The (one-dimensional) m.s. relative dispersion of two particles about their mutual centroid, $\overline{l^2} = \overline{(y_2-y_1)^2}$, is given (see Monin and Yaglom (1967) for instance) by

$$d\overline{l^2}/dt = 4\overline{v^2}\int_0^t[R_L(t-\xi) - R_{2L}(t,\xi)]d\xi \ , \tag{32}$$

where $R_{2L}(t,\xi) = \overline{v_1(t)v_2(\xi)}/\overline{v^2}$, the subscripts refer to the two particles, $t>\xi$, and R_L is the single-particle autocorrelation as previously defined. The importance of this equation lies in its relation to the

N-particle cluster problem, for it can be shown that the average $\overline{l^2}$ over all possible particle pairs in a diffusion cluster equals just twice the relative dispersion, $2\sigma_{yR}^2$. Because any two particles move independently in the RFM, the effect of their initial spatial correlation, $\overline{v_{10}v_{20}}/\overline{v^2}$, on relative diffusion decreases exponentially. It has however been argued, since the relative diffusion studies by Brier (1950) and Batchelor (1950) and, in effect, in Richardson's (1926) study, that R_{2L} should depend on the cluster size, or particle-pair separation, because it is reasonable to expect that spatial correlations will affect particle motions physically in the same way at all stages of cluster growth. Sawford (1982a) reviewed these arguments and devised a model of $\overline{l^2}$ based on integration over an assumed Gaussian cluster.

Unfortunately nothing very specific can be said about R_{2L}. Since, as Sawford (1983) points out, R_{2L} must in general involve the instantaneous, rather than ensemble-averaged value of $l(t)$ at each stage, the problem is much more difficult than for the case of independent particle motions. Attempts, such as those by Sawford (1982a) and Durbin (1980), to resolve this fundamental difficulty have necessarily required additional assumptions. Experimental data, particularly in the atmosphere, are also inadequate to define R_{2L}. Further experimental work on this problem would be very desirable.

The correct evaluation of the relative diffusion σ_{yR}, is especially important in the problem of concentration fluctuations, since it determines the partitioning of the total diffusion into (instantaneous) relative, cluster spreading and centroid "meandering" through Eq. (12). The chief qualitative difference between Sawford's curve of $\sigma_y R$ and that of the RFM, Eq. (10), is that the latter is displaced toward smaller values of T in the range $T < 20$, by an amount that is small there but reaches a factor of 3 (in T) at $T = 0.1$. Figure 3 illustrates these two relative diffusion curves for the initial value equal to .00214, which corresponds to one of the curves of Sawford's (1982a) Figure 1. The difference is well within the scatter of the existing atmospheric diffusion data points, which consequently do not particularly favor either curve. The two relative diffusion curves coincide at small values of T, where source size effects tend to dominate the diffusion process, and at large values of T, in the asymptotic $T^{1/2}$-range, where particle motions are certainly independent.

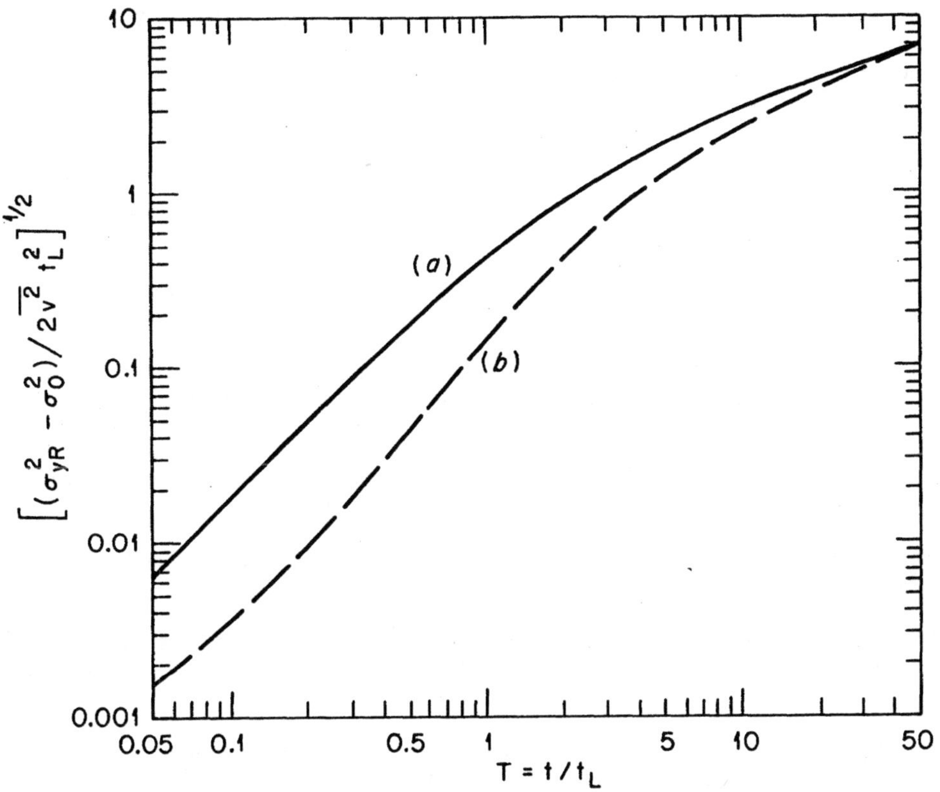

Figure 3. Non-dimensional relative dispersion curves, σ_{yR}, based (a) on Eq. (10), and (b) on Sawford's (1982a) curve for η_0 = .001 (where $\sigma_0 = 2.143\eta_0$).

6. Acknowledgments

The writer is deeply indebted to Drs. J. T. Lee, G. L. Stone, S. Barr, F. B. Smith, H. Tennekes, and H. A. Panofsky for many useful comments at various stages of the work reported above. Support by the U.S. Department of Energy, Office of Health and Environmental Research, and the U.S. Army, Atmospheric Sciences Laboratory, through contracts with the University of California, Los Alamos National Laboratory, is also gratefully acknowledged.

7. REFERENCES

Angell, J. K.: 1960, 'Analysis of operational 300-mb transosonde flight from Japan in 1957-58', J. Meteorol. 17, 20-35.

Barr, S.: 1983, 'The random force theory applied to regional scale tropospheric diffusion', Los Alamos Nat. Lab., ESS-7 Div. MS.

Batchelor, G.K.: 1950, 'The application of the similarity theory of turbulence to atmospheric diffusion', Quart. J. R. Met. Soc. 76, 133-146.

Bigg, E.K., Ayers, G.P., and Turvey, D.E.: 1978, 'Measurement of the dispersion of a smoke plume at a large distance from the source', Atmos. Environ. 12, 1815-1818.

Brier, G. W.: 1950, 'The statistical theory of turbulence and the problem of diffusion in the atmosphere', J. Meteorol. 7, 283-290.

Carras, J. N., and Williams, D. J.: 1981, 'The long-range dispersion of a plume from an isolated point source', Atmos. Environ. 15, 2205-2217.

Chadam, J.: 1962, 'On a theory of turbulent diffusion', AFCRL Research Note, AFCRL-62-1107, vii and 26 pp. U.S. Air Force.

Charney, J.: 1971, 'Geostrophic turbulence', J. Atmos. Sci. 28, 1087-1095.

Durbin, P. A.: 1980, 'A stochastic model of two-particle dispersion and concentration fluctuations in homogeneous turbulence', J. Fluid Mech. 100, 279-302.

Gifford, F. A.: 1982, 'Horizontal diffusion in the atmoshpere: a Lagrangian, dynamical theory', Atmos. Environ. 16, 505-512.

Gifford, F. A.: 1983, 'Atmospheric diffusion in the mesoscale range: the evidence of recent plume width measurements', Preprint volume, 6th Symp on Turbulence and diffusion, March 22-25, Boston, Am. Meteor. Soc., pp 300-304.

Golitsyn, G.: 1973, 'Introduction to the dynamics of planetary atmospheres', Gidrometeoizdat, Leningrad, 103pp, NASA trans. No. TT F-15, 627, Dec. 1974.

Hanna, S. R.: 1979, 'Some statistics of Lagrangian and Eulerian wind fluctuations', J. Appl. Meteorol. 18, 518-525.

Krasnoff, E. and Peskin, R. L.: 1971, 'The Langevin model for turbulent diffusion', Geophys. Fluid Dynam. 2, 123-146.

Kao, S. K.: 1962, 'Large-scale diffusion in a rotating fluid with applications to the atmosphere', J. Geophys. Res. 67, 2347-2359.

Kao, S. K.: 1973, 'Basic characteristics of global scale diffusion in the troposphere', Proc. Symp. on Turbulent Diffusion in Environmental

Pollution, April 8-14, 1973, Charlottesville, Virginia, Adv. in Geophys. 18B, 15-32, (1974).

Lee, J. T., and G. L. Stone: 1983, 'The use of Eulerian initial conditions in a Lagrangian model of turbulent diffusion,' Preprint volume, 6th Symp. on Turbulence and Diffusion, March 22-25, 1983, Boston, Am. Meteorol. Soc., pp. 13-17.

Levin, A. V.: 1971, 'Random processes and the Richardson-Obukhov law', Trudy Ukrainsk. Nautchna-Issled. Gidrometeorol. Inst. 106, 3-12.

Levin, A. V.: 1973, 'Statistical model of atmospheric diffusion with two relaxation times', Trudy Ukrainsk. Nautchna-Issled. Gidrometeorol. Inst. 125, 150-157.

Ley, A. J. and Thompson, D. J.: 1982, 'A random walk model of dispersion in the diabatic surface layer', to be pub. in Quart. J. R. Met. Soc.

Li, W.-W. and Meroney, R. N.: 1983, 'Laboratory support for the determination of atmospheric diffusion parameters from field estimates of the Lagrangian scale', Rep. no. CEP83-84WWL-RNM-3, Dept. of Civil Engrg., Colorado State Univ., Ft. Collins, CO 80521.

Lilly, D. K.: 1983, 'Stratified turbulence and the mesoscale variability of the atmosphere', J. Atmospher. Sci. 40, 749-761.

Lin, C. C. and W. H. Reid: 1963, 'Turbulent flow', in Handbuch der Physik VIII/2, 438-523, Springer,Berlin.

Monin, A. S.: 1972, 'Weather forecasting as a problem in physics', MIT Press, Cambridge, MA.

Monin, A. S. and Yaglom, A. M.: 1967, 'Statistical fluid mechanics', Nauka, Moscow (Trans. MIT Press, Cambridge, MA, 1971).

Novikov, E. A.: 1963, 'Random force method in turbulence theory', Soviet Phys. JETP 17, 1449-1454.

Obukhov, A. M.: 1959, 'Description of turbulence in terms of Lagrangian variables', Adv. in Geophys. 6, 113-116.

Reid, J. D.: 1979, 'Markov chain simulation of vertical dispersion in the neutral surface layer for surface and elevated releases', Bound.-Layer Met. 16, 3-22.

Richardson, L. F.: 1926, 'Atmospheric diffusion shown on a distance-neighbor graph', Proc. R. Soc. Lond. 110A, 709-737.

Sawford, B. L.: 1982a, 'Comparison of some different approximations in the statistical theory of relative dispersion', Quart. J. R. Met. Soc. 10B, 191-206.

Sawford, B. L.: 1982b, 'Lagrangian Monte Carlo simulation of the turbulent motion of a pair of particles', Quart. J. R. Met. Soc. 108, 207-213.

Sawford, B. L.: 1983, 'The effect of Gaussian particle-pair distribution functions in the statistical theory of concentration fluctuations in homogeneous turbulence', Quart. J. R. Met. Soc. 109, 339-354.

Smith, F. B.: 1968, 'Conditioned particle motion in a homogeneous turbulent field', Atmos. Environ. 2, 491-508.

Taylor, G. I.: 1921, 'Diffusion by continuous movements', Proc. London Math. Soc. 20, 196-211.

Tennekes, H.: 1978, 'The exponential Lagrangian correlation function and turbulent diffusion in the inertial subrange', Atmos. Environ. 13, 1565-1567.

Tennekes, H., and Lumley, J. L.: 1972, 'A first course in turbulence', xii and 300 pp., MIT Press, Cambridge, MA.

Thompson, D. J.: 1983, 'Random walk modelling of diffusion in inhomogeneous turbulence', UK Met. Office, Met. O. 14, T.D.N. No. 144.

Yaglom, A. M.: 1977, 'Semi-empirical equations of turbulent diffusion in boundary layers', Fl. Dynamics Trans., Polish Acad. Sci., 7, 99-144.

Questions, answers and comments

S.P.S. ARYA: Many observations of σ_Y in the P.B.L. show that it varies with the distance from the source as $\sigma_Y \prec x$ initially, but approaches $\sigma_Y \prec x^{0.5}$ behavior in a relatively snort distance of a few kilometers. How would you reconcile this with your results for the long-range diffusion which indicates a power law being valid dor very long distances?

F.A. GIFFORD: My feeling is that what has been observed in short-range diffusion experiments reflects the limited sampling time of most experiments. Instanteneous plume spreading seems to be at a rate such that $\sigma_Y \prec x^p$ $p \geq 1$, up to quite large distances of the order of hundred kilometers.

J.C.R. HUNT: Your model assumes that the turbulence is stationary. I believe that situations of practical interest may arise when this assumption is not valid. For example, between over land and over sea. Then the Langevin Equations need to be modified as our recent work shows (Van Dop, Nieuwstadt, Hunt, 1984).

F.A. GIFFORD: Yes, the R.F.M. is certainly not any more exempt from this kind of problems than any other models. The R.F.M. I described has the property that the turbulence time scale is quite large and controlled by the earth's rotation. Thus, I would expect that the contribution to diffusion from large-scale eddies, and consequently the diffusion at long times will tend to be less affected by small-scale surface differences than the near-field diffusion is. This seems to be what is found in the diffusion data, where the relative scatter of σ_Y points about their mean tends to <u>decrease</u> with diffusion time. Of course, we have too few such observations at present to be very sure.

MASS CONSISTENT MODELS FOR WIND DISTRIBUTION IN COMPLEX TERRAIN - FAST ALGORITHMS FOR THREE DIMENSIONAL PROBLEMS

Nathan Dinar
Israel Institute for Biological Research
P.O.Box 19, Ness-Ziona 70450, Israel

ABSTRACT

Mass consistent models for wind distribution in complex terrain are extremely useful and readily applied to many practical situations, such as the siting of wind turbines, or as input in the estimation of diffusion and transport of pollutants in complex terrain. These models are based on the numerical solution of the steady state three dimensional continuity equation for the mean wind components. The momentum and energy equations are not solved explicitly, but considered indirectly using parametric relations and wind data. In practical applications the equations must be solved numerous times (e.g. for each time interval). Standard techniques for numerical solution of three dimensional problems are frequently very expensive and thus not suitable for practical needs. In the present work, great emphasis is given to the development of fast algorithms, and techniques based on the multigrid approach are shown. Two mass consistent programs are described, the first based on the parametric representation of one of the wind components, and the second based on a few wind measurements and a variational principle. To verify the reliability of the variational approach, a measurement program related to a project of wind energy, is being performed at Har-Ahim, a site located in the Galilee.

I. INTRODUCTION

Mass consistent models for the estimation of the windfield in complex terrain can be regarded as an useful method for practical needs, such as the search for appropriate sites for wind turbines (e.g.Endlich et al., 1982; Walton et al., 1980; Liu and Yocke, 1980; Endlich, 1984) as an input in the estimation of transport and diffusion in complex terrain (e.g.Sherman, 1978; Dickerson et al., 1984), or as the assessment of spread of fires in forests (Liu et al., 1974).

The models are suitable for such practical problems because of their physical simplicity, and because the input data they require is relatively easy to obtain.

Boundary-Layer Meteorology **30** (1984) 177–199. 0006–8314/84/0302–0177$03.45

In spite of their simplicity, considerable numerical efforts are still needed in order to solve the equations derived from the model. This is due to the fact that the real world with complex topography is regretfully three dimensional in space, and no symmetry considerations can be generally applied in order to reduce the dimensionality of the problem. Therefore the equations, although linear, are usually three dimensional and with complicated boundary conditions.

The implementation of the mass consistent approach in the practical problems described above, usually requires an intensive use of the algorithms which are dedicated to the numerical solution of the equations derived from the model. Therefore, in order to exploit the advantages of mass consistent models, it is extremely important to develop fast and efficient solvers.

In the field of the atmospheric boundary layer, usually not enough efforts are invested in developing accurate and efficient algorithms for numerical solutions of the equations derived from the models. This is due to the fact that people are obviously interested in getting results in the simplest way possible, and complicated algorithms, involving elaborate programming, are usually avoided. Grid sizes are often specified for resolution purposes, but no grid size considerations with respect to the discrete approximations of the continuous problems are usually mentioned. The accuracy of the discrete equations near boundaries is often not consistent with the accuracy far from boundaries. Therefore an increased interest in numerical methodology is important and is certainly needed both in research programs for academic purposes and in practical problems. But in the latter case, methodology is a crucial factor which may determine the success or failure of the method.

One of the purposes of this work is to describe a methodology which can be efficiently and successfully applied to the numerical solution of all kinds of equations used in the field of the atmospheric boundary layer, including general flow problems, diffusion, turbulence, etc. Section II will include a short and general description of the Multigrid Method. This method is used in the solution of the equations derived from the mass consistent models which are described in Section III. A description of an experimental program related to a project of wind energy and aimed at verifying the reliability of a mass consistent model, is given in Section IV.

II. Multigrid methods

Multigrid methods were first developed for elliptic boundary value problems. A complete and extensive description of multigrid methods for these problems is given by Brandt (1977). Today they are applied in a variety of problems including steady state incompressible Navier-Stokes equations, Brandt and Dinar (1979); Compressible flows, Brandt (1981). General ideas for the multigrid approach in time-dependent problems are outlined by Dinar (1979).

The multigrid method is an iterative method which uses several levels of discretization of the continuous equation, so that the differential character of the problem is always retained (see Fig. 1).

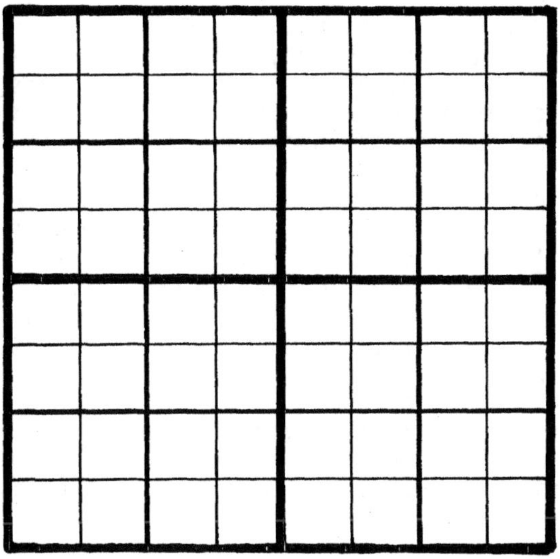

Fig. 1. Multigrid method. Several levels of discretization.

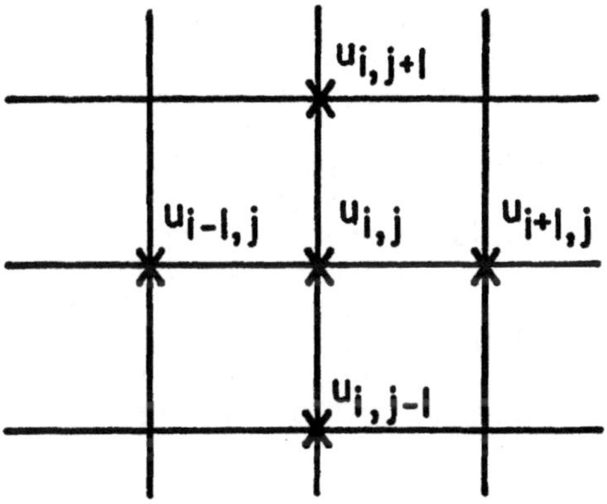

$$u_{i,j} = \tfrac{1}{4}(u_{i+1,j} + u_{i-1,j} + u_{i,j+1} + u_{i,j-1})$$

Fig. 2. Relaxation of Laplace equation.

On the other hand however, in standard techniques, once the discrete equivalents of the equations are set, all the numerical efforts are usually invested in the solution of algebraic equations, and their differential origin is not exploited. This is an inefficient way to solve the problems for various reasons,such as:

- The number of arithmetical operations is usually very large, of the order of N^2, or even N^3, depending on the method of solution. N is the total number of unknowns, proportional to h^{-d}, where h is a typical grid length, and d is the dimension of the problem.
- There is no indication whether the grid is fine enough for convergence purposes. Checking convergence implies normally the repetition of the algorithm in a finer grid.
- As a consequence, when the algebraic equations are solved iteratively, there is no way of knowing when to stop the process,thus resulting in the investment of unnecessary efforts to obtain very accurate solutions to the discrete algebraic equations, which may in fact be relatively poor approximations to the true solutions of the differential equations.

In the multigrid technique, on the other hand, solutions are obtained by using a number of arithmetical operations which is proportional to the number of unknowns (or even less than that when using extrapolation techniques), and in that respect the multigrid method is optimal. This is a general property of multigrid methods which can be proved rigorously, and is independent of the dimension of the problem and the shape of the domain. In all practical problems the coefficient of proportion is small, so that the total number of operations is small as well. In addition, multigrid algorithms contain intrinsic information that allows to stop iterations when errors are of the same order of magnitude as truncation errors, and it can make use of adaptive discretizations, using finer grids in regions where this is needed, and coarser grids in regions with smooth behaviour. An example of the use of adaptive grids is shown in Section III.

The multigrid method is an iterative method consisting of several processes, performed in a given order, and defining a multigrid iterative cycle. The main processes involved are the following:

- Relaxation
- Fine grid to coarse grid transfers
- Coarse grid to fine grid transfers.

Relaxation

Relaxation is a local operator in which the solution is usually adjusted sequentially at each grid point. For example, in a typical Laplace equation in two dimensions, five grid points are usually related in a single discrete equation, and the value of the central grid point approximant is corrected (see Fig.2). Because of the localness of the operator, its influence is restricted to the immediate neighborhood of the grid point relaxed. Therefore, in a single relaxation sweep over all the points of the domain, only high frequency errors, of the order of the grid size, are efficiently reduced, while low frequency errors are mostly not reduced. This is actually the reason for the slow convergence found in relaxation methods, and everyone who has used relaxation is familiar with the number of thousands of relaxations which are often needed in order to get reasonable approximations to solutions. The multigrid technique does not use relaxation to solve the equations. The use of relaxation is just for reducing the high frequency errors and to get approximations with smooth errors. Once this purpose is achieved, relaxations are discontinued.

Fine grid to coarse grid transfers

This process consists of a weighted transfer of the residuals, (i.e., the errors in the equations which remain after the relaxation process) from a fine grid to a coarser grid. By doing so, it is possible to define a discrete problem similar to the original one, but in the coarse grid. This problem can again be treated by the same multigrid process, (i.e., relaxation and transfer of residuals to a still coarser grid, etc.). This transfer of residuals is justified by the fact that since errors after relaxation are smooth, a coarse grid representation of the problem will probably be a good one.

Coarse grid to fine grid transfers

When some convergence criteria are applied in a coarse grid, the corrections to the solutions are transferred to a fine grid by interpolation techniques, and a new approximation in the finer grid is obtained. In doing so, a new source of errors is introduced by the interpolation. But these are obviously high frequency errors, which are again efficiently reduced by subsequent relaxations.

As an illustration let us consider a typical problem given by

$$LU = f \quad \text{in the domain } \Omega , \qquad U|_{\Gamma} = g \tag{1}$$

Γ is the boundary domain. L is a linear operator, U is the solution to equation (1), and f and g are given functions. Suppose we seek a discrete approximation U^h in a grid with mesh size h. The discrete equivalent is given by

$$L_h U^h = f^h \tag{2}$$

where L_h is a discrete operator approximating L, and f^h are the values of f at grid points. Suppose we have some approximation u^h to the exact solution U^h of (2). We define the error V^h by

$$V^h = U^h - u^h. \tag{3}$$

This error will satisfy the following correction equation with zero boundary conditions:

$$L_h V^h = r^h, \tag{4}$$

where r^h is the residual and it is given by

$$r^h = f^h - L_h u^h \tag{5}$$

If the approximation u^h was obtained by a few relaxation sweeps, then V^h is expected to show a smooth behaviour, and therefore it is reasonable to assume that it will be fairly represented by solving its discrete equation (4), but in a coarse grid with mesh size 2h, i.e.,

$$L_{2h} V^{2h} = I_h^{2h} r^h, \tag{4'}$$

where I_h^{2h} represents the fine to coarse transfer operator from grid h to grid 2h. This equation (4') has exactly the same structure as the original equation (2) in the fine grid h, and therefore can be treated identically,i.e., relax the equation, define a similar correction equation in a grid with meshsize 4h, and so on. If this process is continued we will eventually reach a very coarse grid in which the number of unknowns of the discrete equations is sufficiently small to allow a direct solution of the equations.

Suppose we have satisfactorily solved the correction equation (4') in the 2h-grid. A new approximation in the fine grid $u^h{}_{NEW}$ is obtained by

$$u^h{}_{NEW} = u^h + I_{2h}^h V^{2h}, \tag{6}$$

where I_{2h}^h represents the coarse to fine transfer operator from grid 2h to h. This completes the multigrid iterative cycle.

A typical Multigrid algorithm uses 5-6 levels of discretization, and the switching from one level to another, coarser or finer, is controlled either by some fixed strategy determined in advance, or by an accommodative strategy relying on internal checks performed during the solution process.

The amount of numerical work required in a multigrid algorithm, in order to solve the equations to the level of truncation errors, is frequently less than 10 work units for a typical problem, where a work unit represents the amount of numerical work needed in order to make a relaxation sweep in the finest grid. This is about one or two order of magnitude faster than well known standard techniques like Relaxation, SOR, ADI, etc. FFT (Fast Fourier Transform) methods are almost as fast as multigrid methods, but they are usually restricted to very simple domains, while multigrid can be efficiently applied in principle in any given domain.

One of the disadvantages attached to the multigrid methods is its relatively complicated programming. But today software is available for a variety of problems, and continous efforts are invested in the software aspects of multigrid (Ophir, 1979). Two multigrid algorithms were developed for the treatment of the two mass-consistent models described in this work.

III. MASS CONSISTENT MODELS

Mass consistent models for the estimation of the mean wind field in complex terrain consider explicitly only the equation of conservation of mass, while energy and momentum equations are taken into account indirectly, by means of parametrization techniques and/or field measurements. This is, of course, a disadvantage of these methods. But sophisticated models, involving more physical variables, will also need usually some parametrization. (e.g. diffusivities definitions or other closure assumptions). Moreover, the more elaborate is the model, the more reliable are the data needed in order to exploit the advantages of the model. So there must always be a balance between the complexity of the model and the quality of the input data available. This is particularly true when the model is to be implemented in purely quantitative estimations, as in wind energy estimates. In fact, many references can be found in which mass consistent models are strongly recommended for these purposes. See for example, Hiester and Pennell(1981).

Two mass-consistent approaches are considered in this work, the first is based on a parametric representation of one of the wind components and follows ideas by Liu and Yocke (1980). The second one is derived from a variational principle, the equations being similar to Endlich et al., (1982), and Sherman (1978). As mentioned, the emphasis in this work is the development of efficient algorithms that have to be used intensively. Therefore, descriptions of models and methods of solution are given below in some detail.

Method A

In general, the mass consistent approach involves the solution at the stationary continuity equation

$$u_x + v_y + w_z = 0, \qquad (7)$$

where $U = (u,v,w)$ is the mean wind vector.

The region Ω in which (7) has to be satisfied can be defined as a three-dimensional box-shaped region which is bounded by the topography at the bottom and by the boundary layer height at the top. (see Fig. 3). The normal component of the wind must usually be specified at the boundaries, providing the necessary boundary conditions.

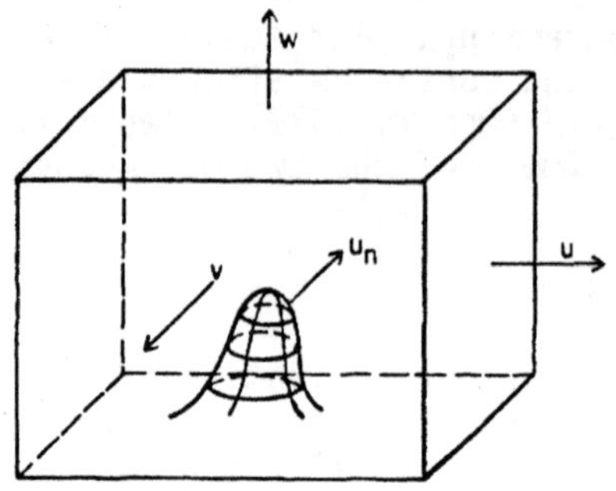

Fig. 3 Box-shaped three dimensional domain

These conditions are given by

$$\left.\frac{\partial u}{\partial n}\right|_{\text{boundaries}} = g(x,y,z), \qquad (8)$$

where g = o at the topography and take some given values at the free boundaries.
The boundary function g must satisfy an additional compatibility condition imposed by the divergence theorem. This condition is given by

$$\int_S gdS = 0 \qquad (9)$$

where S means boundary surfaces. This condition is not mentioned and is apparently disregarded in the current formulation of mass-consistent models. Even if boundary conditions are defined properly and the condition is satisfied in the continous equation, it will be generally not satisfied in the discrete equivalent of the equations, unless g is a given linear function. Another source of errors which may affect this compatibility condition is input boundary data subject to experimental errors. If (9) or its discrete form does not hold, the problem is ill-posed and iterative methods would not converge, although errors may usually remain bounded. To get a correct well posed problem, an appropriate scalar constant C can be added to the right hand side of (7). It can be shown that C is given by

$$C = (1/V) \int_{S^h} gdS, \qquad (10)$$

where V denotes the volume of Ω and $\int_{S^h}$ represent the discrete integral operator.

The equations (7) to (10) do not determine a unique solution, because only one equation is defined for the three unknowns u,v,w. In order to close the system, two more equations are needed.

The idea is to assume a parametric relation for one of the wind components, say w,given by some function

$$w = F\,(x,y,z), \qquad (11)$$

and to assume some value for the vertical component of vorticity

$$u_y - v_x = A \tag{12}$$

Defining A = 0 as in Liu and Yocke (1980), implies horizontal potential flow at each plane z = constant. Endlich (1984) uses terrain-following coordinates, F=0, and A a vorticity wind field interpolated from measurements.

Using the formalism of Liu, a sequence of two dimensional problems for given values of z are defined and represented by the following Cauchy-Riemann system.

$$\begin{aligned} u_x + v_y &= -\frac{\partial F}{\partial z} = -Q(x,y) \\ u_y - v_x &= 0 \end{aligned} \tag{13}$$

These equations are defined in a horizontal plane which may intersect topography, and therefore equations (13) are defined in the domain B outside the topography. (see Fig. 4).

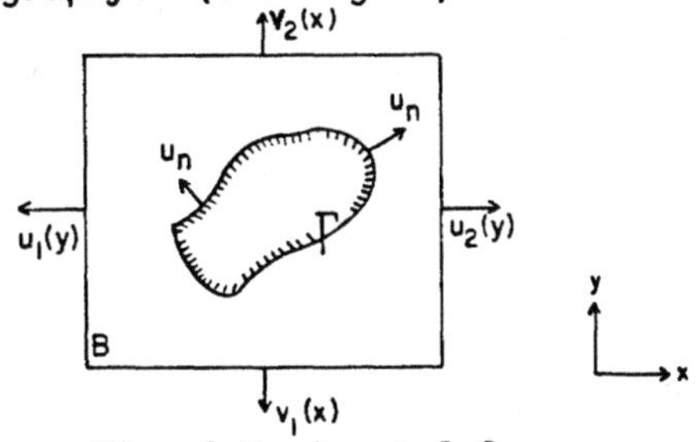

Fig. 4 Horizontal layer

Boundary conditions must satisfy a two dimensional compatibility condition given by

$$\iint_B -Q(x,y)dxdy = \int[u_2(y)-u_1(y)]dy + \int[v_2(x)-v_1(x)]dx - \int_\Gamma u_n d\Gamma \tag{14}$$

where u_1, u_2, v_1, v_2, and are shown in Fig. 4. If there is a hole in the domain due to topography, an additional condition must be added in order to take into account that there are no sources or sinks inside. This equation (Stokes Theorem) is given by

$$\int_S u_s \, ds = 0, \tag{15}$$

where s is some closed curve surrounding the hole.

Condition (14) is not mentioned in the work by Liu and Yocke (1980). Condition (15) is satisfied in Liu's work since the system (13) is replaced by an equivalent representation, involving a two-dimensional potential function ϕ , defined by

$$u = \frac{\partial\phi}{\partial x} \quad ; \quad v = \frac{\partial\phi}{\partial y} \tag{16}$$

The equivalent equation is a Poisson equation

$$\nabla^2\phi = - Q(x,y) \tag{17}$$

and Neumann conditions $\frac{\partial\phi}{\partial n}$ must be satisfied at boundaries.

At a first look it seems preferable to use (17) instead of (13), since a single equation is involved. But Neumann conditions in difficult and irregular boundaries are quite difficult to discretize properly, i.e., without losing the truncation accuracy obtained inside the domain, far from boundaries. The usual replacement of irregular boundaries by nearby grid lines may cause large numerical errors. In this work the representation given by (13) is used.

The parametric representation of the right side of the equation involved is based, as a first approximation, on linearity of solutions and superposition of various physical effects that are important in complex terrain. Thus, when it is assumed that $Q_1, Q_2, \ldots, Q_N$ are the functions defined by the parameterization relationships, then the right hand side of (13), or (17) is given by $\alpha_1 Q_1 + \ldots + \alpha_N Q_N$, where α_i are tunable parameters describing the relative importance of each single effect. Alternatively, denoting by U_i the wind field which is obtained when only Q_i is considered, then the superposed wind field is defined by $\alpha_1 U_1 + \ldots \alpha_N U_N$. This distinction will become clear further on in the text.

Three separate effects are included in the algorithm developed in this work. These are

- Effects of topography
- Effects of boundary layer
- Thermal effect, mountain valley flows.

The effect of topography on the flow can certainly be dominant in complex terrain, and flow will strongly depend on stability conditions. In a very stable atmosphere, streamlines will have a tendency to be horizontal and surround the hills, i.e., $w = o$, while in neutral or unstable situations, streamlines will climb the hill and will be approximately confined to vertical planes parallel to the main flow direction. The idea here is to get solutions for the above two extreme conditions, and then to superpose the solutions. For extremely stable conditions, the equations are given by

$$\begin{aligned} u_x^{(1)} + v_y^{(1)} &= 0 \\ u_y^{(1)} - v_x^{(1)} &= 0 \end{aligned} \tag{18}$$

These equations are solved repeatedly for a sequence of horizontal layers $z_o, z_1 \ldots, z_H$, and a three dimensional windfield

$U_1(x,y,z) = (u_1^{(1)}, v_1^{(1)}, 0)$ is calculated.

For the other extreme condition, the x-axis is defined parallel to the main flow direction, and therefore v = o. The equations are given by

$$u_x^{(2)} + w_z^{(2)} = 0,$$
$$u_z^{(2)} - w_x^{(2)} = 0,$$

and are solved for a sequence of vertical layers $y_1 \ldots y_M$ defining a three dimensional wind field $U_2(x,y,z)=(u^{(2)}, o, w^{(2)})$.

The solution for a non-extreme situation is defined by

$$U = p\,U_1 + q\,U_2 \qquad (19)$$

where p,q are positive, and p + q = 1. It is easily shown that U defined in (19) is a solution of the three dimensional continuity equation (7). p must be dependent on a stability parameter, like the Froude number. Of course, these solutions are appropriate for potential flow only, and situations which are characterized by the effects of flow separation cannot be handled by this approach.

Effects of boundary layer and mountain valley winds are incorporated in the present algorithm in the same way as Liu and Yocke (1980). The parametrizations are defined for the vertical wind component and therefore solutions are calculated in horizontal planes intersecting the topography. For the boundary layer effects, the following equation is used:

$$w_i = \frac{u_{i-1}}{\delta} \left(\frac{f_i}{f_i} - 1\right) \qquad (20)$$

where i represents the z_i layer, f_i are dimensionless velocities defined by Businger et al.(1971);

$$f_i = f(z_i) = \ln \frac{z_i}{z_o} + 4.7 \frac{z_i - z_o}{L} \qquad (21)$$

for stable conditions, and

$$f_i = \ln \frac{B(z_i)}{B(z_o)} + 2 \text{ arctg } [A(z_i)] - 2 \text{ arctg } [A(z_o)] + \frac{\pi}{2} \qquad (22)$$

for unstable conditions, where

$$B(z) = \frac{A(z)-1}{A(z)+1} \; ; \; A(z) = \left(1 - 15 \frac{z}{L}\right)^{1/4}$$

z_o is the roughness length, L the Monin-Obukhov length and δ the horizontal length scale.

Mountain valley flows are thermal local effects which are independent of the mean flow. The vertical component is defined by

$$w = U_H \text{ tg}\alpha \qquad (23)$$

where α is the average slope in the height interval $[z_i, z_{i+1}]$ and U_H is a horizontal wind component imposed as a boundary condition in the intersection with topography. U_H is given by

$$U_H = \frac{[gh(T_C - T_E) \sin \alpha]^{1/2}}{C_D T_C \cos \alpha} \tag{24}$$

where g is the gravity acceleration, T_C the local temperature and T_E the ambient air temperature.

The numerical solution of the equations described above, proceed by a multigrid algorithm which solves two dimensional Cauchy Riemann equations, like equations (13), in a general irregular domain. The construction of the algorithm requires an appropriate definition of the discrete equations and an appropriate relaxation scheme. The discretization uses a staggered grid and the function errors through the domain are $O(\Delta x^2)$, where $\Delta x = \Delta y$ is the mesh size.

Fig. 5 shows the grid structure. u-components are defined on the vertical grid lines, v-components on horizontal grid lines. Right hand sides -Q are defined on the cell centers. The first equation (continuity) is defined relative to the cell centers, and the second equation (vorticity) is defined relative to the nodes. The discrete equations are

$$u_x + v_y \simeq \frac{u_2 - u_1}{\Delta x} + \frac{v_2 - v_1}{\Delta x} = -Q \tag{25}$$

$$u_y - v_x \simeq \frac{u_4 - u_3}{\Delta x} - \frac{v_4 - v_3}{\Delta x} = 0 \tag{26}$$

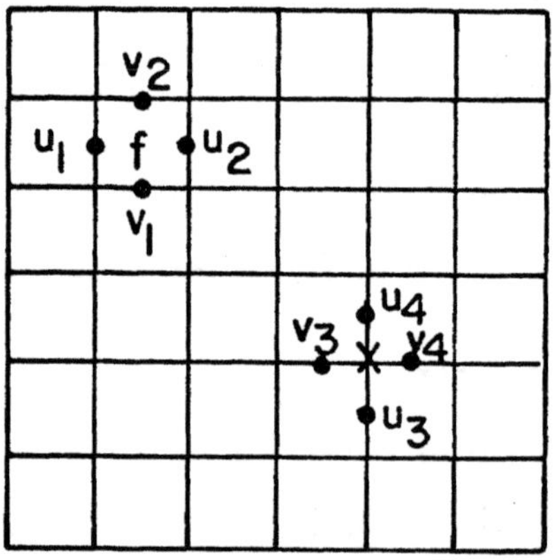

Fig. 5. Discretization of the Cauchy-Riemann equations in staggered grids

Equation (25) and (26) are for regular points far from irregular boundaries. Instead of (25) we can use an integral discrete form, based on the divergence theorem, given by

$$\Delta x^2 \text{ div } U = -\Delta x^2 Q = \Delta x u_2 + \Delta x v_2 - \Delta x v_1 - \Delta x u_1 \qquad (27)$$

(See Fig. 6). Equation (27) is identical with (25) for regular points, but for irregular points (near boundaries) it can be easily generalized (see Fig. 7), resulting in the following discretization:

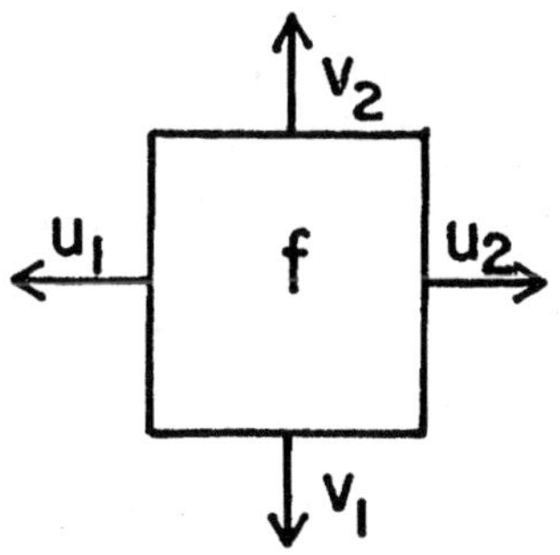

Fig. 6. Discretization using the divergence theorem. (Regular cell)

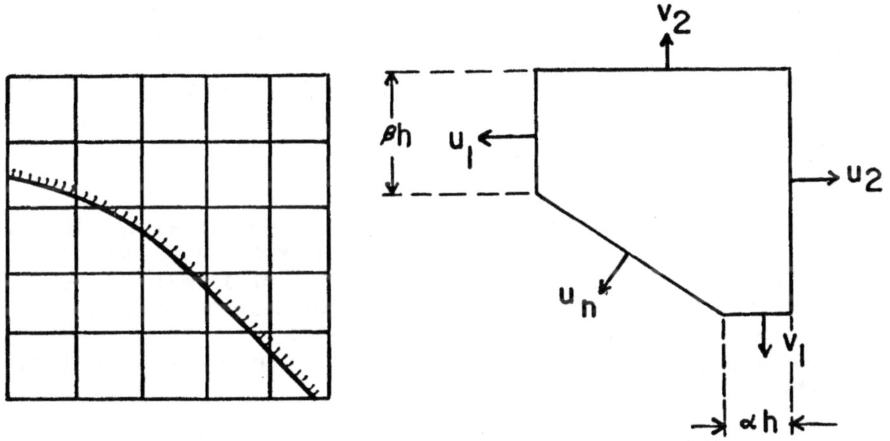

Fig. 7. Discretization using the divergence theorem. (Irregular boundary cell)

$$\frac{u_2 - \beta u_1}{\Delta x} + \frac{v_2 - \gamma v_1}{\Delta x} - \frac{u_n}{\Delta x} \sqrt{(1-\gamma)^2 + (1-\beta)^2} = -[1-(1-\gamma)(1-\beta)] \qquad (28)$$

No special treatment of (26) to irregular points used in the present algorithm . The relaxation scheme used is the same as described by Dinar (1979). Separate relaxation sweeps are performed on the two equations (25) or (28), and (26). This relaxation scheme is called a distributive relaxation scheme and has the important property that in the process of the relaxation of any one of the equations, the residuals of the other equation are not affected.

This multigrid algorithm is very efficient. Three dimensional wind fields are obtained by solving a sequence of two dimensional problems, and this is the main advantage of this approach.

The main disadvantages of this method from a practical point of view, are the following:

- Model tuning and verification is difficult in general topography, because the normal wind components must be prescribed in vertical planes. Reliable data of this kind are difficult to obtain.

- In general topography, horizontal planes may be very irregular, resulting in a domain with several holes. When the fraction of boundary to regular cells is relatively large, the accuracy of the numerical solution will be seriously affected.

- Flow separation and other common effects cannot be treated.

Therefore this method is especially suitable for regions surrounded by planes, like islands, in which terrain is not too complex and wind profiles in the boundaries are easily obtained or calculated. Liu and Yocke (1980) used these ideas in Hawaii in the island of Maui. For wind energy purposes in Israel, Method B described below should be preferred.

Method B

In this method, a solution to the three-dimensional continuity equation (7) is obtained. The additional information needed in order to define a complete set of equations consists of a small number of surface wind measurements, which are relative easy to obtain.

The method is composed of two main steps:

- Interpolation, in which an interpolated mean wind field is obtained from the existing wind data.

- Mass consistency enforcement, in which the interpolated wind field is used as a first guess, and then it is adjusted with minimal corrections in order to satisfy the continuity equation.

The first guess is obtained by interpolating the surface wind data measured at a small number of stations, and synoptic data obtained through meteorological services or by means of kites or pilot balloons.

The interpolation is performed in two steps, vertical interpolation and horizontal interpolation. Vertical interpolation is based on usual power or logarithmic laws, for instance (Endlich, 1982):

$$U(z) = U(z_m) + (U(H) - U(z_m)) \frac{\log (z/z_m)}{\log (H/z_m)} \tag{29a}$$

where $U = (u,v,w)$, is the wind vector, H is the boundary layer height, and z_m is the measurement height. Horizontal interpolation is usually linear, with weight coefficients that are inversely proportional to the square of the distance from a given point to the measured point.

$$u(x,y) = \frac{\Sigma f_i \, u(x_i,y_i)}{\Sigma f_i} \tag{29b}$$

where (x_i,y_i) are the measurement coordinates, and

$$f_i = \frac{1}{(x-x_i)^2 + (y-y_i)^2} \tag{30}$$

The wind field obtained by interpolation, $U_o = (u_o,v_o,w_o)$, does not usually satisfy the equation of mass conservation, i.e.,

$$\frac{\partial u_o}{\partial x} + \frac{\partial v_o}{\partial y} + \frac{\partial w_o}{\partial z} \neq 0.$$

The idea is to correct this wind field in order to enforce mass consistency, without changing it too much. This approach can be stated by the following variational problem :

Find a vector $U = (u,v,w)$ that minimizes the functional J defined by

$$J = \int_V [\alpha_1(u-u_o)^2 + \alpha_2(v-v_o)^2 + \alpha_3(w-w_o)^2] \, dxdydz \tag{31}$$

subjected to the following constraint:

$$u_x + v_y + w_z = 0,$$

where α_i are tunable parameters describing the relative changes between horizontal and vertical motion. The problem can be put in an equivalent form using a Lagrange multiplier λ:

Minimize $I(u,v,w,\lambda)$ where

$$I(u,v,w,\lambda) = \int_V [\alpha_1(u-u_o)^2 + \alpha_2(v-v_o)^2 + \alpha_3(w-w_o)^2 + \lambda(u_x+v_y+w_z)]dxdydz \tag{32}$$

without any additional constraints. For this problem, an equivalent system of Euler equations of four equations with four unknowns, can be derived:

$$\lambda_x = 2\alpha_1 \, (u - u_o) \tag{33a}$$

$$\lambda_y = 2\alpha_2 \, (v - v_o) \tag{33b}$$

$$\lambda_z = 2\alpha_3 (w - w_o) \tag{33c}$$

$$u_x + v_y + w_z = 0 \tag{33d}$$

Appropriate boundary conditions can be defined by

$$\begin{aligned} u_n &= 0 \quad \text{on the topography,} \\ \lambda &= 0 \quad \text{on free boundaries} \end{aligned} \tag{34}$$

There are some existing codes to solve these equations, for example MATHEW (Sherman,1978), and COMPLEX(Endlich 1982),but in order to apply this method efficiently, development of fast solvers is essential. The complex boundary imposed by topography can be handled numerically in two different ways. One is by a special numerical treatment of boundaries, like in Method A of this work and in MATHEW. This has the advantage that the original set of equations is not altered. The second approach is using a mapping of the irregular domain to a regular domain, by "terrain following coordinates", like in COMPLEX. The regular domain obtained is a three dimensional box, and the transformation is given by

$$\sigma = \frac{z-h(x,y)}{H-h(x,y)} \tag{35}$$

where $h(x,y)$ defines the topography and H is the boundary layer height. The price paid for the simplification of the working domain is generally, of course, the complication of the differential equation, but using the following transformation of the wind field,

$$\begin{aligned} u^* &= u(H-h) \\ v^* &= v(H-h) \\ w^* &= w-(1-\sigma)(uh_x + vh_y), \end{aligned} \tag{36}$$

it can be shown that

$$u^*_x + v^*_y + w^*_\sigma = 0 \tag{37}$$

which is similar to (33d).

Euler equations (33) are not convenient for the numerical treatment and equivalent equations can be derived. One possibility is to eliminate u,v,w from (33) getting a single equation for λ , given by

$$\frac{\partial^2\lambda}{\partial x^2} + \frac{\partial^2\lambda}{\partial y^2} + \frac{\alpha_1}{\alpha_3}\frac{\partial^2\lambda}{\partial z^2} = -2\alpha_1 \left(\frac{\partial u_o}{\partial x} + \frac{\partial v_o}{\partial z} + \frac{\partial w_o}{\partial z}\right) \tag{38}$$

where $\alpha_1 = \alpha_2$. The desired wind field (u,v,w) is then obtained using (33a,b,c). This approach was adopted in MATHEW. A slightly different approach was used in COMPLEX, λ is eliminated from equations (33) and three uncoupled equations are defined for the wind components.

In vector notation these equations are:

$$LU = LU_o - \text{grad div } U_o \tag{39}$$

where $U = (u,v,w)$; $U_o = (u_o,v_o,w_o)$ and

$$L = \frac{\partial}{\partial x^2} + \frac{\partial}{\partial y^2} + \frac{\alpha_1}{\alpha_3} \frac{\partial}{\partial z^2} . \tag{40}$$

In the present work, a multigrid algorithm was constructed which solves (40), together with (33d) and using the transformation (35) and (36). Adaptive grid refinements are implemented in the code, resulting in finer discretization near the topography, where sharp wind profiles may occur. (See Fig. 8). A line relaxation was used in this code, instead of the usual point relaxation outlined in Fig. 2, for a similar two dimensional operator. It can be shown by Fourier analysis that line relaxation is more efficient in reducing high frequency errors, due to the small values used for α_1/α_3. A detailed description of the algorithm and code is given by Givati and Dinar (1984).

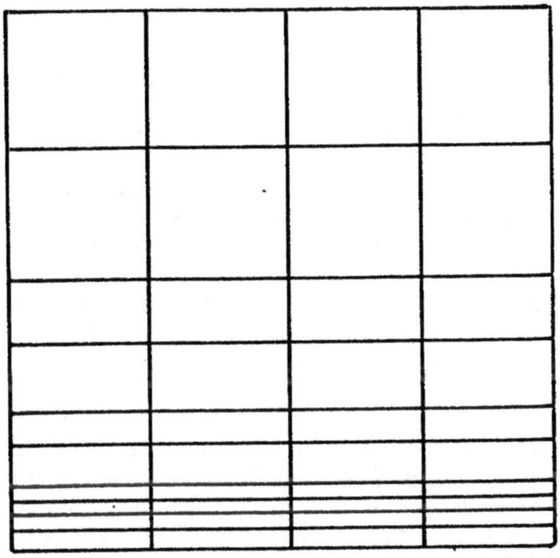

Fig. 8. Adaptive grid refinements

The present algorithm allows the intensive use of the code at very low cost. A typical problem involving about 20000 equations is solved in about 10 seconds in a CDC Cyber. With similar existing software the same problems may require 30-100 times longer to solve.

IV A MODEL TUNING PROGRAM

One of the important applications of the mass consistent models described in this work, is the estimation of the wind field with relation to the problem of siting of large wind turbines. The axis of large turbines may be located at a height of the order of 100 m, and therefore a reliable method which can estimate winds at 100 m, using cheap measurements near ground level will be extremely useful.

In order to check the applicability of the model, a tuning program is being conducted at Har-Ahim, a site located in the Galilee. The aim of this program is to tune the model, i.e., to choose the appropriate weighing parameters, to check sensitivity and reliability of the model in predicting wind energy estimates at 100-300 m altitudes, and to determine the density of experimental data needed for a good representation of the wind field in the area. The measurement program consists of one-hour mean wind measurements (u,v,w) for a 24 hr period, once a month, during a year. The first measurements were performed in October 1983. The main topographic characteristics of the site are shown in Fig. 9. The eight circles represent the location of the surface stations (6 m height).

The following strategy is used to compare measurements to model predictions. The experimental data are split into two groups. The first group (usually five or six stations) is used to feed the model. Prediction of the model are compared with the data of the remaining stations. In addition, data obtained from a pilot balloon are also compared with model predictions, and used to check model performance at 100-300m altitudes. Partial evaluation of the experiments performed shows good model performance. As an example, a comparison between measured and predicted wind velocities, at the two surface stations denoted by I and II in Fig. 9, is shown in Fig. 10. Comparison between model predictions and pilot balloon measurements (circles) at the highest point in the area is shown in Fig. 11. Prediction seems satisfactory. The sharp gradients are resolved by the algorithm, due to the adaptive grid refinement technique mentioned in Section III. A complete evaluation of the tuning program is given in Givati and Dinar (1984).

V SUMMARY

Two different mass consistent algorithms were developed and described in this work. Mass consistent models are very suitable in practical applications. Fast solvers are essential to enhance the usefulness of these models, and the multigrid method is probably the best approach for this purpose.

Some changes in the mass-consistent philosophy, directed towards introducing more physics in the basic assumptions, but retaining the fundamental simplicity of the approach, is certainly desirable and may result in substantial improvements.

ACKNOWLEDGEMENTS

I would like to thank Mr. Reuven Givati for his contribution in Sections III and IV, and to Dr. Hadassah Kaplan for the helpful dis-

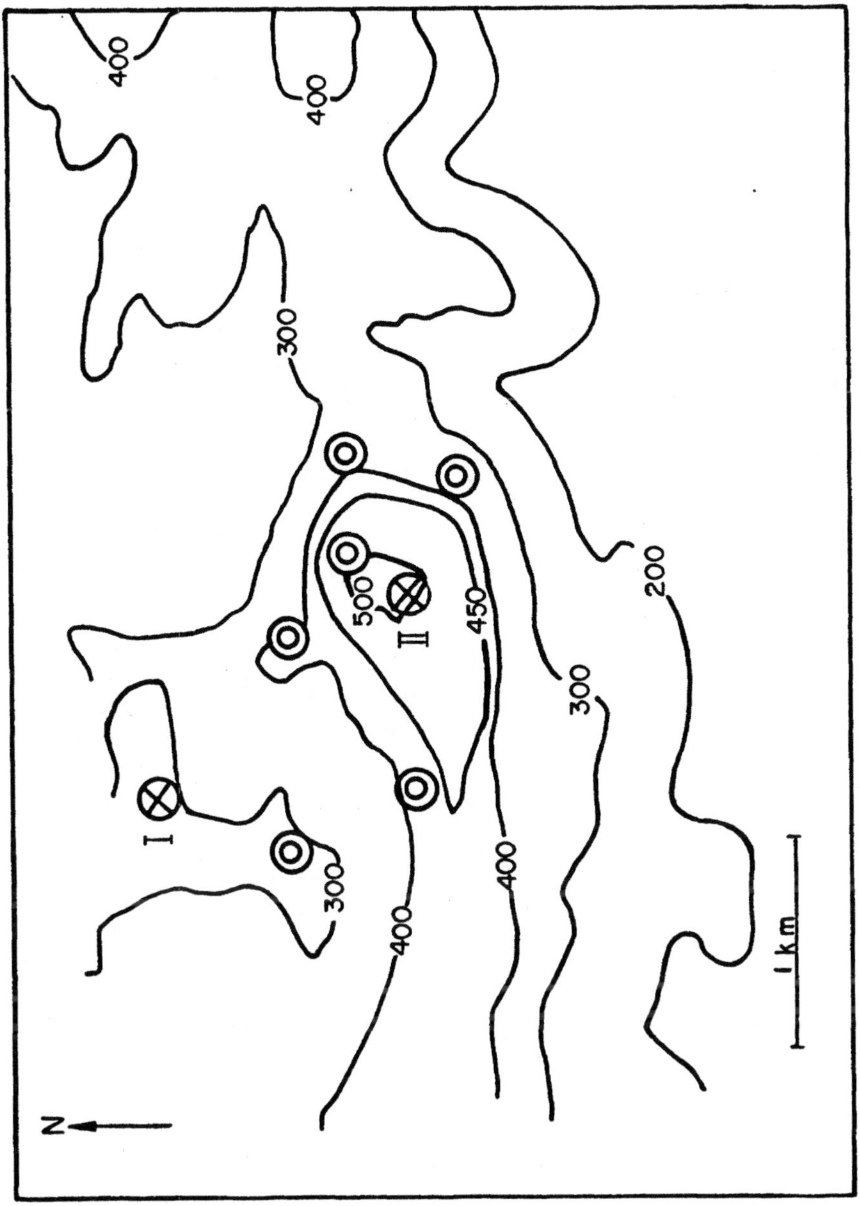

Fig. 9. Height contours (m.) near the Har-Ahim site. Circles represent surface stations (6m height). I and II represent surface stations used for checking model performance.

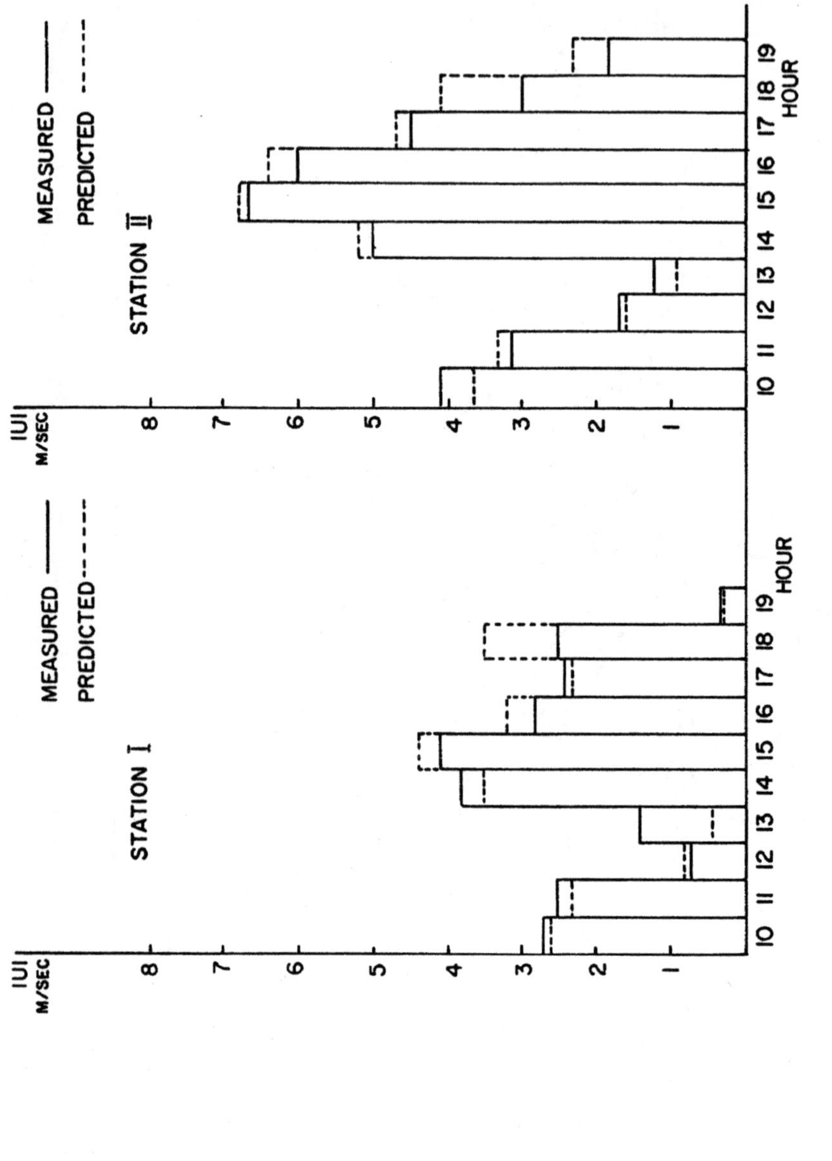

Fig. 10. Evaluation of model performance at two surface stations ($\alpha_1/\alpha_3 = 10^{-4}$).

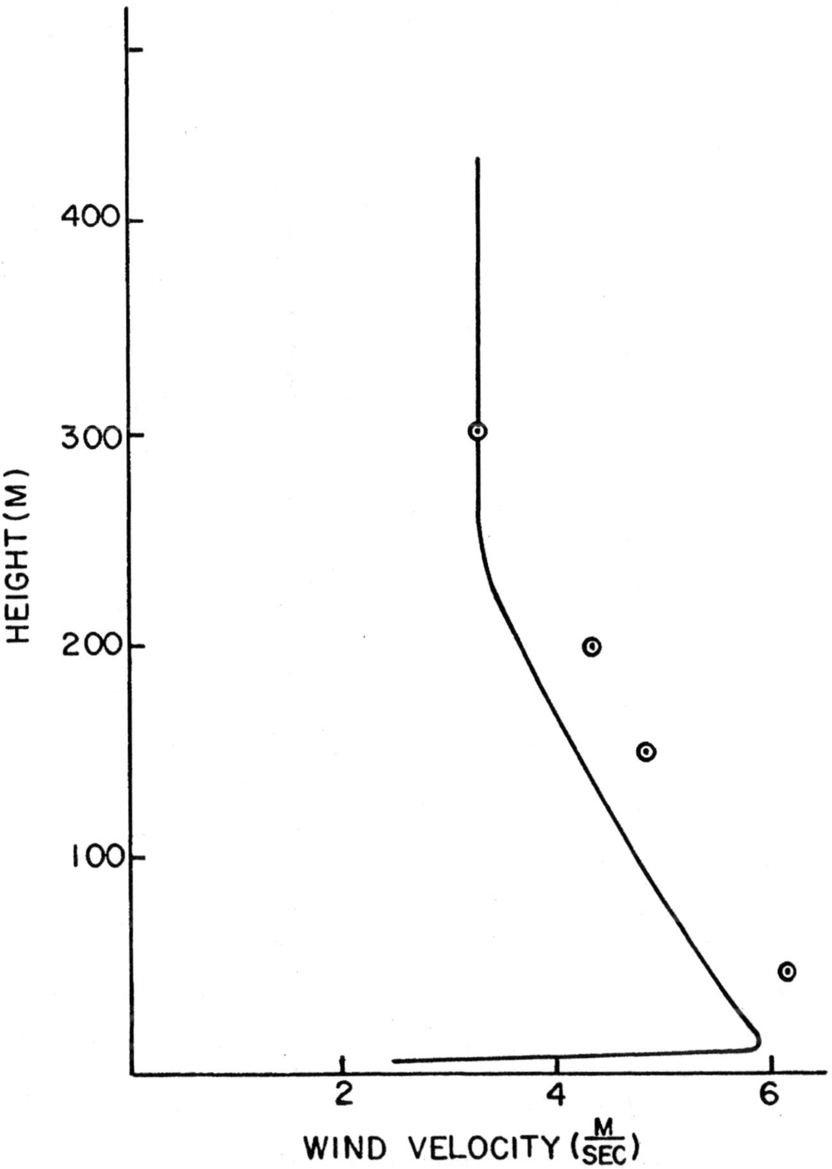

Fig. 11. Comparison between model predictions (solid line), and pilot balloon measurements (circles),at the top of the hill (near station II in fig. 9).

cussions and comments during the preparation of this work.

The tuning program described in Section IV is supported by the Israel Ministry of Energy and Infrastructure, Research Contract No. 82-5-06.

REFERENCES

Brandt, A.: 1977. 'Multilevel Adaptive Solutions to Boundary Value Problems'. Mathematics of Computation, 31, 353-390.

Brandt, A.: 1981. 'Multigrid Solutions to Steady State Compressible Navier-Stokes Equations'. Proc. Fifth International Symposium on Computing Methods in Applied Sciences and Engineering. Versailles, France.

Brandt, A. and Dinar, N.: 1979. 'Multigrid Solutions to Elliptic Flow Problems'. In "Numerical Methods for Partial Differential Equations". (S. Parter, Ed.). Academic Press, New York. 53-147.

Businger, J.A., Wyngaard, J.C., Izumi, Y. and Bradley, E.F.: 1971. 'Flux-Profile Relationships in the Atmospheric-Surface Layer'. J. Atmos. Sci., 28, 181-189.

Dickerson, M.H., Foster, K.T. and Gudiksen, P.H.: 1984. 'Experimental and Model Transport and Diffusion Studies in Complex Terrain With Emphasis on Tracer Studies'. Proc. of the 29th. OHOLO Conference (Kaplan and Dinar, Eds.), Reidel, Holland.

Dinar, N.: 1979. 'Fast Methods for the Numerical Solution of Boundary Value Problems'. Ph.D. Thesis, Weizmann Institute of Science, Rehovot, Israel.

Endlich, R.M., Ludwig, F.L. and Bhumralkar, C.M.: 1982. 'A Diagnostic Model for Estimating Winds at Potential Sites for Wind Turbines'. Journal of Appl. Meteorol., 21, 1441-1454.

Endlich, R.M.: 1984. 'Wind Energy Estimates by Use of a Diagnostic Model'. Proc. of the 29th. OHOLO Conference (Kaplan and Dinar, Eds.), Reidel, Holland.

Givati, R. and Dinar, N.: 1984. 'A Mass Consistent Model for the Estimation of the Wind Field in Complex Terrain, Related to a Wind Energy Siting Program. (To be published).

Hiester, T.R. and Penell, W.T.: 1981. 'Meteorological Aspects of Siting Large Wind Turbines'. Batelle Pacific Northwest Labs., 048335000, 9500022.

Liu, M.K. and Yocke, M.A.: 1980. 'Siting of Wind Turbine Generators in Complex Terrain'. J. Energy, 4, 10-16.

Liu, M.K., Mundkur, P. and Yocke, M.A.: 1974. 'Assessment of the Feasibility of Modelling Wind Fields Relevant to the Spread of Brush Fires'. R74-15, Forest Fire Lab., Dept. of Agriculture, Riverside, Ca.

Ophir, D.: 1979. 'Language for Processes of Numerical Solutions to Differential Equations'. Ph.D. Thesis, The Weizmann Institute of Science, Rehovot, Israel.

Sherman, C.A.: 1978. 'A Mass Consistent Model for Wind Fields Over Complex Terrain'. J. Appl. Meteorol., 17, 312-319.

Walton, J.J., Sherman, C.A. and Knox, J.B.: 1980. 'Wind-Power Site Screening Methodology, Final Report'. Lawrence Livermore Nat. Lab. California.

Questions, answers and comments

L. BERKOFSKY: In limited area modeling in numerical weather prediction, instabilities arise on the internal boundaries due to time dependence, even though space derivatives are mass-consistent. In your multigrid method you did not have instabilities, possibly because you deal with a steady-state problem. Would you care to comment on the application of your multigrid method to time dependent problems?

N. DINAR: The mass-consistent approach used in this work implies the solution of a steady-state three dimensional elliptic system of equations for the one-hour mean wind field in a finite space domain. Large temporal low frequency changes in the wind field can eventually result in large inaccuracies. However, these changes are usually associated with low wind speed, or restricted to short time intervals. Relatively strong and stable winds are the most important for wind energy estimates, and therefore this approach seems to be appropriate.

Most of the numerical methods known in the literature for this kind of problem are stable, including of course the multigrid method. Multigrid ideas can be easily implemented in time dependent problems. One obvious and straightforward way to do it is to replace the method of solution used in a single implicit time-step by an appropriate multigrid solver. Another, more sophisticated, multigrid approach is based on explicit time-steps performed on coarse and fine grids, as shown for example in Dinar (1979).

SPECTRAL METHODS FOR ATMOSPHERIC DIFFUSION MODELING

Ruwim Berkowicz
National Agency of Environmental Protection
Air Pollution Laboratory
Risø National Laboratory
DK-4000 Roskilde, Denmark

ABSTRACT

The rate of diffusion of a cloud depends on cloud dimensions. As the cloud enlarges, larger eddies come into play and the rate of diffusion increases. The turbulent diffusion process is scale-dependent. The gradient-transfer theory (K-theory) is only appropriate when the dimensions of the dispersed material are much larger than the size of the turbulent eddies. Introduction of a spectral turbulent diffusivity function (STD) makes it possible to treat the diffusive transport in a Eulerian system, with diffusivity effectively dependent on the actual size of the concentration distribution. The basic innovation is that diffusion is treated in the Fourier space and the diffusion coefficient is dependent on the wave number of the Fourier components of the concentration distribution. It is shown that the concept of the wave-number-dependent diffusivity leads to a non-local flux-gradient relation.

1. INTRODUCTION

The gradient-transfer theory (K-theory) is widely used for the description of turbulent diffusion. One of the most important advantages of this theory is that it makes it possible to describe the diffusive transport in a Eulerian framework. The basic assumption of the gradient-transfer theory is that the flux of material is proportional to the local gradient of the concentration.

$$F = K \cdot \text{grad} \cdot c \qquad (1)$$

where the diffusivity coefficient K is in general a tensor.

The earliest attempts to use the gradient-transfer approximation for atmospheric diffusion were made by Schmidt (1925) and, since then, numerous studies have been undertaken (see e.g. Pasquill, 1974). The use of the K-theory, however, is limited by serious restrictions. The theory is based on the analogy of turbulent diffusion to molecular diffusion (Fick's law); this analogy is, in fact, a very poor one. Molecular diffusion is

Boundary-Layer Meteorology 30 (1984) 201–220. 0006–8314/84/0302–0201$03.00

caused by microscale motions. Turbulent diffusion, on the other hand, is governed by eddies, usually of a broad range of sizes. The K-theory approximation is only appropriate when the size of the turbulent eddies responsible for the diffusion is small compared with the length-scale of the diffusing distribution. A classic example of the weakness of the K-approach is diffusion of a plume from an elevated point source in the atmospheric boundary layer. In the initial stage of dispersion, the plume size is much smaller than the size of the turbulent eddies, and the use of the gradient-transfer approximation is not justified. Introduction of a diffusivity coefficient as a function of time of travel or distance from a source allows us to retrieve some essential properties of point source diffusion, but at the expense of losing the possibility of using the method in a Eulerian system.

Turbulent diffusion has been treated extensively in terms of statistical theory. Since the pioneering works of Taylor (1921), the statistical theory has been applied with success to diffusion from both continuous and instantaneous point sources (Batchelor, 1953). The statistical theories explain quite well the most important features of turbulent diffusion, and the results of all other theories are usually compared with the results predicted by the statistical theory. Studies of transport of atmospheric pollutants, however, often require treatment of non-homogeneous and non-stationary flows. Emission from several sources and physical and chemical transformations must also be accounted for. Such problems are difficult or even impossible to model in terms of Taylor's statistical theory, and description based on a Eulerian model is more suitable.

Several sophisticated methods for description of turbulent diffusion have been proposed. One can mention methods based on higher-order closure approximation (Lewellyn and Teske, 1976; Enger, 1983), direct-interaction approximation (Roberts, 1961), random-walk model (Monin and Yaglom, 1965) or renormalized perturbation methods (Phythian and Curtis, 1978). These theories, however, are not frequently applied in practical studies of diffusion, e.g. in air pollution modeling, possibly because of their high level of complexity.

The spectral turbulent diffusivity theory (Berkowicz and Prahm, 1979; Prahm et al., 1979; Berkowicz and Prahm, 1980), which is reviewed here, arose from the need for a practical approach in air pollution modeling. The theory reveals the essential features of turbulent diffusion, but also preserves the simplicity and flexibility of the K-theory formulation.

The basic innovation is that diffusion is treated in the Fourier space and the diffusion coefficient is made dependent on the wave number of the Fourier components of the concentration distribution. Introduction of a spectral turbulent diffusivity function (STD) makes it possible to treat the diffusive transport in a Eulerian system, but with diffusivity effectively dependent on the actual size of the concentration distribution. The concept of a wave-number-dependent diffusivity leads to a non-local flux-gradient relation, i.e. just contrary to the basic assumption (Eq. 1) of the gradient-transfer theory. The gradient-transfer theory appears as a small-scale limit of the more general STD-theory.

The STD-theory is especially easy to apply in modeling turbulent transport when a numerical technique based on the so-called pseudo-spectral method (Christensen and Prahm, 1976; Zlatev et al., 1983) is used to solve the diffusion equation.

2. THE SPECTRAL TURBULENT DIFFUSIVITY CONCEPT

The basic concept of the spectral turbulent diffusivity was presented by Berkowicz and Prahm (1979). However, Monin (1955, 1956) and Schönfeld (1962) have previously suggested diffusion models which are closely related to the STD-approach.

The rate equation of a contaminant diffusing in a turbulent flow is given by

$$\frac{\partial c}{\partial t} = -\mathrm{div}(\bar{v} \cdot c) - \mathrm{div}\, F + S + Q \tag{2}$$

where $\bar{v}$ is the vector of the mean flow velocity, and S and Q are terms describing sinks and sources, including physical and chemical processes. Equation (2) is undetermined and an additional relationship between the turbulent flux F and the mean concentration c is required. In terms of the gradient-transfer approximation, the so-called "first order closure" is used, i.e. it is assumed that flux is proportional to the gradient of the concentration (Eq. 1).

Let us consider the case of diffusion from an instantaneous point source (puff). Restricting ourselves to only one dimension, we can write:

$$\frac{\partial c}{\partial t} = K \frac{\partial^2 c}{\partial y^2} \tag{3}$$

The solution of (3) for an instantaneous point source of a strength Q at y = 0 is the well-known Gaussian distribution,

$$c(y,t) = \frac{Q}{(2\pi)^{\frac{1}{2}}\sigma} \exp\left(-\tfrac{1}{2}\frac{y^2}{\sigma^2}\right) \tag{4}$$

where the standard deviation σ is given by

$$\sigma = (2Kt)^{\frac{1}{2}} \tag{5}$$

It is well known that the square root time dependence of a diffusing cloud's dimension postulated by (5) is not valid until the cloud-size does not become larger than the size of the turbulent eddies. Berkowicz and Prahm (1979) suggested that the Fourier transform of the diffusion equation (3) be considered:

$$\frac{\partial}{\partial t} \tilde{c}(k,t) = -k^2 K\, \tilde{c}(k,t) \tag{6}$$

Because the range of eddies able to diffuse a concentration distribution represented by a given Fourier mode will be larger for a "long wave" than for a "short wave", it is logical to assume that the diffusivity coefficient should in fact be different for different Fourier modes of the concentration distribution. The generalized diffusivity equation now reads:

$$\frac{\partial}{\partial t} \tilde{c}(k,t) = -k^2 K(k) \tilde{c}(k,t) \tag{7}$$

where K(k) is the spectral turbulent diffusivity coefficient (STD).

The STD-coefficient depends only on the properties of the turbulent flow, and as such is thus a real Eulerian parameter. On the other hand, the relative distribution of the Fourier components depends on the actual shape of the diffusing cloud. Because the STD-coefficient depends on the wave number k, the rate of diffusion also becomes dependent on the size of the cloud. This is the main feature of the STD-approach; i.e., it allows modeling of the turbulent diffusion in a Eulerian system, and at the same time, it takes the scale dependence into account.

Equation (7) was derived for the case of a homogeneous turbulence field. As (7) indicates, the rate of diffusion of a single Fourier mode does not depend on the other Fourier modes. This is not a surprising result because an interaction among various Fourier modes would violate the requirement of invariance of the diffusion equation toward a homogeneous translation in space. However, Eq. 7 is also local in time. The The more general form, valid for homogeneous diffusion, is

$$\frac{\partial}{\partial t} \tilde{c}(k,t) = \hat{H}(k,t,t') \tilde{c}(k,t') \tag{8}$$

where $\hat{H}(k,t,t')$ is an integral operator in the time space. The STD-formulation is a special case of (8) with

$$\hat{H}(k,t,t') = -k^2 K(k) \int_0^t dt' \delta(t-t') \tag{9}$$

An equation similar to (8) was presented by Roberts (1961) (see also the discussion given in the Appendix of Berkowicz and Prahm, 1979). The work by Roberts is based on the so-called Direct Interaction Approximation theory developed by Kraichnan (1959). An excellent review is given by Leslie (1973). The STD-theory thus appears to be an approximation of (8) which neglects the time-dependence, while the gradient-transfer formulation is a further approximation which also neglects the k-dependence.

3. THE INTEGRAL EQUATION OF DIFFUSION

The spectral diffusion equation (7) can be converted to real space, resulting in

$$\frac{\partial}{\partial t} c(y,t) = \frac{\partial}{\partial y} \int_{-\infty}^{\infty} D(y-y') \frac{\partial}{\partial y'} c(y',t) dy' \tag{10}$$

where

$$D(y-y') = \frac{1}{2\pi} \int_{-\infty}^{\infty} K(k) \exp(ik(y-y')) dk \tag{11}$$

is the turbulent diffusivity transfer (TDT) function introduced by Berkowicz and Prahm (1979). From (10), it follows that the turbulent flux of concentration is given by

$$F = -\int_{-\infty}^{\infty} D(y-y') \frac{\partial}{\partial y'} c(y',t)dy' \tag{12}$$

Thus, according to the STD-formulation, the turbulent flux is related to the mean concentration gradient by an integral equation. This contradicts the main assumption of the gradient-transfer theory that the flux is proportional to the local gradient of concentration. The non-local character of the turbulent diffusion is the main feature of this type of diffusion. Batchelor and Townsend (1956) were the first to suggest that this non-localness may be best expressed by an integro-differential equation. Schönfeld (1962) has presented an integro-differential equation which also is non-local in time. He does not, however, relate this equation to a spectral formulation,which is the basis of the derivation of Eq. (10).

It is easy to show that the gradient-transfer approximation is just a limiting case of the integro-differential Eq. (10). Neglecting the k-dependence of the STD-coefficient, i.e. assuming that $K(k) = \text{const} = K_0$, we obtain from (11) $D(y-y') = K_0\delta(y-y')$. In this case, Eq. (10) converts to the usual diffusivity equation. The non-local character of the new diffusion equation given by (10) is thus entirely the consequence of introduction of the k-dependent spectral diffusivity coefficient K(k). The STD-theory leads to a diffusion equation which is non-local in space, but still local in time. Because the integro-differential diffusion equation is more appropriate for treating a scale-dependent diffusion, the STD-theory can be considered a significant improvement of the gradient-transfer approach in view of its application in Eulerian models.

4. SPECTRAL DIFFUSIVITY FOR A CLOUD

The spectral diffusivity coefficient K(k) must depict the properties of the turbulent flow. In Berkowicz and Prahm (1979), the form of K(k) was deduced from the known behaviour of an evolving cloud in a homogeneous, stationary turbulent field. Pasquill (1974) shows that as long as the dimensions of a cloud are smaller than the size of the most energetic eddies, the standard deviation of the concentration distribution is given by

$$\sigma = \alpha\varepsilon^{1/2}t^{3/2} \tag{13}$$

where ε is the turbulent energy dissipation rate and α is a dimensionless constant of the order of one. From the relation,

$$K = \frac{1}{2}\frac{d\sigma^2}{dt} \tag{14}$$

it follows that

$$K = \frac{3}{2}\alpha^{2/3}\varepsilon^{1/3}\sigma^{4/3} \tag{15}$$

Here, the diffusivity coefficient is expressed in terms of the cloud's linear dimension (σ). Implicitly, the diffusivity is treated as a function of time. Eqs. (13) and (15) express only diffusion of a cloud in the case of an instantaneous point source. This is the approach used in Lagrangian models.

Berkowicz and Prahm (1979) suggest replacing the σ-dependence in (15) with a wave-number dependence assuming that $\sigma \sim k^{-1}$. When the cloud dimension enlarges and becomes greater than the size of the most energetic eddies, the diffusivity coefficient tends toward a constant value. On the basis of the imposed limiting behaviour of the STD-coefficient, the following form of K(k) is proposed:

$$K(k) = \frac{K_0}{1+B(k/k_m)^{4/3}} \qquad (16)$$

where K_0 is the diffusivity of the long-time diffusion limit, k_m is the wave number corresponding to the largest turbulent eddies and B is a dimensionless constant. Prahm et al. (1979) show that the STD-coefficient of the form given by (16) results in the same time-dependence of diffusion of a cloud as given by Eq. (13) when an instantaneous point source is considered. One must note that the spectral turbulent diffusivity approach can be applied for an arbitrary source configuration and initial conditions.

5. DERIVATION OF THE SPECTRAL DIFFUSIVITY EQUATION

In Berkowicz and Prahm (1980), the STD-formulation is derived from the concept of eddy transport. The procedure presented there is, to some extent, based on Prandtl's (1925) "mixing length" concept. The turbulent field is considered to be made up of eddies of a finite size and with some characteristic transport velocity. Let us say that the probability density of occurrence of an eddy of length ℓ is $f(\ell)$ and that the transport velocity of such an eddy is $v(\ell)$. An eddy of length ℓ can transport material over a distance ℓ, or smaller. Only those eddies which are situated in the region $(y-\ell, y+\ell)$ can thus contribute to the flux at y (Fig. 1). In the case of homogeneous turbulence, the eddies must be distributed equivalently in the space. The probability that an eddy of length ℓ is in a region of length $d\rho$ is $d\rho/\ell$. The flux at y, caused by eddies of length ℓ and originating at $y+\rho$ and $y-\rho$, is thus

$$-v(\ell)c(y+\rho)f(\ell)\frac{d\rho}{\ell} + v(\ell)c(y-\rho)f(\ell)\frac{d\rho}{\ell} \qquad (17)$$

In order to find the total flux at y, we must integrate (17) with respect to ρ (from 0 to ℓ) and over all values of ℓ.

$$F(y) = \int_0^\infty \left(\int_0^\ell v(\ell)f(\ell)\{c(y-\rho)-c(y+\rho)\}\frac{d\rho}{\ell}\right)d\ell \qquad (18)$$

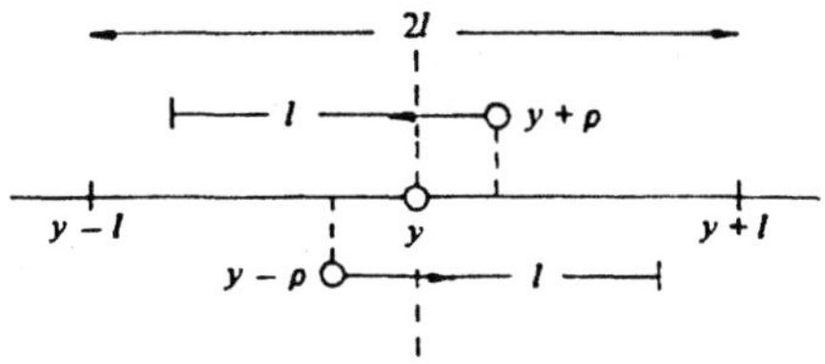

Figure 1. Diagram illustrating the contribution of eddies of length ℓ to the eddy flux at point y. Eddies starting from points y+ρ and y-ρ are shown. Only eddies originating in the region (y-ℓ,y+ℓ) contribute.

Using the continuity equation,

$$\frac{\partial c}{\partial t} = - \frac{\partial}{\partial y} F(y) \tag{19}$$

and, after differentiation with respect to y and integration with respect to ρ, we obtain

$$\frac{\partial c}{\partial t} = \int_0^\infty \frac{v(\ell)f(\ell)}{\ell} \{c(y+\ell) + c(y-\ell) - 2c(y)\} d\ell \tag{20}$$

Using a polynomial approximation of the concentration field and keeping only terms to second order with respect to ℓ, we obtain

$$\frac{\partial c}{\partial t} \simeq [\int_0^\infty v(\ell)\ell f(\ell)d\ell] \frac{\partial^2 c}{\partial y^2} \tag{21}$$

Comparing (21) with (3), one can immediately see that a second order approximation of the concentration distribution leads to the gradient-transfer formulation with a diffusivity given by

$$K = \int_0^\infty v(\ell)\ell f(\ell)d\ell \tag{22}$$

Instead of polynomial approximation, we use now a Fourier representation of the concentration distribution in (20)

$$c(y) = \int_{-\infty}^\infty \tilde{c}(k)\exp(iky)dk \tag{23}$$

Substituting (23) into (20), we obtain

$$\frac{\partial}{\partial t} \tilde{c}(k) = -k^2[\int_0^\infty v(\ell)\ell f(\ell) \frac{\sin^2 \frac{1}{2}k\ell}{(\frac{1}{2}k\ell)^2} d\ell] \tilde{c}(k) \tag{24}$$

Comparing (24) with (7), we can see that

$$K(k) = \int_0^\infty v(\ell)\ \ell f(\ell)(\frac{\sin\frac{1}{2}k\ell}{\frac{1}{2}k\ell})^2 d\ell \tag{25}$$

While the gradient-transfer formulation corresponds to a second-order approximation of the space derivatives of the concentration distribution, the STD-formulation results from an exact evaluation of the derivatives by means of Fourier transformation.

6. EVALUATION OF THE STD-FUNCTION FROM ENERGY SPECTRA

In order to determine $K(k)$, both $f(\ell)$ and $v(\ell)$ have to be specified. In Berkowicz and Prahm (1980), a rather lengthy method was presented, and here, we will only state the main results. The probability of occurrence of eddies of length ℓ, $f(\ell)$, is connected to energy spectrum $E(k)$ by the relationship

$$\tfrac{1}{2}\int_0^\infty v^2(\ell)f(\ell)d\ell = \int_0^\infty E(k)dk = \sigma_v^2 \qquad (26)$$

Assuming that there is a unique relationship between the eddy size and the spectral wave number k,

$$k = \pi/\ell \qquad (27)$$

we can write

$$\tfrac{1}{2}v^2(\ell)f(\ell) = \frac{\pi}{\ell^2}E(\pi/\ell) \qquad (28)$$

6.1. Inertial Sub-Range Form.

A simple form of the energy spectrum is adapted here. It is assumed that turbulent energy is supplied at a wave number $k_m = \pi/\ell_m$, where ℓ_m is the size of the largest eddies; it is also assumed that the energy is transferred toward the shorter eddies according to the $k^{-5/3}$- law (inertial sub-range). No energy is transferred to eddies larger than ℓ_m. This means that a sharp cut-off exists at the small wave-number range of the spectrum (Fig. 2).

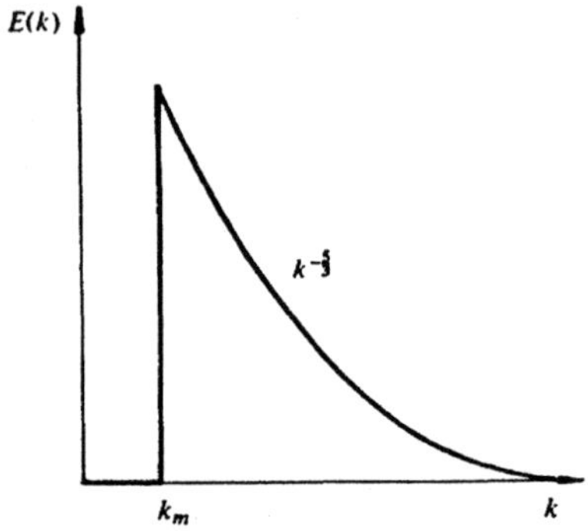

Figure 2. Schematic representation of the energy spectrum $E(k)$. The inertial sub-range form ($k^{-5/3}$) is extended from k_m to infinity; k_m is the wave number at which the turbulent energy is produced. No energy is assumed to be transferred to wave numbers smaller than k_m.

From these arguments, and using (26), we can write

$$\tfrac{1}{2}v^2(\ell)f(\ell) = \begin{cases} 2/3\ \sigma_v^2\ \ell_m^{-2/3}\ \ell^{-1/3} & \text{for } \ell \leq \ell_m \\ 0 & \text{for } \ell > \ell_m \end{cases} \tag{29}$$

In the inertial sub-range, the energy is transferred by a cascade process between eddies of differing size. The rate at which the transfer process takes place is constant for all eddies. The lifetime (or the relaxation time) for an eddy of length ℓ is

$$\tau(\ell) \propto \ell/v(\ell) \tag{30}$$

As the energy is proportional to $v^2(\ell)$, the requirement of a constant dissipation rate ε, leads to

$$\varepsilon \propto v^2(\ell)/\tau(\ell) = v^3(\ell)/\ell = \text{const} \tag{31}$$

From (31), it appears that

$$v(\ell) \propto \ell^{1/3} \tag{32}$$

Assuming that the largest eddies of length ℓ_m have a transport velocity v_m, we can write

$$v(\ell) = v_m(\ell/\ell_m)^{1/3} \tag{33}$$

The transport velocity v_m of the largest eddies can also be interpreted as the maximum velocity fluctuation (Prahm et al., 1979). According to Markee (1963), the maximum observed velocity fluctuation is about six times the standard deviation σ_v. Prahm et al. (1979) and Berkowicz and Prahm (1980) estimated the proportionality constant to be 2π, thus

$$v(\ell) = 2\pi\sigma_v(\ell/\ell_m)^{1/3} \tag{34}$$

Substituting (29) and (34) into (25), we finally obtain

$$K(k) = \frac{2}{3\pi}\ \sigma_v \ell_m^{-1/3} \int_0^{\ell_m} \ell^{1/3} \left(\frac{\sin\frac{1}{2}k\ell}{\frac{1}{2}k\ell}\right)^2 d\ell \tag{35}$$

This is the form of the STD-coefficient derived for the inertial sub-range of the energy spectrum. For $k = 0$, we have

$$K(0) = K_o = \frac{1}{2\pi}\ \sigma_v \ell_m \tag{36}$$

Combining (36) with (35), we obtain

$$K(k) = K_o \int_0^1 \frac{4}{3}\ s^{1/3} \left(\frac{\sin\frac{1}{2}\pi s(k/k_m)}{\frac{1}{2}\pi s(k/k_m)}\right)^2 ds \tag{37}$$

where $s = \ell/\ell_m$ and $k_m = \pi/\ell_m$.

K(k) computed from (37) is shown in Fig. 3. For $k \gg k_m$, K(k) becomes proportional to $(k/k_m)^{-4/3}$. It is the behaviour postulated by Berkowicz and Prahm (1979) mainly on the basis of dimensional arguments (Eq. 16). In Fig. 3, the formula (16) is shown with B = 0.87. The agreement with the curve corresponding to (37) is very good for $k > 3k_m$. The somewhat larger disagreement for smaller wave numbers is, however, not practically important because the product $k^2K(k)$ appears in the diffusion expressions, making the deviation less significant. The simple formula (16) might thus be useful for practical applications. Another simple expression, which fits (37) very well, was suggested by Berkowicz and Prahm (1980).

$$K(k) = K_0 \; [1-\exp(-B^{-1}(k/k_m)^{-4/3}] \qquad (38)$$

with B = 0.87. The curve corresponding to (38) is also shown in Fig. 3.

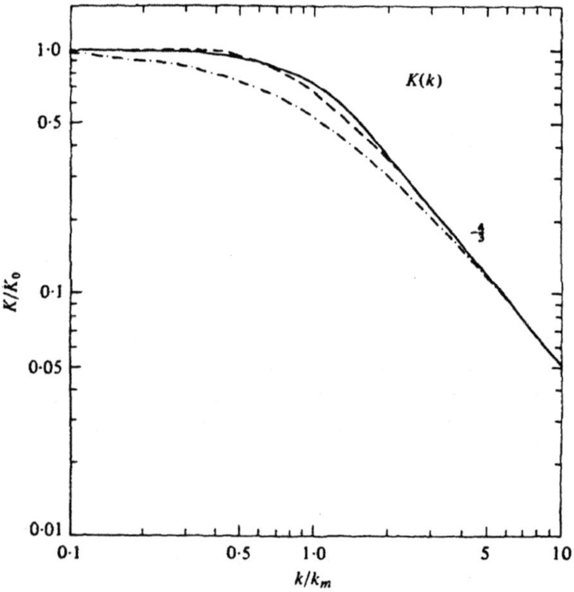

Figure 3. The spectral turbulent diffusivity K(k) estimated from (37). Note that K(k) decreases like $k^{-4/3}$ for large k values. The expression given by (16) is plotted with B = 0.87. K(k) is shown to be well approximated by the simple exponential expression (38), also with B = 0.87. ——, equation (37); —·—·—·, equation (16); ----, equation (38).

6.2. General Form

Deriving the form of K(k) for the inertial sub-range, we have postulated the form of the energy spectrum E(k) and $v(\ell)$. The probability density of occurrence of eddies of length ℓ, $f(\ell)$, was not specified. In fact, $f(\ell)$ is, according to (29) and (35), given by

$$f(\ell) = \begin{cases} (3\pi^2\ell)^{-1} & \text{for } \ell \leq \ell_m \\ 0 & \text{for } \ell > \ell_m \end{cases} \tag{39}$$

In the case of homogeneous and stationary turbulence, eddies are at equilibrium and are equivalently distributed in space. In a volume element of space (a linear element, in the case of one-dimensional turbulence), the number of eddies of length ℓ is proportional to ℓ^{-1}. In general, thus,

$$f(\ell) = \alpha\ell^{-1} \tag{40}$$

where the proportionality constant depends on the range of eddies over which the turbulence is distributed.

The transport velocity $v(\ell)$ can be obtained from (28) and (40):

$$v(\ell) = [2\pi E(\pi/\ell)/f(\ell)]^{\frac{1}{2}}/\ell =$$
$$= [2\pi\alpha E(\pi/\ell)/\ell]^{\frac{1}{2}} \tag{41}$$

Substituting (40) and (41) into (25), we get

$$K(k) = \int_0^\infty [2\pi\alpha^3\ E(\pi/\ell)/\ell]^{\frac{1}{2}} \left(\frac{\sin\frac{1}{2}k\ell}{\frac{1}{2}k\ell}\right)^2 d\ell \tag{42}$$

When the energy spectrum is known, the STD-coefficient can be computed from (42). For large k, i.e. $k \gg 2\pi/\ell_m$, the integral in (42) can be well approximated by

$$K(k) \simeq \int_0^{2\pi/k} [2\pi\alpha^3\ E(\pi/\ell)/\ell]^{\frac{1}{2}}\ d\ell \tag{43}$$

Changing the integration from ℓ to $\kappa = \pi/\ell$, we get

$$K(k) = \gamma \int_{k/2}^\infty [E(\kappa)\kappa^{-3}]^{\frac{1}{2}}\ d\kappa \tag{44}$$

where γ is a new dimensionless constant.

Expression (44) is, in fact, identical with the expression for the spectral eddy viscosity introduced by Heisenberg (1948) in his theory of turbulent energy transport.

We have previously shown that when the energy spectrum exhibits the $k^{-5/3}$-dependence, the STD-coefficient has, for large k, a shape proportional to $k^{-4/3}$. The $k^{-5/3}$-behaviour is characteristic for the unstable stratified turbulent flow in the atmospheric boundary layer. In the case of large-scale turbulence (synoptic scale) and turbulence in the stable stratified boundary layer, the turbulence is mainly characterized by a constant relaxation time τ, i.e.

$$\tau = \ell/v(\ell) = \text{const}$$
or
$$v(\ell) \propto \ell \tag{45}$$

The energy spectrum, according to (28), becomes

$$E(k) \propto k^{-3} \tag{46}$$

Substituting (46) into (44), we obtain for large k,

$$K(k) \propto k^{-2} \tag{47}$$

7. EXPERIMENTAL VERIFICATION OF THE STD-THEORY

The main result of the STD-theory is that in the case of diffusion in a field of homogeneous and stationary turbulence, the time-dependence of the Fourier coefficients of the concentration distribution is given by

$$\tilde{c}(k,t) = \tilde{c}(k,o)\exp[-k^2K(k)t] \tag{48}$$

Prahm et al. (1979) suggest that this relationship be used for a direct test of the STD-theory. The question which must be answered is: Does $\tilde{c}(k,t)$ decrease exponentially with time? Furthermore, by studying the k-dependence of the exponent in (48), the spectral turbulent diffusivity coefficient can be determined. Such work has been carried out at the Department of Fluid Mechanics, Technical University of Denmark. Particle concentration measurements in a plume were made using a Laser Doppler Anemometer (Pedersen et al., 1984). Measurements were carried out in a wind tunnel with grid-generated turbulence. Concentration profiles were measured at different distances from the particle source. Due to the very high resolution of the Laser Doppler Anemometer measurements, Fourier analysis of the profiles was possible (Pedersen and Kristensen, 1982).

In Fig. 4, the Fourier components of the concentration profiles are shown as function of x, the distance from the source. In the log-linear plot used here, the Fourier compounds depict clearly a linear dependence on x. Because the source was not exactly a point source (the diameter was about 3 mm), the lines do not cross at x = 0, but at a distance of about 30 mm from the source. In addition, the turbulence could not be considered exactly stationary in the immediate vicinity of the source. The diffusivity coefficient was determined from the shape of the lines shown in Fig. 4.

In Fig. 5, K(k) is shown as function of the wave number k. It is clearly seen that the diffusivity is here a decreasing function of k. The line corresponding to the $k^{-4/3}$-shape is also shown, but the curve of K(k) does not extend far enough into the range of large k-values to be able to determine the shape. Besides, the measurements were performed on a plume, and as shown by Prahm et al. (1979), the shape of K(k) in this case should approach k^{-1}-behaviour rather than $k^{-4/3}$. Furthermore, the value of k_m is here computed from the frequency spectra of the longitudinal velocity component. The transversal component should preferably be used, but unfortunately, it could not be measured. The value of k_m used in Fig. 5 is thus probably too small.

The study of Pedersen and Kristensen (1982) is the first direct, although qualitative, verification of the STD-theory.

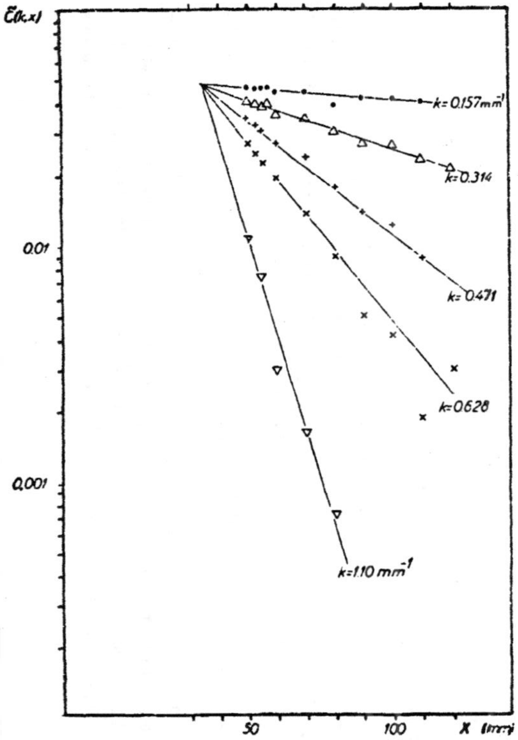

Figure 4. Fourier components of the concentration distribution as function of the distance from the source. The values of k are indicated in the figure (from Pedersen and Kristensen, 1982).

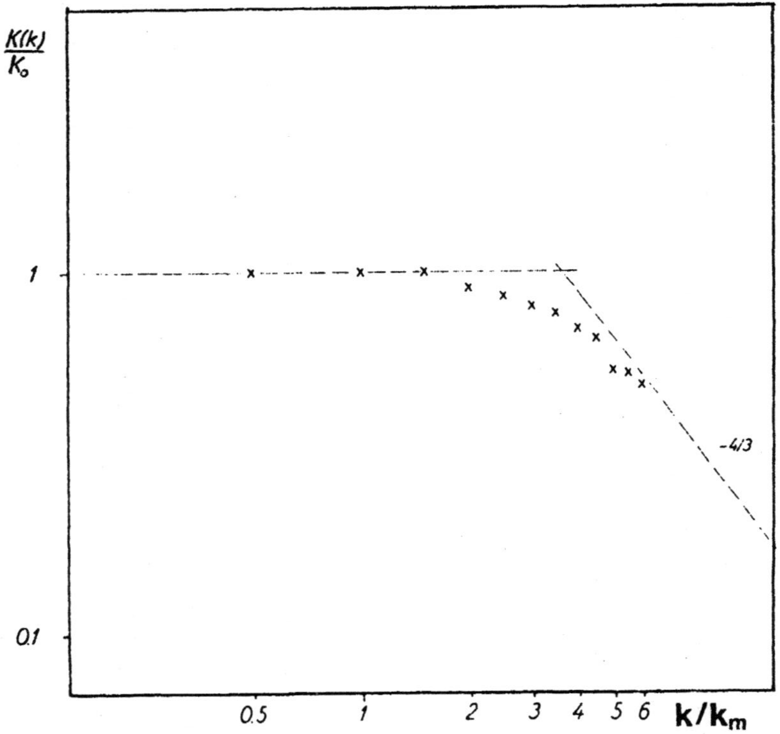

Figure 5. Normalized STD-function as derived from the shapes of the Fourier components shown in Fig. 4 (from Pedersen and Kristensen, 1982).

8. APPLICATION OF THE STD-APPROACH IN NUMERICAL MODELS

The spectral turbulent diffusivity theory is especially easy to apply when a numerical technique based on the so-called pseudo-spectral method is used to solve the transport equation (2) (Christensen and Prahm, 1976; Prahm and Christensen, 1977). The method is based on use of truncated Fourier series for evaluation of the space derivatives of the concentration field. If Eq. (2) is to be numerically solved on a discrete grid where y_i denotes the i-th grid point, we write

$$c(y_i) = \sum_{j=0}^{N} \tilde{c}(k_j) \exp(i\ k_j\ y_i) \tag{49}$$

where

$$k_j = \frac{2\pi \cdot j}{N\ \Delta y} \tag{50}$$

N is the number of grid points and Δy is the space interval.

The space derivative of c is approximated by

$$\frac{\partial c}{\partial y} \simeq \sum_{j=0}^{N} i\ k_j\ \tilde{c}(k_j) \exp(i\ k_j\ y_i) \tag{51}$$

Fast Fourier Transform (FFT) are used to perform both forward and backward Fourier transformations. The space derivatives evaluated by means of (51) are substituted into Eq. (2) and all other operations are performed in the real space. Especially fast and exact methods for time integration are now developed (Zlatev et al., 1982). The diffusion term of Eq. (2) is computed by multiplying Eq. (51) by the diffusivity coefficient K and the space derivative of the flux can now be evaluated. In order to make use of the STD-theory, the constant diffusivity is just replaced by a k-dependent diffusivity.

$$\frac{\partial}{\partial y}\left(K\frac{\partial c}{\partial y}\right) \rightarrow -\sum_{j=0}^{N} k_j^2\ K(k_j)\tilde{c}(k_j) \exp(ik_j y_i) \tag{52}$$

No additional operations are required, just as is the case with use of a constant diffusivity.

When diffusivity is constant in space and does not vary too much in time, it is useful to split the transport equation (2) into advective and diffusive parts and, after each time step, evaluate the change in time of the Fourier components which is due to diffusion only:

$$\tilde{c}(k_j, t+\Delta t) = \tilde{c}(k_j, t) \exp(-k_j^2\ K(k_j)\Delta t) \tag{53}$$

Fourier components computed by means of (53) can now be used to evaluate the space derivatives appearing in the advection terms. The method was used by Zlatev et al. (1983) for simulation of long-range transport of air pollution in Europe. Here, however, the k-dependence of the diffusivity is neglected because the grid-spacing used in the model is 150 km, which is much larger than any typical length scale of turbulence relevant in this case.

The STD-theory has been applied to model transport of pollutants on a global scale in a two-dimensional troposphere-stratosphere model (Berkowicz et al., 1979). Here, use of a proper, scale-dependent diffusivity is of importance. The scale of eddies ranges from a hundred to several thousand kilometers and application of a conventional K-theory is, in this case, very doubtful. The STD-theory in that study was adapted for non-homogeneous turbulence, but in a somewhat primative way. The spectral diffusivity was given by an expression similar to (16), but with K_0 being a function of the space coordinates. The value of ℓ_m used to compute k_m was 10,000 km for the horizontal diffusivity and was set equal to the height of the tropopause for the vertical diffusivity. The model was tested on the radioactive tracer C^{14} originating from the nuclear detonation tests in the stratosphere. When the gradient-transfer approach was used, the model predicted a distribution of the radioactive cloud which was significantly smoothed compared with the measurements. With the STD-approach, the cloud still remained compact several months after detonation, which was also in better agreement with the measurements.

9. FUTURE DEVELOPMENT: NON-HOMOGENEOUS DIFFUSION

The STD-theory presented here was derived under the assumption that the turbulence is homogeneous and stationary. Extension of this theory to also account for non-homogeneous diffusion requires additional work. Some general features can, however, be stated already on the basis of the present results. Considering the integral form of the diffusion equation (Eq. 10), one can see that a straightforward extension of this equation for the non-homogeneous case is to assume that the turbulent diffusivity transfer function is not only a function of the distance in the space, but also a function of the space coordinates. Referring to vertical diffusion in the atmospheric boundary layer, we can thus write

$$\frac{\partial}{\partial t} c(z,t) = \frac{\partial}{\partial z} \int_0^\infty D(z,z') \frac{\partial c}{\partial z'} dz' \tag{54}$$

The non-homogeneous STD-function is not only a function of k, but also dependent on z, and what is more, it is a complex quantity:

$$K(k,z) = K^0(k,z) + i\, K'(k,z) \tag{55}$$

The imaginary part of the complex STD-function can be related to a k-dependent velocity:

$$w(k,z) = k\, K'(k,z) \tag{56}$$

The imaginary part of the non-homogeneous STD-function results from asymmetry of the TDT-function; that means that $D(z,z') \neq D(z',z)$. This behaviour is characteristic for vertical diffusion in the convective boundary layer. There, the diffusive transport is by large convective eddies extending through the whole boundary layer. The upward and downward motions are not equally probable due to skewness of the vertical velocity distribution. Plume diffusion has, under these conditions, a quite peculiar behaviour. In the case of a groundlevel source, the locus of maximum concentration rises, while a decrease is observed for an elevated source. These effects were demonstrated in a series of laboratory experiments by Willis and Deardorff (1976) and also in a numerical model by Lamb (1978). Recently, a Lagrangian Monte Carlo model was constructed by Bærentsen and Berkowicz (1984), and it was demonstrated that when using proper distribution functions for the vertical velocity fluctuation, a good agreement can be obtained with the laboratory experiments of Willis and Deardorff. A Eulerian model can be constructed using the STD-approach for non-homogeneous diffusion. A qualitative agreement with the previously mentioned laboratory experiments was obtained using a complex STD-function (Berkowicz and Prahm, unpublished). However, only the asymmetry due to gradients of the vertical turbulence was accounted for and the skewness effects were not considered. The non-homogeneous energy spectra used to derive the STD-function were obtained from spectral representation of the vertical velocity fluctuations in the convective boundary layer (Berkowicz and Prahm, 1984). It is our opinion that by using these spectra and taking the skewness of the vertical velocity into account, it will be possible to construct a Eulerian model of diffusion in the convective boundary layer.

ACKNOWLEDGEMENTS

The author is indebted to P. N. Pedersen for discussion of the wind-tunnel experiments and for making the results available to us. We wish to thank Mary Bille for help in preparing and typing the manuscript.

REFERENCES

Batchelor, G. K.: 1953, The Theory of Homogeneous Turbulence, Cambridge University Press, pp.197.

Batchelor, G. K. and Townsend, A. A.: 1956, 'Turbulent diffusion', in Surveys in Mechanics, ed. G. K. Batchelor and R. M. Davies, Cambridge University Press, 352-399.

Berkowicz, R. and Prahm, L. P.: 1979, 'Generalization of K-theory for turbulent diffusion. Part I. Spectral turbulent diffusivity concept', J.Appl.Met. 18, 266-272.

Berkowicz, R. and Prahm, L. P.: 1980, 'On the spectral turbulent diffusivity theory for homogeneous turbulence', J.Fluid Mech. 100, 433-448.

Berkowicz, R. and Prahm, L. P.: 1984, 'Spectral representation of the vertical structure of turbulence in the convective boundary layer', Quart.J.Roy.Meteorol.Soc. 110 (in press).

Berkowicz, R., Prahm, L. P. and Louis, J. F.: 1979, 'Global 2-D spectral dispersion model', WMO Rep. no. 538, Proc. WMO Symp. on the Long-Range Transport of Pollutants and its Relation to General Circulation Including Stratospheric-Tropospheric Exchange Processes, Sofia.

Bærentsen, J. H. and Berkowicz, R.: 1984, 'Monte Carlo simulation of plume dispersion in the convective boundary layer', Atm. Env. 18, (in press).

Christensen, O. and Prahm, L. P.: 1976, 'A pseudospectral model for dispersion of atmospheric pollutants', J.Appl.Meteorol. 15, 1284-1294.

Enger, L.: 1983, 'Numerical boundary layer modeling with application to diffusion', A Two Dimensional Higher Order Closure Model, Department of Meteorology, University of Uppsala, Report No. 70-71, pp.54.

Heisenberg, W.: 1948, 'Zur statistischen Theorie der Turbulenz', Z.Phys. 124, 628-657.

Kraichnan, R. H.: 1959, 'The structure of isotropic turbulence at very high Reynolds numbers', J.Fluid Mech. 5, 497-543.

Lamb, R. G.: 1978, 'A numerical simulation of dispersion from an elevated point source in the convective planetary boundary layer', Atm.Env. 12, 1297-1304.

Leslie, D. C.: 1973, Developments in the Theory of Turbulence, Clarendon Press, pp.368.

Lewellen, W. S. and Teske, M. E.: 1976, 'Second-order closure modeling of diffusion in the atmospheric boundary layer', Boundary-Layer Meteorol. 10, 69-90.

Markee, E. H.: 1963, 'On the relationships of range to standard deviation of wind fluctuations', Mon.Weath.Rev. 91, 83-87.

Monin, A. S.: 1955, 'Equation of turbulent diffusion', Dokl.Akad.Nauk SSSR 105, 256-259.

Monin, A. S.: 1956, 'Horizontal mixing in the atmosphere', Isv.Akad.Nauk SSSR, Ser.Geofiz. 3, 327-345.

Monin, A. S. and Yaglom, A. M.: 1965, Statistical Fluid Mechanics of Turbulence, Part I. Washington, Trans-Joint Publication Research Service, pp.769.

Pasquill, F.: 1974, Atmospheric Diffusion, 2nd edition, Wiley, pp.425.

Pedersen, P. N. and Kristensen, H. S.:1982, Experimental Investigation of Diffusion of Particles in Turbulent Flows, AFM 82-04, Department of Fluid Mechanics, Technical University of Denmark, pp.46 (Danish)

Pedersen, P. N., Tambo, M. and Kristensen, H. S.: 1984, Concentration Measurements using a Laser Doppler Anemometer, AFM 81-8, Department of Fluid Mechanics, Technical University of Denmark, pp.23.

Phythian R. and Curtis, W. D.: 1978, 'The effective long-time diffusivity for a passive scalar in a Gaussian model fluid flow', J.Fluid Mech. 89, 241-250.

Prahm, L. P. and Christensen, O.: 1977, 'Long-range transport of pollutants simulated by the 2-D pseudospectral dispersion model', J.Appl. Met. 16, 896-910.

Prahm, L. P., Berkowicz, R. and Christensen, O.: 1979, 'Generalization of K-theory for turbulent diffusion. Part II. Spectral diffusivity model for plume dispersion', J.Appl.Met. 18, 273-282.

Prandtl, L.: 1925, 'Bericht über Untersuchungen zur ausgebildeten Turbulentz', Z.Angew.Math.Mech. 5, 136-139.

Roberts, P. H.: 1961, 'Analytical theory of turbulent diffusion', J.Fluid Mech. 11, 257-283.

Schmidt, W.: 1925, 'Der Massenaustausch in freier Luft und Verwandte Erscheinungen', Probleme der Kosmischen Physic, Hamburg, Verlag Von Henri Grand, pp.176.

Schönfeld, J. C.: 1962, 'Integral diffusivity', J.Geophys.Res. 67, 3187-3199.

Taylor, G. I.: 1921, 'Diffusion by continuous movements', Proc. London Math.Soc. 20 (Ser. 2), 196-211.

Willis, G. E. and Deardorff, J. W.: 1976, 'A laboratory model of diffusion into the convective planetary boundary layer', Quart.J.Roy.Meteorol.Soc. 102, 427-445.

Zlatev, Z., Berkowicz, R. and Prahm, L. P.: 1982, 'Choice of time-integration scheme in pseudospectral algorithm for advection equations', Numerical Methods for Fluid Dynamics, eds. K. W. Morton and M. J. Baines, Academic Press, London, 303-321.

Zlatev, Z., Berkowicz, R. and Prahm, L. P.: 1983, 'Numerical treatment of the advection-diffusion equation. Part 6, Special treatment of the diffusion terms', MST LUFT - A72, Air Pollution Laboratory, National Agency of Environmental Protection, Risø National Laboratory, DK-4000 Roskilde, Denmark, pp.35.

Zlatev, Z., Berkowicz, R. and Prahm, L. P.: 1983, 'Testing subroutines solving advection-diffusion equations in atmospheric environments', Computer and Fluids, 11, 13-38.

Questions, answers and comments

F. B. SMITH: The spectral method you describe recognizes the importance of the size of the eddies relative to the distribution of concentration. In this it meets one of the objections to conventional K-theory. However it does not appear to meet the other major objection, namely that the rate of change of concentration at a point with time depends on the concentration distribution at all y, implying that knowledge of the concentration at distant points is transmitted to the point when $\frac{\partial c}{\partial t}$ is being evaluated infinitely quickly. Could you comment on this?

R. BERKOWICZ: It is true that the S.T.D. method does not take the finite transport time into account. As such, the method must be considered as an approximation to turbulent diffusion, but a better approximation than the ordinary K-theory formulation. The aim of our work was to construct a method which can be applied in a Eulerian-framework, which is as simple as the K-theory approach, but avoids the most important drawbacks of this approach. In my opinion, the time non-locality is less important than the space non-locality. The time non-locality can be taken into account by replacing in equation (17) and further on $c(y \pm \varsigma, t)$ by $c(y \pm \varsigma, t - \frac{\varsigma}{v(l)})$. Inclusion of time non-locality would complicate the model dramatically. The S.T.D. theory can be expected to perform well when the difference in space is larger than the difference in time, considering the most significant eddies.

J.C.R. HUNT: How do you intend to generalize this approach to inhomogenous turbulence? Do you have any systematic way of defining $D(y-y')$ in such flows?

R. BERKOWICZ: We don't have any general method of defining the turbulence-diffusivity transfer function for non-homogenous turbulence. The function will depend on the particular structure of the turbulence. Some overall features of the TDT-function in the case of non-homogenous turbulence are discussed in the paper.

S.P.S. ARYA: I have a conceptual difficulty in understanding how you define the spectrum of the distribution of an ensemble mean concentration.

R. BERKOWICZ: The spectrum of the distribution used in the S.T.D. method is just a Fourier decomposition of the mean concentration profile. It has nothing to do with spectra of concentration fluctuations.

PART III

WIND DISTRIBUTION IN COMPLEX TERRAIN

STRATIFIED AIRFLOW OVER ONE OR TWO HILLS

J.C.R. Hunt
Department of Applied Mathematics and Theoretical Physics
University of Cambridge, Silver St. Cambridge CB3 9EW, U.K.
K.J. Richards
Institute of Oceanographic Sciences,
Wormley, Godalming, Surrey, GU8 5UB, U.K.

ABSTRACT

Methods for calculating, interpolating and idealising air flow in complex terrain are reviewed. Then the general structure of stratified airflow over a single hill of height H and length L_1 is studied in detail and shown to be determined by the upwind velocity profile, the magnitude of a characteristic Froude number and the dimensions of the hill. Let $N(L_1)$ be the buoyancy frequency upwind at a height L_1, and u_* and U_0 be the upwind friction and mean velocity respectively; then the flow is effectively neutral if $u^*/NL_1 >> 1$. But if $u_*/NL_1 < 1$ and $U_0/NL_1 >> 1$, the stratification is weak enough to affect the upwind turbulence and velocity profile but not the dynamics of the flow over the hill. If $U_0/NL_1 < 1$ but $U_0/NH > 1$ the buoyancy forces are strong enough to affect the mean flow over the hill but not strong enough to prevent it passing over the top. In this regime the flow is very sensitive to the form of the upwind temperature profile. If $U_0/NH < 1$, much of the flow passes round the hill. A similar classification, with different flow patterns, is appropriate for unstably stratified flows. When the wind is weak enough, local slope winds can dominate. Results from the analysis of these different regimes are described and compared with laboratory of field measurements where possible. It is shown how some of these results can be extended to groups of hills.

1. INTRODUCTION

Many studies of air flow near the earth's surface are directed towards helping with practical problems such as the making of estimates of air pollution dispersion, or of the potential for wind-energy conversion systems in different sites, or of forces on structures, etc. (see for example the recent book reviewing these applications by Plate (1982). Because of the practical importance of these air flows and because they are strongly affected by the variations in their kinematical and thermodynamical boundary conditions, i.e. the elevation, roughness (defined as the distribution of surface obstacles on scales less than 10 m), and temperature of the surface, the study of these airflows has become an

Boundary-Layer Meteorology 30 (1984) 223–259. 0006-8314/84/0302-0223$05.55

active area of research in fluid mechanics as well as microscale and mesoscale meteorology.

There are two other reasons for the importance of this research. The first is that these variations of surface conditions can significantly change the useful characteristics of these air flows (a change that may or may not be beneficial); for example, wind-energy potential on a hill top may be greater than on a level plain, but air pollution concentration may be worse on the surfaces of hills. The second reason is that in the most populated regions of the world, the natural variations of surface conditions are particularly marked; most major cities are close to boundaries between land and water, and many cities, particularly those whose air is severely polluted, are close to mountains. Of course the cities themselves also change the surface conditions.

The scientific study of the movement of air flow (and heat and water vapour) over complex terrain has three main objectives. (i) To understand the different mechanisms dominating in different circumstances, (ii) to calculate, compute or estimate the air-flow, using the equations of motion, in a given region and given some upwind boundary conditions, and (iii) to develop methods of interpolation (not necessarily based on solving the equations of motion) of the airflow between a number of ground level or remotely-sensed measurements above ground. Of course these latter two objectives are not entirely different because some methods of computation may use as much data as possible for specifying boundary conditions and others use very little. An important reason for developing interpolation methods is that they require much less computer time, but certain assumptions have to be made, for example, that mass is conserved and that, depending on the stable stratification, the air flows round mountains or over them (Sherman 1978). In other words these methods incorporate some basic physical mechanisms about the structure of the flow in complex terrain so the better these mechanisms are understood, the better the interpolation. At present the latter methods are often more reliable than large-scale computations, even with rather sparse data. (This paragraph is largely based on the papers and discussions held at the Euromech 173 at Delphi, which are summarised in Hunt et al (1984).

In this review, we discuss a few idealisations of air flow over complex terrain and describe for each of these (a) the location and scale of the main flow regions which are dominated by different physical processes, (b) the main scaling laws, and (c) results of laboratory modelling and field measurements. We shall use linear analyses wherever possible to help with tasks (a) and (b), because by then applying superposition, many combined effects of changes in elevation roughness and temperature can be estimated. However, we shall also show that there are many important situations or regions of flows where linear analyses are quite inappropriate.

The style of the review reflects the belief of mathematical and experimental researchers that the salient characteristics of flows over "idealised" complex terrain have simple dynamical explanations which can be quantified. Such a belief often leads to useful practical formulaes, but it also leads to interesting hypotheses which can be tested by further analysis or experiment. In that (Popperian) sense it ensures that what we are doing is science!

These analyses are also useful as a means of exploring the limitations of the two main computational methods in current use, namely (i) the hydrostatic model, such as developed by Mahrer & Pielke (1975), in which vertical accelerations are neglected compared with the vertical hydrostatic pressure gradient, and in which turbulent shear stresses and fluxes are approximated rather simply (e.g. eddy diffusivities); (ii) the full dynamical equations combined with more complex but still, approximate equations for the turbulence, e.g. equations of the second-order moments of velocity and temperature, etc. Inevitably the greater the complexity of the model the simpler the flow that can be studied. Such methods hafe, as far as I am aware, been used only for flows over changes in surface roughness (Rao et al. 1974) and for flows over a single hill, in both cases in neutral conditions. For applications to the atmospheric boundary layer these "higher-order" turbulence models (and also large eddy simulation methods) are still mostly being developed for uniform terrain.

In defining idealised forms of airflow over complex terrain, one has to begin by defining the scale for the idealisation. Thus flow over hilly terrain can be thought of as flow over a series of isolated hills (on a scale < 10 km) with ideal upwind conditions or with each flow determined by complex upwind boundary conditions, determined by the other hills. Or one could consider the hills as pairs whose air flow interacts. The shape of the hills may be idealised as some smooth curve by ignoring the small-scale undulations on the hill (Walmsley et al. 1982). The purpose of these idealisations is to help the understanding, calculations or interpretations of flow around one or two hills. They do not help the understanding of air flow over and around a whole area of hilly terrain, perhaps on a scale of tens or hundreds of kilometers.

Large scales of air flow over complex terrain have to be idealised differently. The most appropriate form of this idealisation depends critically on the prevailing meteorological conditions; for example in strong winds with neutral or moderately stable conditions, when air flow largely passes <u>over</u> hills, they can probably be idealised as rough surface undulations (Blumen 1965). In strongly stable conditions when the air flows tends to travel in horizontal planes around hills, they can be idealised as a distribution of resistance within the flow extending up to the tops of the mountains (Newley et al. 1984; Hunt et al. 1983).

There are other meteorological conditions when a range of hills primarily acts as a region of strong heating or cooling of the atmosphere; for example at night the difference in temperature between the centres and the sides of the valleys drives katabatic winds which extend well outside the range of hills and can continue all night often intermittently.

These winds may develop a strong gravity current bounded by a marked front (Blumen 1983). An interesting idealisation of how temperature differences induce large-scale flows (e.g. sea breezes) in a stratified rotating atmosphere by generating large internal waves is developed by Rottuno (1983). These are just some of the examples of how local idealisations differ from the regional idealisations made by meso- and synoptic meteorologists.

This review is restricted to an example of local idealisations of complex terrain, namely flow over and around one and two hills. But there is now a growing interest in studying air flow in complex terrain over distances ranging from the local to the mesoscale because of regional air pollution problems and the need to find areas of high wind energy.

2. STRATIFIED FLOW OVER A HILL WITH LOW SLOPE

2.1. Defining the flow regions

This section develops a general method of calculating or estimating the flow of a turbulent boundary layer with various temperature and velocity profiles over a hill which may have various shapes or scales. Provided certain assumptions are satisfied, which correspond to a wide range of natural conditions of practical interest, the analyses lead to useful approximate formulae and understanding of the velocity field.

We assume that the length of the hill L_1 is much larger than its height H, and also larger than the vertical scale h_m of the region of significant shear in the upwind velocity profile.

As usual we take the turbulent velocity to be determined by the surface shear stress $\rho\tau_o$ and to be of the same order as $\sqrt{\tau_o} = u_*$; the surface friction velocity (ρ is the density). It is assumed that u_* is much less than the mean velocity in the boundary layer $U_o(z)$. This is why any changes in surface roughness, shear stress or temperature over a hill are confined to an inner region of thickness ℓ which is much less than L_1 (Fig. 1 and Section 2.2). In the lower part of this region the inertial and shear stresses are in balance. Inner regions form and have similar structure whether there is a surface change in elevation, roughness or temperature (Hunt & Simpson 1982).

Because this layer is thin, the hill displaces the flow and the temperature profile abote it, leading to accelerations and buoyancy forces in the outer region which extends upwards to a distance L_1 and even higher when internal waves are generated in stratified flows.

Much of this Section is drawn from papers by Hunt, Leibovich & Richards (1984) and Hunt, Richards & Brighton (1984) denoted hereafter by HLR and HRB. These papers develop the earlier work of Jackson & Hunt (1975), Sykes (1980), Brighton (1977).

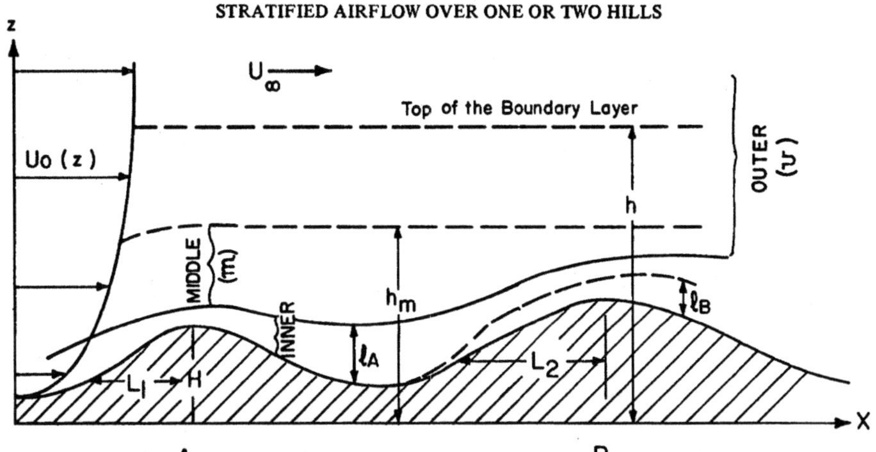

Fig. 1. Schematic diagram of air flow over two hills showing the flow regions for the upwind hill (A). ($\mathcal{M}$) and ($\mathcal{U}$) denote the middle and upper layers of the outer region. $U_0(z)$ is the mean velocity profile in the upwind region. U_∞ is the wind above the boundary layer. h is the boundary layer height. ℓ_A, ℓ_B are the height of the inner region of hill A and B respectively. The inner region of hill B grows within the wake of inner region of hill A. h_m is the height of the middle layer. L_1, L_2 are the length of hill A and B respectively and H is the height of the hill.

2.2. Inner Region Analysis

Governing equations. In the inner regions over hills with low slopes (i.e. $H/L_1 \ll 1$), to the first approximation the air flow follows the surface of the hill ($z = Hf(x)$). So in a perturbation analysis it is natural to let the mean horizontal and vertical velocity components, the mean shear stress $\rho\tau$, the mean potential temperature Θ and the heat flux $\overline{w\Theta}$ be expressed as the displaced upwind profile plus perturbations to it, viz.

$$\bar{u}(x,z) = U_0(\Delta z) + \Delta\hat{U}(x,\Delta z), \quad \bar{w} = H\, f'(x)\, [U_0 + \Delta\hat{U}] + \Delta\hat{W}$$

$$\rho\tau = \rho(\tau_o + \Delta\hat{\tau}), \quad \Theta = \Theta_o(z_o) + \Delta\Theta_o(\Delta z) + \Delta\hat{\Theta}(x,\Delta z) \tag{2.1}$$

$$\overline{w\Theta} = (w\Theta)_o + \Delta\overline{w\Theta}$$

where $\Delta z = z - H\cdot f(x)$. But in a stratified flow the first approximation to the pressure is the pressure at the same height upwind plus a perturbation correction. We shall denote by $p(x,z)$ the pressure over the hill divided by ρ (the air density), and the upwind pressure divided by ρ as $p_o(z)$.

$$p(x,z) = p_o(z) + \Delta\hat{p}(x,\Delta z) \text{ where } \frac{dp_o}{dz} = g\,\frac{\Theta_o + \Delta\Theta_o(z)}{\Theta_o}$$

and then the governing equations (to first order in $\Delta U/U_0$ and H/L_1) are

$$U_0 \partial\Delta\hat{U}/\partial x + \Delta\hat{W}\, \partial U_0/\partial\Delta z = -(\partial\Delta p/\partial x)_{\Delta z} - (\partial\Delta\hat{p}/\partial\Delta z).\ \partial\Delta z/\partial x + \partial\Delta\tau/\partial\Delta z \quad (2.2a)$$

$$U_0 \partial\Delta\hat{W}/\partial x = -\ \partial\Delta\hat{p}/\partial\Delta z + (g/\Theta_0)(\Delta\Theta_0 + \Delta\hat{\Theta} - \Delta\Theta_0(z)) \quad (2.2b)$$

$$\partial\Delta\hat{U}/\partial x + \partial\Delta\hat{W}/\partial\Delta z = 0 \quad (2.2c)$$

$$U_0\ \partial\Delta\hat{\Theta}/\partial x + \Delta\hat{W}\partial\Delta\Theta_0/\partial\Delta z = \partial\Delta\overline{w\Theta}/\partial\Delta z \quad (2.2d)$$

Note that the term $(-\partial\Delta p/\partial\Delta z)\partial\Delta z/\partial x$ only has to be considered in strongly stratified flows when perturbations to hydrostatic pressure gradients may be significant. For the problems considered here it is negligible; it is $O(H^2/L_1^2)$. The perturbation in temperature $\Delta\Theta$ is shown by (2.2) to be caused by convergence or divergence of the streamlines or by a change in the surface flux.

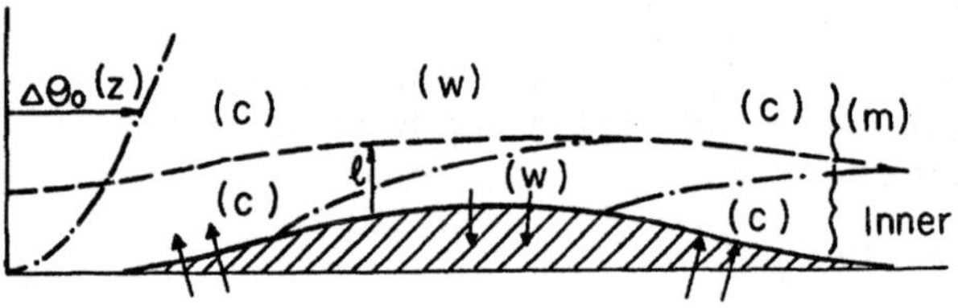

Fig. 2. Perturbation temperature field $\Delta\hat{\Theta}$ over a hill, in the inner and middle regions when the upwind air flow is stable and the ground is cooling. Zones of warmer ($\Delta\hat{\Theta}>0$) or cooler ($\Delta\hat{\Theta}<0$) air are denoted by (w) and (c). Surface arrows denote the changes in the perturbation surface heat flux induced by changes in $\Delta\hat{\tau}$.

<u>Flux gradient relations.</u> The turbulence near the surface in the inner region is in local equilibrium so that changes in shear stress $\Delta\hat{\tau}$ and heat flux $\Delta\overline{w\Theta}$ can be estimated from Monin-Obukhov surface layer similarity results. (Since the shear stress and heat flux tend to vary rapidly (Fig. 2), the use of M-O similarity is only justifiable in the lowest part of this layer). Since

$$\frac{\partial u}{\partial z} = \frac{\tau^{\frac{1}{2}}}{kz}\ (1 + \alpha z/L_{Mo}) = \frac{\tau^{\frac{1}{2}}}{kz}\ \left(1 + \frac{\alpha z g\ \overline{w\Theta} k}{\tau^{3/2}\Theta_0}\right)$$

by substituting the expression in (2.1) and using binomial expansions, we find

$$\Delta\hat{\tau} = \frac{2ku_* \, \Delta z \, \partial\Delta\hat{U}/\partial\Delta z}{(1-2\alpha \, \Delta z/L_{Mo})} - \frac{2u_*^2\alpha\Delta z}{L_{Mo}} \frac{\Delta\overline{w\Theta}}{(\overline{w\Theta})_o} \quad (2.3)$$

where L_{Mo} is the upwind Monin-Obukhov length, $u_*^2 = \tau_o$, and $\alpha(\simeq 5)$, $k(\simeq 0.4)$ are universal constants. The denominator of the first term is caused by $\Delta\hat{\tau}$ increasing the local L_{Mo}.

We can estimate the change in the heat flux $\Delta\overline{w\Theta}$ from the known variation of surface heat flux $\overline{w\Theta}$ (o) with surface stress $\tau(o)$ and the angle n between the sun and surface normal. From Thom's (1975) and Venkatram's (1980) correlation of stable surface-layer data.

$$\Delta\overline{w\Theta}(o) = \overline{w\Theta}(o) \left(\chi_\tau \frac{\Delta\tau(o)}{\tau(o)} - \chi_R \frac{\delta\hat{n}}{\hat{n}_o} + \chi_\Theta \frac{\Delta\hat{\Theta}(o)}{\Theta_o} + \chi_q \, \Delta q/q_o\right) \quad (2.4)$$

where the first term is the shear term, the second term is the radiation term, the third term is a temperature term and the fourth term is the humidity term.

In unstable neutral conditions χ_τ varies between about 0.3 and 0 depending on the amount of transpiration from plants. In stable conditions, where the flux is directly proportional to mechanical transfer, $\chi_\tau \simeq -\frac{1}{2}$ (Venkatram 1980). $\chi_R, \chi_\Theta, \chi_q$ depend on atmospheric conditions, and χ_Θ, χ_q also depend on plant cover. Estimates of these terms have not been tested in complex terrain.

Laboratory measurements show that the change in water vapour flux from saturated undulating surfaces is proportional to $\Delta\hat{\tau}/\tau(o)$ (Dawkins & Davies 1981; Verma & Cermak 1971).

Suppose that $\Delta\overline{w\Theta}(o) = \overline{w\Theta}(o) \cdot \chi_\tau \dfrac{\Delta\hat{\tau}(o)}{\tau(o)}$

we get by combining (2.3) and (2.4) the perturbation shear stress $\Delta\hat{\tau}$ which can be expressed in terms of the local perturbation velocity gradient as

$$\Delta\hat{\tau}(x, \Delta z) = 2k\hat{\phi} \, (\Delta z/\ell, \ \ell/L_{Mo}) \, u_*\Delta z \, \partial\Delta\hat{U}/\partial\Delta z \quad (2.5a)$$

where $$\hat{\phi} = [1-2(1-\chi_\tau)\alpha\Delta z/L_{Mo}]^{-1} \quad (2.5b)$$

when $\Delta z/\ell << 1$, $\alpha\ell/L_{Mo} << 1$

$\Delta\hat{\tau}$ can be approximated by its surface value and then:

$$\Delta\hat{U} \simeq \frac{1}{2} \frac{\Delta\hat{\tau}(x,z_0)}{ku_*} \left(\ln \Delta z/z_o - 2(1-\chi_\tau)\alpha\Delta z/L_{Mo}\right) \quad (2.5c)$$

This means that near any surface, whether hilly or not, a given increase in shear stress so alters the turbulence and the stratification that the increase in the mean velocity gradient is greater in unstable than in stable conditions (Fig. 3a). In other words the perturbation eddy viscosity increases with the Richardson number. (This result is of course equally valid in the inner region at a change in roughness.

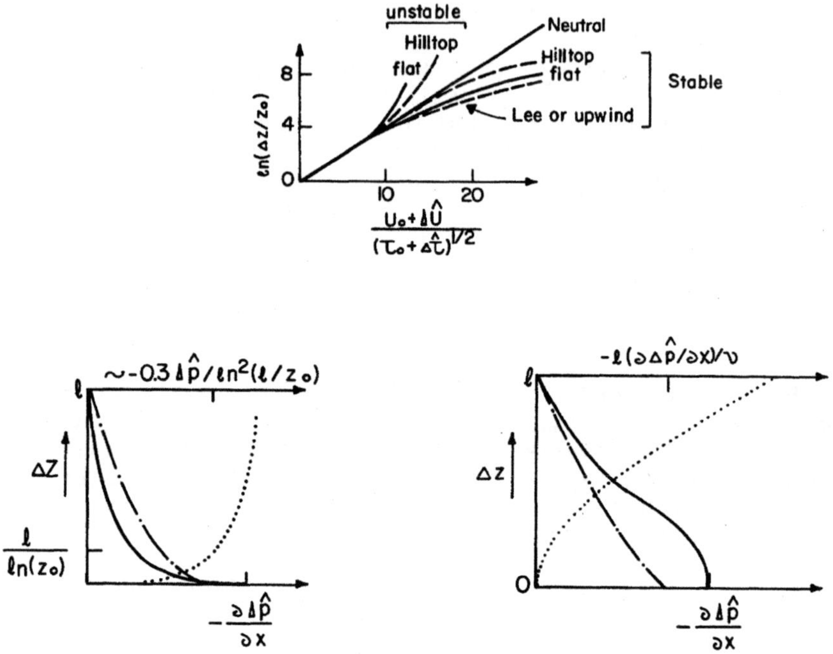

Fig. 3. Structure of the inner region over a hill. a) Mean velocity profiles: (b) Profiles of a perturbation shear stress $\Delta\hat{\tau}$ and perturbation velocity $\Delta\hat{U}$ for (i) turbulent, and (ii) laminar (constant eddy viscosity).

———— $\partial\Delta\hat{\tau}/\Delta z$

—·—·—· $\Delta\hat{\tau}$

········ $U_o(\Delta z)\partial\Delta\hat{U}/\partial x$

In the outer part of the inner layer the perturbation shear stress becomes very small, so that the turbulence structure is determined by upwind conditions and by the straining of the turbulence caused by changes in the mean flow (Britter et al. 1981). In other words, the turbulence structure does not depend on the local changes in heat flux very near the surface. A suitable hypothesis is that the perturbation eddy diffusivity is approximately the same as upwind so

$$\hat{\phi}(\ell) = 1/(1 + \alpha\ell/L_{Mo})$$

This is consistent with laboratory and field observation (e.g. Laver et al. 1981) that ℓ is reduced in stable conditions.

Distribution of $\Delta\hat{\tau}$ and $\Delta\hat{U}$. An understanding of how the perturbation shear stress $\Delta\tau$ is determined in general terms by the modelling of the turbulence, the nature of the outer region flow and the effects of stratification in the inner region, can be derived from an equation for $\Delta\hat{\tau}$ obtained from (2.2a) and (2.5). By differentiating (2.2a) w.r.t Δz and then multiplying by $2ku_*\Delta z\hat{\phi}/U_0$, $\Delta\hat{\tau}$ is described by a kind of diffusion equation with a source term

$$\frac{\partial\Delta\hat{\tau}}{\partial x} - \frac{2ku_*\hat{\phi}\Delta z}{U_0(\Delta z)}\frac{\partial^2\Delta\hat{\tau}}{\partial\Delta z^2} = -2ku_*\hat{\phi}\Delta z\Delta\hat{W}U_0''/U_0(\Delta z) - \frac{2ku_*\Delta z\hat{\phi}}{U_0(\Delta z)}\frac{\partial^2\Delta p}{\partial x\partial\Delta z} \quad (2.6)$$

From the l.h.s the thickness $\ell(x)$ of the inner layer, for changes of elevation, roughness and temperature, following Jackson & Hunt (1975), can be estimated as

$$d\ell/dx \simeq 2k^2\hat{\phi}(\ell/L_{Mo})/\ln(\ell/z_0) + \alpha\ell/L_{Mo}) \quad (2.7a)$$

so that an average value over a hill is given by

$$\ell \sim 2k^2L_1\ \hat{\phi}(\ell/L_{Mo})/(\ln(\ell/z_0) + \alpha\ \ell/L_{Mo} \quad (2.7b)$$

For a typical hill in neutral conditions $\ell/L_1 \sim 1/20$. Examples of other inner layers are given by Hunt & Simpson (1982).

From (2.2a) the boundary condition for $\Delta\hat{\tau}$ at the surface $\Delta z = z_0$, where $\Delta\hat{U} = \Delta\hat{W} = 0$ is

$$\partial\Delta\hat{\tau}/\partial\Delta z(z_0) = \partial\Delta\hat{p}/\partial x \quad (2.8a)$$

Thence a general relation between $\Delta\hat{p}$ and $\Delta\hat{\tau}$ can be obtained by dividing through by $\Delta z\phi/Uo$ and integrating (2.6) with respect to Δz from z_0 to $\ell(x)$ using the continuity equation. We use the outer boundary condition that $\Delta\hat{\tau}\to o$ with all its derivatives when $\Delta z/\ell >> 1$ and when $x/L_1 \to \infty$

$$\frac{\partial}{\partial x}\int_{z_0}^{\ell}\{Uo(\Delta z)\ \frac{\Delta\hat{\tau}}{\phi} + \Delta z U_0''\ 2ku_*\int_0^{\Delta z}\Delta\hat{U}dz'\}\ \frac{d\Delta\tau}{\Delta\tau} = -\ 2ku_*\ \frac{\partial\Delta\hat{p}}{\partial x}\ (\Delta z = \ell) \quad (2.8b)$$

Since the integrand is greatest where $\frac{\Delta z}{\ell} << 1$ and

where $\Delta\hat{U} = \frac{1}{2} \frac{\Delta\hat{\tau}(o)}{ku_*} \ell n \frac{\Delta z}{z_o}$ and $U''_o = -\frac{u_*}{k\Delta z^2}$

then (2.8b) reduces to

$$\int_{z_o}^{\ell} \ell n \frac{\Delta z}{z_o} \left(\frac{\Delta\hat{\tau}}{\phi} + G\Delta\hat{\tau}(o)\right) \frac{d\Delta z}{\Delta z} = -2k^2\Delta\hat{p}(\ell) \tag{2.9}$$

where

$$G = (\ell n \frac{\Delta z}{z_o} - 1) / \ell n \frac{\Delta z}{z_o} \simeq 1 \text{ when } \ell n \frac{\Delta z}{z_o} >> 1.$$

Since $\ell n \frac{\ell}{z_o} >> 1$ the integral of (2.9) can be expressed as

$$\Delta\hat{\tau}(x,z_o) = -\frac{2k^2\Delta\hat{p}(\ell)}{\ell n^2(\frac{\ell(x)}{z_o})(1+\lambda)} \tag{2.10}$$

where $\lambda = 0\left[\frac{\Delta\hat{\tau}(\frac{\ell}{2})}{\Delta\hat{\tau}(z_o)} \cdot \frac{1}{\hat{\phi}(\frac{\ell}{L_{Mo}})}\right]$

The general result (2.10) has many implications. (i) In order that (2.8a) and (2.10) should be consistent, it means that the vertical gradient of $\Delta\hat{\tau}$ near the surface is much larger than $\Delta\hat{\tau}/\ell$, since

$$\Delta\hat{\tau} \sim -\Delta\hat{p}/\ell n^2(\ell/z_o), \text{ (2.8a) and (2.7b) imply}$$

$$\partial\Delta\hat{\tau}/\partial\Delta z\ (z_o) \sim \Delta\hat{\tau}/\ell . \ell n(\ell/z_o)$$

In other words $\partial\Delta\hat{\tau}/\partial\Delta z$ is $O(\ell n(\ell/z_o))$ greater near the surface than through the bulk of the inner layer, as Sykes (1980) pointed out, while $\partial\Delta\hat{U}/\partial x$ is much smaller near the surface. The schematic diagram of Fig. 3 shows how $\Delta\hat{\tau}$ and $\Delta\hat{U}$ vary in the layer. An important consequence is that where $\Delta\hat{\tau}$ decreases monotonically, the effect of $\Delta\hat{\tau}$ on $\Delta\hat{U}$ is largely confined to the lowest part of the layer. This is not necessarily true in the wake where $\Delta\hat{\tau}$ increases and then decreases.
(ii) If the flow in the inner region is described by a constant viscosity (or eddy viscosity) ν, then approximately,

$$\partial\Delta\hat{\tau}/\partial x \simeq (\nu/U_o(\ell))\partial^2\Delta\hat{\tau}/\partial\Delta z^2 \tag{2.11a}$$

In fact $\partial^2\Delta\hat{\tau}/\partial\Delta z^2$ is zero as $\Delta z \to 0$, so

$$\Delta\hat{\tau}(z_0) \simeq -\left(\frac{\partial\Delta\hat{\tau}}{\partial z}\right)\ell \simeq \frac{\partial\Delta\hat{p}}{\partial x} \qquad (2.11b)$$

so the surface shear stress is in phase with the pressure gradient and $\partial^2\Delta\hat{\tau}/\partial\Delta z^2$ is greatest at the outer edge of the inner region.

(iii) Thus in stably stratified flows over hills, where large gradients of surface pressure are induced by the outer region flows, there are large qualitative differences between the behaviour of the surface stresses in laminar or turbulent flows. Note that even if the inner region becomes stably stratified, it does not tend to behave like a laminar (or constant eddy viscosity) flow, because near the surface (where the integral is weighted), the flow remains turbulent.

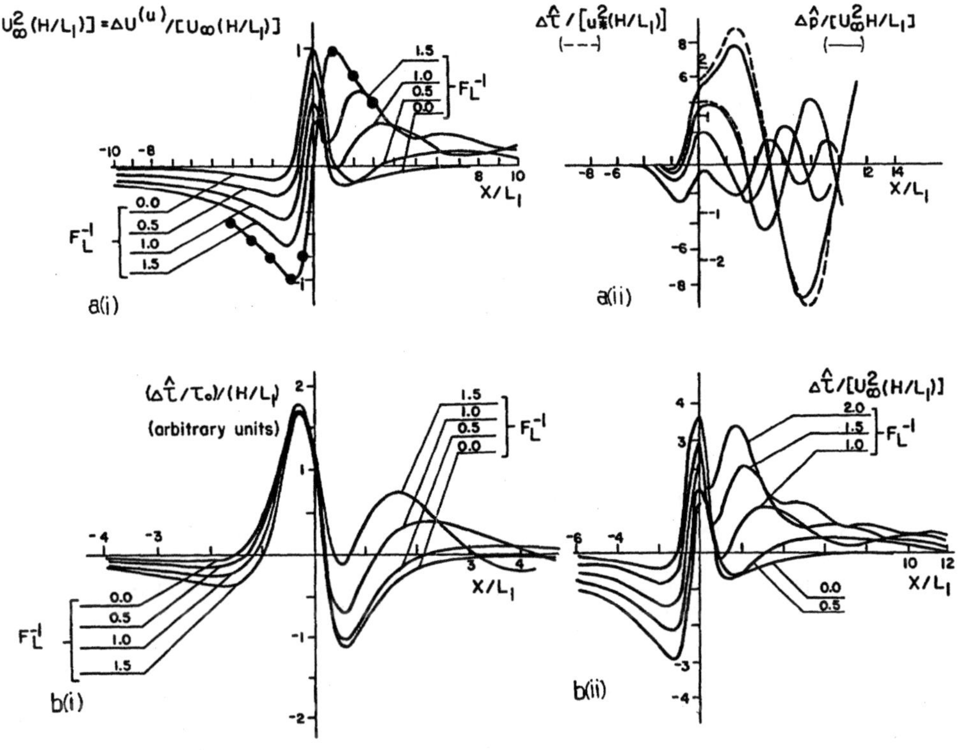

Fig. 4. Stably stratified flow over a hill. a) Surface pressure $\Delta p(x,o)$ which is proportional to the perturbation velocity ΔU in the upper layer, for uniform stratification (i) in an unbounded flow (—●— denotes the asymptotic solution $F_L\Delta UL_1/(H\ U_\infty)$) and (ii) with a strong inversion at $z=3L_1$, (b) Perturbation surface shear stress $\Delta\hat{\tau}$ for laminar flows (i) and turbulent flows, (ii) (unbounded uniform stratification); a (ii) also includes surface shear stress (----). (F_L is the Froude number).

(iv) The magnitude of $\Delta\hat{\tau}(x,z_0)$ and its distribution, (using (2.7a) for $\ell(x)$, taking $\ell(x)=0$ where $\Delta\hat{p}$ changes sign), is reasonably well estimated by (2.10) for a variety of flows when the inner region is effectively neutral, even though the outer region is not (i.e. $\ell/L_{Mo} << 1$). In Fig. 5 we compare for neutral flow over a wavy surface the estimate of (2.10) with the solution of the approximate equation for $\Delta\hat{\tau}$ obtained by omitting the term in $\Delta\hat{w}\partial U_0/\partial\Delta z$ in (2.2a), (Jackson & Hunt 1975), and with computations of the full equations (2.2a-c) using the closure (2.5) (Richards & Taylor 1980). Note how the phase lead by $\Delta\hat{\tau}$ decreases as the roughness length z_0 decreases.

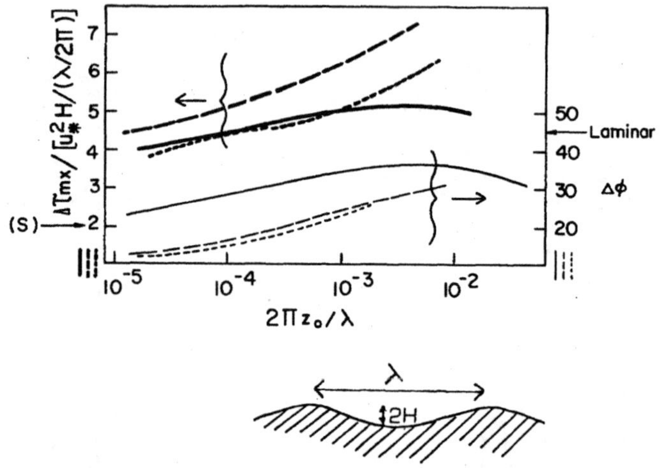

Fig. 5. Different calculations for maximum perturbation surface shear stress $\Delta\hat{\tau}_{mx}$ and the phase difference between the surface shear stress and the surface $\Delta\phi$. ——— computation (Richards & Taylor 1980); - - - approximate innerlayer solution (Jackson & Hunt 1975) ······· estimates from (2.10).

(v) Some surface shear measurements over hills have been made in a wind tunnel by Meroney et al. (1978) and compared with surface pressure measurements. On the upwind face they found that $\Delta\hat{\tau}$ was approximately proportional to $\Delta\hat{p}$, and that $\Delta\hat{\tau}/(\Delta\hat{p}(u_*/U(L))^2)\simeq 2{\cdot}2$. The theoretical value from (2.10) is approximately 3, the appropriate value of z_0 being rather uncertain. It is interesting to note that the result (2.10) gives a good estimate for the surface shear stress distribution $\hat{\tau}(x)$ in most turbulent boundary layers (outside regions of separated flow), for example aerofoils and bluff bodies. (See for example the measurements on a circular cylinder by Achenbach 1968).

(vi) Near the surface within the inner region $\Delta\hat{U}$ is given by (2.5c), so that if $\Delta\hat{U}$ increases monotonically, as it does over the top of a hill, the expression (2.10) for $\Delta\hat{\tau}$ leads to an estimate for

$$\Delta\hat{U}\ (x,\ell) \sim \frac{-\Delta\hat{p}\ k}{u_* \ell n(\ell/z_o)} \sim -\Delta\hat{p}/U_o(\ell) \qquad (2.12)$$

We shall show from the outer region analysis in the next section that this estimate is also valid in stratified flows.

Lee side and wake flows. On the lee slopes of hills in neutral or stable conditions, Δp may be positive, or slightly negative, while $\partial\Delta\hat{p}/\partial x$ is large and positive. Then $|\Delta\hat{\tau}|$ and $\Delta\hat{U}$ do not vary monotonically with Δz within the inner region (Fig. 3). Typically $\Delta\hat{\tau}$ and $\Delta\hat{U}$ have maximum values above the surface. In such situations if the slope is greater than about 1/10, the turbulence should be generated and controlled by this elevated shear layer rather than by shear at the surface. This is probably the reason why the perturbed surface layer model for shear stress in (2.3) overpredicts $\Delta\hat{\tau}(z_o)$ and underpredicts ΔU (Britter, Hunt & Richards 1981). The model is not as appropriate as a model based on the turbulence in a free shear layer (Plate & Lin 1965). Hunt, Leibovich & Lumley (1983) have suggested how a suitable model might be constructed which includes these two limiting cases, and could be applied to hills which have large or small slopes.

Further downwind in the wake the turbulence structure is also not determined by local production and dissipation. The assumption of a local eddy viscosity is not physically justifiable. Nevertheless, useful estimates for the wake structure have been made on this basis (Counihan et al. 1974): their limitations have been pointed out by Castro (1979). The effects of stable stratification on the wakes of surface obstacles and hills depends largely on the outer region flow which we can consider.

2.3. Outer region

Governing equations and boundary conditions. Since in the outer region of flows over low hills ($H \ll h_m$) with low slope ($H/L_1 \ll 1$) the mean velocity is equal to the upwind velocity $U_o(z)$ at the same elevation above level ground, it is natural to express $\bar{u}$ in terms of perturbations to U_o.

The perturbed flow in the outer region is driven by the vertical displacement of the flow in the inner region. Far above the hill this displacement tends to zero and this suggests that the vertical displacement in the outer region is of the same order as the horizontal scale in the inner region, L_1. Therefore the perturbation velocity in the outer region is expressed as function of x/L_1, so

$$\bar{u} = U_o(z) + \Delta U(x/L_1, z/L_1),\ \bar{w} = \Delta W(x/L_1, z/L_1) \qquad (2.13)$$

where

$$\Delta U \ll U_o,\quad \Delta W \ll U_o.$$

If the height of the hill is comparable with that of the boundary layer, it is more appropriate to express $\bar{u}$, $\bar{w}$ in terms of (2.1).

The extensive numerical, analytical and experimental work on neutral flows supports the order-of-magnitude analysis that, when $z>\ell$, the perturbed flow is essentially unaffected by shear stresses. Even in strongly convective conditions in the outer region, the gradient of the perturbation shear stress is negligible compared with the acceleration of the perturbation flow, i.e.

$$\frac{\partial \Delta\tau}{\partial z} \; \left(\sim \Delta\tau/L_1 \sim \frac{u_*^2 \Delta U}{L_1 U_o}\right) << U_o \; \frac{\partial \Delta U}{\partial x} \; \left(\sim \frac{U_o \Delta U}{L_1}\right) \tag{2.14}$$

Observations indicate that, even when the mean temperature distribution is unstable ($z/L_{MO}<0$), the turbulence and mean flow over a hill do not exhibit any singular growth and downwind of the hill the flow returns to its upwind form. (Linear stability analysis indicates that the only modes which can remain stationary relative to the hill are decaying. The growing modes travel past; these are the turbulent fluctuations over the hill).

Another important aspect of the outer region is that the gradient of the perturbation heat flux $\partial \Delta\overline{w\theta}/\partial z$ is negligible compared with the rate of change in temperature produced by the deflection of the streamlines $\Delta\hat{w}\partial\Delta\Theta_o/\partial z$.

Consequently the mean flow is governed by the equations for perturbations to stably or unstably stratified flows where shear stresses are negligible and the heat flux between streamlines is negligible. Over a single two-dimensional hill, ΔW satisfies

$$[\partial^2/\partial x^2 + \partial^2/\partial z^2 + S] \quad \Delta W = 0 \tag{2.15a}$$

where

$$S = [N^2/U_o^2 - d^2U_o/dz^2/U_o]; \; \pm N^2 = \frac{g}{\Theta_o} \cdot \frac{d\Delta\Theta_o}{dz} \tag{2.15b}$$

and for a three-dimensional hill

$$\left[\frac{\partial^2}{\partial x^2} \quad (\nabla^2 - d^2U_o/dz^2/U_o) + N^2/U_o^2 \, \nabla_H^2\right]\Delta W = 0$$

where

$$\nabla^2 = \partial^2/\partial x^2 + \partial^2/\partial y^2 + \partial^2/\partial z^2; \; \nabla^2_H = \partial^2/\partial x^2 + \partial^2/\partial y^2 \tag{2.16}$$

and S may be positive or negative (Scorer 1949); HRB).

Boundary conditions. As $z/L_1 \to 0$, ΔW is determined by matching with the vertical velocity of the displaced profile of (2.1), so for near-neutral profiles the appropriate boundary conditions is that

$$\Delta W(z/L_1 \to 0) \simeq (H/L_1)\, f'(x/L_1)\, U_0(z \simeq \ell) \qquad (2.17a)$$

The boundary condition when $z/L_1 >> 1$ depends very much on the atmospheric conditions. In neutral flows it is sufficient to apply the condition

$$z/L_1 \to \infty\ ,\ \Delta W \to 0 \qquad (2.17b)$$

but in stably stratified flows when $gd\Delta\Theta/dz/\Theta_0 > 0$ then the appropriate radiation condition must be applied, namely that the waves radiate away from the hills (Smith 1980). The general form of the far field solutions to (2.15) for various S(z) are given by Bois (1983).

In unstable conditions, vertical perturbations are likely to be limited by an elevated inversion layer at $z = h_i$ (as in Bradley's (1980) experiments). The variation of the inversion height at h_i is analysed later. Solutions to (2.15) when $N^2 < 0$ are given by HRB.

Defining the middle and upper layers. Inspecting the governing equations (2.15) for the outer region, shows that the perturbed flow near the surface is generally quite different to the flow at and above the top of the boundary layer. Near the surface, but above the inner region, $d^2U_0/dz^2 \propto 1/z^2$, so $\partial^2\Delta W/\partial z^2$ must be large enough to balance $S\ \Delta W$. Typically $\partial^2\Delta W/\partial z^2$ is of order $\Delta W/h_m$ where h_m^2 is the thickness of this middle layer (M). Consequently if the length of the hill L_1 is large compared with h_m,

$$\partial^2\Delta W/\partial x^2 << \partial^2\Delta W/\partial z^2, \text{ and } \partial^2\Delta W/\partial x^2 << S\ \Delta W \qquad (2.18a)$$

Thus h_m can be estimated as $S(h_m) \sim 1/L_1^2$, whence

$$h_m \sim L_1/\ \ell n^{\frac{1}{2}}(L_1/z_0) \qquad (2.18b)$$

for a neutral boundary layer, such that $h \gtrsim L_1$. In (M), the middle layer equations (2.15) (2.16) can be simplified by ignoring $\partial^2\Delta W/\partial x^2$ and $\partial^2\Delta W/\partial y^2$.

A further simplification is often possible by ignoring the effects of stratification in the middle layer.

The surface layer similarity theory of Monin-Obukhov shows that the ratio of the two terms in the Scorer parameter is, when $z/L_{Mo} > 1$,

$$U_0''/U_0/(N/U_0)^2 \simeq 4(\sigma_w/N)\ (1 + \sigma_w/zN)^2/z \qquad (2.19a)$$

$$\text{Thus}\quad U_0''/U_0 >> (N/U_0)^2 \quad \text{when} \quad z \lesssim 4(\sigma_w/N) \qquad (2.19b)$$

where σ_w^2 is the variance of the vertical turbulence component.

In recent field measurements over rolling terrain near the Rocky Mountains on the 300 m tower at the Boulder Atmospheric Observatory, Hunt et al. (1983) found that over a wide range of stable conditions, with L_{Mo} varying from 30 to 300m,

$$|d^2U_o/dz^2/U_o| \gtrsim 5\ (N/U_o)^2 \tag{2.19c}$$

Consequently if (2.19b) or (2.19c) are valid, (2.15) and (2.16) reduced in the middle layer to

$$\partial^2\Delta\hat{W}/\partial z^2 - (d^2U_o/dz^2/U_o)\Delta\hat{W} = 0 \tag{2.20}$$

For higher hills (that are still of low slope) the air flow in the middle layer essentially flows over the hill, so perturbation velocities defined in (2.1) are expressed in terms of x and Δz. Thence the middle-layer equation for ΔW is

$$(\partial^2/\partial\Delta z^2 - \frac{\partial^2(U_o + \Delta\hat{U})}{\partial\Delta z^2} /(U_o + \Delta\hat{U}))(\Delta\hat{W} + (H/L_1)f'(U_o + \Delta\hat{U})) = 0 \tag{2.21}$$

The nonlinear terms of order $(U_o\Delta\hat{W}/L_1{}^2)(\frac{HL_1}{h_m{}^2})$ are retained because they are larger than the streamline derivative terms of $O(U_o\Delta\hat{W}/L_1{}^2)$.

At the top of the boundary layer the effects of shear are often negligible compared with the effects of stratification, so we define an upper layer above $z = h_m$ where (2.15) and (2.16) reduce to

$$(\partial^2/\partial x^2 + \partial^2/\partial z^2 + N^2/U_o{}^2)\Delta W = 0 \text{ over two-dimensional hills} \tag{2.22a}$$

$$\text{and } (\partial^2/\partial x^2\nabla^2 + N^2/U_o^2\nabla_H^2)\ \Delta W = 0 \text{ over three-dimensional hills} \tag{2.22b}$$

The height h_m which divides these two layers is from (2.18) of order h if the hill is long enough that $L_1 >> h$, but is less than h for shorter hills $L_1 << h$.

General form of middle-layer solutions. From the governing middle-layer equation (2.20), the boundary conditions of (2.17) and the equations for Δp and ΔU in (2.2), it follows that

$$\Delta W^{(m)} = (H/L_1)f'(x/L_1)U_o(z) + \frac{\partial\Delta p^{(m)}(x)}{\partial x} \quad U_o(z) \int_0^z \frac{dz'}{U_o{}^2(z')} \tag{2.23a}$$

$$\Delta U^{(m)} = -H\ f\ (x/L_1)\ U_o'(z) - (\Delta p^{(m)}/U_o(z)).\ J(z) \tag{2.23b}$$

where $\Delta p^m(x)$ is the perturbation pressure, which does not vary with height, in the middle layer and is determined by the flow in the upper layer. The velocity profile function is

$$J(z) = 1 + U_o(z)\, U_o'(z) \int_0^z U_o^{-2}(z')\, dz'.$$

Note that $J(z)\to 1$ when $U_o'\to 0$ so $J(z)\simeq 1$ when $z>h_m$. Although for most profiles in neutral conditions $J\simeq 1$ (with error of order $u_*/U_o(h)$), $J>1$ in stable conditions in complex flows such as wakes of hills.

An important implication of (2.23a) is that at the top of the middle layer

$$\Delta W^{(m)} = (H/L_1) f'(x/L_1)\, U_o(h_m) \tag{2.23c}$$

with an error of $O(h_m/L_1)$ which is small if $h_m \ll L_1$.

This middle-layer analysis is similar to that of Stewartson & Williams (1969) and Brighton (1977) for laminar flows. For higher hills a similar solution to (2.23) can be derived for $\Delta\hat{W}$, $\Delta\hat{U}(x,\Delta z)$ in the middle layer using (2.21).

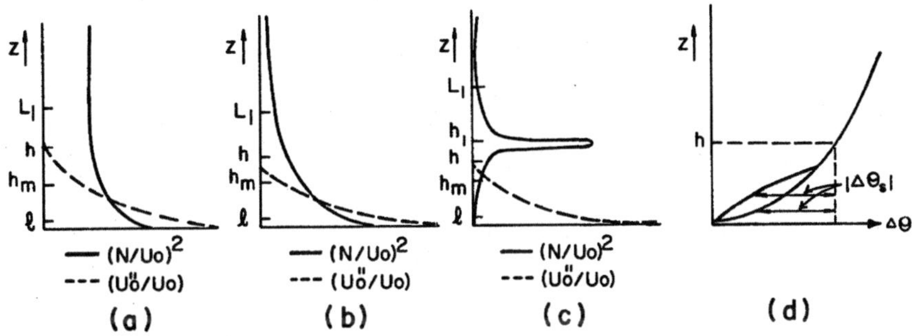

Fig. 6. Typical profiles of the stratification and velocity shear parameters that determine the middle and upper layer in stably stratified flow.
(a) N and U_0 are constant in the upper layer.
(b) (N/U_0) decreases continuously to zero in the upper layer.
(c) There is a strong elevated inversion at $z=h_i$ in the upper layer.
(d) Typical temperature profiles in a stable surface inversion layer, defining $|\Delta\Theta_s|$ relative to the temperature at the top of the boundary layer, and $\Delta\Theta$ relative to the surface temperature.

When stable (or unstable) layers form near the ground, N^2 increases by a factor of 20 or more (Fig. 6d) in the 50 m nearest the ground. Let the difference between the temperature in this surface layer and at the top of the middle layer (h_m) be $\Delta\Theta_s$. Provided N^2/U_o^2 is still small compared with U_o''/U_o, the changes to the mean velocity components can be expressed in terms of $\Delta\Theta_s$ and $U_o(z)$. Denoting the new values U with the superscript (M,s) it is found that

$$\Delta U^{(M,s)} = \Delta U^{(m)} - \frac{Hfg}{\Theta_o(o)}\left[\frac{\Delta\Theta_s(z)}{U_o(z)} + U_o'(z) \int_0^z \frac{\Delta\Theta_s(z')dz'}{U_o^2(z')} \right] \qquad (2.24)$$

Thus the correction to ΔU for a stable inversion layer is similar to that of shallow water flow at high Froude number ($U_o^2/Hg>>1$) over a hump in that the increase in ΔU is reduced over the hill top. Note that the correction is independent of the thickness of the layer, provided it is thin. The effects of stratification in the upper layer are incorporated in $\Delta p^{(m)}$ which helps determine $\Delta U^{(m)}$.

Upper layer solutions for neutral flows. In the upper layer in neutrally stratified flows, N = 0, so from (2.21) and (2.22) the perturbation velocity is governed by Laplace's equation and the boundary condition is given by (2.23), namely

$$\Delta W(z\sim h_m) = H/L_1 f'(x/L_1)\, U_\infty$$

Thence at $z\sim h_m$ for two-dimensional hills, $\Delta p^{(u)}$ the pressure perturbation in the upper region is given

$$\Delta p^{(u)}(x,z\sim h_m) = \Delta p^{(m)}(x) = -U_o(h_m)\ \Delta U(x,h_m) \qquad (2.25)$$

where

$$\Delta p^{(u)}(x,h_m) = -\frac{U_\infty}{\pi}\int_{-\infty}^{\infty} \frac{\Delta W^m(x',z\sim h_m)}{(x-x')}dx' = -U_\infty^2(H/L_1)\Sigma\ (x/L_1)$$

$$\Sigma = \sigma_o(x/L_1) + O(h_m/L_1) \qquad (2.26a)$$

and the hill shape function is

$$\sigma_o(x/L_1) = \frac{1}{\pi}\int_{-\infty}^{\infty} \frac{f'(x'/L_1)dx'}{x-x'} \qquad (2.26b)$$

For example, the variation of Δp with x for a bell-shaped hill defined by

$$f = \frac{1}{1+(x/L_1)^2},$$

$$\sigma_o = \frac{1-(x/L_1)^2}{(1+(x/L_1)^2)^2},$$

see Fig. 4a . Similarly for three-dimensional hills

$$\Delta p^{(u)}(x,y,z\sim h_m) = -U_\infty^2\cdot(H/L_1)\cdot\sigma_o(x/L_1,y/L_1) \tag{2.27a}$$

where

$$\sigma_o(x/L_1,y/L_1) = \frac{1}{2\pi}\int_{-\infty}^{\infty}\int_{-\infty}^{\infty}\frac{f'(x'/L_1,y'/L_1)(x-x')dx'dy'}{((x-x')^2+(y-y')^2)^{3/2}} \tag{2.27b}$$

For a bell-shaped hill where

$$f = \frac{1}{(1+(x^2+y^2)/L_1^2)^{3/2}}\ ,\ \sigma_o = \frac{1-2(x/L_1)^2+(y/L_1)^2}{(1+(x/L_1)^2+(y/L_1)^2)^{5/2}} \tag{2.28}$$

Note that the perturbation velocity and pressure in the upper layer are independent (to $O(h_m/L_1)$) of the velocity profile of the middle layer.

Middle-layer solution for different upwind profiles. We can now use (2.23) to calculate the perturbation velocity within the middle layer, for different kinds of upwind velocity profile. If

$$U_o = (u_*/k)\ \ell n\ (z/z_o) \tag{2.29a}$$

or

$$U_o = U_\infty(z/h)^n \tag{2.29b}$$

then

$$\Delta U^{(m)} = -U_o'\ \ Hf + (H/L_1)\cdot\sigma_o(x/L_1)\cdot(U_o^2(h)/U_o(z))J(z) \tag{2.30}$$

where $J(z) = (1-n)/(1-2n)$ for profile (2.29b) (note $n<<1$), and

$$J(z) = 1 + 0\left(\frac{1}{\ell n(z/z_o)}\right) = 1 + O(u_*/U_\infty)$$

for the logarithmic profile (2.29a). Note that the first term is the reduction of velocity at a given height caused by the upward displacement of the upwind profile. The second term is caused by the speedup in the upper layer induced by this upward displacement of the flow. This is the same as the perturbation wind speed at a given displacement, Δz. For large shear $\Delta\hat{U}$ can be more accurately calculated by the solution to (2.21), namely

$$\Delta\hat{U}(x, z) = H/L_1\ \sigma_o\ (x/L_1)\ \frac{U_o^2(h)}{U_o(\Delta z)}\ \ J(\Delta z) \tag{2.31}$$

This reduces to the particular expressions of Britter et al. (1981), (2.5), derived from Jackson (1977), when $\Delta z<<L_1$, and of Dawkins & Davies (1981)

In the limit of strong stable stratification, or when a hill lies in the wake of another one, U_o may be approximated by a linear profile $U_o=U_\infty(1-\Omega+\Omega(z/h))$. Substitution in (2.23) or a direct analysis of equations (2.20), (2.22a) shows that if $\Omega H/h<<1$ and $L_1>>h$,

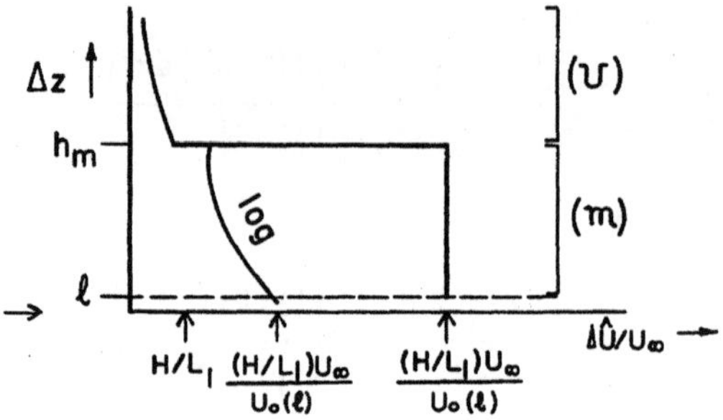

Fig. 7. Vertical profile of the displaced perturbation velocity $\Delta\hat{U}$ over a hill for logarithmic and linear upwind mean velocity profiles.

$$\Delta U^{(m)} = -U_0' \;\; Hf + (H/L_1)\sigma_0(x/L_1)\; U_0{}^2(h)/U_0(z_0) \tag{2.32}$$

or $J(z) = U(z)/U(z_0)$ (for details see HLR). (2.33)

The distributions of $\Delta\hat{U}$ for the log and linear profiles are plotted in Fig. 7. Both show how $\Delta\hat{U}$ is generally greater in the middle layer than the upper layer, but that $\Delta\hat{U}$ increases nearer the surface for the log or power law profiles. The explanation is that the difference between $\Delta\hat{U}$ and the upper-layer perturbation velocity $\Delta U(h)$ is caused by displacement of the vortex lines of the upwind profile. If $\Delta U^{(u)}$ increases, these are displaced downwards and consequently increases $\Delta\hat{U}$. Since the vorticity is greatest near the surface in the power law and log profiles, the increase in $\Delta\hat{U}$ is greatest near the surface; the vorticity in the linear profile is constant, so $\Delta\hat{U}$ is constant. Note that even when the slope is large, the effects of uniform upwind shear produce similar effects (e.g. Fraenkel 1961).

Note also that (2.31) is also valid for three-dimensional flows. Several field experiments have confirmed its validity. In some cases it needs to be modified for small-scale undulations on the hill (Mason & Sykes 1979; Walmsley et al. 1982; Hunt & Simpson 1982).

Upper-layer solutions for uniform stable stratification. The two extremes of ideal upwind conditions for stably stratified flows are where N and U_0 are constant in the upper layer (Fig. 6a) and where they are discontinuous (Fig. 6c). For either case, the solutions to (2.22a) and (2.22b) subject to (2.23c) and the appropriate radiation conditions can be obtained by Fourier transform techniques. For uniform N and U_0 typical streamlines when the Froude number $F_L = U_\infty/(NL_1)$ is of the order of unity, which indicate the lee-waves, are shown in Fig. 8a, and computations (given in HRB) for the distribution and the maximum surface winds over two and three-dimensional hills are given in Figs. 4a and 8b. Some useful analytic results can be obtained in certain limits.

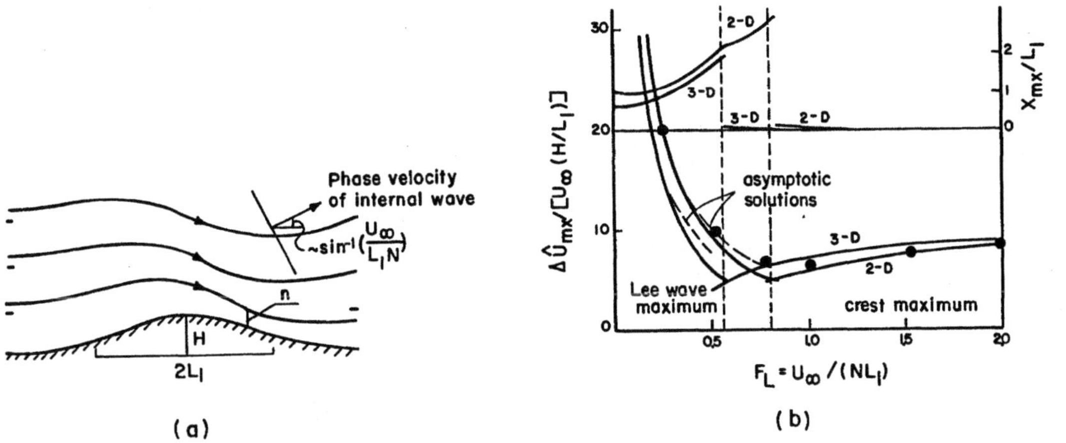

Fig. 8. Stably stratified air flow over a hill; (uniform unbounded stratification). (a) Typical streamlines, (b) maximum perturbation velocity in the upper layer for two- and three-dimensional bell-shaped hills. --- asymptotic solution of (2.35) and • approximate estimate in (2.36).

In weak stratification, when $F_L \gg 1$, over a bell-shaped hill, where $(x/L_1) \ll 1$

$$-\Delta p^{(u)}/U = \Delta U^{(u)}(x,h) = (1-3(x/L_1)^2-\tfrac{1}{2}F_L^{-2}\ln F_L+O(F_L^{-2})).(U_\infty H/L_1) \quad (2.34)$$

showing that a small increase in stability decreases the magnitude of the upper layer velocity and pressure perturbations; but has a small effect on the distribution of $\Delta U^{(u)}$ over the hill as is also shown in Fig. 4a.

The other limit is when $F_L \ll 1$ which we term "moderate stratification", because we still assume that $F_H = U_0(H)/NH \gg 1$. Then the upper layer velocity at the top of the boundary layer and the perturbation pressure through the boundary layer can be expressed as general integrals in terms of the shape of the hill, (Drazin and Su, 1975; HRB), viz.

$$\Sigma(x/L_1,y/L_1) = F_L^{-1}\,\sigma_\infty(x/L_1,\,y/L_1) \qquad (2.35a)$$

where $\sigma_\infty = 1/\pi \int_{-\infty}^{\infty} \frac{f(x'/L_1)dx'}{x-x'}$ <u>for two-dimensional hills</u> (2.35b)

$\sigma_\infty = \frac{1}{2\pi}\int_{-\infty}^{\infty} \frac{f(x',y')(x-x')dx'dy'}{((x-x')^2+(y-y')^2)^{3/2}}$ <u>for three-dimensional hills</u> (2.35c)

Thus the amplification is very much greater than in neutral flow, by a factor F_L^{-1}, but the distribution of $\Delta U^{(u)}$ is also different; in fact it is the same as neutral potential flow over a surface

$$\hat{f} = \int_{-\infty}^{x/L_1} f(x'/L_1)d(x'/L_1)$$

<u>whose slope</u> is equal to the local height of the hill. Alternatively σ_∞ can be calculated by integrating σ_o, since

$$\sigma_o = \partial\sigma_\infty/\partial x \quad \text{for 2 or 3 dimensions.} \qquad (2.35d)$$

These provide easy ways of visualising the forms of these stratified flows (Drazin & Su, 1975); HRB.

Fig. 8 shows that these asymptotic results for bell-shaped hills agree well with the exact computations for $F_L \lesssim 1$. Fig. 8b also shows that a simple useful formula can be suggested for interpolating between the limits of weak and moderate stratification for the maximum value of $\Delta U^{(u)}(x,h)$:

$$\Delta U_{mx}^{(u)} \simeq (H/L_1)U_\infty\left(\frac{(\sigma_o)_{mx}}{1+F_L^{-2}} + (\sigma_\infty)_{mx}/F_L\right) \qquad (2.36)$$

where $(\ )_{mx}$ indicates the maximum values of σ_o and σ_∞.

<u>Upper layer flow with an elevated inversion</u>. At the other extreme of ideal upwind conditions there is an elevated inversion layer at $z = h_i$ where the mean temperature has a jump $\Delta\Theta_i$. Then $(N/U_o)^2$ is a delta function:

$$(N/U_o)^2 = U_\infty^{-2}\,\Delta S\,\delta(z-h_i) = F_i^2/h_i \cdot \delta(z-h_i)$$

where $\Delta S = g\Delta\Theta_i/\Theta_o(h_i)$ and F_i is the Froude number for the inversion layer. Typically such inversions occur about 300m over hills longer than 1 km, but much lower than 1 km, for this reason and for mathematical simplicity, we assume $h_m \ll h_i \ll L_1$. There may also be a stable layer <u>above</u> the inversion, in which case

$$(N/U_o)^2 = (F_i^{-2}/h_i)\ \delta(z-h_i) + (L_1F_L)^{-2}\ \mathcal{H}(z-h_i) \qquad (2.37)$$

where $\delta(\quad)$ and $\mathcal{H}(\quad)$ are Dirac and step functions respectively and $F_L = U_\infty/NL_1$.

A convenient way of expressing the perturbation to the flow in this case is by calculating the difference, n_i, between the height of the inversion layer $h_i + \Delta h_i$, which is deflected an amount Δh_i by the flow, and the local height of the hill Hf (Fig. 9), so

$$n_i = h_i + \Delta h_i - Hf$$

The solution to (2.22a), (2.37) subject to (2.17) and the radiation condition is (HRB)

$$n_i = h_i + \frac{Hf}{F_i^2-1} - \frac{h_iF_i^2}{F_i^2-1}\{H/L_1 . \Sigma(x/L_1, F_{L_1} h_i) - \frac{h_i}{L_1}(\frac{H}{L_1}\frac{f''}{6} + \frac{1}{3}\frac{\Delta h_i''}{L_1})\} \qquad (2.38)$$

where $U_\infty(H/L_1)\,\Sigma$ is the velocity perturbation ΔU just above the inversion, which is calculated in terms of Δh_i by the same method as for the upper layer in uniform stratification over a hill. When $F_i{>>}1$, $\Sigma=\sigma_o$ when $F_L{>>}1$, as defined by (2.26b), and $\Sigma=F_L^{-1}\sigma_\infty$ when $F_L<1$.

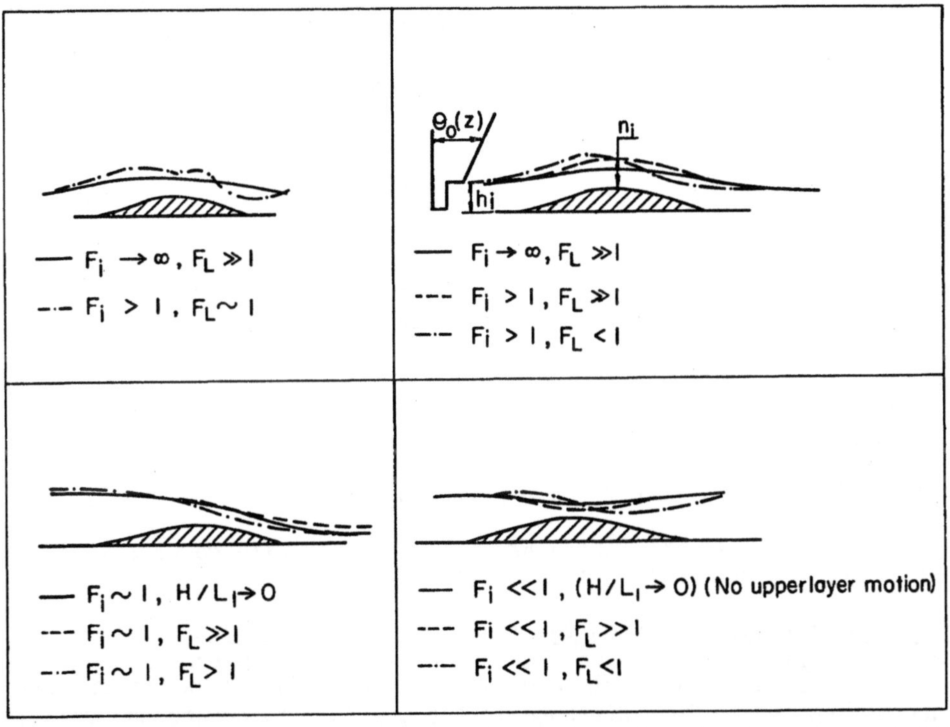

Fig. 9. Deflection of the inversion layer when the air flow over a hill is neutrally or stably stratified and the air below neutral (or unstable), and $h_i{<<}L_1$.

The first conclusion is that the elevated inversion only affects the flow when $F_i^2<(L_1/h_i)/\Sigma$. This means that an increase in stable stratification in the upper layer, which increases Σ, effectively reduces the relative effect of an elevated inversion on the air flow. If $F_i>>1$, whatever the value of F_L the effect of the inversion is to decrease ΔU over a hill top; in fact the hill top may be a position of minimum velocity, as in high Froude number hydraulic flow over a wavy surface. Stable upper-layer stratification induces the inversion to be deflected downwards on the lee side and large wave-like deflections in the lee of the hill. Recently Carruthers & Choularton (1982) made a field and computational study of the flow of a thin, strongly stable layer below slightly stable upper air over 500 m hills in Northern England, for a parameter range $2<F_i^2<8$, while $0.4\lesssim F_L\lesssim 0.7$. Their results broadly agree with the linear theory.

For a strong elevated inversion when $F_i^2<<1$, the inversion is deflected downwards and the velocity is a maximum at the hill top.

The flow above the inversion layer only affects n_i by $O(F_i^4h_iH/L_1)$ when $F_L>>1$ and by $O(F_i^4h_iHN/U_\infty)$ when $F_L\lesssim 1$. Consequently calculations which ignore non-hydrostatic pressure gradients induced above the inversion layer, such as have been made to include the nonlinear perturbation when $F_i\sim 1$, are reasonably accurate, as shown by the recent laboratory and computational work by Lamb & Britter (1984).

Upper-layer flow with uniform stratification below a strong inversion. In this situation where

$$(N/U_o)^2=\frac{F_i^{-2}}{h_i}\,\delta(z-h_i)+(L_1^2F_L^{-2})\,\mathcal{H}(h_i-z) \qquad (2.39)$$

trapped resonant waves occur below the inversion, which can induce very large surface winds when $F_i<(h_i/L)/\pi$ (Klemp & Lilly, 1975). An estimate for U can be derived from continuity; since $|\Delta U/\Lambda|\sim|\Delta W/h_i|$ where Λ is the wave length of the internal waves below the inversion.

Using the value for Λ obtained from linear theory and the condition that $\Delta W\sim U_\infty(H/L_1)$, it follows that, if $F_i<<1$ and $F_L<(h_i/L)/\pi$.

$$\Delta U\sim(H/L_1)U_\infty 2\pi F_L/[(h_i/L_1)(1-\pi^2F_L^2(L_1/h_i)^2)^{\frac{1}{2}}] \qquad (2.40)$$

of course this estimate (which has been compared with computations by HRB) is quite wrong when F_L is close to h_i/L.

Effects of stratification on the flow in the middle layer. Even when the direct effects of buoyancy forces on the mean flow can be neglected in the middle layer, the action of weak buoyancy forces on the turbulence acting over a long distance in the upwind flow change the mean velocity profile of the wind approaching the hill (or other complex terrain). Over level terrain, it is found that in the surface layer,

for stable conditions,

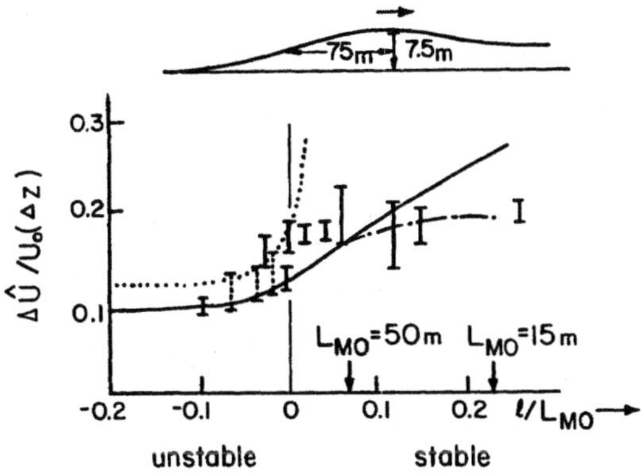

Fig. 10. Variation of speed up at height ℓ over the top of a low hill with the stability of the upwind flow. Measurements by Bradley (1983). I z_0 = 3 mm; z_0 = 2 mm. Theory: ----- only allowing for upwind velocity profile (from (2.42) (2.44) ($F_L^{-1}=0$, $\Sigma=\sigma_0=0.7$)
·-·-· also allowing for mean buoyancy forces.
···· upwind velocity profile taking $h_m \approx L_1$.

$$U_0(z) = (u_*/k)(\ell n(z/z_0) + \alpha z/L_{Mo});\alpha \simeq 5;\ L_{Mo} > 0 \qquad (2.41a)$$

and for unstable conditions,

$$U_0(z) = (u_*/k) \int_{z_0}^{z} \frac{dz}{z(1-\beta z/L_{Mo})^{1/4}} \ ;\ \beta \simeq 15;\ L_{Mo} < 0 \qquad (2.41b)$$

These profiles can be substituted into the general formulae (2.23) for the middle layer velocity perturbations at a given displacement Δz using (2.26a) for the pressure deflection,

since $\Delta\hat{U} = \Delta U - Hf\ U_0'$, then

$$\Delta\hat{U} = (H/L_1)\ U_0^2(h_m)\ \Sigma\ (x/L_1,\ y/L_1,\ F_L)/U_0(\Delta z) \qquad (2.42)$$

where Σ is defined by the upper-layer analysis. For hills long enough that $L_1 >> h$ and Δz near enough to the surface that $\Delta z < L_{Mo}$, $U_0(h_m) \simeq U_\infty$. Thus in slightly stable conditions over the top of a bell-shaped hill, the relative increase in wind speed can be estimated as

$$\frac{\Delta\hat{U}}{U_0(\Delta z)} \approx (H/L_1)(1 - \frac{F_L^{-2}}{2}\ \ell n F_L)\left(\frac{\ell n\ h/z_0 + \alpha h/L_{Mo}}{\ell n \Delta z/z_0 + \alpha\Delta z/L_{Mo}}\right)^2 \qquad (2.43)$$

so a small increase in stability increases this speed up, by $O(2\alpha(h/L_{Mo})/\ln(h/z_0)) \sim O(Nh/u_*/\ln(h/z_0))$ because the vorticity of the upwind boundary layer flow is increased. But the stability decreases the speed-up by

$$O(F_l^{-2}\ln F_l) \sim O((NL_1/U_\infty)^2 \ln(NL_1/U_\infty))$$

by the action of buoyancy forces in the upper layer. For example in a weakly stratified flow over a hill 1 km long, where

$$N(h) \sim .0035^{-1},\ h \sim 500\text{ m},\ L_{Mo} \sim 300\text{ m},\ u_* \sim .2\text{ms}^{-1},\ U_\infty \sim 10\text{ ms}^{-1},$$

$$z_0 \sim .01\text{ m},\ (Nh/u_*)/\ln(h/z_0) \simeq .7,\ (U_\infty/U_0(\Delta z))^2$$

increases by 2.9, while $NL_1/U_\infty \sim .3$, so Σ decreases by 0.5.

For short hills, where $L_1<h$, the height of the middle layer h_m is also affected by the change in the upwind profile which reduces U_0''/U_0. It follows from (2.18) that in stable conditions

$$h_m \sim L_1/(\ln(h_m/z_0) + 5h_m/L_{Mo})^{1/2} \qquad (2.44)$$

Recent field measurements by Bradley (1983) of the speed up over a small unsymmetrical hill ($H \simeq 7.5$ m, $L_1 \simeq 75$ m, $\sigma_0 \simeq 0.7$) have shown how $\Delta\hat{U}/U_0(\Delta z)$ changes with stability.[1] As conditions change, $\Delta\hat{U}/U_0(\Delta z)$ measured at the height of the inner layer of about (3 m) increased from about .12 to .17 in neutral conditions (for different roughness lengths from 2 to 4 mm), to about 0.2 to 0.3 in stable conditions when $L_{Mo} \simeq 50$ m. A further decrease of L_{Mo} to 15 m produced no further increase. The effect of unstable stratification produced a sudden decrease in speed up with a decrease in $(-L_{Mo})$ to -20 m. A further decrease had less effect. In Fig. 10, the measurements and theoretical results based on (2.42) and (2.44) (with $\Sigma = \sigma_0$ and Σ corrected for the weak buoyancy forces) are plotted along with graphs based on the assumption that $h_m = L_1$. Note that if h_m is estimated as being of order L_1 as suggested previously by Jackson & Hunt (1975), then $\Delta\hat{U}$ can be severely overestimated. The effect of moderate buoyancy forces begin to reduce $\Delta\hat{U}$ directly when L_{Mo} is as low as 15 m; then $F_L \simeq 1.3$ and Figs. 4a and 8 shows that Σ is reduced by about 30% for a two-dimensional hill.

Bouwmeester (1978) and others at Colorado State University have also found in their studies of wind speeds over hills in a stratified wind tunnel this increase in speed up in stable conditions over hill tops (see Hunt 1981), and increased sheltering on the upwind slopes. These studies also confirmed the implication of (2.24) that weak stratification does not affect the distribution of $\Delta\hat{U}(x/L_1)$ over the hill, only its magnitude. This distribution is only affected when buoyancy forces affect the mean flow in the upper flow.

Effects of the inner region on the outer region. An essential assumption of this analysis is that the vertical velocity outside the inner layer is determined by the shape of the hill and the upwind velocity, i.e. the boundary condition (2.17). The reason is that the change in the vertical velocity by the perturbed flow in the inner region is negligible for flow over low hills if the inner region flow is turbulent and if the flow does not separate. In analyses with constant viscosity or eddy viscosity, the perturbed flow in the inner region may displace as much flow as the hill itself, and in that all three regions have to be calculated interactively. This is also true in turbulent flows if the flow on the lee side separates; but such calculations have not yet been done. For example in Bouwmeester's (1978) wind-tunnel studies the speed up over steep hills was approximately half what it would have been without separation.

Nonlinear corrections. The general result (2.42) shows that the criterion for linear analysis to be valid in the middle and inner layers is that

$$(H/L_1)\ U_o^2\ (h_m)/U_o^2(\Delta z) << 1 \quad \text{in neutral and weakly stable conditions} \tag{2.45a}$$

and $$F_L^{-1}(H/L_1)U_o^2(h_m)/U_o^2(\Delta z) << 1 \quad \text{in moderately stable conditions} \tag{2.45b}$$

So if the velocity gradient in the middle layer is large enough that $U_o^2(h_m)/U_o^2(\Delta z) >> 1$, the slope (H/L_1) must be smaller in the middle layer than in the upper layer, in order that the linear analysis should be valid, i.e.

$$(H/L_1) << (U_o(\ell)/U_o(h_m))^2 \tag{2.45c}$$

However, provided H/L_1 is small enough for linear analysis to be valid in the upper layer, and provided $L_1/h_m << 1$, a simple nonlinear analysis of the middle layer is possible based on the solution to (2.21), basically because the pressure is still determined by the upper layer. If $\Delta\hat{U}^{(o)}$ is the value of $\Delta\hat{U}$ calculated by linear theory (2.24), then the value corrected for the nonlinear terms $\Delta\hat{U}^{(1)}$ can be derived, namely

$$\frac{\Delta\hat{U}^{(1)}}{U_o(\Delta z)} \simeq -1 + [1 + 2\Delta\hat{U}^{(o)}\ (\Delta z)/U_o(\Delta z)]^{\frac{1}{2}} \tag{2.46}$$

(In fact this is the same as obtained from Bernoulli's equation, with given Δp). Thus over a hill where $H/L_1 \simeq 0.4$, $\Delta\hat{U}$ is typically of the order of 0.8 in neutral conditions; corrected for nonlinearity, $\Delta\hat{U}$ decreases to 0.6, a reduction of 25%, whereas the correction for nonlinearity in the upper layer would only be about 15%.

3. STRONGLY STRATIFIED FLOW OVER A HILL

3.1. Order-of-magnitude analysis

When stable stratification in the boundary layer is strong enough that the Froude number based on the hill height is small, i.e.

$$F_H = U_0(H)/N(H).H) \lesssim 1$$

the analysis of Section 2 shows that there must be a large perturbation to the incident flow. Clearly the incident flow does not have enough kinetic energy for all of it to pass over the hill.

$$(\text{i.e. } \tfrac{1}{2}\rho U_0^2 < \tfrac{1}{2}\rho N^2H^2).$$

Therefore there is a region $[H_1]$ extending upwards from the ground where either the flow must travel round the hill in approximately horizontal planes (Fig.11). But within a height H^* of the summit of the hill there is a region [T] where the kinetic energy $\tfrac{1}{2}\rho U_0^2$ is just enough for the flow to develop the potential energy $\tfrac{1}{2}\rho N^2H^{*2}$ to pass over the top of the hill, so

$$H^* = H.F_H \tag{3.1a}$$

and the height

$$H_c = H-H^* \tag{3.1b}$$

is the dividing height between regions [T] and $[H_1]$. If N varies with z, the energy argument suggests

$$\int_{H_c}^{H}(H-z')N^2(z')dz' = \tfrac{1}{2}U_0^2 \tag{3.1c}$$

Thus the streamlines at or below a critical height H_c impinge on the windward force of the hill, while those above H_c pass over the hill.

If the hill is two-dimensional, the upwind flow below H_c stagnates. But the flow is not really well defined in that situation because the upwind extent of this stagnant region keep increasing until other effects are significant, e.g. shear stresses or thermal effects.

If the shear in the upwind flow is strong enough, the difference in stagnation pressure at different heights within $[H_1]$ is large enough to induce a downward deflection of the streamlines by $O(U_0dU_0/dz/N^2)$. This can be of the order of H_c in many circumstances because intense stratification is often associated with strong shear (Snyder et al. 1984).

The vertical displacements within region [T] produce weak lee waves with displacements of order (U_0/N). These propagate downwards, weakening as they do so, into the region $[H_1]$ and upwards into $[H_2]$.

A number of laboratory experiments and recent field measurements around an isolated three-dimensional hill have confirmed this estimate for H_c in (3.1) originally due to Sheppard (1956) (see Snyder et al. 1980; Lavery et al. 1982).

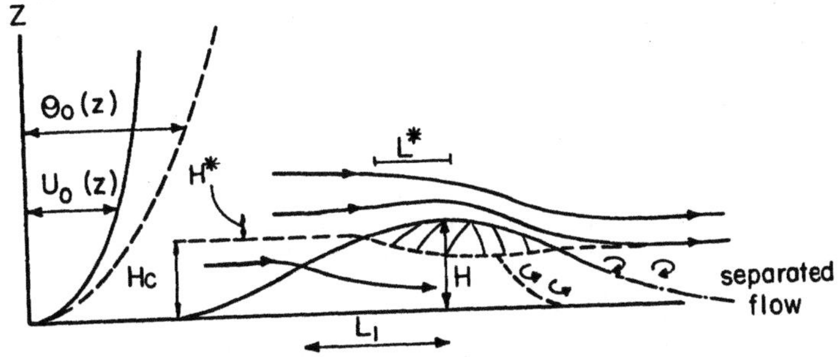

Fig. 11. Strongly stable flow over a three-dimensional hill, when F_H 1. The flow above the critical height H_c is similar to that over a 'cut off' hill (height H^*, and half length L^*), indicated by shading.

3.2. Theoretical models

In the limit of $F_H << 1$, for hills of arbitrary slope the nearly horizontal flow in region [H_I] was first analysed by Drazin (1961). Corrections due to shear and Coriolis acceleration were added by Brighton (1978). However, neither analysis included the effects of waves propagating downwards from region [T] , nor the fact that the flow separates. So to lowest order in F_H the solutions were those of potential flow in horizontal planes.

There is no consistent theory or even computation for the summit region [T] which matches with the flow in [H_1], because even when $F_H << 1$ the flow is three-dimensional and nonlinear. However, an approximate theory can be constructed for this region by basing it on the results of flow visualisation experiments (Hunt & Snyder 1980). The experiments showed that the interface between regions [H_1] and [T] is approximately horizontal when N is constant with height (this is an even better approximation when N decreases with height). So the flow in [T] appears to be similar to the flow over a 'cut off' hill, of height $H^* = H - H_c$, placed on a horizontal plane at height H_c. The half length of this cut off hill L^* is, for a rounded hill, large compared with H^*. So

$$F_{H^*} = U_0(H)/NH^* \simeq 1 \text{ but } F_L^* = U_0(H)/(NL^*) << 1.$$

Therefore the results of the theory for strongly stratified flow over hills of low slope given in equations (2.35) and Figs. 4a and 8 can be used to estimate the flow in this region. For example (see Fig. 11), this explains why the streamlines come close to the surface on the lee side of the region [T], thus causing locally the highest wind speeds and the highest ground-level concentrations from sources of pollution released upwind at heights greater than H_c. Because of this large increase in wind speed on the lee side, the streamlines dip below the dividing plane H_c, and generally in this region the model is only qualitative (see HRB).

3.3. Effects of stratification on separation

For reasons given in Section 2 in flows over hills of low slope, the surface perturbation shear stress is largely determined by the local perturbation pressure, when the boundary layer is turbulent, but is proportional to the local pressure gradient when the boundary layer is laminar. Experiments suggest that this is a useful approximation for the total surface shear stress even when the perturbations are large, and can be used to indicate the onset of separation.

In turbulent stratified flows the perturbation pressure is determined by the hill shape and the upwind profiles of mean velocity and density gradient. Figs. 4a,b show that with uniform stratification in the upper layer, Δp_x and $\Delta\tau$ are negative on the lee slope (defined by $0 < x/L_1 < 2$) when $F_L \gtrsim 1$. Consequently if H/L_1 and/or $U_0(h_m)/U_0(\ell)$ are large enough, and if $F_L \gtrsim 1$, the usual boundary-layer separation occurs on the lee slopes of a bell-shaped hill. As F_L decreases to about 1, the boundary induced acceleration flow on the lee side reduces the length of any separated region. When $F_L \lesssim 1$ if $F_H > 1$, the pressure is largely determined by the buoyancy forces in the upper layer. Figs. 4a and 9 show how the pressure distribution is related to upper-layer stratification, and in particular to the local wavelength of internal waves Λ. In general the distance x_{sep} from the hill top to separation location is approximately $\Lambda/2$ (Hunt & Snyder 1980; Brighton 1977).

If the stratification is strong enough that $F_H < 1$, then separation occurs around the sides of a three-dimensional hill in region $[H_1]$ up to the critical height $H_c = H(1-F_H)$.

Note that in stratified flows the flow on the upwind slope tends to separate since Δp is negative, when $F_L \gtrsim 1$ separation can occur on these slopes while being suppressed on the lee slopes. When $F_H \lesssim 1$, the flow upwind of two-dimensional hills continues to be stagnant while upwind of three-dimensional hills it is not.

3.4. Summarising flow over a single hill

Whether a hill has a low or high slope, the theory has shown how the factors determining a typical perturbation of wind speed (or deflection of streamlines) can be expressed schematically as

$$\Delta U = U_\infty \, (H/L_1) \cdot A_{vp}(z/h_m, \frac{u_*}{h_m N}) \quad \cdot \quad A_D\,(F_o)$$

slope x effect of upwind velocity profile x dynamic effects of stratification

where F_o is an overall Froude number, defined according to the particular stratification profile. This expression leads to a useful classification for different kinds of stratified flow over a hill.

<u>Neutral flow</u> $u_*/(h_m N) >> 1;\ F_L >> 1,\ F_H >> 1$

The flow depends on the hill shape and the neutral upwind velocity profile, and its structure is independent of the wind speed at a given height, say U_∞. Typically $\Delta U \sim U_\infty^2\,(H/L_1)/U_o(\ell)$

<u>Weakly stratified flow</u> $u_*/(h_m N) \sim 1,\ F_L >> 1,\ F_H >> 1$

The gradients of the upwind velocity profile is increased by weak stable stratification and decreased by unstable stratification, acting on the turbulence upwind of the hill. The buoyancy forces over the hill are negligible so the upper layer is largely unaffected, but the changed upwind profile affects the middle layer and inner region. Typically

$$\Delta\hat{U} \sim U_\infty^2\,(H/L_1)/U_o(\ell)$$

<u>Moderately stratified flow</u> $F_L \lesssim 1,\ F_H > 1$

As well as the upwind velocity profile being changed, buoyancy forces affect the flow in the upper layer, well above the hill. The same structure of flow regions is appropriate. Typically $\Delta\hat{U} \sim U_\infty^2\,(H/L_1)F_L^{-1}/U_o(\ell)$

<u>Strongly stratified flow</u> $F_L << 1,\ F_H \lesssim 1$

Now the buoyancy forces are strong enough to affect the flow at the height of the hill. A new pattern of flow regions emerges. Typically

$$\Delta U \sim U_o(H)$$

<u>Heating and cooling of the slopes</u>

Typically if $\ell/|L_{Mo}| > 1/5$ stratification significantly affects the structure of the inner region. The interaction between thermally-driven slope winds and the incident flow have not yet been systematically investigated. But see Fitzjarrald (1984). For recent work on slope winds, see Manins & Sawford (1981).

4. FLOW OVER TWO RIDGES

4.1. Complex flows with hills

4.1.1. Roughness and elevation changes

The linear analysis for flow over a single hill can be used to estimate many kinds of flows over complex distributions of changes of surface elevation, roughness and temperature. The best example where two such effects have been studied is the flow over escarpments where the surface roughness increases on the escarpment or slope (e.g. Jensen & Petersen 1978). There the effects of the roughness changes are confined to the inner region while the elevation change effects all three regions. A laboratory study of a sudden reduction in roughness over a hill ($H/L_1 = 0.4$) also showed that the mean perturbation velocity caused by roughness change could be superposed on that produced by the hill, but only on the windward slope and the top of the hill. The effect of a roughness change had a large effect on the wake, in this case suppressing separation (Britter et al. 1981). In general separated flow on hills of moderate slope are quite sensitive to changes in local slope, roughness or heating (Scorer 1955).

Various analyses suggested that only the scales of turbulence that are smaller than the scale of the inner layer, ℓ, are locally determined and can be determined from the superposition of the mean flow. Large scales take longer to adjust and are either determined by advection from upwind or by distortion in the middle layer (Hunt 1981; Panofsky et al. 1982).

4.2. Two hills

Neutral conditions. The calculations and measurements of flow over a single hill can be used to infer the flow over two low two-dimensional hills, A,B. In neutral conditions the influence of a hill on the flow upwind of it is small (see Fig. 4a), so the main interaction is in the effect of the wake of the upwind hill flowing onto the downwind hill (Fig. 1). So the upwind velocity profile impinging onto hill B has a strong shear up to height ℓ_w caused by the wake of hill A, above which the profile is approximately that of the undisturbed boundary layer. Also the upwind turbulence is large below ℓ_w.

Since ℓ_w is larger than the thickness of the inner layer over hill B, the upwind profile affects the perturbations $\Delta\hat{U}$ in the middle layer and inner region over hill B, which can be calculated by the methods of Section 2 (Hunt, Leibovich & Lumley 1983). A notable feature of these calculations is that despite the velocity defect in the wake of hill A, its vorticity amplifies $\Delta\hat{U}$ (by the mechanisms outlined in Section 2), so the net effect of the wake of hill A is much reduced.

Wind-tunnel studies of the flow over a chain of hills by Counihan (1974) suggest that the net perturbation to the displaced profile

$\Delta\Theta(\Delta z)$ over the second and subsequent hills is almost zero. (This is unlikely to be true if there is any variation in the height of the hills).

Note that the turbulence over hill B can be somewhat increased by the effect of A's wake. Recent field experiments by Mason & King (1984) have shown that energetic low-frequency motions are also a noticeable feature of valley winds.

Stable conditions. In stable conditions the extent of the upwind influence of a hill steadily increases (Fig. 4a) so unlike the flow in neutral conditions, the flow over hill B is now affected by hill A. Nevertheless, if the slopes of the hill are small (taking into account the amplifications to ΔU by the upwind velocity profile and by buoyancy forces), this interaction can be approximated by superposing solutions for flow over two hills a distance D apart. The form of the outer flow is determined by F_L and by the ratio D/L_1. Also the height and intensity of any elevated inversion plays a most important role, because it controls the wavelength Λ of internal waves.

Two features of the flow of great practical interest are (i) whether the flow stagnates in the valley or whether the valley is ventilated; (ii) the location and magnitude of the greatest surface winds.

Following Scorer (1972) it is natural to expect the largest perturbations when $D \sim \Lambda$. Nonlinear numerical and laboratory experiments by Bell & Thompson (1980) showed strong ventilation for flow between two bell-shaped hills ($H/L_1 \simeq 0.6$, $D/H = 8$, for unconfined stratification when $D \simeq 2\pi U_0/N$ or $F_H = 1.3$. However linear analysis suggests that D/L_1 is as important as D/Λ as a criterion, especially in unconfined stratification.

Fig. 12 shows the result of superposing the solutions of Fig. 4a for ΔU in the upper layer or $-\Delta p$, for $D/\Lambda = 1$ and various values of D/L_1. This shows that there is always a negative perturbation velocity except when $D/L_1 \simeq 6$, and $F_L \simeq 1.0$. For closer or larger spacing than this, the upwind influence of hill B leads to stagnation in the valley. When $F_L \lesssim 1.0$, the largest surface winds always occur on the lee face of hill B.

If there is a strong elevated inversion and the flow is subcritical, then the velocity perturbations are so regularly sinusoidal (Fig. 4a (ii)) because of the strong internal wave motion, that the only criterion would then have to be that $D \sim \Lambda$.

Tampieri (1984) has recently made flow visualisation studies of laminar flow over two hills in the small stratified flow plume at Cambridge (Brighton 1978), and shown how sensitively the onset of ventilation varies with D/L_i. His results for the streamline pattern largely agree with the estimates from linear theory.

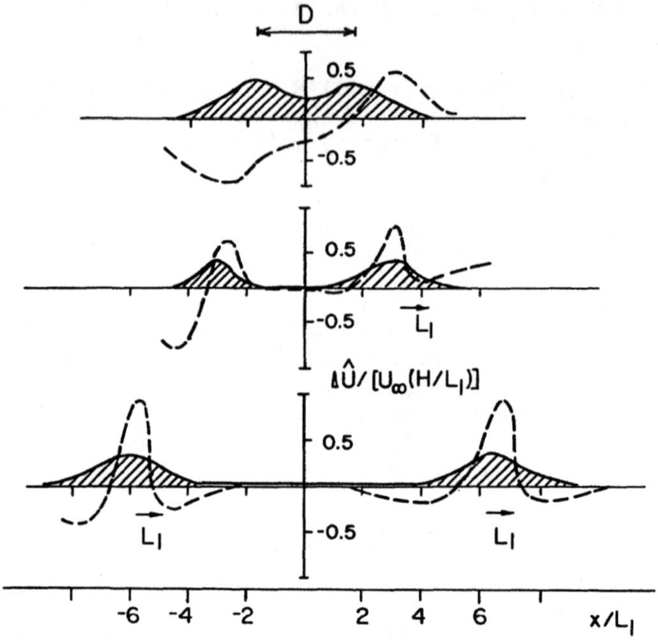

Fig. 12. Stable flow over two hills with low slopes, when the natural wavelength ($2\pi U_0/N$) equals the valley length D. (a) $F_L = 0.5$; $D/L_1 = 3$; (b) $F_L = 1.0$; $D/L_1 = 6$; (c) $F_L = 2.0$; $D/L_1 = 12$.

Acknowledgements

JCRH partly worked on this paper at Flow Analysis Associates of Ithaca, N.Y. with support from the State of Maryland Department of Natural Resources and the U.S. Air Force Geophysics Laboratory. KJR was supported at Cambridge by an Ernest Cook Fellowship of the Royal Society. We are grateful to Phillip Brighton for being the source of many ideas presented here, and to Drs. H. Kaplan and N. Dinar for their careful editing of the paper. Any remaining errors are our responsibility.

References

Achenbach, E.L.: 1968 J. Fluid Mech. 34, 625-639
Bell, R.C. and Thompson, R.O.R.Y.: 1980 J. Fluid Mech. 96, 757-767.

Blumen, W.: 1965 Geofys. Public. Geophys. Norv. XXVI, No. 2.
Blumen, W.: 1983 Monthly Weather Review 111, 1052.
Bois, P.A.: 1983 Proc. Course 'Models for Atmospheric Flows'. C.I.S.M., Udine.
Bouwmeester, R.J.B.: 1978 Ph.D. Dissertation. Colo. State Univ. See also Report CER 77-78 RJB RNM VAS 51.
Bradley, E.F.: 1980 Q.J. Roy. Met. Soc. 106, 101-124.
Bradley, E.F.: 1983 Proc. 6th. Wind Engineering Conf., Brisbane.
Brighton, P.W.M.: 1977 Ph.D. Dissertation. Univ. of Cambridge.
Brighton, P.W.M.: 1978 Q.J. Roy. Met. Soc. 104, 289-307.
Britter, R.E., Hunt, J.C.R. and Richards, K.J.: 1981 Q.J. Roy. Met. Soc. 107,
Carruthers, D.J. and Choularton, T.W.: 1982 Q.J. Roy. Met. Soc. 108, 603-624.
Castro, I.P.: 1979 J. Fluid Mech. 93, 631-659.
Counihan, J.: 1974 C.E.R.L. Report RD/L/N57/74.
Counihan, J., Hunt, J.C.R. and Jackson, P.S.: 1974 J. Fluid Mech. 64, 529-563.
Dawkins, R.A. and Davies, D.R.: 1981 J. Fluid Mech. 108, 423-442.
Drazin, P.G.: 1961, Tellus.
Drazin, P.G. and Su, C.H.: 1975 J. Atmos. Sci. 32, 437.
Fitzjarrald, D.: 1984 J. Atmos. Sci. (in press).
Fraenkel, L.E.: 1961 J. Fluid Mech. 11, 400-406.
Hunt, J.C.R.: 1981 Symp. on 'Designing with the wind'. Centre Scient. Technique du Baltiment, Nantes, France, June 1981.
Hunt, J.C.R., Kaimal, J.C., Gaynor, J., Korrel, A.: 1981 NOAA Env. Lab. Report Boulder, Colo.
Hunt, J.C.R., Leibovich, S., Lumley, J.L.: 1983

Hunt, J.C.R., Lalas, D.P. & Asimakopoulos, M.: 1984 J. Fluid Mech. 142, 201-216.
Hunt, J.C.R., Leibovich, S. and Richards, K.J.: 1984 Q.J. Roy. Met. Soc. (Submitted).
Hunt, J.C.R., Richards, K.J. & Brighton, P.W.M.: 1984 Q.J. Roy. Met. Soc. (Submitted).
Hunt, J.C.R. and Simpson, J.E.: 1982 Chap. 7 of Engineering Meteorology (Ed. E.J. Plate) Elsevier.
Hunt, J.C.R. & Snyder, W.H.: 1980 J. Fluid Mech. 96, 671-704.
Jackson, P.S.: 1977 Proc. 4th. Int. Conf. on Wind Effects on Buildings and Structures, pp. 33-40, (Ed. K. Eaton). Cambridge Univ. Press.
Jackson, P.S. and Hunt, J.C.R. 1975: Q.J. Roy. Met. Soc. 101, 929-956.
Jensen, N.O. and Peterson, E.W. 1978: Q.J. Roy. Met. Soc. 104, 719.
Klemp, J.B. and Lilly, D.K.: 1975 J. Atmos. Sci. 32, 320.
Lamb, V.R. and Britter, R.E.: 1984 J. Fluid Mech. (in press).
Lavery, T.F. et al.: 1982 EPA-600/3-82-036. April 1982.
Mahrer, Y. and Pielke, R.A.: 1975 J. Atmos. Sci. 32, 2144-2155.
Manins, P.C. and Sawford, B.L.: 1979 Q.J. Roy. Met. Soc. 105, 1011.
Mason, P.J. and King, J.C.: 1984 Q.J. Roy. Met. Soc. July 1984.
Mason, P.J. and Sykes, R.I.: 1979 Q.J. Roy. Met. Soc. 105, 393-395.
Meroney, R.N., Sandborn, V.A., Bouwmeester, R.J.B., Chien, H.C. and Rider, M.: 1979 Colo. State Univ. Report CER77-78 RJB RNM-VAS50G.

Newley, T.J., Pearson, H.J. and Hunt, J.C.R.: 1984 (submitted for publication).
Panofsky, H.A., Larks, D., Lipshutz, R., Stone, G., Bradley, E.F., Bowen, A.J. and Hojstrup, P.J.: 1982 Q.J. Roy. Met. Soc. 108, 215.
Plate, E.J. and Lin, C.Y.: 1965. Colo. State Univ. Report CEP65, EJP14.
Plate, E.J. (Ed.) 1982. Engineering Meteorology, Elsevier.
Rao, K.S., Wyngaard, J.C. and Cote, O.R. 1974: J. Atmos. Sci. 31, 738.
Richards, K.J. and Taylor, P.A.: 1980 Geophys. J.R. Astr. Soc.
Rottuno, R.: 1983 J. Atmos. Sci., 40, 1999-2009.
Scorer, R.S.: 1949 Q.J. Roy. Met. Soc. 75, 41-56.
Scorer, R.S.: 1955 Q.J. Roy. Met. Soc. 81, 340.
Scorer, R.S.: 1972 Clouds of the World. David & Charles (Newton Abbott).
Sheppard, P.A.: 1956 Q.J. Roy. Met. Soc. 82, 528-9.
Sherman, C.A.: 1978 J. Appl. Met. 17, 312-319.
Smith, R.B.: 1980 Tellus 32, 348-364.
Snyder, W.H., Britter, R.E. and Hunt, J.C.R.: 1980 Proc. 5th. Int. Conf. on Wind Engineering. Pergamon.
Snyder, W.H., et al.: 1984 J. Fluid Mech. (Submitted).
Stewartson, K. and Williams, P.G.: 1969 Proc. Roy. Soc. A. 321, 181-206.
Sykes, R.I.: 1980 J. Fluid Mech. 101, 647-670.
Tampieri, F.: 1984 (Submitted for publication).
Thom, A.: 1975 Vegetation and the Atmosphere (Ed. J.L. Monteith). pp. 57-110 Acad. Press.
Venkatram, A.: 1980 B.L. Met. 19, 481.
Verma, S.B. and Cermak, J.E.: 1971 Colo. State Univ. CER 71-72 SBV-JEC 11.
Walmsley, J.L., Salmon, J.R. and Taylor, P.A.: 1982 B.L. Met. 23, 1746.

Questions, answers and comments

P. ALPERT: You have shown a few examples of linear solutions to flow over complex terrain. Could you please summarize common features to flows where linear theory works well. There are forced boundary flows which have a nice linear solution although velocity perturbations are quite high.

J.C.R. HUNT: It is difficult to know exactly when the linear solutions for an air flow over hills are inaccurate. If the linear theory is used and large perturbations are calculated, then the results should be treated with caution. A correction for non-linear effects over a hill is included in our paper. Linear calculations appear to over-predict turbulent shear-stresses in the wake when the slopes are greater than about 0.1.At low Froude -Numbers the velocity perturbations are often large. But in general it is hard to define the limitations of linear models without having made non-linear calculations.

L. BERKOFSKY: The calculations you presented were based on steady-state models. Are you investigating unsteady flows over complex terrain?

J.C.R. HUNT: Yes. Sea breezes or gravity currents impinging on hills. Drs. Asimakopoulos and Lalas are studying these effects over Mt. Hymmetas near Athens. There are many other problems of unsteady flows over topography which certainly need to be studied.

PHYSICAL MODELLING OF FLOW AND DISPERSION OVER COMPLEX TERRAIN

J. E. Cermak
Director, Fluid Dynamics and Diffusion Laboratory and
Professor-in-Charge, Fluid Mechanics and Wind Engineering Program
Colorado State University
Fort Collins, Colorado 80523 U.S.A.

ABSTRACT

Atmospheric motion and dispersion over topography characterized by irregular (or regular) hill-valley or mountain-valley distributions are strongly dependent upon three general sets of variables. These are variables that describe topographic geometry, synoptic-scale winds and surface-air temperature distributions. In addition, pollutant concentration distributions also depend upon location and physical characteristics of the pollutant source. Overall fluid-flow complexity and variability from site to site have stimulated the development and use of physical modelling for determination of flow and dispersion in many wind-engineering applications. Models with length scales as small as 1:12,000 have been placed in boundary-layer wind tunnels to study flows in which forced convection by synoptic winds is of primary significance. Flows driven primarily by forces arising from temperature differences (gravitational or free convection) have been investigated by small-scale physical models placed in an isolated space (gravitational convection chamber). Similarity criteria and facilities for both forced and gravitational-convection flow studies are discussed. Forced-convection modelling is illustrated by application to dispersion of air pollutants by unstable flow near a paper mill in the state of Maryland and by stable flow over Point Arguello, California. Gravitational-convection modelling is demonstrated by a study of drainage flow and pollutant transport from a proposed mining operation in the Rocky Mountains of Colorado. Other studies in which field data are available for comparison with model data are reviewed.

NOMENCLATURE

B_T = dimensionless coefficient

B_U = dimensionless coefficient

C_p = specific heat at constant pressure

Boundary-Layer Meteorology 30 (1984) 261–292. 0006-8314/84/0303-0261$04.80

g	=	gravitational acceleration
h_{mn}	=	mean difference in elevation
k	=	thermal conductivity of fluid or concentration coefficient
k_c	=	molecular diffusivity of pollutant
k_s	=	equivalent sand roughness
K	=	Kármán constant
K_c	=	eddy diffusivity for pollutant
K_M	=	eddy diffusivity for momentum
L	=	reference length or Monin-Obukhov length
L_{u_x}	=	integral scale of longitudinal turbulence component
L_x	=	longitudinal distance from beginning of boundary layer
p	=	pressure
Q	=	pollutant emission rate
R	=	gas constant
t	=	time
T	=	mean temperature
T_*	=	friction temperature
ΔT	=	temperature difference
u_*	=	shear velocity $\sqrt{\tau_o/\rho}$
U	=	mean velocity
x	=	longitudinal coordinate
y	=	lateral coordinate
z	=	vertical coordinate
z_o	=	aerodynamic roughness length
α	=	wind direction
δ	=	boundary-layer thickness
δ_{ij}	=	Kronecker delta
ε_{ijk}	=	permutation tensor
θ	=	temperature fluctuation
τ_o	=	surface shear stress
ν	=	kinematic viscosity
ρ	=	mass density
σ_U	=	root-mean-square of longitudinal turbulence component
τ_o	=	surface shear stress
ϕ	=	turbulent energy dissipation function
χ	=	concentration of pollutant
Ω	=	angular velocity of coordinate system

Subscripts

i	=	ith component
j	=	jth component
k	=	kth component

m = quantity associated with model
o = reference quantity
p = quantity associated with prototype
∞ = quantity of height $z = \delta$

Superscripts

$()^*$ = nondimensional quantity
$\overline{()}$ = time average
$()'$ = fluctuation from time average

1. INTRODUCTION

Atmospheric flow over complex terrain is generally a combination of forced convection generated by a synoptic-scale wind field and natural convection driven by air-density differences. During periods of strong synoptic-scale winds air flow over ridges and through valleys is dominated by flow in the planetary boundary layer; however, when the synoptic-scale motion becomes weak, or ceases, local temperature gradients cause forces that control the flow.

Physical modelling of atmospheric flow over complex topography, until recently, has been confined to forced-convection flows. Wind tunnels, ranging from a simple (0.8 m wide x 0.35 m high x 0.9 m long) uniform-flow test section stratified with dry ice used by Abe (1929) to a special meteorological boundary-layer wind tunnel having a 2 m x 2 m x 30 m test section with heating and cooling capabilities for stratification (Cermak, 1958 and 1981; Plate and Cermak, 1963) have been used for this purpose. In some cases water has been used as the modelling fluid (Long, 1959; Snyder, 1983). On the other hand, physical modelling of flow over complex terrain that is dominated by gravitational forces, mountain and valley winds, is still in the exploratory stage of development. Stimulated by the ever growing need to predict concentration distributions of pollutants released from sources located in complex terrain, efforts have been initiated in the Fluid Dynamics and Diffusion Laboratory to simulate winds with stable stratification forced by synoptic-scale pressure gradients as well as mountain-valley winds generated by local temperature gradients. Studies of the former type include flow and dispersion over Point Arguello, California; simulation of mountain lee waves (Lin and Binder, 1967); flow and dispersion over Elk Mountain, Wyoming, (Kitabayshi, Orgill and Cermak, 1971); flow and dispersion over the Eagle River Valley, Colorado (Cermak, Orgill and Grant, 1970; Orgill and Cermak, 1972); and flow and dispersion near the Westvaco Paper Mill, Maryland (Petersen, Cermak and Ghavam, 1980; Weil, 1979; and Weil, Cermak and Petersen, 1981). Studies of the latter category of flow include drainage flows at geothermal power plant sites in the Geysers Field of northern California (Petersen and Cermak, 1978 and 1980a) and a molybdenum mine site in the Rocky Mountains near Crested

Butte, Colorado; (Petersen and Cermak, 1980b; and Petersen, Cermak and Nelson, 1980).

Similarity requirements for physical modelling of flow and dispersion in the atmospheric boundary layer (ABL) have been presented in several publications (Chuang and Cermak, 1967; Mery, 1969; Cermak and Arya, 1970; Cermak, 1971; Snyder, 1972; Cermak, 1975). However, a review of the requirements with emphasis on modelling flow over complex terrain is presented in the following section.

The primary purpose of this paper is to review techniques by which physical modelling of atmospheric motion and dispersion over complex terrain can be accomplished for different meteorological conditions. Meteorological flows considered are neutral ABL's, stratified ABL's and pure gravitational convection. An objective is to examine the agreement between information obtained from measurements in the small-scale and full-scale flow fields. Consequently, physical-modelling examples cited are primarily those for which some comparable field data are available.

2. SIMILARITY CRITERIA

Basic criteria for similarity of two flow fields differing in geometric, kinematic, dynamic and thermic scales are derived from inspectional analysis (Ruark, 1935) of the equations for conservation of mass, momentum and energy. The nondimensional equations obtained by scaling of all dependent and independent variables are also used to establish laminar flow similarity criteria for stable flow over very small-scale models.

2.1 General (Gross Flow) Similarity Criteria

The results of this analysis (Cermak, 1971), where scaled variables are

$$\bar{U}_i^* = \frac{\bar{U}_i}{U_o},\ (u_i')^* = \frac{u_i'}{U_o},\ (\theta')^* = \frac{\theta'}{T_o},\ x_i^* = \frac{x_i}{L_o},\ t^* = \frac{tU_o}{L_o},$$

$$\rho^* = \frac{\rho}{\rho_o},\ \Omega_j^* = \frac{\Omega_j}{\Omega_o},\ \bar{P}^* = \frac{\bar{P}}{(\rho_o U_o^2)},\ \bar{T}^* = \frac{\bar{T}}{T_o},\ \overline{\Delta T}^* = \frac{\overline{\Delta T}}{\Delta T_o},\ g^* = \frac{g}{g_o}$$

and ϕ^* = dimensionless dissipation function, (1)

are as follows:

$$\frac{\partial \rho^*}{\partial t^*} + \frac{\partial(\rho^* U_i^*)}{\partial x_i^*} = 0 \tag{2}$$

$$\frac{\partial \bar{U}_i^*}{\partial t^*} + \bar{U}_j^* \frac{\partial \bar{U}_i^*}{\partial x_j^*} + \left[\frac{L_o \Omega_o}{U_o}\right] 2\varepsilon_{ijk}\Omega_j^* \bar{U}_k^* = - \frac{\partial \bar{P}^*}{\partial x_i^*}$$

$$- \left[\frac{\Delta T_o}{T_o} \frac{L_o g_o}{U_o^2}\right] \frac{\overline{\Delta T^*}}{\bar{T}^*} g^* \delta_{i3} + \left[\frac{\nu_o}{U_o L_o}\right] \frac{\partial^2 \bar{U}_i^*}{\partial x_k^* \partial x_k^*} + \frac{\partial(-\overline{u_i'^* u_j'^*})}{\partial x_j^*} \tag{3}$$

and

$$\frac{\partial \bar{T}^*}{\partial t^*} + \bar{U}_i^* \frac{\partial \bar{T}^*}{\partial x_i} = \left[\frac{k_o}{\rho_o C_{po} \nu_o}\right]\left[\frac{\nu_o}{L_o U_o}\right] \frac{\partial^2 \bar{T}^*}{\partial x_k^* \partial x_k^*}$$

$$+ \frac{\partial(-\overline{\theta'^* u_i'^*})}{\partial x_i^*} + \left[\frac{\nu_o}{U_o L_o}\right] \frac{U_o^2}{C_{po} T_o} \phi^* . \tag{4}$$

The conservation of mass statement given by Eq. 2 is invariant for this transformation in which geometrical similarity ($x_i^* = x_i/L_o$) and kinematic similarity ($\overline{U_i^*} = \bar{U}_i/U_o$ and $t^* = tU_o/L_o$) are preserved. Conservation of linear momentum is expressed by Eq. 3 with the Boussinesq approximation ($\overline{\Delta T} << T_o$) where the equation of state for the fluid is taken as

$$P = \rho RT . \tag{5}$$

Equation 3 states that requirements for dynamic similarity are that the following dimensionless parameters be equal for two systems (model and prototype): Rossby number, $Ro = U_o/(L_o\Omega_o)$; Richardson number, $Ri = (\Delta T_o/T_o)(L_o g_o/U_o^2)$; and Reynolds number, $Re = U_o L_o/\nu_o$. Thermic similarity, according to the conservation of energy statement (Eq. 4), requires equality for two systems of the Prandtl number, $Pr = \nu_o \rho_o C_{po}/k_o$; the Eckert number, $Ec = U_o^2/(C_{po}T_o)$; and Re.

The requirements for geometric, dynamic, and thermic similarities obtained by inspectional analysis must be augmented by additional specifications if requirements for "exact" similarity are to be complete. These external conditions are the following: (1) similarity of surface roughness and temperature distributions at ground level; (2) similarity of flow structure above the ABL; (3) similarity of the horizontal pressure gradient; and (4) sufficient upwind fetch to establish equilibrium of the ABL consistent with external conditions.

Even a casual review of the requirements for "exact" similarity of the ABL, when a wind-tunnel boundary layer is one of the flow systems, leads to the conclusion that physical modelling of the ABL must be considered on the basis of "approximate" similarity. The primary compromises that must be made is a relaxation of the requirements for equal Reynolds numbers and equal Rossby numbers. The degree to which the similarity requirements can be achieved has been discussed by Cermak (1971, 1975) and will be summarized here.

With no rotation of the wind tunnel about a vertical axis, Rossby number for the wind-tunnel flow will be smaller than for the atmosphere by a factor approximately equal to the length scale. Accordingly, for flow over a plane boundary, the mean velocity will not change direction with height as in the atmosphere. Local mean velocities in flow around and immediately above highly irregular boundary geometry such as rough complex terrain are governed by shear stresses and local pressure gradients induced by the geometry. These effects are strong compared to Coriolis effects. Accordingly, relaxation of Rossby-number equality is considered as an approximation that does not introduce significant error in simulation of flow over complex terrain.

The ratio of Reynolds numbers for the wind tunnel and the atmosphere will also be approximately equal to the length scale. Fortunately, as shown in Fig. 1, boundary-layer flow characteristics over a rough surface of length, L_x, become independent of Reynolds number at values readily attainable in a wind-tunnel with test-section length of 20-30 m (Cermak; 1958, 1981, 1982). Through selection of proper combinations of wind-tunnel length, surface roughness and ambient wind speed; boundary-layer flows can be realized that model the micrometeorological features of the high Reynolds number ABL. A criterion for local Reynolds-number independence; i.e., the surface is aerodynamically rough, is given by $u_* k_s/\nu > 70$ (Schlichting, 1968). This Reynolds-number criterion is valid for neutral and, with lesser restriction, to unstable thermally stratified flow. However, further research is needed to establish a corresponding criterion for stably stratified flow. When the geometric scaling is 1:2000, small physical roughness elements may scale down to become submerged in viscosity dominated flow near the surface ("laminar sublayer") and result in an aerodynamic smooth model. In such cases, an aerodynamically rough boundary can be achieved by exaggerating the roughness element heights to protrude above the nominal viscous-zone depth ($5\nu/u_*$) by its scaled height (Chaudhry and Cermak, 1971). Flow over complex terrain is influenced by a larger roughness than k_s--the difference in height between high and low points in the area. When the terrain is a complex of valleys and ridges varying in direction over the area, a reasonable criterion for Reynolds number independence is $u_* h_{mn}/\nu > 70$ (h_{mn} = mean difference in elevation of ridges and valleys). Additional research is required to determine the generality of this criterion.

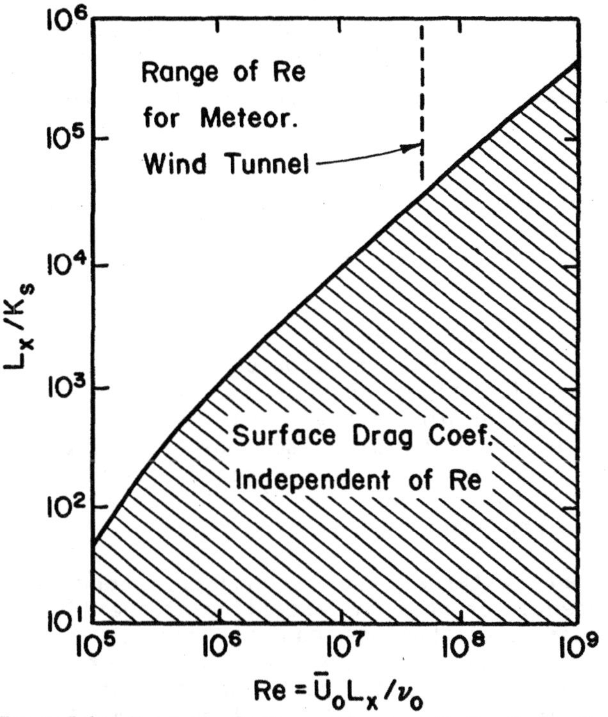

Figure 1. Reynolds Number Independence of Surface Drag Coefficient as Function of Relative Roughness in Neutral Flow (Schlichting, 1968)

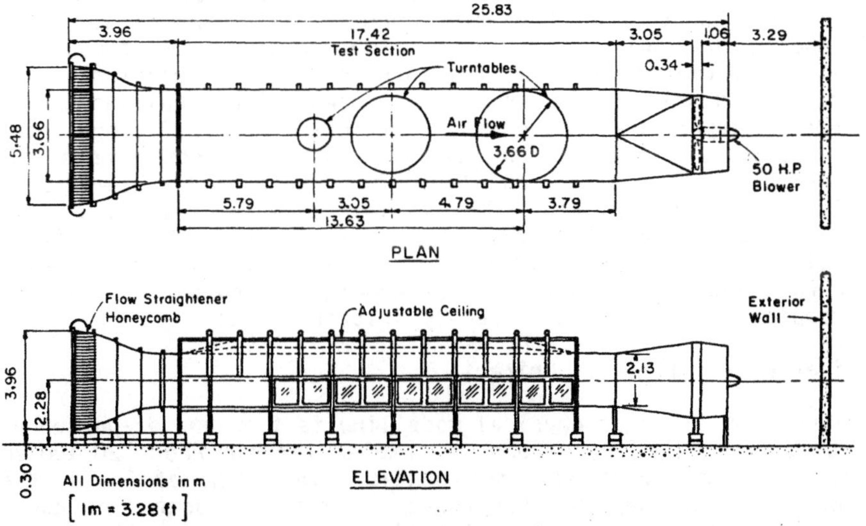

Figure 2. Open-circuit BLWT, Fluid Dynamics and Diffusion Laboratory, Colorado State University.

Typical gross Richardson numbers for the atmospheric surface layer are in the range $-1 < Ri < 1$. In order to achieve Richardson numbers of this magnitude in the BLWT, large values of ΔT_o and small values of U_o must be obtainable to equalize the small length-scale ratio. Richardson-number equality is essential if the BLWT is to be used for a wide range of diffusion studies. A noteworthy observation is that the square root of Ri^{-1} is a generalized bulk Froude number that is commonly used as the essential similarity parameter for stratified flow over complex terrain by both forced and gravitational convection.

Prandtl-number equality is automatically satisfied if air is used as the wind-tunnel fluid. The Eckert number, when Richardson numbers are equal for stratified flow, will be approximately one order-of-magnitude smaller for the wind tunnel than for the ABL. Since the Eckert number can be expressed as a Mach number squared, it will be small compared to unity for both the atmosphere and the BLWT. Therefore, little distortion of the flow field can be attributed to inequality of the Eckert numbers in this context.

The associated external, or boundary, conditions impose additional requirements for physical modelling of the ABL. Similarity of surface roughness distributions can be achieved by addition of roughness elements to the model surface, while surface temperature distributions can be realized by addition of heating and cooling elements. Longitudinal pressure variation can be controlled through slight divergence of the wind-tunnel test-section walls, or installation of a flexible ceiling, or both, that can be adjusted for specific surface roughnesses or models. To accommodate a variety of flow structures above the boundary layer, flow entering the test section should be uniform in mean velocity with low turbulence intensity. Addition of grids at the entrance can add turbulence of a desired scale and intensity over the test area. The requirement that the boundary layer be in equilibrium with external conditions (primarily surface roughness and temperature) does not impose any limitations on simulation capabilities, but suggests the need for a reasonably long test section. This is particularly true if a long length of equilibrium boundary layer is required for either fundamental flow studies or diffusion tests. The extent to which upwind terrain should be modelled under modelling flow over a selected area of complex terrain is addressed by Britter (1982).

2.2 Laminar Flow Similarity Criteria

An important class of physical modelling is flow and disperion of stably stratified flow over areas of complex terrain about 20-100 km in diameter. Since existing wind tunnels have widths of 2-4 m, the peak geometric scale required for these large areas range from about 1:5,000 to 1:50,000 and smaller for larger areas. Equality of Ri requires a low wind speed $\bar{U}_o$ and a large temperature difference ΔT. This usually results in a laminar flow over the model (Cermak and Peterka, 1966).

The last term in Eq. 3 can be approximated by taking

$$- \overline{u_i' u_j'} \simeq K_m \, \partial \bar{U}_i / \partial x_j \tag{6}$$

to give for the last two terms in Eq. 3

$$\left[\frac{\nu_o}{U_o L_o} \right] \frac{\partial^2 \bar{U}_i^*}{\partial x_j^* \partial x_j^*} + \frac{\partial(-\overline{u_i'^* u_j'^*})}{\partial x_j^*} \simeq \left[\frac{\nu_o}{U_o L} + \frac{K_m}{U_o L} \right] \frac{\partial^2 \bar{U}_i^*}{\partial x_j^* \partial x_j^*} \tag{7}$$

For the model flow $K_m \simeq 0$ and for the atmosphere $\nu_o << K_m$ ($K_m \simeq 10^4 - 10^5 \, \nu_o$); therefore, similarity of flow ($\bar{U}^*_{i_{model}} = \bar{U}^*_{i_{atm}}$) requires

$$(U_o L/\nu_o)_{model} \simeq (U_o L/K_m)_{atm} \tag{8}$$

and geometric and thermic similarities are satisfied. Equation 8 will be satisfied if $L_{model}/L_{atm} \simeq 1/10{,}000$ to $1/100{,}000$. Similarity of this type has been employed by Fultz, 1951; Cermak and Peterka, 1966; and Hertig, 1983.

An inspectional analysis of the equation for conservation of mass and use of a turbulent mixing coefficient K_c in the approximation of turbulent diffusion terms gives

$$\frac{\partial \bar{\chi}^*}{\partial t^*} + \bar{U}_j^* \frac{\partial \bar{\chi}^*}{\partial x_j^*} = \left[\frac{k_c}{U_o L_o} + \frac{K_c}{U_o L_o} \right] \frac{\partial^2 \bar{\chi}^*}{\partial x_j^* \partial x_j^*} \tag{9}$$

Accordingly, with $(\bar{U}_j^*)_{model} \simeq (\bar{U}_j^*)_{atm}$ and similar mass sources, the requirement for similarity of the mean concentration field ($\chi^*_{model} \simeq \chi^*_{atm}$) is

$$(U_o L_o/k_c)_{model} \simeq (U_o L_o/K_c)_{atm} \, . \tag{10}$$

3. FLOW MODELLING FACILITIES

Air, water, "Freon" and other fluids may be used for modelling flow over complex terrain. However, air at atmospheric pressure is most convenient, requires the least expensive facility and offers great flexibility of applications without sacrifice in meeting

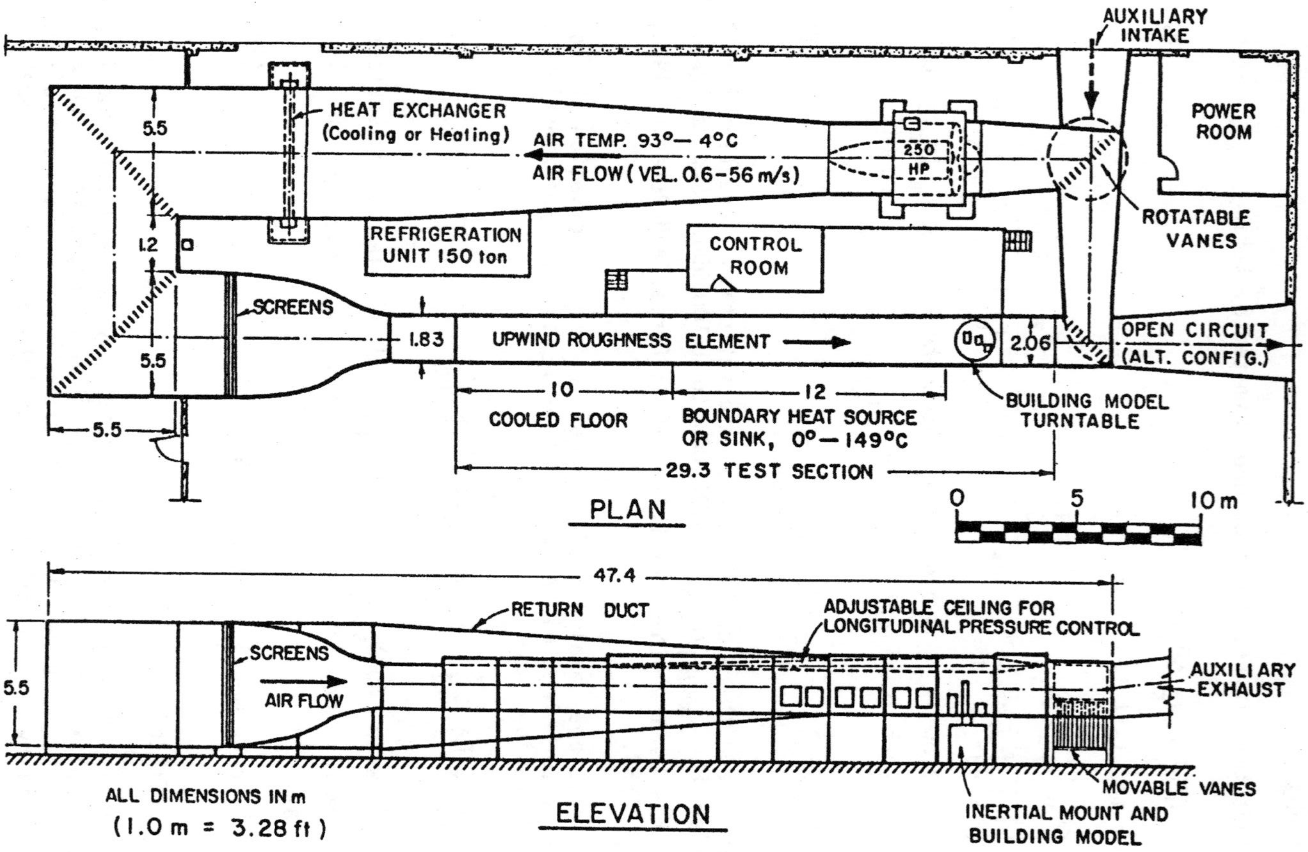

Figure 3. Meteorological Wind Tunnel, Fluid Dynamics and Diffusion Laboratory, Colorado State University

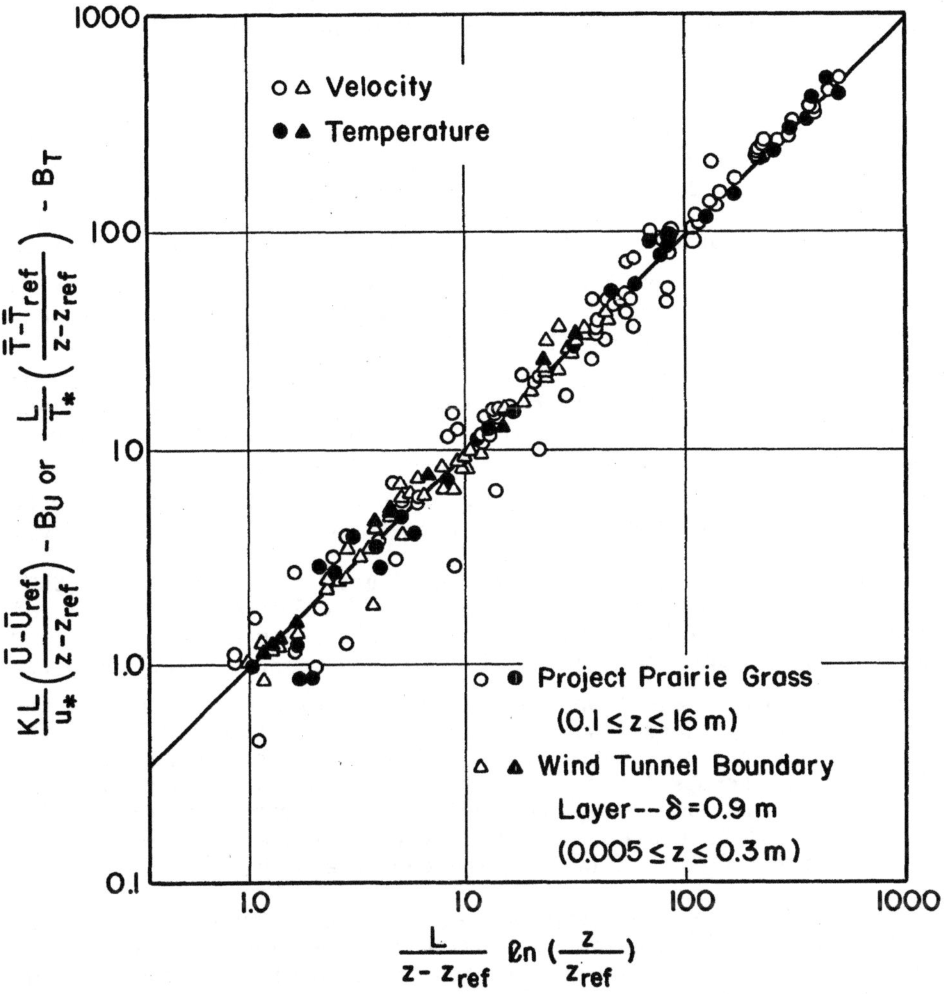

Figure 4. Mean Velocity and Temperature from Meteorological BLWT (Chuang and Cermak, 1967) and Atmosphere (Barad, 1952) Compared to Log-linear Profile

similarity requirements. Therefore, the following descriptions are limited to low-speed boundary-layer wind tunnels for modelling forced convection flows and an "office-size" chamber filled with air at atmospheric pressure for modelling pure gravitational convection.

3.1 Boundary-layer Wind Tunnels (BLWT)

Two basic types of low-speed BLWT are used for physical modelling of flow and dispersion over complex terrain--open-circuit and closed-circuit configurations. An open-circuit BLWT such as illustrated in Fig. 2 is used primarily for simulation of a neutral ABL. This facility has a maximum wind speed U_0 of 10 m/s, an entrance flow turbulence intensity of 1.0% and a boundary-layer thickness over a smooth floor of 0.5 m at 15 m downwind from the test-section entrance when U_0 = 10 m/s. By adding spires or vortex generators and fences at the test-section entrance, the boundary-layer thickness may be increased to 1.5 m. The effect of these devices on mean velocity and turbulence properties of the simulated ABL is reported by Cermak (1982). By introducing a bank of dry ice (solid CO_2) over about a 4 m length of the test-section entrance, an open-circuit BLWT can be used to simulate a stably stratified ABL (Cermak, Orgill and Grant; 1970).

Figure 3 illustrates the earliest closed-circuit BLWT (meteorological wind tunnel) designed specifically for simulation of thermally stratified ABL's (Cermak, 1958; Plate and Cermak, 1963; and Cermak, 1981). Over the range of U_0 from 0.5 to 37 m/s the speed can be maintained with less than ± 1% variation from any setting. At the lower wind speed, thermal stratification over the lowest 0.1 m of the boundary layer can be controlled to give values of Ri ranging from -1.0 to +1.0. Across the test-section entrance mean velocities vary by less than 1% and turbulence intensity is approximately 0.05%. Elevated inversions can be simulated by cooling the upstream 10 m of the test-section floor and warming the downstream 10 m.

Mean velocities and temperature, Reynolds stresses, turbulence energy spectra and turbulent diffusion in the simulated ABL's have been found to be in good agreement with corresponding micrometeorological data (Cermak and Arya, 1970; Cermak, 1975 and 1982; Poreh and Cermak, 1984). Mean velocities and temperatures measured in the simulated ABL and in the atmosphere are compared with log-linear relationships for stable, neutral and unstable flow in Fig. 4.

3.2 Convection Chamber

Physical modelling of gravitation convection over complex terrain caused only by Archimedian forces requires that the model be placed in an isolated space free from air disturbances. Since a model used for this purpose will ordinarily be used also for studies of forced convection in a wind tunnel, the model size will be determined by the wind-tunnel size. For the wind-tunnel shown in Fig. 2 the model size would be about 3.6 m wide and 6 m long.

The model should be centered in the chamber with sufficient distance between model and the chamber walls to reduce wall effects on flow near the model edges. This distance is not well defined but for the size model referred to, a distance of about 1 m was observed to be satisfactory (Cermak and Petersen, 1981). The chamber ceiling height above the model should correspond to the scaled height of an inversion layer. The foregoing considerations suggest that a thermally insulated room about 5.5 m wide, 8 m long and 2.5 m high is satisfactory for models of the size indicated (Cermak and Petersen, 1981). A ceiling that can be raised and lowered would be a useful feature. Provisions for remote movement of temperature and velocity sensors over the cooled and/or heated model should be made. Figure 5 shows the interior of a convection chamber configured for cooling an aluminum shell model to simulate drainage flows.

4. MODELS OF FLOW AND DISPERSION OVER COMPLEX TERRAIN

Experiences in physical modelling of flow and dispersion over complex terrain now include forced convection by flow with neutral, unstable and stable stratification and drainage flows resulting from gravitational convection. Only those cases for which some full-scale meteorological and/or concentration data are available are addressed in the following paragraphs. Flow and dispersion studies over isolated two- and three-dimensional hills has been addressed by Snyder (1983) but, since these geometries should not be classified as complex terrain, they are not included in the following discussion.

4.1 Neutral ABL Flow

Flow characteristics over the Rakaia River Gorge region of New Zealand have been studied by measurements over a physical model and in the field (Meroney, 1980). The undistorted 1:5,000 scale model covered an area 6,100 m wide and 18,300 m long centered on the gorge which has an average elevation difference between river and gorge top of about 200 m and a width that varies from 500 m to 2,600 m. The University of Canterbury BLWT having a 1.22-m square cross section and 12.2-m length was used. Running with U_0 equal to 13 m/s the approach mean wind profile followed a power-law exponent of 0.14 with a depth that scaled to 500 m and a value of z_0 that scaled to 4.5 cm. Mean wind speed was measured with a cup anemometer at a height of 10 m for several locations.

Correlation coefficients for the field and model wind speeds at corresponding locations were found to be as large as 0.95. The high correlation was for a model surface with steps between laminations

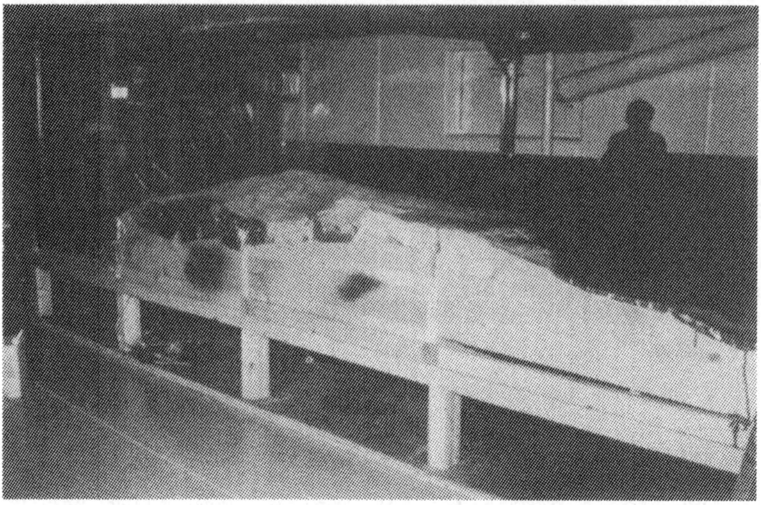

Figure 5. Interior of Convection Chamber with Platform for Placement of Model Over Dry Ice Bed (Petersen and Cermak, 1980)

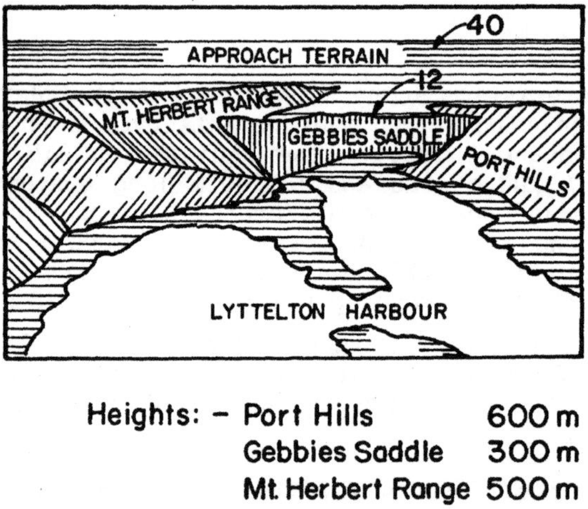

Figure 6. Terrain Features of the Gebbies Pass Region--View from the NE (Neal, 1983)

filled, then roughened to give a full-scale z_0 of 50 cm and shelter belts added. Correlations with the stepped (terraced model) gave wind-speed correlations of only 0.7.

Two significant conclusions are derived from the Rakaia River Gorge study:

1. An upwind fetch of modelled terrain is necessary to develop an equilibrium boundary layer over the site in question.

2. Local surface details (geometry, vegetation, etc.) must be accurately reproduced in order to obtain a good simulation of winds in the surface layer ($z < 100$ m).

Flow characteristics over 1:4,000 and 1:8,000 scale models and in the field over Gebbies Pass, New Zealand, shown in Fig. 6 were measured by Neal (1983). The University of Canterbury wind tunnel (1.2-m square cross section and a length of 12.2 m) was used. The approach flow boundary layer scaled to 640 m and the model surface was smoothed to remove lamination steps. Surface roughness was added to reproduce 20-m high shelter belts and scrub vegetation 3-5 in height. Field measurements included mean wind speed at a height of 10 m at a variety of locations, mean wind velocity to a height of 300 m (measured by Tala kites) and turbulence data at several heights up to 20 m.

Correlations coefficients of 0.91 to 0.96 between mean velocities at 12 m and 40 m for the two models indicate that 6-7% blockage for the 1:4,000 scale model had negligible effect on the modelled wind field. Figures 7 and 8 show a good degree of agreement between mean wind speed and turbulence intensity profiles for the model and full-scale data. Both models gave an integral length scale for the longitudinal turbulence component L_{u_x} of about 200 m for point 40 (upwind of pass) but gave 350 m and 800 m for the 1:4,000 and 1:8,000 scale models, respectively, at point 12 (on top of pass).

Important conclusions from the Gebbies Pass study are the following:

1. Wind-tunnel blockage of 7% has negligible effect on flow in the lower 50% of the boundary layer.

2. Modelled wind characteristics are in excellent agreement with corresponding field data in the lower 100 m.

3. Accurate reproduction of hill-top geometry becomes more important as the model scale is reduced.

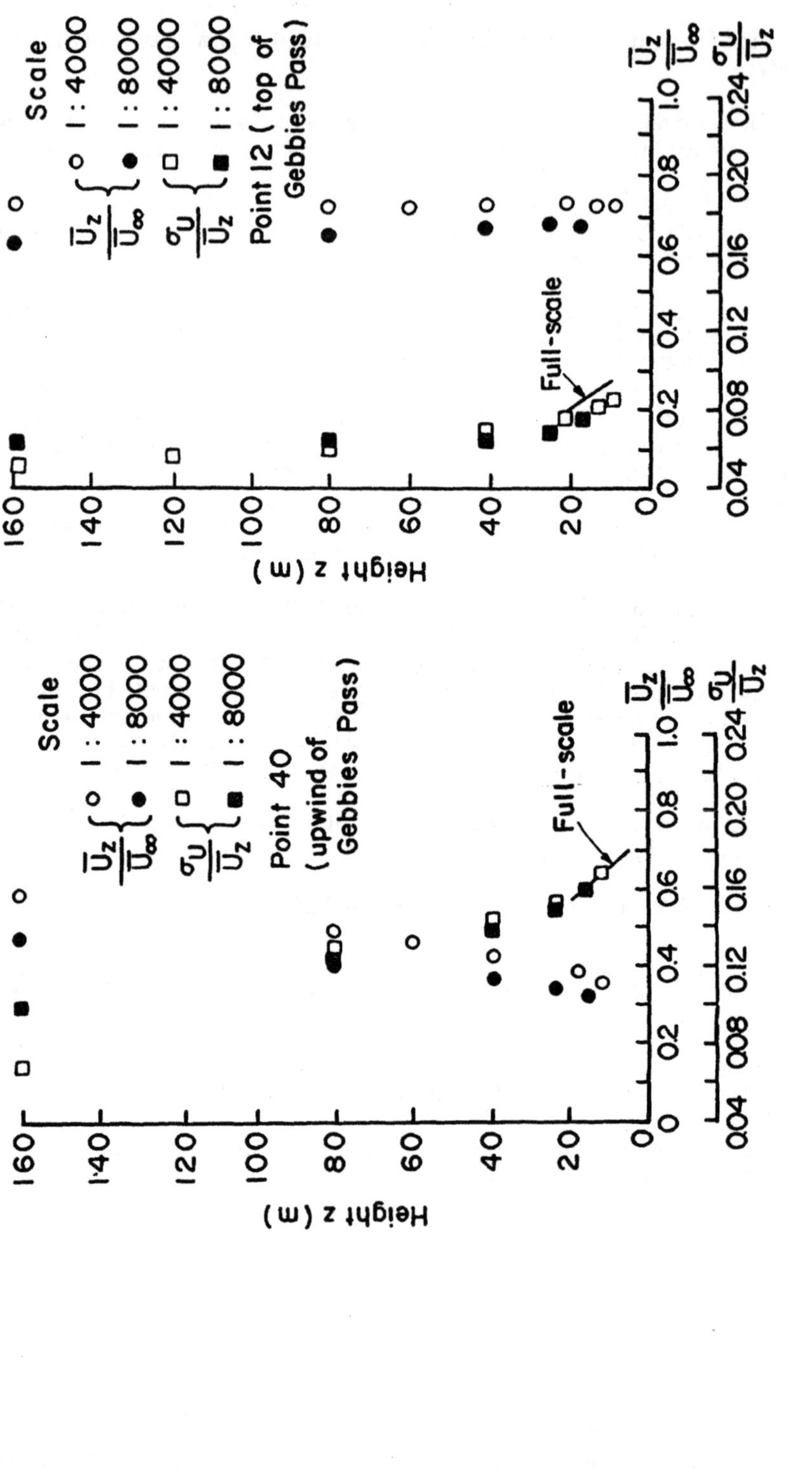

Figure 7. Mean Wind Speed and Local Turbulence Intensity Profiles over Gebbies Pass Region (Neal, 1983)

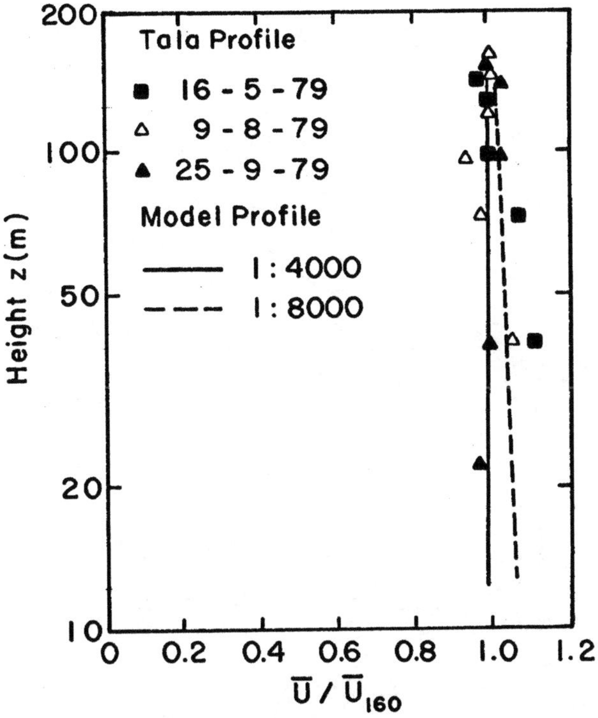

Figure 8. Mean Wind Speeds at Point 12 (at top of Gebbies Pass) Measured by Talakites and Over Models (Neal, 1983)

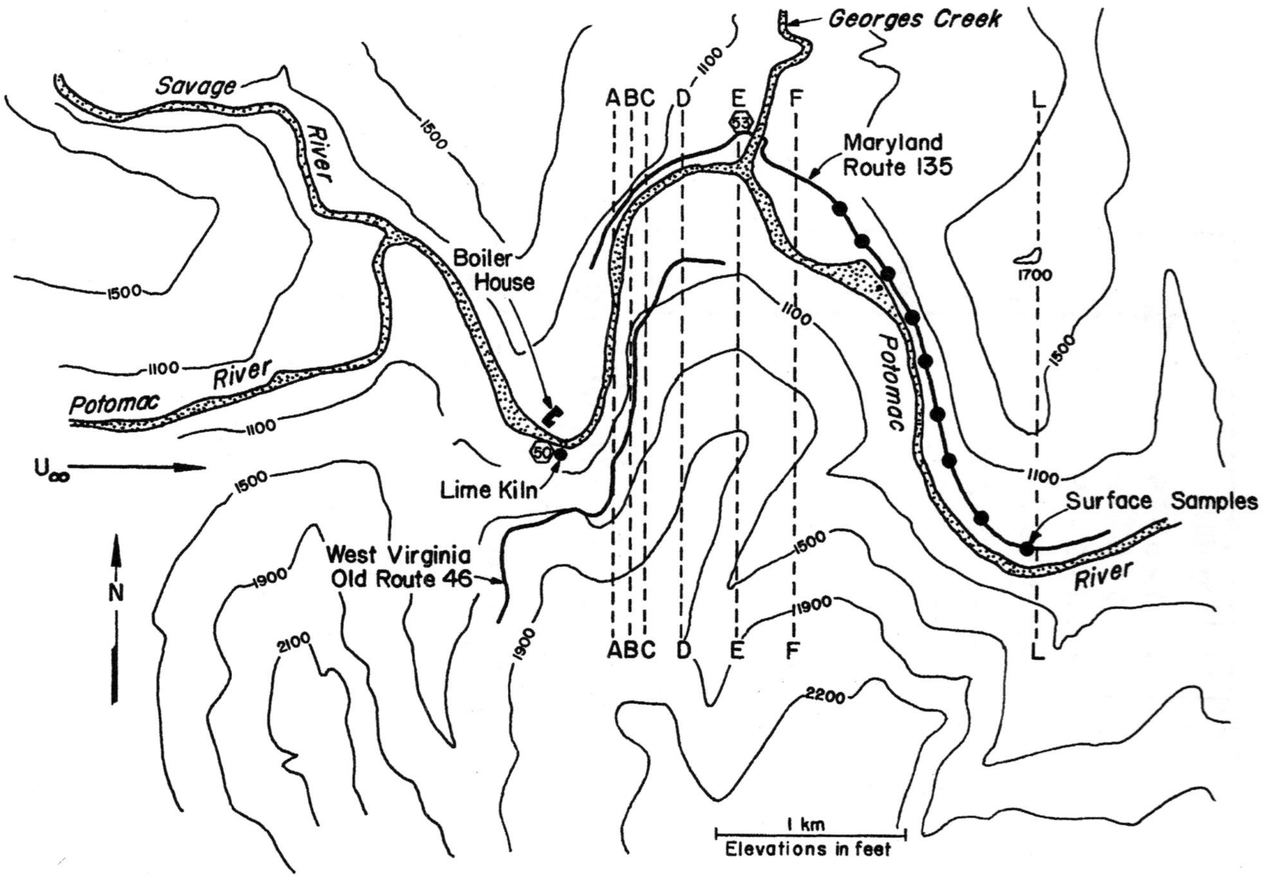

Figure 9. Source Locations and Ground-level Concentration Sampling Positions in Field and Wind Tunnel. Heavy Solid Lines Denote Mobile Van Sampling Routes Used in Field; Dashed Lines and Closed Circles Show Sampling Locations in Wind Tunnel (Petersen, Cermak and Ghavam; 1980)

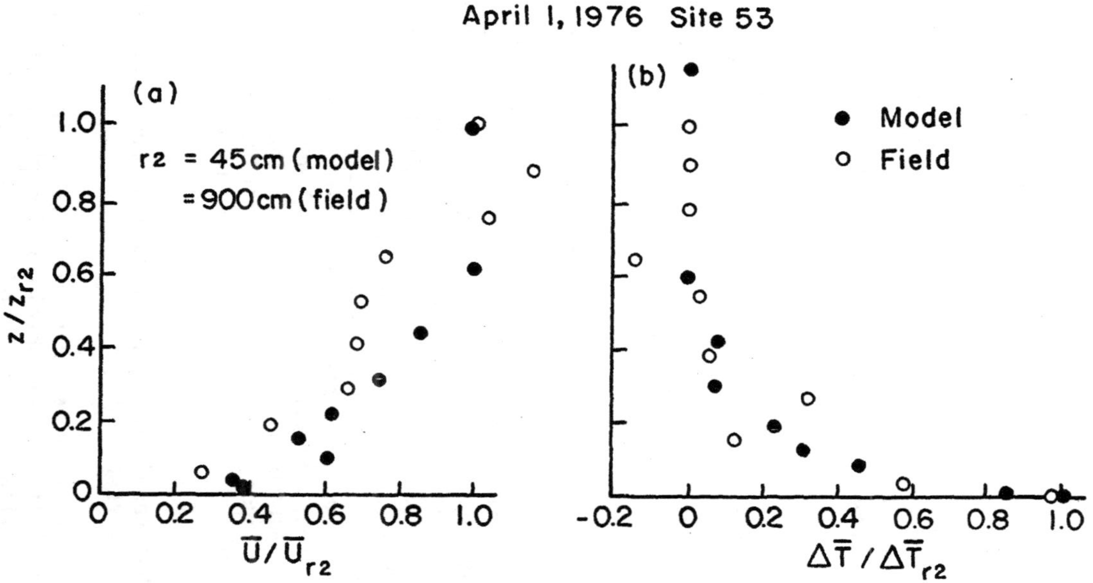

Figure 10. Wind Tunnel and Field Measurements during Unstable Conditions for (a) Vertical Profiles of Wind, and (b) Vertical Profiles of Temperature (Weil, 1979)

4.2 Thermally Stratified ABL Flow

Flow and dispersion over the Westvaco Paper Mill site shown in Fig. 9 were studied by means of physical modelling and field measurements of meteorological variables and SO_2 concentrations (Petersen, Cermak and Ghavam, 1980; Weil, 1979; Weil, Cermak and Petersen, 1981). A 1:2,000 scale aluminum (heavy aluminum foil) shell model shaped by pressing over a previously constructed styrofoam topographic model was placed 10.4 m from the meteorological wind tunnel test-section entrance shown in Fig. 3. The aluminum shell permitted heating and cooling of the surface by heat transfer to the heated or cooled wind tunnel floor (stimulated by fans between the surfaces). The surface was roughened by adding 0.3 cm stone particles to represent vegetation over the area. Concentration measurements were made for emission from the boiler-house stacks in neutral, stable and unstable boundary-layer flows. Bulk Richardson numbers for points 50 and 53 (Fig. 9) based on mean temperature and mean velocity differences between 1 cm and 9 cm heights over the model and in the field study are given in Table 1.

TABLE 1. Bulk Richardson Numbers

	Prototype (1 April 1976)	Wind Tunnel	
Test	Point 53	Point 53	Point 50
Neutral	~0	0	0
Unstable	-0.29	-0.27	-0.45
Stable	--	+0.15	+0.59

A comparison of mean velocity and temperature profiles over site 53 shown in Fig. 10 for unstable flow in the atmosphere and over the model reveals that good similarity was achieved. Figures 11 and 12 compare cross wind ground-level concentration distributions measured along routes 46 and 135 (see Fig. 9), respectively, for unstable and neutral (route 135 only) flow. Field and model concentrations are in excellent agreement along route 46. On route 135 the model gave concentrations smaller than those measured in the field. The differences are due, at least in part, to the model Reynolds number based on ridge height being approximately 4,400 rather than a recommended value of about 20,000 (Snyder, 1972). Unfortunately, no field data are available for stable flow that can be used to establish model-field agreement.

This study has resulted in the following conclusions (Weil, 1979; Weil, Cermak and Petersen, 1981):

1. Good agreement of meteorological variables and concentration distributions can be obtained for physically modelled and field realizations of unstable forced-convection flows.

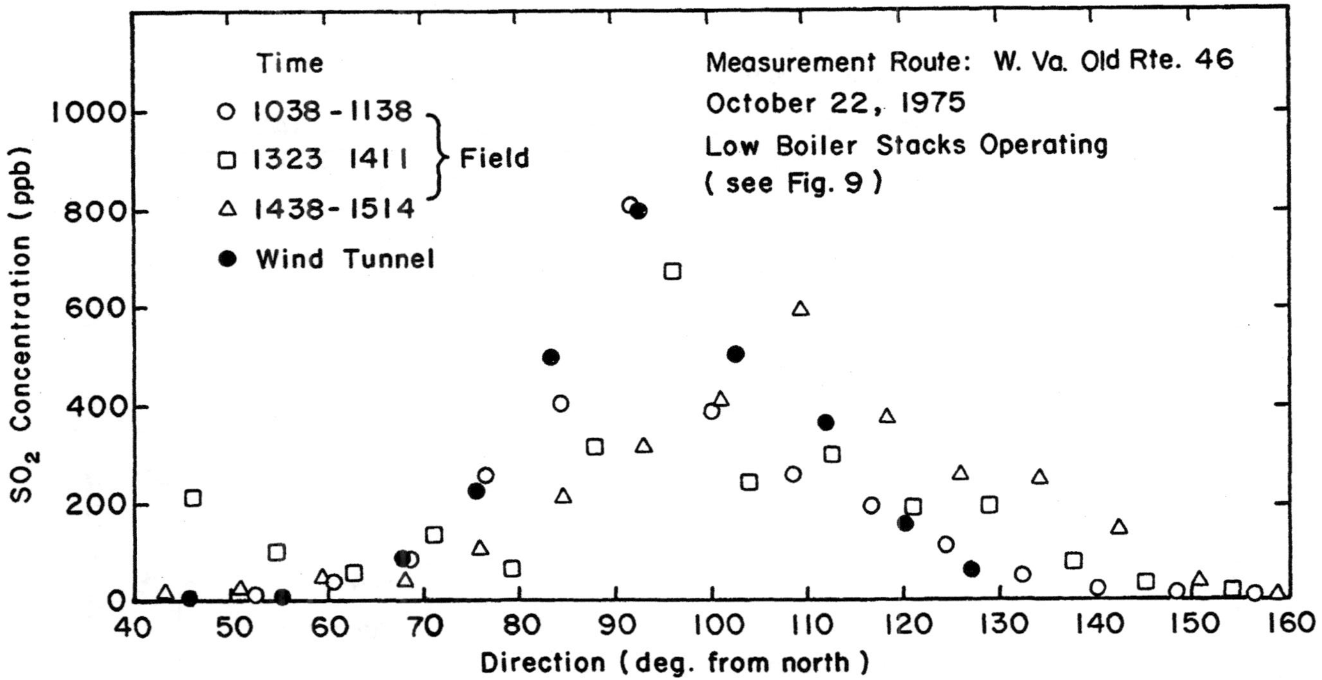

Figure 11. Crosswind Ground-Level Concentration Profiles Measured on Elevated Terrain in Field and Wind Tunnel during Unstable Conditions (Weil, 1979))

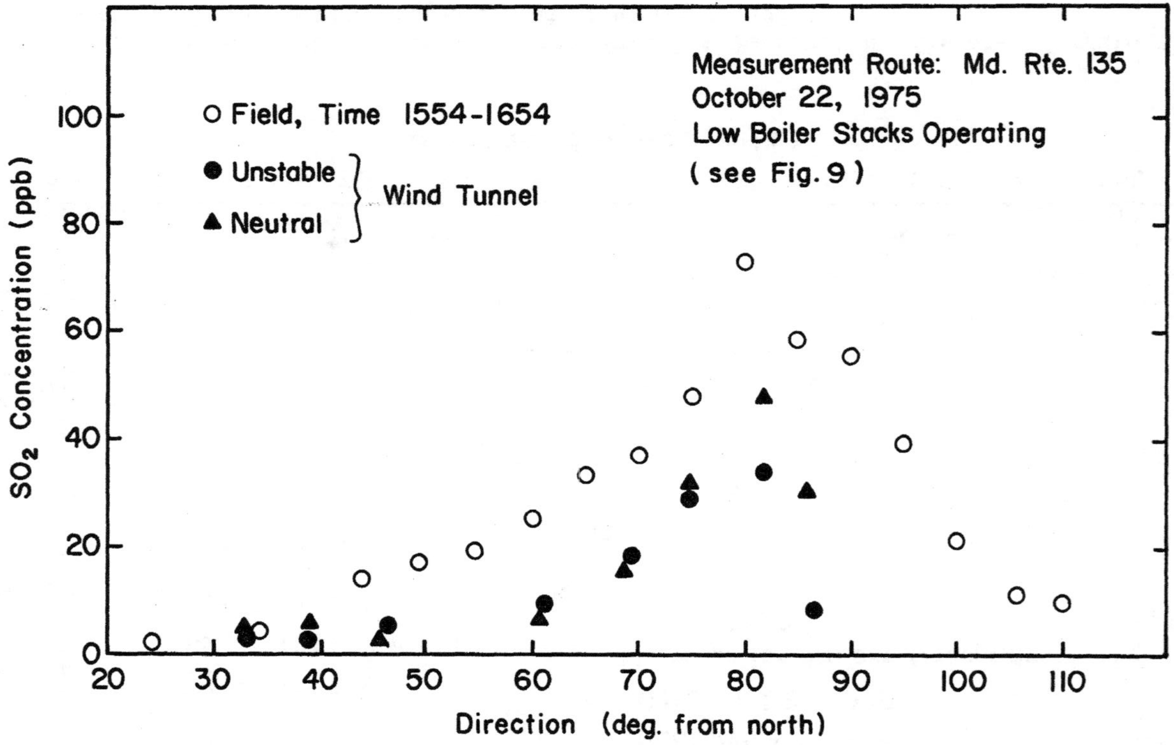

Figure 12. Ground-Level Concentration Profiles Measured in Field and Wind Tunnel during Neutral to Unstable Conditions. Measurements Obtained in Valley on Lee Side of Hill to the East of Westvaco Plant (Weil, 1979)

Figure 13. Visualization of Advective Dispersion over Point Arguello, California, by Stably Stratified Flow (Cermak and Peterka, 1966)

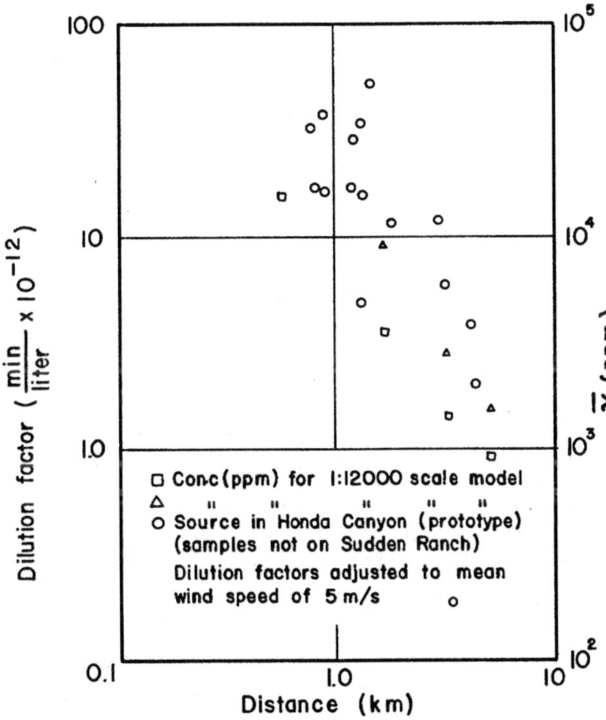

Figure 14. Attenuation of Maximum Ground-Level Concentration Downwind from Point Source over Model of Point Arguello and Prototype (Cermak and Peterka, 1966)

2. The model-field correspondence supports use of physical modelling in place of field measurements and for testing mathematical models designed to predict dispersion over complex terrain.

3. Observations of stable flow in the physical model show that a three-dimensional mathematical model is required for dispersion prediction over complex terrain.

A study confirming the usefulness of laminar-turbulent flow similarity (Eqs. 8 and 10) was made on a 1:12,000 scale model of stable flow and dispersion over Point Arguello, California (Cermak and Peterka, 1966; Cermak, 1971). The model was inserted into the meteorological wind tunnel (Fig. 3) near the downstream end of the test section. Cooling of the wind-tunnel floor to 40°C below ambient air temperature gave a gross Richardson number of 0.2 over the lower 5 cm of a 35 cm deep laminar boundary layer (near values of Ri for wind approaching over the Pacific Ocean) with U_0 equal to 1.5 m/s. The mean velocity characteristics near ground level was observed by movement of constant-level balloons released in the field. Flow features observed in this manner were also observed in flow over the model. Concentration measurements were made for a source of tracer material located at the smoke source shown in Fig. 13 over the model and in the field. As shown in Fig. 14 the decay rate of ground-level concentration were found to be similar.

This study demonstrates that physical modelling of very stable flow and dispersion over complex terrain with model scales of 1:10,000 and smaller is possible. In these flows dispersion is dominated by advective transport in the highly nonuniform flow field that results from the complex geometry.

4.3 Drainage Flows Resulting from Gravitational Convection

Drainge flow down Coal Creek Canyon (Fig. 15) was investigated by physical modelling with an aluminum shell model placed in the convective chamber shown in Fig. 5. The 1:1,920 scale model was cooled by circulating (by 120 small fans) gas cooled with two tons of dry ice placed in an insulated space below the model. Mean velocity and temperature profiles were measured at sensor locations in the canyon. Typical profiles for point G are shown in Fig. 16. Table 2 lists model flow parameters and two sets of data from field measurements. The Reynolds number listed is $Re_x = \bar{U}x/\nu_{(z=0)}$ and the Richardson number Ri is for the layer $0 \leq z \leq z_{(\bar{U}=\bar{U}_m)}$.

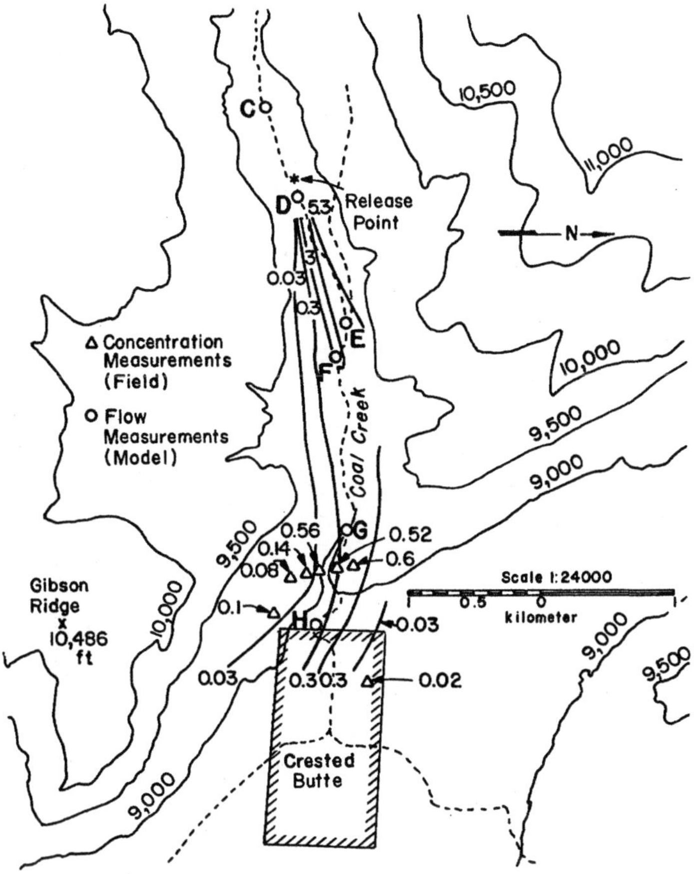

Figure 15. Isopleths of Ground-Level Normalized Concentration ($\bar{\chi}\bar{U}_m/Q \times 10^5$) for a 6.1 m Release Height (Petersen and Cermak, 1980b)

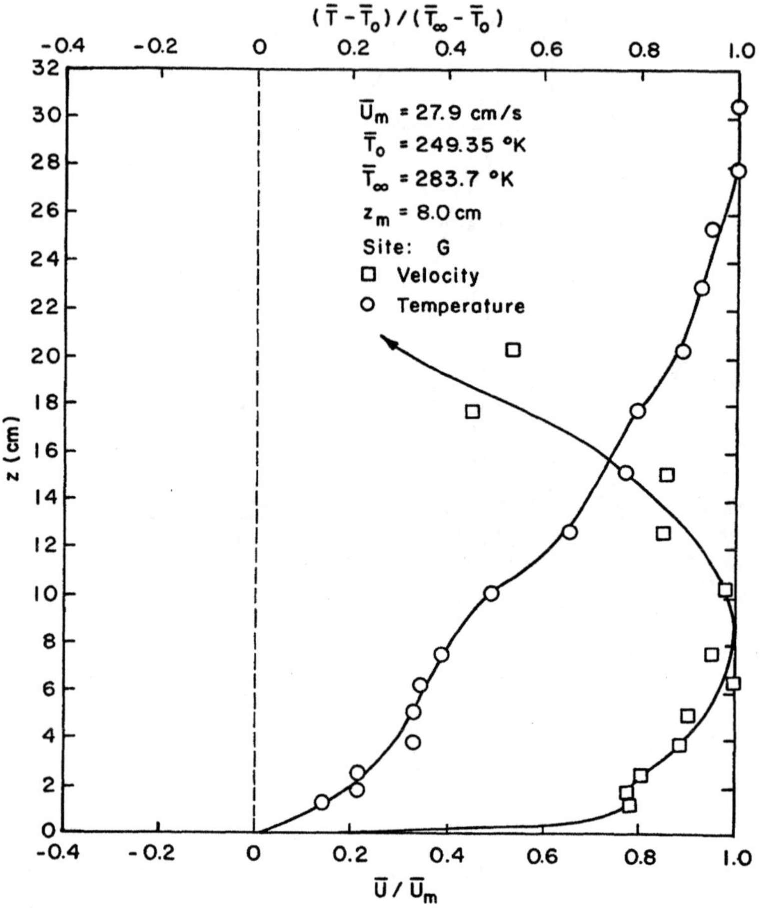

Figure 16. Velocity and Temperature Profiles Taken at Concentration Sampling Tap 16 (G) for the Coal Creek Drainage Flow Tests--Ri = 0.54 (Petersen and Cermak, 1980b)

TABLE 2. Mountain-Wind Characteristics for Model (M) and Prototype (P)

Location	Down-Valley Distance x (km)	Mountain-Wind Depth (m) M	Mountain-Wind Depth (m) P	Height to $\bar{U}_m$ (m) M	Height to $\bar{U}_m$ (m) P	$\bar{U}_m$ (m/s) M	$\bar{U}_m$ (m/s) P	Re_x M	Re_x P	Ri M	Ri P
A	2.75	461	-	49	-	0.18	-		-		-
B	4.25	384	-	146	-	0.18	-		-	1.9	-
C	5.15	461	-	196	-	0.20	-		-		-
D	5.85	346	-	122	-	0.21	-		-	1.1	-
E	6.90	-	200	-	100	-	3			-	0.8
F	7.15	480	-	146	-	0.25	-		-		-
G	8.40	422	~200	154	105	0.28	5	10^5	$2x10^9$	0.6	0.7
H	9.25	461	-	37	-	0.27	-		-		-

The magnitude of $\bar{U}_m$ was found to increase as the square root of distance from the canyon origin. Davidson and Rao (1963) observed a similar behavior in their field observations.

If Reynolds-number independence is assumed and all boundary conditions are similar, the relationship between $(\bar{U}_m)_m$ and $(\bar{U}_m)_p$ should be determined on the basis of Richardson-number equality. This condition (as shown in Table 2 for location G, this was clearly satisfied) gives the following relationship:

$$\bar{U}_p = (L_p/L_m)^{1/2} (\overline{\Delta T}_p/\overline{\Delta T}_m)^{1/2} \bar{U}_m .$$

Assuming that this relationship is valid for the flow being studied $\bar{U}_p$ can be calculated for location G since $(Ri)_m \simeq (Ri)_p$. For this calculation $L_p/L_m = 1920$ and $\overline{\Delta T}_p/\overline{\Delta T}_m = 4/21$; therefore,

$$\bar{U}_p \simeq 19\bar{U}_m .$$

This result predicts that $\bar{U}_p = 19x0.28$ or 5.3 m/s. The measured prototype value listed in Table 2 is 5 m/s. This close agreement gives strong evidence that physical modelling of drainage flows can be achieved in a practical manner.

One difficulty in physical modelling of drainage flows is that an appropriate value of Richardson number for the prototype is not known. Hence, the physical model must be designed to operate at some estimated value of Ri_p. For a particular site, the estimate can be refined by measurements of mean temperature and air speed at some convenient location in the drainage system. In the present study no data were available for design of the model; therefore, the fact that $(Ri)_m \simeq (Ri)_p$ was indeed fortunate.

An application of the physically modelled drainage flow is determination of pollutant concentrations resulting from release in valleys and on the slopes of complex terrain. Reid (1979) concluded that application of Gaussian models to dispersion in mountain-valley winds was futile. Therefore, physical modelling of the flow field and measurement of concentrations of tracer gases released from potential sources provides a method for not only obtaining information for a specific pollution problem but information that can be used to formulate realistic numerical models. Dispersion in Coal Creek Canyon was investigated by means of the physical model (Petersen and Cermak, 1980b). Isopleths of ground-level concentrations ($\bar{\chi}\bar{U}_m/Q \times 10^5\ m^{-2}$) are shown in Fig. 15 for the model as well as some field measurements of concentrations. The agreement is sufficiently good to encourage further development of this type of physical modelling.

From this investigation the following conclusions may be derived:

1. Physical modelling of gravitational convective flow and associated dispersion over complex terrain can be realized in a practical manner.

2. More general gravitational convective modelling should be possible by use of an inverted topographic model heated with electrical heaters in a manner to develop a variety of non-uniform surface temperature distributions.

5. CONCLUSIONS

The results obtained by physical modelling of flow and dispersion over complex terrain have been validated by field measurements for a wide variety of meteorological conditions. Excellent model-field agreement is found for forced-convection flows with neutral and unstable thermal stratification; however, further model-field comparisons are needed to attain the same level of confidence for flow with stable stratification. Laminar-turbulent similarity is a useful modelling concept for forced-convection flows with strong stable stratification over complex terrain modelled to scales of 1:10,000 and smaller. Physical modelling of drainage flow (gravitational convection) can be achieved with sufficient accuracy to provide useful predictions of air pollutant dispersion. Successful modelling of pure drainage flow suggests that techniques for modelling gravitational convection should be extended to permit modelling of more complex mountain-valley wind systems.

REFERENCES

Abe, M.: 1929, Mountain Clouds, Their Forms and Connected Air Currents, Bull. Cent. Meteor. Obs. Vol. 7, No. 3, Tokyo, 1929.

Barad, M. L., Ed.: 1952, Project Prairie Grass--A Field Program in Diffusion, Geophysical Research Paper 59, Geophysical Research Institute, Bedford, Massachusetts.

Britter, R.: 1982, Modeling Flow Over Complex Terrain and Implications for Determining the Extent of Adjacent Terrain to be Modeled, Proceedings of the International Workshop on Wind Tunnel Modeling Criteria and Techniques in Civil Engineering Applications, Gaithersburg, Maryland, pp. 186-196.

Cermak, J. E.: 1958, Wind Tunnel for the Study of Turbulence in the Atmospheric Surface Layer, Fluid Dynamics and Diffusion Laboratory, Technical Report CER58-JEC42, Colorado State University, Fort Collins, Colorado.

Cermak, J. E. and Peterka, J. A.: 1966, Simulation of Wind Field over Point Arguello, California, by Wind-Tunnel Flow over a Topographic Model, Fluid Dynamics and Diffusion Laboratory, Technical Report CER65JEC-JAP64, pp. 86.

Cermak, J. E. and Arya, S. P. S.: 1970, Problems of Atmospheric Shear Flows and Their Laboratory Simulation, Int'l. Jl. of Boundary-Layer Meteorology, Vol. 1, pp. 40-60.

Cermak, J. E., Orgill, M. M. and Grant, L. O.: 1970, Laboratory Simulation of Atmospheric Motion and Transport over Complex Topography as Related to Cloud Seeding Operations, Proceedings of Second National Conference on Weather Modification, Santa Barbara, California, 6 April 1970, pp. 59-65.

Cermak, J. E.: 1971, Laboratory Simulation of the Atmospheric Boundary Layer, AIAA Jl., Vol. 9, No. 9.

Cermak, J. E.: 1975, Applications of Fluid Mechanics to Wind Engineering--A Freeman Scholar Lecture, ASME Jl. of Fluids Engineering, Vol. 97, Ser. 1, No. 1.

Cermak, J. E.: 1981, Wind-tunnel Design for Physical Modeling of Atmospheric Boundary Layers, ASCE Jl. of Engineering Mechanics, Vol. 107, No. EM3, pp. 623-642.

Cermak, J. E. and Petersen, R. L.: 1981, Physical Modeling of Downslope Mountain Wind and Atmospheric Dispersion, Proceedings of the Fourth U.S. National Conference on Wind Engineering Research, University of Washington, Seattle, Washington, pp. 371-381.

Cermak, J. E.: 1982, Physical Modeling of the Atmospheric Boundary Layer (ABL) in Long Boundary-layer Wind Tunnels (BLWT), Proceedings of the International Workshop on Wind Tunnel Modeling Criteria and Techniques in Civil Engineering Applications, Gaithersburg, Maryland, Cambridge University Press, pp. 97-125.

Chaudhry, F. and Cermak, J. E.: 1971, Simulation of Flow and Diffusion Over An Urban Complex, Proceedings of the American Meteorological Society--Air Pollution Control Association Conference on Air Pollution Meteorology, Raleigh, North Carolina, 7 pp.

Chuang, H. and Cermak, J. E.: 1967, Similarity of Thermally Stratified Shear Flows in the Laboratory and Atmosphere, Physics of Fluids (supplement) Vol. 10, Part II, No. 9, pp. S255-S258.

Davidson, B. and Rao, P. K.: 1963, Experimental Studies on the Valley-Plain Wind, Int. Jl. Air Met. Poll. Vol. 7, 907.

Fultz, D.: 1951, Experimental Analogies to Atmospheric Motions, Compendium of Meteorology, American Meteorological Society, Boston, Massachusetts, pp. 1235-1248.

Hertig, J.-A. and Liska, P.: 1983, Simulation of Regional Atmospheric Flows on Small Scale Topographical Models, Proceedings of the Sixth International Conference on Wind Engineering, Gold Coast, Australia, 21-25 March 1983, Journal of Wind Engineering and Industrial Aerodynamics, Vol. 15, Nos. 1-3, pp. 77-89.

Kitabayshi, K., Orgill, M. M. and Cermak, J. E.: 1971, Laboratory Simulation of Airflow and Atmospheric Transport-Dispersion over Elk Mountain, Wyoming, Fluid Dynamics and Diffusion Laboratory, Technical Report 70-71KK-MMO-JEC65, Colorado State University, Fort Collins, Colorado, 90 pp.

Lin, J. T. and Binder, G. J.: 1967, Simulation of Mountain Lee Waves in a Wind Tunnel, Fluid Dynamics and Diffusion Laboratory, Technical Report CER67-68JTL-GJB24, Colorado State University, Fort Collins, Colorado, 120 pp.

Long, R. R.: 1959, A Laboratory Model of Air Flow Over the Sierra Nevada Mountains, The Rossby Memorial Volume, pp. 372-380.

Meroney, R. N.: 1980, Wind-Tunnel Simulation of the Flow Over Hills and Complex Terrain, Journal of Industrial Aerodynamics, Vol. 5, pp. 297-321.

Méry, P.: 1969, Reproduction en Similitude de la Diffusion dans la Couche Limite Atmospherique, La Houille Blanche, Vol. 3169, pp. 327-344.

Neal, D.: 1983, The Influence of Model Scale on a Wind-Tunnel Simulation of Complex Terrain, Journal of Wind Engineering and Industrial Aerodynamics, Vol. 12, pp. 125-143.

Orgill, M. M. and Cermak, J. E.: 1972, Atmospheric Flow and Diffusion over Moutainous Terrain, AIAA 5th Fluid and Plasma Dynamics Conference, Paper No. 72-648, Boston, Massachusetts, 11 pp.

Petersen, R. L. and Cermak, J. E.: 1978, Atmospheric Transport of Hydrogen Sulfide from Proposed Geothermal Power Plant (Unit 19), Predictions by Physical Modeling in a Wind Tunnel, Fluid Dynamics and Diffusion Laboratory, Technical Report CER78-79RLP-JEC10, Colorado State University, Fort Collins, Colorado.

Petersen, R. L. and Cermak, J. E.: 1980a, Atmospheric Transport of Hydrogen Sulfide from Proposed Geothermal Power Plant (Unit 20), Predictions by Physical Modeling in a Wind Tunnel, Fluid Dynamics and Diffusion Laboratory, Technical Report CER80-81RLP-JEC9, Colorado State University, Fort Collins, Colorado.

Petersen, R. L. and Cermak, J. E.: 1980b, Physical Modeling of Plume Dispersion at Alkali and Coal Creek, near Crested Butte, Colorado, Technical Report CER79-80RLP-JEC43, Fluid Dynamics and Diffusion Laboratory, Fort Collins, Colorado.

Petersen, R. L., Cermak, J. E. and Ghavam, H. R.: 1980, Westvaco Paper Mill Dispersion Study, Fluid Dynamics and Diffusion Laboratory, Technical Report, CER79-80RLP-JEC-HRG57, Colorado State University, Fort Collins, Colorado, 393 pp.

Petersen, R. L., Cermak, J. E. and Nelson, R. A.: 1980, Simulation of Pollutant Dispersion in the Mountain Wind, Proceedings of Second Joint Conference on Applications of Air Pollution Meteorology and Second Conference on Industrial Meteorology, New Orleans, Louisiana, pp. 540-545.

Plate, E. J. and Cermak, J. E.: 1963, Micro-meteorological Wind-Tunnel Facility, Fluid Dynamics and Diffusion Laboratory, Technical Report CER68EJP-JEC9, Colorado State University, Fort Collins, Colorado, 65 pp.

Poreh, M. and Cermak, J. E.: 1984, Wind Tunnel Simulation of Diffusion in a Convective Boundary Layer, Proceedings of the 29th OHOLO Conference on Boundary Layer Structure--Modelling and Application to Air Pollution and Wind Energy, 25-28 March 1984, Zichron Ya'acov, Israel.

Reid, J. D.: 1979, Studies of Pollutant Transport and Turbulent Dispersion Over Rugged Mountainous Terrain Near Climax, Colorado, Atmos. Environment, Vol. 13, 23.

Ruark, A. E.: 1935, Inspectional Analysis, Journal Elisha Mitchell Science Society, Vol. 51, pp. 127-133.

Schlichting, H.: 1968, Boundary-Layer Theory, Sixth Edition, McGraw-Hill Book Co., New York, 748 pp.

Snyder, W. H.: 1972, Similarity Criteria for the Application of Fluid Models to the Study of Air Pollution Meteorology, Boundary Layer Meteorology, Vol. 3, pp. 113-134.

Snyder, W. H.: 1983, Fluid Modeling of Terrain Aerodynamics and Plume Dispersion--A Perspective View, Preprint Volume of Extend Abstracts, Sixth Symposium on Turbulence and Diffusion, American Meteorological Society, Boston, Massachusetts, pp. 317-320.

Weil, J. C.: 1979, Modeling of Buoyant Plume Dispersion in Complex Terrain, Technical Report Environmental Center, Martin Marietta Corporation, Baltimore, Maryland, 85 pp.

Weil, J. C., Cermak, J. E. and Petersen, R. L.: 1981, Plume Dispersion About the Windward Side of a Hill at Short Range: Wind-Tunnel vs Field Measurements, Proceedings Fifth Symposium on Turbulence, Diffusion and Air Pollution, American Meteorological Society, 9-13 March 1981, Atlanta, Georgia, pp. 159-160.

Questions, answers and comments

J.C.R. HUNT: Are there situations where in wind-tunnel simulations it is appropriate or justifiable to use models with distorted vertical scales? I ask because the British Forestry Commission model air flow over hills with a 2:1 distortion.

J.E. CERMAK: The vertical scale of surface roughness may be exaggerated for small-scales of terrain in order to achieve an aerodynamically rough-surface. An increase in roughness elements height by an amount approximately equal to the depth of the viscous zone over the same surface when smooth, has been used with satisfactory results. Distortion of hill geometry by vertical scale exaggeration also distorts the local flow field. This may cause serious errors in the location of flow separation. Therefore, models of primary geometrical features should retain geometric similarity.

ON THE ENHANCED SMOOTHING OVER TOPOGRAPHY IN SOME MESOMETEOROLOGICAL MODELS

P. Alpert* and J. Neumann**
*Dept. of Geophysics and Planetary Sciences, Tel Aviv University, Israel. On leave from CIMMS, Norman, OK.
**Emeritus, Dept. of Atmospheric Sciences, Hebrew University of Jerusalem, Israel. Now visiting Dept. of Met., Helsinki, Finland.

ABSTRACT

An equation is derived for the components of the horizontal (turbulent) frictional force in the σ-coordinate system with special attention to mesometeorological flow models. The starting point is the horizontal equation of motion in its "flux-form" in the σ-system in which we replace (following Reynolds' procedure) the velocity components u,v and $\dot{\sigma}$ as well as other relevant quantities by terms of the form $u = \overline{u} + u', \ldots, \sigma = \overline{\sigma} + \dot{\sigma}'$, etc. ($\overline{u}$ = time average of u; u' = fluctuating part of u.) Next, the equation is averaged with respect to time and terms which we believe are small in mesometeorological flows, are neglected. On expressing $\dot{\sigma}'$ by an appropriate expression that involves w', the result shows the appearance of two new terms which have not been considered previously in the published literature. While the expression earlier used in the literature involved the σ-derivative of $\overline{u'w'}$ alone, the new terms add the σ-derivatives of $\rho\alpha\,\overline{u'^2}$ and $\rho\beta\,\overline{u'v'}$ for the x-component of the force, and the σ-derivatives of $\rho\beta\,\overline{v'^2}$ and $\rho\alpha\,\overline{u'v'}$ for the y-component, where α and β are the slopes of the σ-surfaces in the x- and y-directions, respectively. Further, a few numerical simulations of the sea-breeze over topography are carried out with and without the correction terms. It is shown that when corrections terms are not included the effective smoothing is stronger above the sloping regions and may amount to as high as 50 percent of the convergence with slopes of ∼.04. The inclusion of the new terms does not lead to any special computational difficulties and for that reason there is no compelling reason to neglect them, all the more so because, as is shown, the addition of the new terms results in a consistent apportioning of the 'degree' of horizontal diffusion.

Boundary-Layer Meteorology 30 (1984) 293–312. 0006-8314/84/0303-0293$03.00

1. INTRODUCTION

In mesometeorological studies using the (x,y,σ,t) terrain-following coordinate system, it is customary to write the terms representing the turbulent frictional force in the "flux-form" of the equation of motion in, say, the x-direction as

$$F_x = -\frac{\partial}{\partial x}(\bar{p}_* \overline{u'u'}) - \frac{\partial}{\partial y}(\bar{p}_* \overline{u'v'}) + g\frac{\partial}{\partial \sigma}(\rho\ \overline{u'w'}) \tag{1}$$

see, e.g., Anthes and Warner (1978, Eqs. (13a), (47) and (56) where

$$\sigma = \frac{p}{p_s} \quad \text{or} \quad \sigma = \frac{p-p_t}{p_*},\ p_* = p_s - p_t, \tag{2}$$

and p_s, p_t are the pressures, respectively, at the surface and at the top of the model atmosphere. Actually, in (1) the terms do not have the dimension of force (per unit mass). This is due to the fact that in the "flux-form" of the equation of motion, (3) below, the terms have the dimension of pressure multiplied by force (per unit mass). The terms in (1) present a kind of "modified" force. For the flux-form of the equations of motion in the σ-system, see e.g., Haltiner (1971, p. 193).

It will appear from the analysis in Section 2 that (1) is appropriate to cases where the slope angle of topography (or that of the σ = const. surfaces; $\sigma = 1$ for the ground surface) is rather small. As will be shown below, a treatment that allows for steeper slopes results, as one would expect, in lengthier expressions[1], though we should point out that we exclude very steep slopes. This latter restriction is not so severe in practice for in numerical work one is forced to adopt not too short horizontal grid distances (say, 1 to 5 km), a fact that leads to a reduction of the slope angles in numerical model work. It will be shown that correction terms which contain the slopes of the σ-surface appear.

In Section 3 the importance of these correction terms will be illustrated in numerical simulations of the sea-breeze circulation with topography. The inclusion of such terms reduces the effective smoothing over sloping σ-surfaces.

2. DEVELOPMENT

In developing the turbulent stresses and components of the frictional force, we shall follow the procedure adopted by Osborne Reynolds (see, e.g., Sutton, 1953, pp. 61-64). The vector form of the momentum-flux

[1]The general expressions for any terrain-following coordinate system were mathematically derived by Gal-Chen and Somerville (1975).

equation in the horizontal is

$$\frac{\partial(p_*\underline{V})}{\partial t} + \nabla(p_*\underline{V}\ \underline{V}) + \frac{\partial}{\partial\sigma}(p_*\dot{\sigma}\underline{V}) + p_* f\underline{k} \times \underline{V} + p_*\nabla\phi + p_* \frac{\sigma}{\rho}\nabla p_* + \frac{p_*}{\rho}\nabla p_t = p_* \underline{F}_M \quad (3)$$

where

$\underline{V}$ = vector of horizontal velocity. $\underline{V} = (u,v)$
$\dot{\sigma}$ ≡ $d\sigma/dt$
f = coriolis parameter
$\underline{k}$ = unit vector along the vertical
ϕ = geopotential
ρ = density
$\underline{F}_M$ ≐ molecular friction force

Following Reynolds we write

$$u = \bar{u} + u',\ v = \bar{v} + v',\ \dot{\sigma} = \bar{\dot{\sigma}} + \dot{\sigma}',\ p_* = \bar{p}_* + p_*',\ \phi = \bar{\phi} + \phi' \quad (4)$$

When (4) is inserted into (3) and the terms are averaged with respect to time, a large number of terms result. We think it is justified to assume that terms involving triple covariances of the fluctuating quantities are numerically small relative to the terms containing double covariances and that therefore they can be neglected. Suppose we now have (3) in its form modified by the introduction of (4), averaging and dropping of small terms, then if we subtract from the modified form of (3), Eq. (3) itself for the average quantities ($\bar{p}_*$, $\bar{\underline{V}}$, etc.), and neglecting momentum fluxes by molecular motion, we are left with the equation in the x-direction:

$$\frac{\partial(\overline{p_*'u'})}{\partial t} = -\frac{\partial}{\partial x}(\bar{p}_*\ \overline{u'u'} + 2\ \bar{u}\ \overline{p_*'u'}) - \frac{\partial}{\partial y}(\bar{p}_*\overline{u'v'} + \bar{u}\ \overline{p_*'v'} + v\ \overline{p_*'u'}) - \frac{\partial}{\partial\sigma}(\bar{p}_*\ \overline{u'\dot{\sigma}'} + \bar{u}\ \overline{p_*'\dot{\sigma}'} + \bar{\dot{\sigma}}\ \overline{p'u'}) + f\ \overline{p_*'v'} + \overline{p_*'\frac{\partial\phi'}{\partial x}} - \sigma\ \overline{\frac{\partial\phi'}{\partial\sigma}\frac{\partial p_*'}{\partial x}} \quad (5)$$

A parallel equation holds in the y-direction.

In mesometeorological practice, p_* is usually between about, say, 250 and 1000 mb and the (turbulent) fluctuations are not likely to reach even 1% of $\bar{p}_*$ (p_t is taken to be constant) or

$$\frac{|p_*'|}{\bar{p}_*} \ll 1\ . \quad (6)$$

In contrast, the fluctuations in the velocity components are not usually small compared with the respective means of the components.

Consider, e.g., the expression in parentheses in the first term on the right-hand side of (5). Dividing that expression by $p_*\ \bar{u}\ \bar{u}$, we have the two terms

$$\frac{\overline{u'u'}}{\bar{u}\ \bar{u}} \text{ and } \frac{\overline{p_*'u'}}{\bar{p}_*\bar{u}}\ . \quad (7)$$

In view of (6), the second term in (7) will be negligible relative to the first. Applying parallel consideration to the other terms on the right of (5) we come up with the estimates

$$\left|\frac{\overline{u'u'}}{\bar{u}\,\bar{u}}\right|, \left|\frac{\overline{u'v'}}{\bar{u}\,\bar{v}}\right|, \left|\frac{\overline{u'\dot{\sigma}'}}{\bar{u}\,\bar{\dot{\sigma}}}\right| \gg \left|\frac{\overline{p_*'u'}}{\bar{p}_*\bar{u}}\right|, \left|\frac{\overline{p_*'v'}}{\bar{p}_*\bar{v}}\right|, \left|\frac{\overline{p_*'\dot{\sigma}'}}{\bar{p}_*\bar{\dot{\sigma}}}\right|, \left|\frac{\overline{p_*'\phi'}}{\bar{p}_*\bar{\phi}}\right| . \tag{8}$$

In (8) we have also assumed that $|\overline{p_*'\phi'}|/|\bar{p}_*\bar{\phi}|$ is smaller than the terms on the left-hand side of the inequality.

Neglecting the terms of this type on the right of (8), we get the x and y components of the (modified) horizontal frictional force in the form

$$F_x = -\frac{\partial}{\partial x}(\bar{p}_*\,\overline{u'u'}) - \frac{\partial}{\partial y}(\bar{p}_*\,\overline{u'v'}) - \frac{\partial}{\partial \sigma}(\bar{p}_*\,\overline{u'\dot{\sigma}'}) \tag{9a}$$

$$F_y = -\frac{\partial}{\partial x}(\bar{p}_*\,\overline{u'v'}) - \frac{\partial}{\partial y}(\bar{p}_*\,\overline{v'v'}) - \frac{\partial}{\partial \sigma}(\bar{p}_*\,\overline{v'\dot{\sigma}'}) . \tag{9b}$$

We note that the last terms on the right of (9) contain the expressions $\overline{u'\dot{\sigma}'}$ and $\overline{v'\dot{\sigma}'}$ and not $\overline{u'w'}$ and $\overline{v'w'}$ as in (1). It is of interest to find the relationships between these quantities. It will be seen below that the replacement of the quantities $\overline{u'\dot{\sigma}'}$ and $\overline{v'\dot{\sigma}'}$ by appropriate expressions leads to the appearance of two terms that are not accounted for in (1) and that these terms can be of importance in regions where the slope angle of the terrain is not too small.

We shall now turn to the components of the quantities $-\overline{u'\dot{\sigma}'}$ and $\overline{v'\dot{\sigma}'}$. In the σ-system

$$w = \frac{dz}{dt} = \frac{1}{g}\frac{d\phi}{dt} = \frac{1}{g}\left(\frac{\partial\phi}{\partial t} + \underline{V}\cdot\nabla\phi\right) - \frac{p_*\dot{\sigma}}{\rho g} \tag{10}$$

where g is the acceleration of gravity. Before continuing,we will carry out an order-of-magnitude estimate of the terms on the r.h.s. of (10) for mesometeorological conditions, similar to the estimate offered by Holton (1973, p. 59) for synoptic-scale motions. We think it reasonable to assume that in mesometeorological systems $\rho(\partial\phi/\partial t)$ is of the order of 1-5 mb day^{-1} (in any case, smaller than the order-of-magnitude of 10 mb day^{-1} adopted by Holton for synoptic-scale motions). Further, taking that the terrain slope is not too small (maximum of ∿ 0.1) and that w is, say 1 m s^{-1}

$$\rho\frac{\partial\phi}{\partial t} \approx 1 \text{ to } 5 \text{ mb day}^{-1} \qquad \rho g w \approx 10^4 \text{ mb day}^{-1}. \tag{11}$$

$$\rho\underline{V}\cdot\nabla\phi \approx 5\times10^3 \text{ mb day}^{-1}$$

Thus we neglect the time derivative term relative to the others. Moreover, $\rho g w$ is, or can be, much larger than in usual synoptic practice where w is commonly taken as 1 cm s^{-1}. We therefore can write

$$w \approx \frac{1}{g}\underline{V}\cdot\nabla\phi - \frac{p_*\dot{\sigma}}{\rho g} \tag{12}$$

On putting

$$\underline{V} = \underline{\bar{V}} + \underline{V}', \quad w = \bar{w} + w' \quad \text{and} \quad \dot{\sigma} = \bar{\dot{\sigma}} + \dot{\sigma}' \tag{13}$$

and using the approximation

$$\rho = \bar{\rho}, \quad p_* = \bar{p}_*, \tag{14}$$

we get that

$$\bar{w} + w' = \frac{1}{g}(\underline{\bar{V}} \cdot \nabla\bar{\phi} + \underline{\bar{V}}\,\nabla\phi' + \underline{V}' \cdot \nabla\bar{\phi} + \underline{V}' \cdot \nabla\phi') - \frac{p_*\bar{\dot{\sigma}}}{\rho g} - \frac{p_*\dot{\sigma}'}{\rho g} \tag{15}$$

Averaging the latter with respect to time and writing $\bar{\bar{w}} = \bar{w}$, $\bar{\bar{\dot{\sigma}}} = \sigma$, we find that

$$\bar{w} = \frac{1}{g}(\underline{\bar{V}} \cdot \nabla\bar{\phi} + \overline{\underline{V}' \cdot \nabla\phi'}) - \frac{p_*\bar{\dot{\sigma}}}{\rho g}. \tag{16}$$

Subtracting the latter from (15) and rearranging,

$$\dot{\sigma}' = -\frac{\rho g}{p_*} w' + \frac{\rho}{p_*}\underline{\bar{V}} \cdot \nabla\phi' + \frac{\rho}{p_*}\underline{V}' \cdot \nabla\bar{\phi} - \frac{\rho}{p_*}\overline{V' \cdot \nabla\phi'} \tag{17}$$

Multiplying the latter by u' and averaging, the result is

$$\overline{u'\dot{\sigma}'} = -\frac{\rho g}{p_*}\overline{u'w'} + \frac{\rho}{p_*}\left(\overline{u'^2}\frac{\partial\bar{\phi}}{\partial x} + \overline{u'v'}\frac{\partial\bar{\phi}}{\partial y} + \bar{u}\,\overline{u'\frac{\partial\phi'}{\partial x}} + \bar{v}\,\overline{u'\frac{\partial\phi'}{\partial y}} + \overline{u'u'\frac{\partial\phi'}{\partial x}} + \overline{u'v'\frac{\partial\phi'}{\partial y}}\right) \tag{18}$$

Presumably, in regions of topography the terms $|\partial\bar{\phi}/\partial x|$ and $|\partial\bar{\phi}/\partial y|$, $\phi = gz$, are larger than $|\partial\phi'/\partial x|$ and $|\partial\phi'/\partial y|$. (In more or less level areas these relations are likely to be reversed.) Neglecting the terms containing the derivatives $\partial\phi'/\partial x$ and $\partial\phi'/\partial y$, we have

$$\overline{u'\dot{\sigma}'} = \frac{\rho g}{p_*}\left[-\overline{u'w'} + \overline{u'^2}\frac{\partial\bar{z}}{\partial x} + \overline{u'v'}\frac{\partial\bar{z}}{\partial y}\right] \tag{19}$$

Similarly

$$\overline{v'\dot{\sigma}'} = \frac{\rho g}{p_*}\left[-\overline{v'w'} + \overline{v'^2}\frac{\partial\bar{z}}{\partial y} + \overline{v'u'}\frac{\partial\bar{z}}{\partial x}\right] \tag{20}$$

where $\bar{z}$ is the average height of the σ surface and thus $(\partial\bar{z}/\partial x)$ and $(\partial\bar{z}/\partial y)$ are the tangents of slope angles (slopes, for brief), respectively, in the x- and y-directions of the σ-surface. Writing

$$\frac{\partial\bar{z}}{\partial x} = \alpha \quad \text{and} \quad \frac{\partial\bar{z}}{\partial y} = \beta, \tag{21}$$

then using (19) through (21), (9a) and (9b) take the form

$$F_x = -\frac{\partial}{\partial x}(\bar{p}_* \overline{u'^2}) - \frac{\partial}{\partial y}(\bar{p}_* \overline{u'v'}) + g\frac{\partial}{\partial \sigma}(\rho \overline{u'w'}) - g\frac{\partial}{\partial \sigma}(\alpha\rho \overline{u'^2}) - g\frac{\partial}{\partial \sigma}(\beta\rho \overline{u'v'}) . \quad (22a)$$

$$F_y = -\frac{\partial}{\partial x}(\bar{p}_* \overline{u'v'}) - \frac{\partial}{\partial y}(\bar{p}_* \overline{v'^2}) + g\frac{\partial}{\partial \sigma}(\rho \overline{v'w'}) - g\frac{\partial}{\partial \sigma}(\beta\rho \overline{v'^2}) - g\frac{\partial}{\partial \sigma}(\alpha\rho \overline{u'v'}) \quad (22b)$$

On comparing (22a) with (1), we note that two new terms have arisen and, of course, the same applies to (22b).

We wish now to consider the new terms appearing in (22). Usually, $\overline{u'^2}$ is several times larger than $|\overline{u'w'}|$. For instance $|\rho\overline{u'w'}|$ might be 1 dyne cm^{-2} while $\overline{u'^2}$ may amount to 10^4 cm^2 s^{-2} or $\rho\overline{u'^2}$ may amount to about 10 dyne cm^{-2} at elevations which are not much above mean sea level. Thus, $\rho\overline{u'^2}$ can be about one order-of-magnitude larger than $|\rho\overline{u'w'}|$. Taking that the numerical value of the slope of the σ-surface is not too small, say it is 0.05-0.1, the arguments $|\alpha\rho\overline{u'^2}|$ and $|\beta\rho\overline{v'^2}|$ under the differentiation symbol $\partial/\partial\sigma$ in (22a) and (22b) can be of a comparable magnitude with those of $\rho\overline{u'w'}$ and $\rho\overline{v'w'}$. It is probably reasonable to assume that in mountainous terrain the variation with height (or, more precisely, with σ) of the aforementioned quantities is appreciable and, under such conditions $|g(\partial/\partial\sigma)(\alpha\rho\overline{u'^2})|$ will not necessarily be small relative to $|g(\partial/\partial\sigma)(\rho\overline{u'w'})|$, etc. Hence we must not ignore the terms $g(\partial/\partial\sigma)(\alpha\rho\overline{u'^2})$, etc., in mesometeorological models including topography.

For the following numerical simulations let us consider the two-dimensional form of the frictional force in the x-direction as formulated by the eddy exchange coefficient theory (Austausch formulation). In 2-D, the frictional force is

$$F_x = -\frac{\partial}{\partial x}(p_* \overline{u'^2}) + g\frac{\partial}{\partial \sigma}(\rho\overline{u'w'}) - g\frac{\partial}{\partial \sigma}(\alpha\rho\overline{u'^2}) \quad (23)$$

we have assumed uniformity in the y-direction. The Austausch formulation for $\overline{u'^2}$ and $\overline{u'w'}$ yields[2,3]

$$\overline{u'^2} = -K_H \frac{\partial u}{\partial x}\Big|_z = -K_H\left(\frac{\partial u}{\partial x}\Big|_\sigma + \frac{\partial \sigma}{\partial x}\frac{\partial u}{\partial \sigma}\right) \quad (24a)$$

$$\overline{u'w'} = -K_V \frac{\partial u}{\partial z} \quad (24b)$$

---

[2]We are using the common nomenclature for which the factor 2 multiplying K_H for the diagonal stress terms is incorporated within K_H, see e.g. Mason and Sykes (1978). In this notation, the actual horizontal diffusivity is represented by half the value of K_H.

[3]We assume $\bar{u}$ to decrease in the direction of increasing x, or, in other words, we assume a transfer down the gradient. It is only transfers of this nature that lend themselves (in an approximate way) to representation by K-theory.

where the chain rule has been used for the transformation from z- to σ-surfaces and K_H, K_V represent the horizontal[4] and vertical eddy viscosity coefficients respectively. It should be pointed out that the transformation to K-formulation is justified only in case the scale of the diffusive motion is small compared to the gradient of the diffused element. Otherwise, countergradient motions may exist yielding negative viscosities. By the substitution of (24a) and (24b) into eq. (23) one gets, with some algebra

$$F_x = \frac{\partial}{\partial x} K_H \frac{\partial u}{\partial x} + 2K_H \left(\frac{\partial \sigma}{\partial x}\right) \frac{\partial^2 u}{\partial \sigma \partial x} + \left(\frac{\partial \sigma}{\partial x}\right)^2 \frac{\partial}{\partial \sigma} \left(K_H \frac{\partial u}{\partial \sigma}\right) +$$

$$+ \left(\frac{\partial \sigma}{\partial z}\right)^2 \frac{\partial}{\partial z} \left(K_V \frac{\partial u}{\partial \sigma}\right) \tag{25}$$

where we have assumed constant K_H for the hybrid term (the second on the right of (25)) and divide by $\bar{p}_*$ to get the non-flux form. The expressions for $\partial\sigma/\partial z$ and $\partial\sigma/\partial x$ may be shown, through the hydrostatic eq., to obey,

$$\frac{\partial \sigma}{\partial z} = - \frac{\rho g}{p_*} \tag{26a}$$

and

$$\frac{\partial \sigma}{\partial x} = \frac{\rho g \alpha}{p_*} \tag{26b}$$

Notice that in (25) all x derivatives are on the σ-surface. If the horizontal eddy viscosity K_H is assumed constant, (25) could be further simplified to get,

$$F_x = K_H \frac{\partial^2 u}{\partial x^2} - 2K_H \alpha \frac{\partial^2 u}{\partial z \partial x} + K_H \alpha^2 \frac{\partial^2 u}{\partial z^2} + K_V \frac{\partial^2 u}{\partial z^2} \tag{27}$$

Thus, the frictional force in the 2-D σ-system includes the four following expressions (from left to right respectively): horizontal friction on σ-surface, hybrid term, vertical term which is induced by the slope and the pure vertical term. Similar expressions were also derived in soil modelling by Mahrer (1982). Denoting these terms by

---

[4]In view of the fact that some investigators believe that K_H and the notion of horizontal diffusion are but numerical tools, we wish to refer to papers, as, e.g. that of Gifford (1982) showing that field observations of diffusion indicate the physical reality of large eddies which significantly affect the diffusion process. Gifford himself uses the value of $K_H = 5 \cdot 10^4 \ m^2 s^{-1}$, see p. 508 and p. 510 in his paper. Eqs. (30) to (32) in his paper lead to estimates that L, the characteristic turbulence length scale is of something like 10 km or a few tens of kilometers with $K \approx 5 \cdot 10^4 \ m^2 s^{-1}$.

A, B, C and D respectively, it comes out from scaling considerations that terms C and D become comparable when the σ-surface slope α obeys the relation, $\alpha^2 = K_H/K_V$. On the substitution of typical values in mesoscale modelling; $K_V \sim 10\ m^2s^{-1}$ and $K_H \sim 10^4 m^2s^{-1}$, one gets a slope of $\alpha \sim 0.03$ for which C and D are comparable. A sloping σ-surface of 0.03 is not considered uncommon in mesoscale modelling over topography and yet the correction term becomes important. The value of term B is more difficult to assess as this depends on the horizontal and vertical length scales. But, when considering typical length scales for mesoscale modelling it could be shown that for slopes in the region 0.1 to 0.01, term B becomes comparable with the other terms. Table 1 illustrates the order of magnitude values for the terms A to D and for slopes of 0.1, 0.01 and 0.001, where the following typical values have been assumed: $K_H \sim 10^4 m^2s^{-1}$, $K_V \sim 10\ m^2s^{-1}$, $\Delta z \sim 10^2 m$, $\Delta x \sim 4$ km, $u \sim 1\ ms^{-1}$.

α	.1	.01	.001
A	6.10^{-4}	6.10^{-4}	6.10^{-4}
B	5.10^{-3}	5.10^{-4}	5.10^{-5}
C	10^{-2}	10^{-4}	10^{-6}
D	10^{-3}	10^{-3}	10^{-3}

Table 1: The order of magnitude values for the different frictional-force terms A,B,C and D in eq. (27).

Similar expressions could be derived for the y-direction momentum eq., and for the temperature turbulent diffusion. Also, it is important to notice that although the aforementioned derivation was performed for the σ-system eqs., the same arguments may be shown to apply to any terrain following coordinate system. In the next section we shall carry out some numerical simulations of the sea-breeze circulation with topography in order to assess the importance of these correction terms.

3. NUMERICAL SIMULATIONS

The model applied for simulating the evolving fields is a two-dimensional σ-system described by Alpert et al. (1982). The vertical turbulent transports are calculated through a high-resolution boundary-layer formulation by Blackadar (1978). Turning to the numerical side, a very accurate advection scheme by upstream cubic spline interpolation was adopted as shown, e.g. by Mahrer and Pielke (1978). Marchuk's (1974) method of splitting was applied in order to permit the use of an efficient implicit method for the vertical diffusion terms. The numerical grid consists of 65 points in the horizontal with a constant spacing of 4 km and 10 levels in the vertical at altitudes of ∼10, 20, 40 up to ∼2800 m above surface. In the following experiments no large-scale pressure gradient was assumed, and the heating is by prescribing a realistic surface temperature, the same as in the study by Alpert et al. (1982).

The model was originally applied without explicit horizontal diffusion due to the use of a highly selective implicit filter proposed by Long et al. (1978). For discussion of the advantages of this filter relative to explicit filters see Alpert (1981). In the first experiment an explicit horizontal diffusion (see eq. 27) with a constant K of $5 \cdot 10^3\ m^2s^{-1}$ was applied in the model with and without the correction terms. A bell-shaped mountain was introduced with the height of 400 m and a peak at the distance of 32 km from the coast. The half-width of the mountain was chosen as 5 km, which is equivalent to an average slope of 0.04.

Fig. 1 shows the horizontal wind arrows without the correction term at 1300 LST. The simulation has been started at 0800 assuming zero wind fields. As noted by Alpert et al. (1982) the dynamical initialization was in the order of 5 to 7 hours. In the present case which consists of less complex terrain the time necessary for dynamic initialization reduces considerably. At 1300 LST the sea breeze is combined with the mountain wind to generate relatively strong winds uphill, while on the lee side the westerly wind just penetrates and advances toward the weak easterlies. The closed contours represent the vertical wind velocities which reach a maximum value of $\sim 70\ cm\ s^{-1}$ at about 1000m above the slope. A value of $\sim -6\ cm\ s^{-1}$ is noticed in the region of the subsidence, right above the coastal region, and on the lee side of the mountain ($-18\ cm\ s^{-1}$).

The same experiment was repeated with the inclusion of the correction terms. For the difference schemes of the diffusion terms see Appendix. The vertical velocities in this experiment differ slightly from the former and reach a value of about $77\ cm\ s^{-1}$ in the upward cell. In order to better notice these differences, the difference fields between the two simulations corresponding to Figs. 1 and 2, is shown in Fig. 3. Fig. 3 depicts a small convergence field in the upwind region (coastal side) and a slightly stronger divergence field downwind, both of which are missing in the first experiment (Fig. 1) where the correction terms were not included. Here the contour interval is $4\ cm\ s^{-1}$ as compared to an interval of $12\ cm\ s^{-1}$ in Figs. 1 and 2. It is important to notice that the maximum differences of about $-6\ cm\ s^{-1}$ amount to only 5-10 percent of the vertical velocity reached in the experiment without the correction terms. Notice how, as expected, the maximum differences are located in the regions where the σ-surfaces show the steepest slopes. The slopes of the σ-surfaces may be followed through the positions of the wind vectors at the pertinent levels.

In the next experiment we increase the horizontal diffusivity K_H from $5 \cdot 10^3$ to $5 \cdot 10^4\ m^2s^{-1}$. Figs. 4 and 5 show the fields in this case after five hours of simulation, without and with the inclusion of the correction terms. The vertical velocities are considerably reduced relative to the former experiment and reach a maximum value of $\sim 23\ cm\ s^{-1}$. The difference fields between the two simulations with and without the correction terms are shown in Fig. 6. The additional convergence field reaches a maximum value of $10\ cm\ s^{-1}$ and dominates the upwind (coastal) region at the higher levels. At the lower layers we can notice a much weaker divergence field. In this

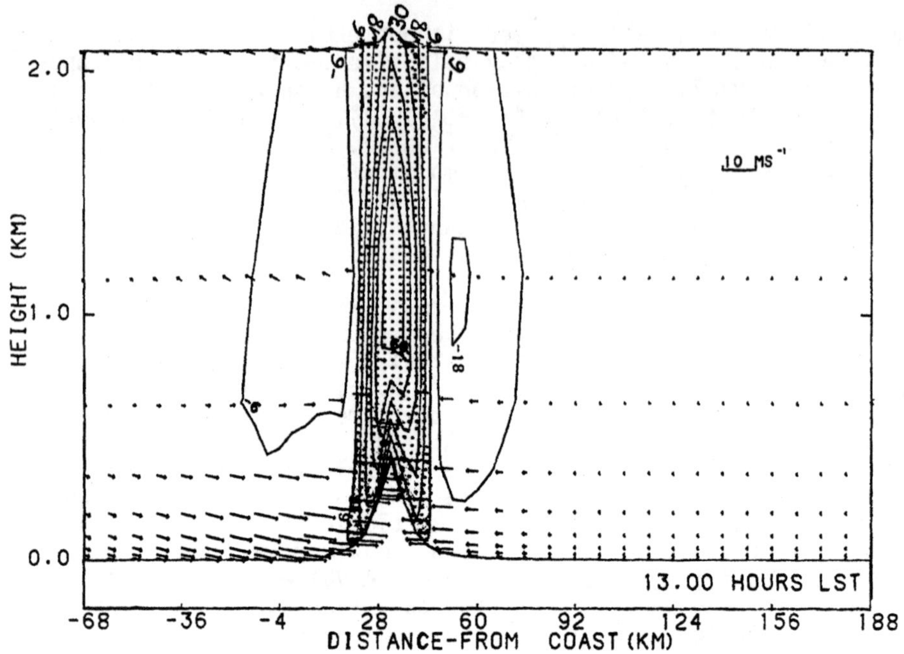

Fig. 1. Horizontal winds, m s^{-1} (arrows), and vertical winds, cm s^{-1} (solid lines), at 1300 LST, after five hours of simulation. Arrows bases are fixed at the level which they represent. Direction of arrows to right represents pure westerly wind, whereas the upward direction indicates southerly wind. The peak of the bell-shaped mountain is located 32 km from the coast. The horizontal wind scale is introduced in the upper right corner. Contour interval for vertical velocity is 12 cm s-1. Upward motion areas (larger than 6 cm s^{-1}) are shaded. Horizontal diffusivity constant of 5.10^3 m^2s^{-1}.

case, the relative contribution of the correction terms is much larger, particularly at the convergence zone, compare Fig. 6 to Figs. 4 and 5. This implies that when the correction terms were not included the effective smoothing became larger by about 50 percent in quite a large domain over the mountain. The zone of divergence in the difference fields corresponds to stronger diffusion with the correction terms but as is noticed from Fig. 6 this region has considerably diminished both in size and magnitude (compare to Fig. 3). This much stronger smoothing of more than 50% over a large domain above the mountain is reached with a constant diffusivity of $5 \cdot 10^4$ m^2s^{-1}. Such values are actually being used in mesoscale simulations, see e.g. Anthes and Warner (1978), although the correction terms are commonly not incorporated in the models.

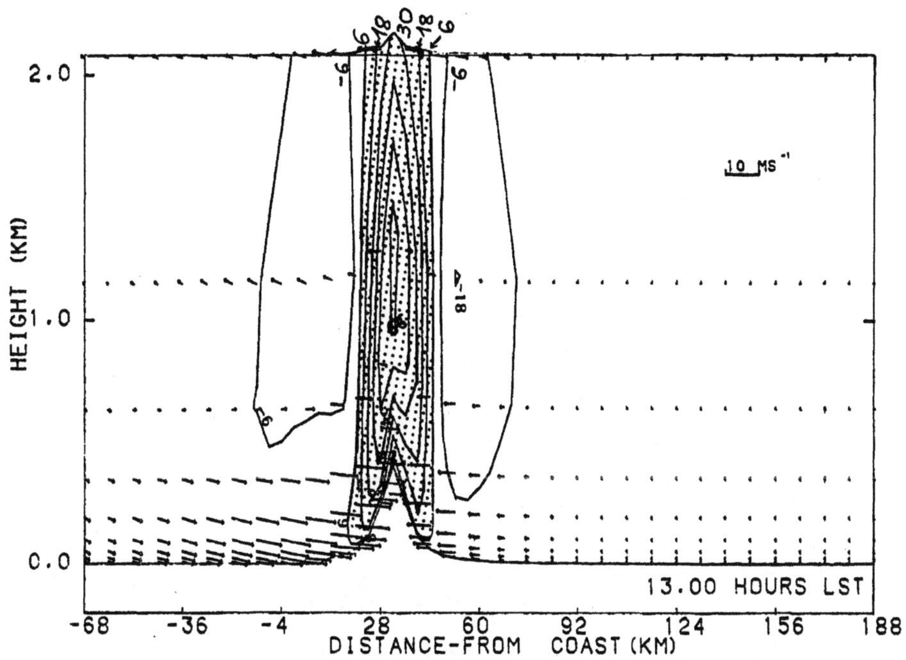

Fig. 2. As in Fig. 1 with the correction term for the diffusion included.

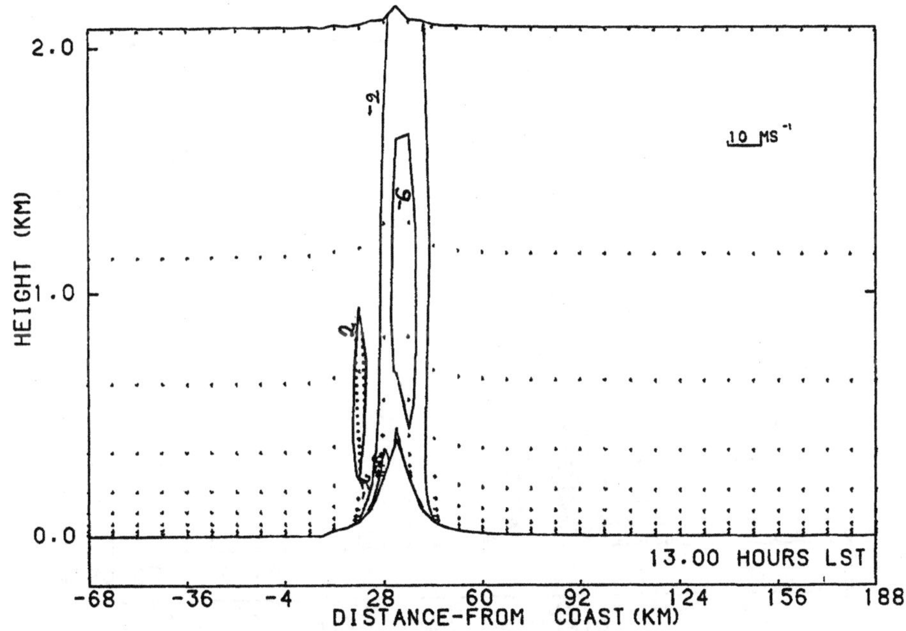

Fig. 3. As in Fig. 1 but with the difference fields between Figs. 1 and 2 (values at Fig. 2 minus the pertinent values at Fig. 1). Contour interval is 4 cm s^{-1}.

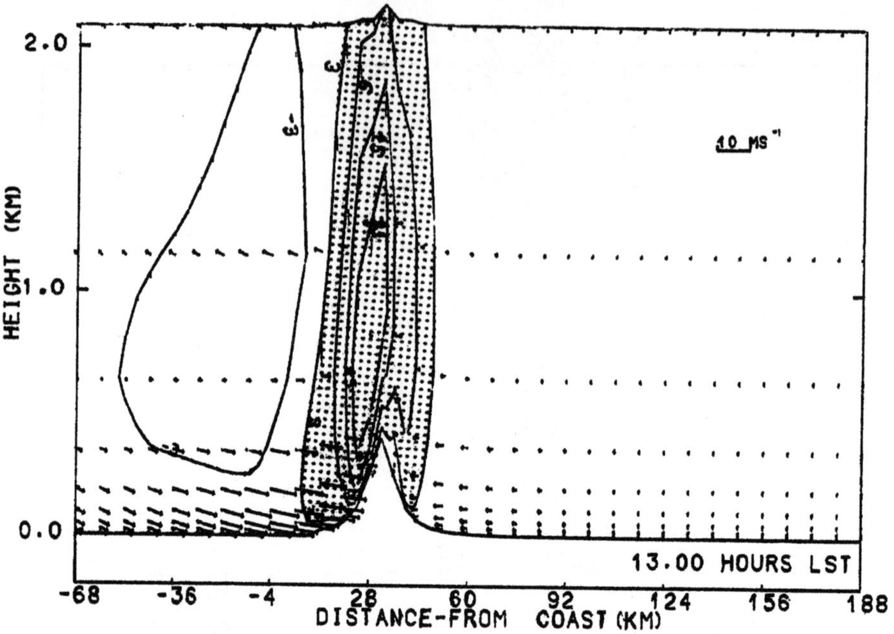

Fig. 4. As in Fig. 1 with the horizontal diffusivity K_H is increased to 5.10^4 m^2s-1. Contour interval of 6 cm s^{-1}.

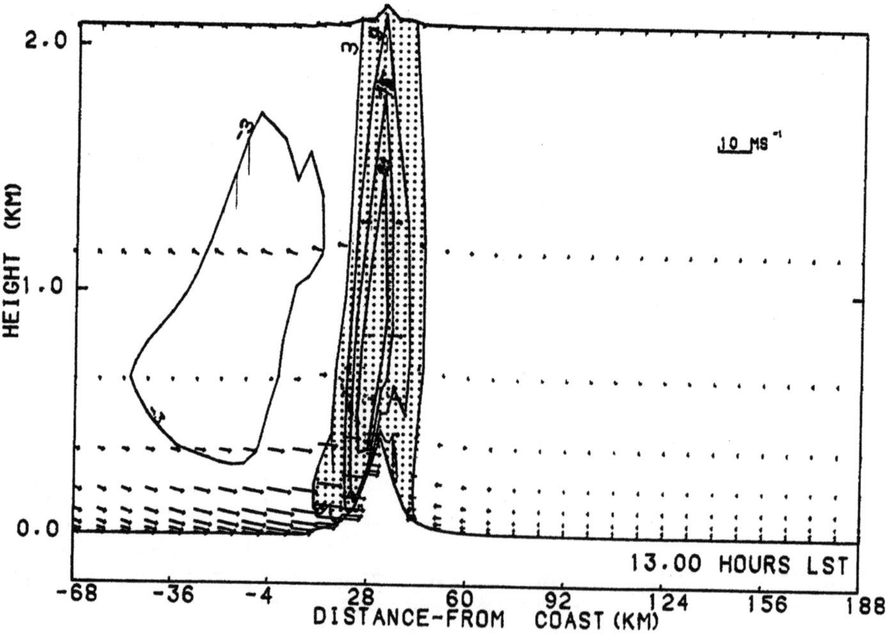

Fig. 5. As in Fig. 4, but with correction terms included.

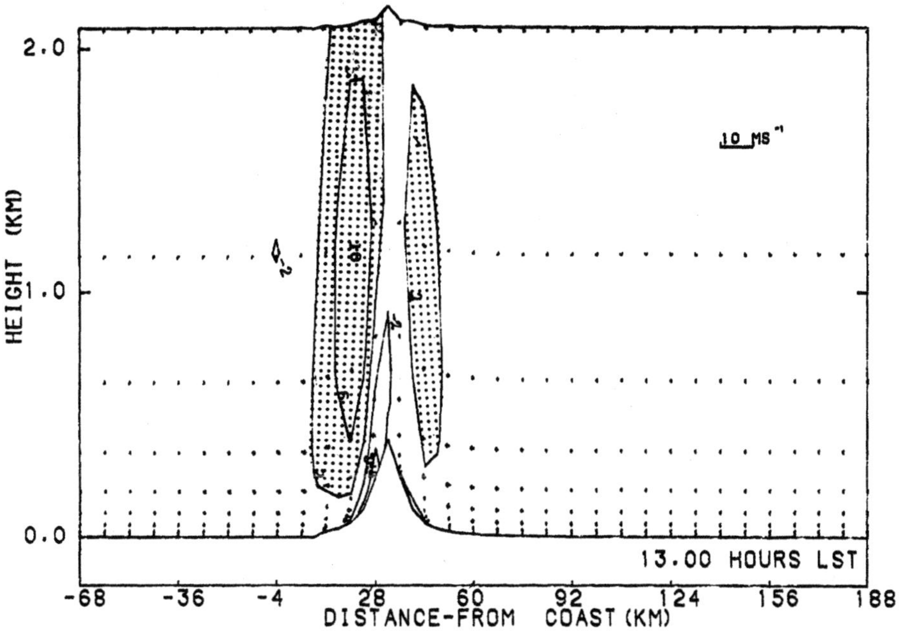

Fig. 6. Difference fields between Figs. 4 and 5.

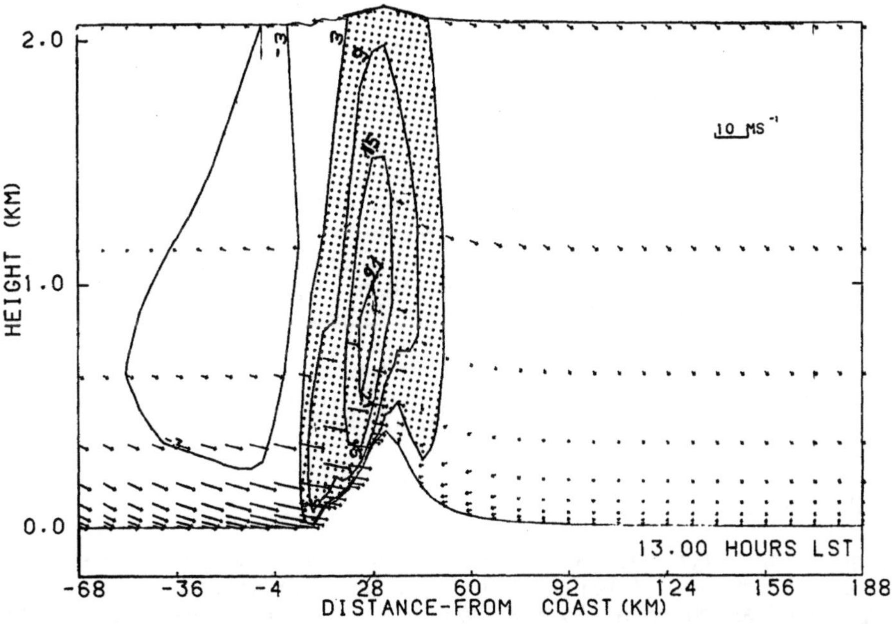

Fig. 7. As in Fig. 4, but with correction terms included and the slope was decreased to 0.02 (half-width mountain of 10 km).

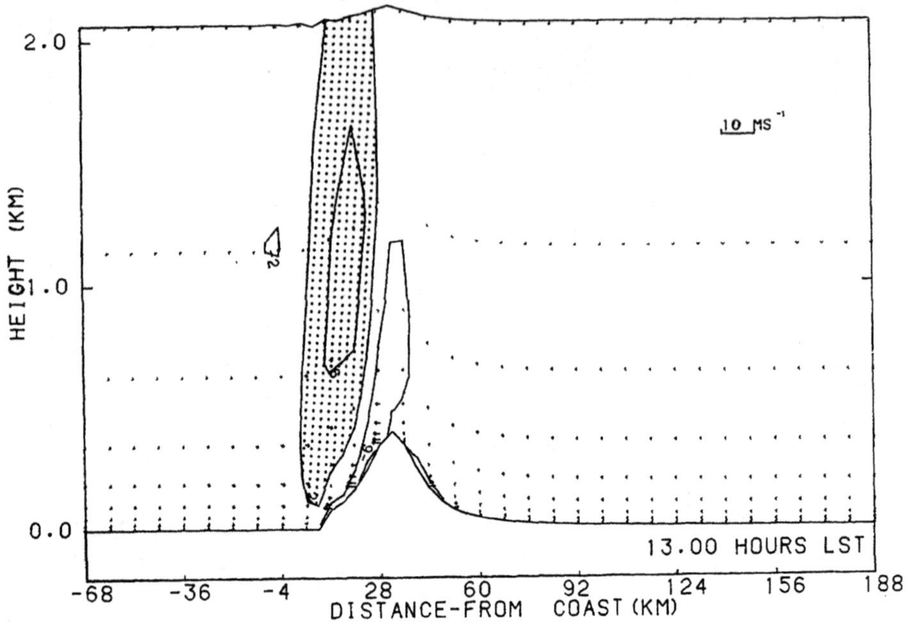

Fig. 8. Difference fields; slope of 0.02, $K_H = 5 \cdot 10^4$ $m^2 s^{-1}$.

In the next experiment we have studied the effect of the slope. The half-width mountain was doubled to 10 km, which is equivalent to the reduction of the average slope by half - to 0.02. The constant diffusivity was kept at the value of $5 \cdot 10^4$ $m^2 s^{-1}$. Fig. 7 illustrates the resulting fields when correction terms were included and it can be noticed that the upward vertical velocity cell is broader and its maximum intensity is reduced. Also, the difference fields in this case of a gentler slope, see Fig. 8, reduce relative to the steeper slope. Nevertheless, the smoothing effect is still considerably high and reaches values of more than 50 percent when 'no-correction' experiment is compared to the experiment in which correction terms were incorporated.

In a final experiment the mountain was removed and in that case the differences become very small and not significant at all. The result for this experiment is not shown here.

4. DISCUSSION

The question which naturally arises when considering these results is whether the numerical horizontal diffusion over mountains is not exaggerated as compared to reality. In a recent letter from A. Simmons at the ECMRWF (European Centre for Medium Range Weather

Forecasting) he stated that actually they have observed an enhanced smoothing due to use of 'horizontal' diffusion along sloping coordinate surfaces in their global prediction model. The diffused temperature in this way has tended to warm mountain tops and cool valleys which in the tropics and summer hemisphere has then triggered unrealistic convective precipitation along mountain chains. This has been corrected in the ECMRWF model by diffusing not the temperature itself but rather some corrected temperature which was equivalent to diffusing along pressure surfaces rather than σ-surfaces.

In this paper we have developed the exact diffusion terms that should be used in order to keep the horizontal diffusion purely horizontal and have found that these correction diffusion terms are much more important in mesoscale as compared to large-scale modelling. In the latter, slopes are typically of the order of 0.001 - 0.01 see e.g. Berkofsky and Bertoni (1955), while in mesoscale the slopes are typically at least one order of magnitude higher, which renders the correction terms comparable to the regular diffusion terms see Table 1.

In a recent paper Pielke and Martin (1981) derive the terms of the hydrostatic equations in a terrain following coordinate system and refer to the subgrid-scale fluxes in the horizontal direction by stating that "as essentially nothing is known about their functional form no purpose is served by writing them".

Even in the opinion of some modellers that those fluxes are included in a model for primarily computational reasons, our study shows that modellers should be cautious when using these fluxes near steep slopes for the following reasons:

(1) with a sloping σ-surface and probably also in any other terrain-following coordinate systems the horizontal diffusion is no longer purely horizontal if correction terms have not been included.

(2) if the correction terms were not included <u>stronger</u> diffusion (or what is equivalent - stronger smoothing) is applied in the sloping regions than in the nearby leveled terrain. The inclusion of the new terms does not lead to any special computational difficulties and for that reason there is no compelling reason to neglect them, all the more that as we have seen the addition of the new terms results in a consistent apportioning of the 'degree' of horizontal diffusion.

Acknowledgements

One of the authors (Alpert) would like to thank the Cooperative Institute for Mesoscale Meteorological Studies for supporting the research of this subject. We wish to thank Profs. R.S. Lindzen and T. Gal-Chen for their constructive comments, and to Dr. Y. Mahrer for his helpful remarks.

Appendix: Difference Schemes for the Diffusion Terms

The four diffusion terms A,B,C and D in Eq. (27) must be calculated by the same scheme and the truncation errors should be minimized in order to get stable and accurate results. This is of special importance for Eq. (27), as the different terms nearly balance each other. In a former experiment for which unequal schemes have been used for the correction terms and for the common diffusion terms we have even got different results on IBM and CDC computers due to the different length of the mantisa(for the real number) on both computers.

Therefore, we have adopted the very stable implicit scheme for the diffusion which has been suggested by Paegle (1976). Originally this scheme was proposed for the vertical diffusion and was extended by Mahrer (1982) for the additional diffusion terms. The difference equation for Eq. (27) takes thus the following form:

$$\frac{U_{ij}^{n+1} - U_{ij}^{n}}{\Delta t} = K_H \left(aDX2_{ij}^{n} + bDX2_{ij}^{n+1}\right)$$

$$- 2K_H \alpha_{ij} \left(aDZX_{ij}^{n} + bDZX_{ij}^{n+1}\right)$$

$$+ \left(K_H \alpha_{ij}^2 + K_v\right) \left(aDZ2_{ij}^{n} + bDZ2_{ij}^{n+1}\right),$$

where the superscript n refers to time, the subscripts i and j refer to vertical and horizontal grid points respectively. The coefficients a and b are 0.25 and 0.75 respectively and,

$$DX2_{ij}^{n} = \frac{U_{ij+1}^{n} + U_{ij-1}^{n} - 2U_{ij}^{n}}{\Delta X^2},$$

$$DZ2_{ij}^{n} = \frac{U_{i+1j}^{n} + U_{i-1j}^{n} - 2U_{ij}^{n}}{\overline{\Delta Z}_{ij}^{2}}$$

$$DZX_{ij}^{n} = \frac{U_{i+1j+1}^{n} + U_{i-1j-1}^{n} - U_{i+1j-1}^{n} - U_{i-1j+1}^{n}}{4\overline{\Delta Z}_{ij}\, \Delta X}$$

$$\alpha_{ij} = \left.\frac{\partial Z_{ij}}{\partial X}\right|_{\sigma}$$

$$\overline{\Delta Z}_{ij} = (Z_{i+1} - Z_{i-1})/2,$$

and parallel expressions for the derivative operators at the time (n+1), namely: $DX2_{ij}^{n+1}$, $DZ2_{ij}^{n+1}$, DZX_{ij}^{n+1}. The horizontal grid interval was 4 km and the time step Δt was 8 s. The altitudes Z_{ij} are calculated at each time step by the height of the σ-surfaces.

Thus, the scheme is fully implicit both in X- and σ- directions and this involves the full numerical domain. Consequently, the solution of the scheme involves a set of 65x10 algebraic simultaneous eqs. The set is solved efficiently by the successive relaxation method and commonly converges after about 5 to 10 over-sweeping of the domain.

References

Alpert, P.: 1981, Implicit filtering in conjunction with explicit filtering, J. Comput. Phys. 44, 212-219.

________, A. Cohen, E. Doron., and J. Neumann.: 1982, A model simulation of the summer circulation from the eastern Mediterranean past Lake Kinneret in the Jordan Valley, Mon. Wea. Rev. 110, 994-1006.

Anthes, R.A., and T.T. Warner.:1978, The development of mesoscale models suitable for air pollution and other mesometeorological studies, Mon. Wea. Rev. 106, 1045-1078.

Batchelor, G.K.: 1956, The theory of homogenous turbulence, Cambridge Univ. Press, 197 pp.

Berkofsky, L. and E.A. Bertoni.: 1955, Mean topographic charts for the entire earth. Bull. Amer. Met. Soc. 36, 350-354.

Blackadar, A.K.: 1978, High resolution models of the planetary boundary layer, Advances in Environmental and Scientific Engineering, Vol.1, Gordon and Breach, 276 pp.

Gal-Chen, T., and R.C.J. Somerville.: 1975, On the use of a coordinate transformation for the solution of the Navier-Stokes equations, J. Comput. Phys. 17, 209-228.

Gifford, F.A.: 1982, Horizontal diffusion in the atmosphere: A Langrangian-dynamical theory, Atmos. Environ. 16, 505-512.

Haltiner, G.J.: 1971, Numerical Weather Prediction, J. Wiley, New York, 317 pp.

Holton, J.R.: 1973, An Introduction to Dynamic Meteorology, Academic Press, New York, 319 pp.

Long, P.E., W.A. Shaffer, J.E. Kemper and F.J. Hicks. 1978: The state of the techniques development laboratory's boundary layer model, 24 May 1977, NOAA Tech. Memo. NWS TDL 66, Silver Spring, MD 63 pp.

Mahrer, Y.: 1982, A theoretical study of the effect of soil surface shape upon the soil temperature profile, Soil Science, 134, 381-387.

Mahrer, Y., and R.A. Pielke.: 1978, A test of an upstream spline interpolation technique for the advective terms in a numerical mesoscale model, Mon. Wea. Rev. 106, 818-830.

Marchuk., G.I.: 1974, Numerical Methods in Weather Prediction, Academic Press. 277 pp.

Mason, P.J. and R.I. Sykes.:1978, On the interaction of topography and Ekman boundary layer pumping in a stratified atmosphere, Quart. J. R. Met. Soc. 104, 475-490.
Paegle, J., W.J. Zdunkowsky and R.M. Walch.: 1976, Implicit differencing of predictive equations of the boundary layer, Mon. Wea. Rev. 104, 1321-1324.
Pielke, R.A., and C.L. Martin.: 1981, The derivation of a terrain following coordinate system for the use in a hydrostatic model, J. Atmos. Sci. 38, 1707-1713.
Sutton, O.G.: 1953, Micrometeorology, McGraw-Hill, New York, 333 pp.
Tennekes, H., and J.L. Lumley.: 1972, A first course in turbulence, MIT Press, 300 pp.

Questions, answers and comments

S.P.S. ARYA: Your parametrization of the velocity variance $\overline{U'^2} = -K_H \frac{\partial \bar{U}}{\partial x}$ is highly questionable, because it would not work in a horizontally homogenous boundary layer over a flat terrain ($\frac{\partial \bar{U}}{\partial x} = 0$ implies infinite K_H). If K_H is considered to be dependent on the terrain slope, then one does not know how to prescribe it. The horizontal diffusivity in the above parametric relation need not have the same meaning or magnitude as the horizontal diffusivity defined from the spreading of a plume.

P. ALPERT: The K-theory formulation is indeed questionable but still used by many modelers as an efficient way of representing sub-grid scale physical processes. It is important to stress that the correction terms proportional to $\alpha \overline{U'^2}$ (α - is the slope of the σ-surface) become important relative to the regular term containing $\overline{U'W'}$ regardless of the K-representation. I have used the K-theory only for the purpose of demonstrating the importance of these correction terms. Regarding the second point, it is correct that we don't know how to express K_H over complex terrain. It has to be included in our model in order to avoid enhanced smoothing (which is involved when the correction terms are not included). The less horizontal diffusion we incorporate in the model, the better is the simulation we are performing.

A. MANES: What is the upper boundary of the model and what effect could this have on the vertical velocity over the hill?

P. ALPERT: The upper boundary of the model was set at top pressure P_t of 700mb. That top pressure is defined by the σ-system formula $\sigma = (p-p_t)/(p_s-p_t)$. Some reflection from the boundary could affect the vertical velocities. In my model this effect is negligible below about 1 km.

J.C.R. HUNT: Can you explain the physical cause of the large horizontal eddy-diffusivity K_H which is many orders of magnitude larger than the vertical eddy-diffusivity. I think it is desirable to understand the causes of this diffusivity before assuming its form in complex terrain.

P. ALPERT: The eddy-diffusivity reflects the subgrid processes which cannot be incorporated in a gridded model. The mesoscale modelers have have used such large K_H values in models with grid distances of about 5 km. We did not make any assumption about the diffusivity form in complex terrain. The only requirement is to keep K_H constant. If we do not include the correction terms, we do not have a constant diffusivity over sloped and level terrain. On the contrary, we get stronger diffusion above slopes because of computational reasons.

F.A. GIFFORD: May I make a comment? I can't explain the physics of this large scale motion either. (The motions mentioned by Dr. Alpert that lead to large values of horizontal diffusivities, $K_H \simeq 5 \cdot 10^4$ $m^2 sec^{-1}$). Motion of this scale is the least studied and the most poorly understood. I can, however, say that they are undoubtedly diffusive in nature, as is clearly shown by the long range diffusion data. They diffuse properties on an instantenous basis at the rate implied by this large K_H.

E. DORON: I would make an additional comment. The diffusion coefficients in these models result from all the processes in a subgrid scale and not only turbulence. It is reasonable to assume that for a larger grid distance, more processes and scales of motion would be involved in the grid scale transports,and this should be reflected in the magnitude of the diffusion coefficient. The horizontal grid distance in mesoscale models, especially hydrostatic ones, is at least an order of magnitude larger than the vertical grid distance, and this leads to larger K values in the horizontal than in the vertical direction.

MESOSCALE FLOW OVER COMPLEX TERRAIN - A FIELD STUDY IN THE LAKE KINNERET AREA

E. Asculai, E. Doron and B. Terliuc
Nuclear Research Centre - Negev, P.O. Box 9001
Beer-Sheva, Israel

A two week observation program was carried out in the summer of 1981 in the Lake Kinneret (Northern Israel) area. The main purpose was to study the mesoscale flow patterns in and around the lake valley and compare them with the results of mesoscale model simulations in the same area. The main effort of the program was directed to the determination of three dimensional trajectories from various points around the valley. For that purpose a new method for the deployment of relatively long term no-lift balloons was used. In addition, surface observations as well as upper air wind observations using pibals were taken at four fixed locations along a straight line across the lake valley. Based on previous studies using surface observations and model results it appeared that the flow regime was determined by the combination of three main mechanisms: the Mediterranean sea breeze, the lake breeze and the mountain-valley wind. This combination results in a daily cycle divided into three distinct flow regimes. The results of the present experiment confirm this basic classification as well as the general structure of the flow for each of the three regimes. The experiment also confirmed the assumption that the large scale synoptic flow has only a minor influence in the valley, and contributes only to the general direction of the winds. In spite of the overall agreement several deficiencies of the model simulations came to light as a result of the experiment. These have to do with the horizontal and vertical resolutions employed in the models, with the fact that all of them were two dimensional (even though several of the models are capable of three dimensional simulations), and with the fact that most of them use the hydrostatic approximation. Due to the lack of appropriate equipment no vertical soundings were performed in order to determine the thermal and humidity stratification. These will have to be completed in subsequent experiments in order to provide the missing data.

Boundary-Layer Meteorology 30 (1984) 313–331. 0006-8314/84/0303-0313$02.85

I. Introduction

In recent years considerable effort has been focused on the subject of the prediction of mesoscale flows over complex terrain. This came about, mainly, as a result of the growing awareness to the problem of mesoscale air pollution, since more and more industrial and power generation complexes are being located in mountaineous areas, especially in river and lake valleys. The study of mesoscale phenomena proceeded in two main directions: The construction and analysis of theoretical models and the collection of observations of the meteorological parameters on that scale. The aim of some of the model simulations was to gain a better understanding of specific phenomena such as the land and sea breeze (Estoque 1961, Neumann and Mahrer 1971, Pielke 1974), the development of mountain and valley winds (Thyer 1966, Manins and Sawford 1979, Rao and Snodgrass 1981) or a combination of both (Doron and Neumann 1976, 1977, Mahrer and Pielke,1977). The purpose of other models was to use such models to study the transport and diffusion of pollutants (Anthes and Warner 1978I, McNider et al. 1981, Segal et al. 1982).

The eastern Mediterranean, and especially the Israeli coastal region, is an ideal location to test the performance of two dimensional mesoscale models (Doron and Neumann 1976). This is due to the long, almost straight, coast which is almost parallel to the mountain range adjacent to it and the Jordan rift valley (Fig.1). In addition the region is also noted for its very steady large scale wind regime in summer. As a result, in recent years, several mesoscale models were used to simulate the flow over the region (Doron and Neumann 1976, Anthes and Warner 1978II, Mahrer and Segal 1979, Alpert et al. 1982). Generally, it was found that all models could simulate, fairly well, the main features of the complex circulation as revealed by the surface observations then available. This was especially true for the two points on the western and eastern shore of the lake for which observations were published (Serruya 1975). Recently a much more comprehensive study, based on a large number of surface observations around the lake and on the slopes surrounding it, was completed (Bitan 1981). It was found, at least as far as surface winds are concerned, that the situation is not as two dimensional as was previously thought, and that caution must be exercised when comparing these results with two dimensional model simulations.

In all model simulations certain interesting features appeared, such as a closed-cell on the western slope in the morning hours and a strong surface convergence zone above the valley to the west. These could not be assessed properly with only surface observations and it was felt that a more detailed observational plan was needed which would include upper air observations as well.

II. Objectives and Methods of Observation

The mesoscale circulation in the Lake Kinneret area is particularly interesting in summer, due to the interaction between the Mediterranean sea breeze and the lake basin local wind system. This local system is mainly a combination of the lake breeze and a mountain valley wind. These two local effects work in the same direction to enhance each other both in the daytime and at night. In addition to the total mesoscale circulation there is, usually, a synoptic scale flow from a general westerly direction. The surface mainfestations of this total flow were observed for quite some time and were summarized in Bitans (1981) work. On the other hand, the vertical structure of the circulation was not investigated at all and the only real information about it was based on results from the numerical simulations. For this reason, we put as the main objective of the present study to try and determine the three dimensional mesoscale flow patterns in the lake valley in summer, especially during the daytime. Several methods to measure upper air wind vectors over complex terrain have been used. Among them were: No lift pilot balloons (Longhetto 1971, Nappo and Snodgrass 1981), tethered balloons fitted with a wind measuring device (Dickerson and Gudiksen 1981), super pressured constant volume balloons (Tetroons, Angell et al. 1972, Wooldridge and Orgill 1978), and single theodolite pilot balloons with constant lift (Hanna et al. 1974, Nappo 1977).

Since we were interested in three dimensional Eulerian wind structure it seems that a dense enough network of tethered balloons would have been the best choice. Unfortunately this is also, by far, the most expensive one, since a three dimensional wind measuring device is needed with each balloon. In addition, this method is very inconvenient when a large percentage of the research area is covered with water. Of the other methods the most widely used was the one with constant volume balloons (Tetroons). These can easily be fitted with a light gas so as to fly at a prescribed isopycnic level and can carry a substantial payload. Their greatest disadvantage is the tendency to stay on a constant density surface. As a result they cannot respond exactly to vertical air motions (Hirsch and Booker 1966, Hanna and Hoecker 1971). This problem may be serious in regions of steep topography such as the area under discussion. Consequently it was decided to use an improved version of the no lift balloons (Terliuc et al. 1983). This method enables the deployment of the balloon from any required level above the ground and can be employed in the field with a minimum of equipment. The main improvement over the previous methods is in allowing relatively long tracking times without the loss of the no lift characteristics.

III. Experiment Set-Up and Observation Program

As already mentioned the aim of the experiment was to compare the actual observed flow with the predicted results of two dimensional models. Consequently, it was decided to concentrate on an east to west cross section of the lake area. Surface synoptic observations and pilot balloon ascents were performed at four points along a more or less straight line. These points are marked by crosses on the map of the region (Fig. 1). The two locations on the western side of the lake were also used as the origin for the no lift trajectory experiments.

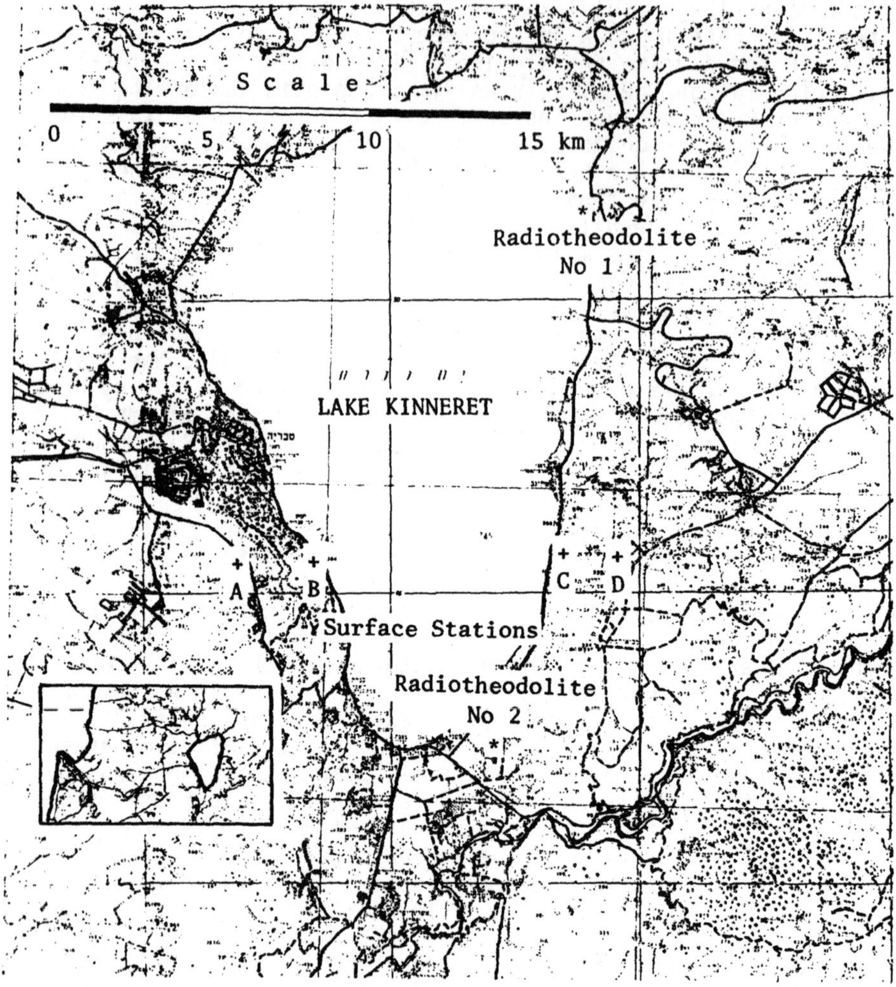

Fig.1: Topographic Map of the Region

1680 MHz radio transmitters were attached to the balloons and they were tracked by two RD-65 radiotheodolites which were positioned along the eastern shore of the lake at the points marked 1 and 2 in Fig.1. We have considered attaching a complete radiosonde to the balloon both for accurate height determination and for atmospheric sounding of temperature, pressure and humidity. However the standard radiosonde available to us is equipped with a baroswitch and does not have a continuous pressure measurement. Since both tracking stations were engaged, most of the time, in tracking the no lift balloon it was not possible to perform separate, conventional, radiosonde ascents. As a result no temperature and humidity measurements other than on the surface were performed.

In addition to the long range no lift balloons, certain no lift experiments were designed to investigate smaller scale phenomena such as closed cells on the valley slopes. These were tracked with optical theodolites.

At the four fixed surface stations, hourly synoptic observations were taken during the entire period of the field study. These included: atmospheric pressure, wind speed and direction, dry and wet bulb temperatures and visual observations such as visibility, cloudiness, and general weather. In addition conventional pilot balloons were released at all four locations simultaneously with the release of the no lift balloon and then repeated one hour later.

The no lift balloons were prepared and launched by the method given in Terliuc et al. (1983). The launch locations and initial flight level were chosen so as to cover as many layers as possible and still enable efficient tracking. A complete list of all no lift launches together with the relevant data is given in Table 1.

Table 1: Launch Data for the Lake Kinneret Experiment

No.	Date(July 1981)	Time	Initial Flight Level (AGL)	Site*
1	6	1745	100	A
2	6	2000	200	A
3	7	0800	100	A
4	7	1300	200	A
5	7	1700	100	A
6	8	0400	100	A
7	8	1400	200	A
8	8	1600	100	A
9	8	1800	200	A
10	9	0700	200	A
11	9	1200	50	A
12	9	1500	50	A
13	9	1700	50	A
14	9	2200	200	A
15	10	0900	20	A
16	10	1200	50	A
17	10	1400	50	A
18	10	1600	100	A
19	12	0900	50	B
20	12	1200	50	B
21	12	1530	50	B
22	12	1900	50	B
23	13	1000	50	A
24	13	1220	50	B
25	13	1400	50	B
26	13	1600	150	B
27	14	1000	50	B
28	14	1300	150	B
29	14	1600	100	B
30	15	1315	50	B
31	15	1600	100	B
32	15	1800	100	B
S1	12	1000	0	B

*see map in Fig. 1.

IV. Features of Summer Flow Structure

In order to be able to put the results of the experiments in the right perspective a short review of our main ideas concerning the flow structure is needed. This is based on a long record of surface observations and the results of two dimensional model simulations.

The main characteristic of the flow is a very pronounced diurnal cycle with three distinct regimes according to the time of day. During the morning hours the flow in the lake valley seemed to be dominated by a combination of the lake breeze with the upslope mountain valley wind. This local circulation prevents the Mediterranean sea breeze from penetrating into the valley. At the meeting zone of these two circulations a strong surface convergence develops with strong upward motion. This leads to a closed circulation cell on the western slopes of the valley. On the eastern slope on the other hand a closed cell does not develop because the large scale westerly winds prevent the return flow (Fig. 2).

Toward early afternoon the lake breeze and the upslope wind weaken, especially on the western slope and the Mediterranean sea breeze plunges suddenly into the valley leading to the afternoon regime. (Fig. 3). This is characterized by a relatively strong wind at the surface leading to quite high waves for a few hours, especially near the eastern shore. Towards evening the breeze weakens and there is a gradual passage into the night regime. This is again dominated by the local breeze and mountain valley winds. This time a closed cell appears on the eastern slope whereas the return flow on the western slope is prevented from developing by the large scale westerlies (Fig. 4). Both closed cells as well as the strong convergence zone were deduced from model results and were not actually observed before.

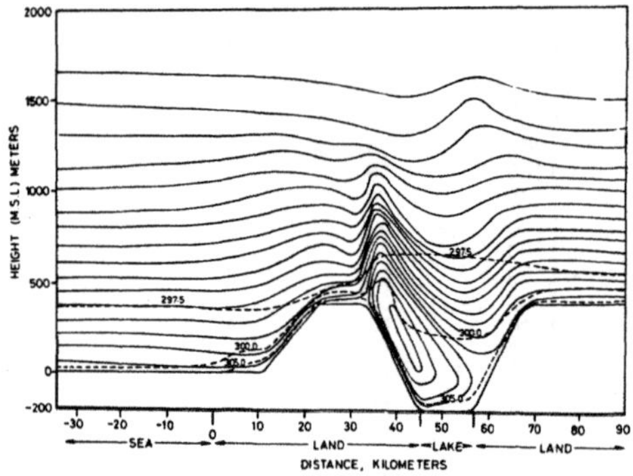

Fig. 2: Streamlines and Isotherms at 0900
After Doron and Neumann (1976).

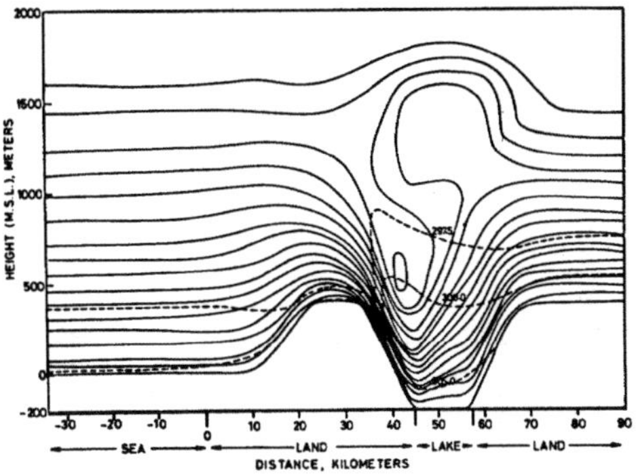

Fig. 3: Streamlines and Isotherms at 1700 After Doron and Neumann (1976).

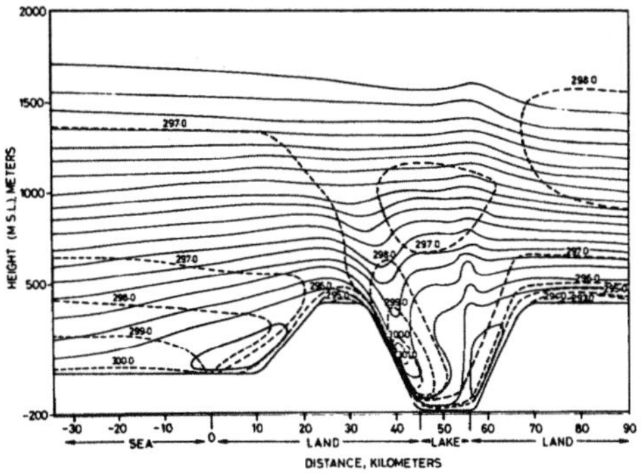

Fig. 4: Streamlines and Isotherms at 0300 After Doron and Neumann (1976).

V. Selected Results

Since a complete review of all the results of the experiment is beyond the scope of the present work only those results most relevant to the description of the previous paragraph will be given. The various experiments to be described will be referred to according to their order in Table 1. Experiment 3 was designed to observe the mid morning regime. The pilot balloon data from the four stations for the same initial time are given in Fig. 5. They indicate a south westerly large scale flow aloft. The horizontal projection of the no lift balloon trajectory is given in Fig. 6 and a height versus time presentation in Fig. 7. After attaining the initial intended level, the balloon continued to rise rapidly whilst traveling in the expected

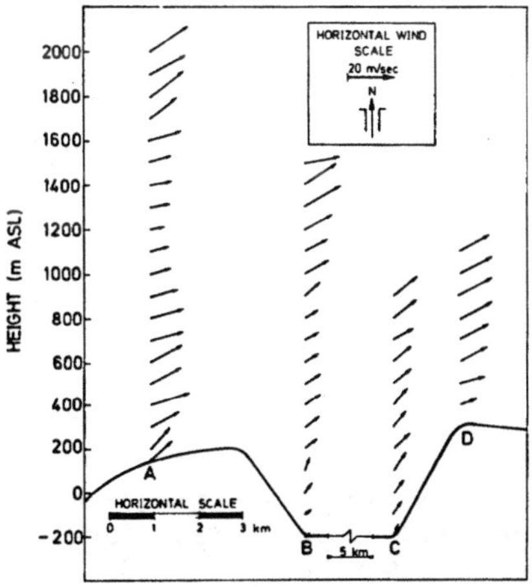

Fig. 5: Horizontal wind as a function of height from pilot balloons at the four surface stations (A,B,C,D). Experiment 3 on July 7 0800.

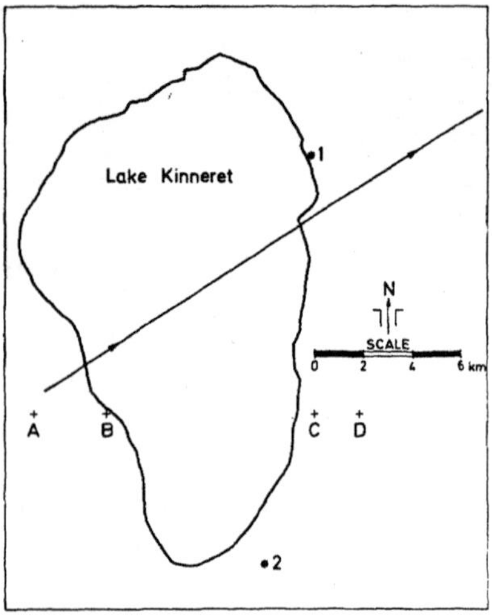

Fig. 6: Horizontal projection of the no lift balloon trajectory. Experiment 3 on July 7 0800.

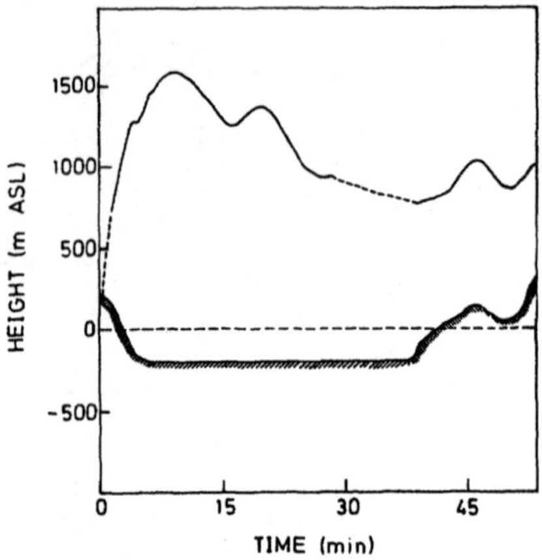

Fig. 7: Height of the no lift balloon and the topography under it as a function of time. Experiment 3 on July 7 0800. Dashed lines represent missing sections of the trajectory.

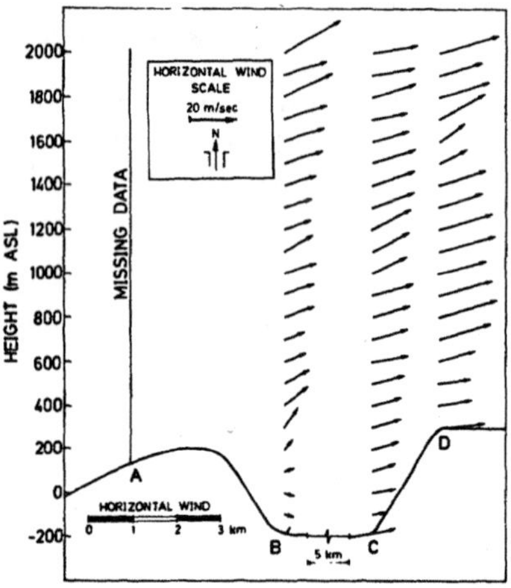

Fig. 8: Horizontal wind as a function of height from pilot balloons at the four surface stations (A,B,C,D). Experiment 4 on July 7 1300.

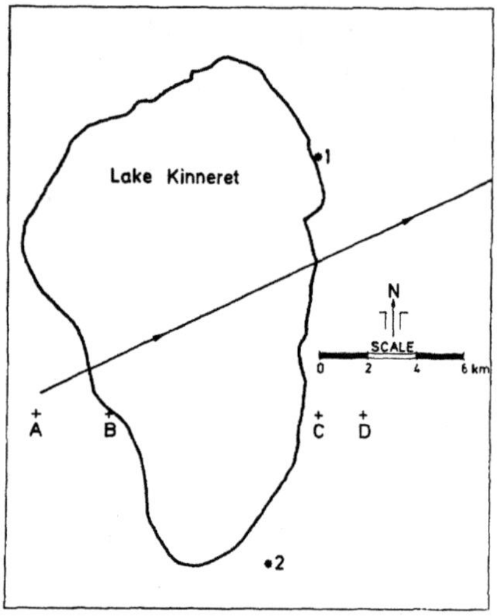

Fig. 9: Horizontal projection of the no lift balloon trajectory. Experiment 4 on July 7 1300.

north easterly direction. At about 1500 ASL it leveled off and started to descend slowly, experiencing an oscillation above the middle of the lake, approximately. After reaching the eastern shore of the lake the balloon started following the contours of the topography, some 800 meter AGL.

Experiment 4 took place on the same day at 1300 LST. The pilot balloon data (Fig. 8) show that at the upper levels the general character of the flow has not changed noticeably from that of the morning. Though the horizontal trajectory of experiment 4 (Fig. 9) resembles closely the one of experiment 3 there was a marked change in the vertical flow pattern over the lake. After rising to a height of about 800 meters (Fig. 10) the balloon leveled off travelling horizontally for some time. It then plunged down into the valley, descending some 700 meters and attaining its lowest point somewhere past the middle of the lake. From there it rose, again rapidly, to about 1200 meters ASL and later continued to follow the topography in much the same way as the previous balloon.

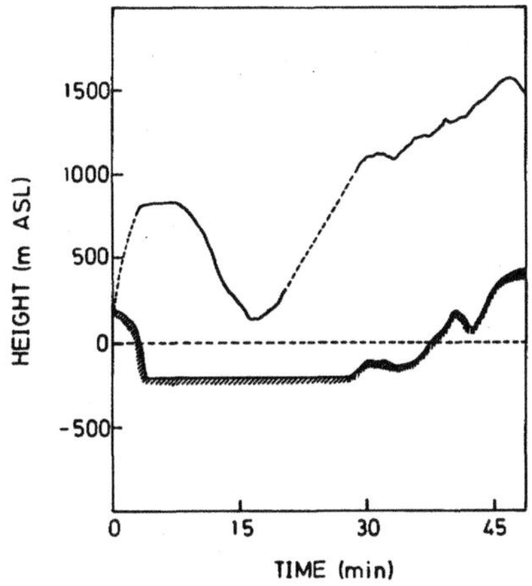

Fig. 10: Height of the no lift balloon and the topography under it as a function of time. Experiment 4 on July 7 1300.

Experiment No. 19 was carried out in order to study the details of the flow near the western shore of the lake. The balloon was launched from point B and was tracked both by radiotheodolites and by two optical theodolites. The initial trajectory was in a northwesterly direction following the contours of the surface (Fig. 12). After travelling up the slope the balloon met the expected convergence zone (see the pilot data in Fig. 11) and changed its direction towards the east. A more detailed presentation of the initial stage is given in Fig. 13, based on the optical theodolite data. After the change of direction the balloon started to rise at a more or less constant rate of about 10 meters per minute indicating a possible failure of the balancing system (Fig. 14). Interesting features of this experiment are the two 90 degree sudden changes in travel direction. The first one occured at an elevation of 2000 meters ASL and later a reversal occurred at 2800 meters ASL. This 800 meter layer was

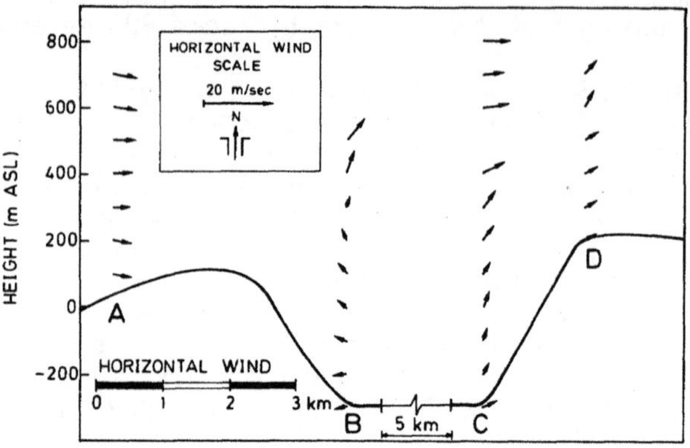

Fig. 11: Horizontal wind as a function of height from pilot balloons at the four surface stations (A,B,C,D). Experiment 19 on July 12 0900.

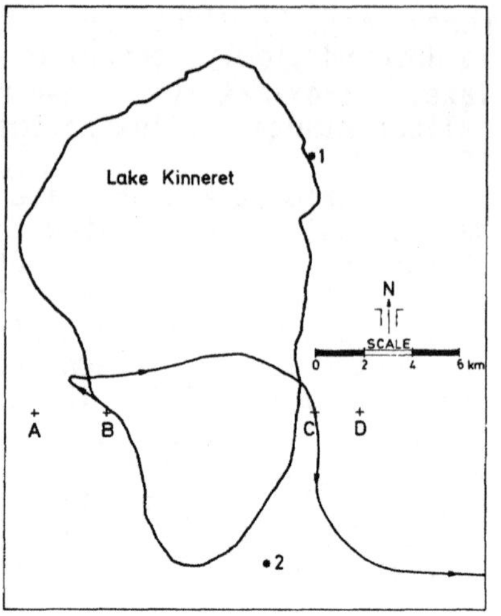

Fig. 12: Horizontal projection of the no lift balloon trajectory. Experiment 19 on July 12 0900.

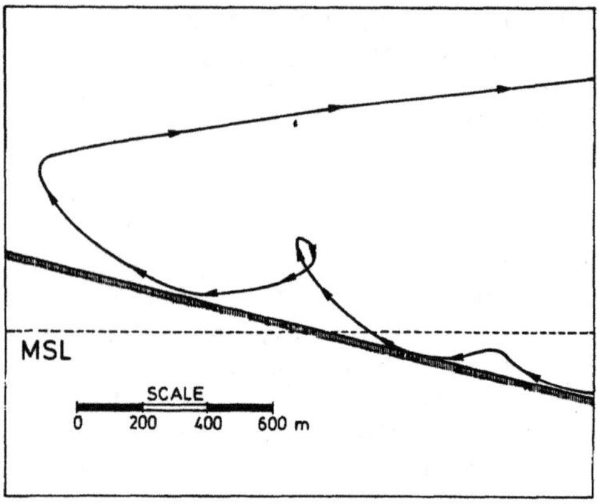

Fig. 13: Projection of the initial part of no lift balloon trajectory on a vertical plane in the NW-SE direction. Experiment 19 on July 12 0900.

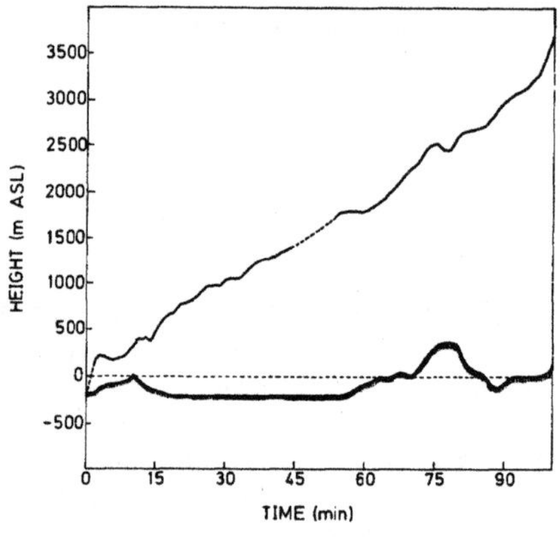

Fig. 14: Height of the no lift balloon and the topography under it as a function of time. Experiment 19 on July 12 0900.

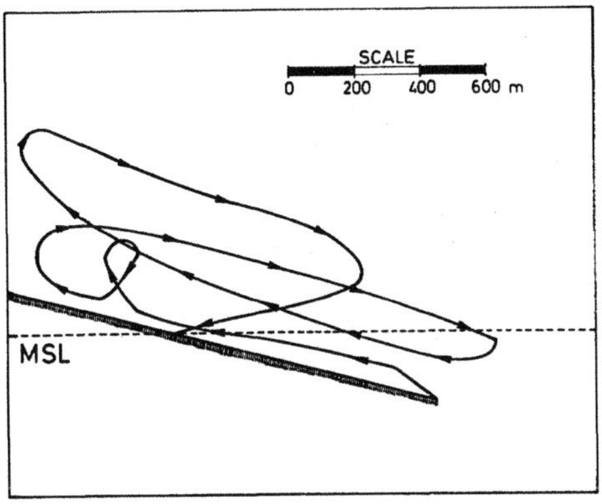

Fig. 15: Projection of the no lift balloon trajectory on a vertical plane in the NW-SE direction. Experiment S1 on July 12 1100.

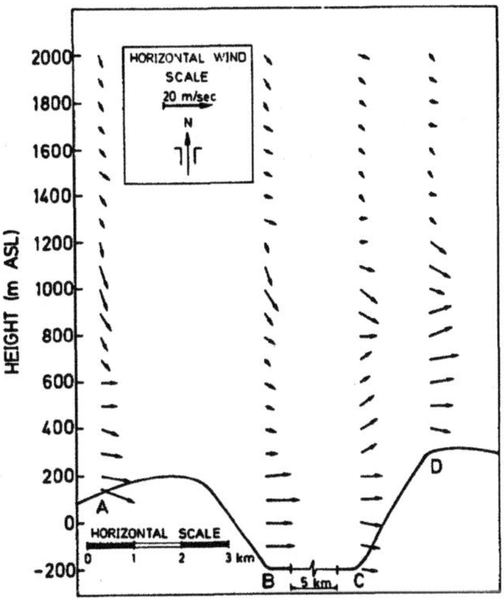

Fig. 16: Horizontal wind as a function of height from pilot balloons at the four surface stations (A,B,C,D). Experiment 21 on July 12 1530.

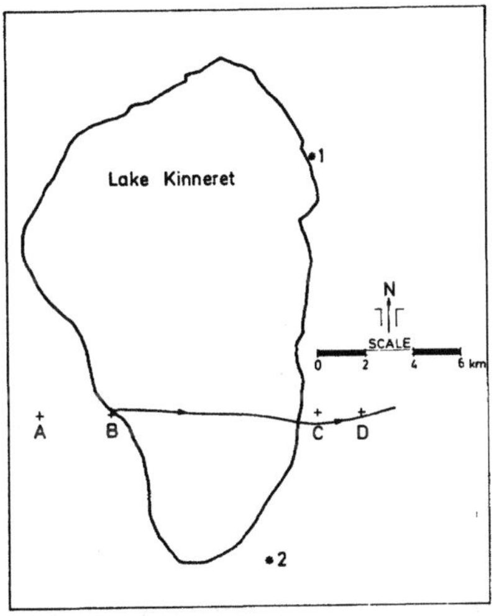

Fig. 17: Horizontal projection of the no lift balloon trajectory. Experiment 21 on July 12 1530.

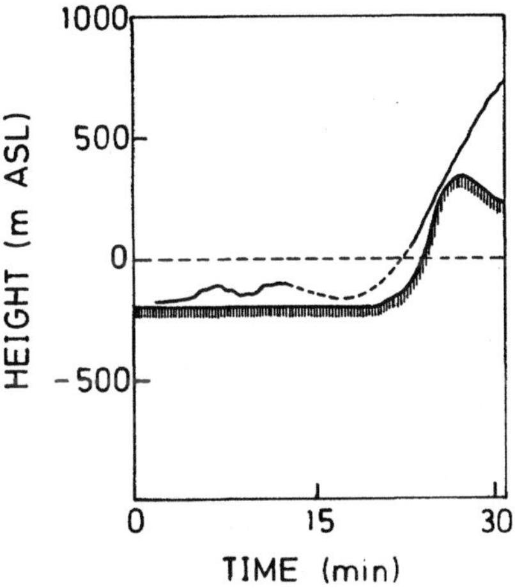

Fig. 18: Height of the no lift balloon and the topography under it as a function of time. Experiment 21 on July 12 1530.

immediately above the cloud base as seen from the ground and is probably associated with an inversion layer acting as a lid to the flow underneath. A special optically tracked launch was carried out two hours after the previous experiment (Labelled S1). In this case the balloon stayed within a closed circulation cell, executing three full loops before hitting the ground (Fig. 15).

An example of a fully developed afternoon flow regime is shown by the results of experiment 21 at 1530 LST. The pilot balloon results (Fig. 16) indicate very strong westerly winds on both sides of the lake. It should be noted that the upper air winds were very weak on that particular day. The no lift balloon was released on the western shore and crossed very rapidly the whole lake from west to east (Fig. 17) close to the water level (Fig. 18). It later rose rapidly over the topography. As an example of the evening and nighttime regime we present the results of experiment 22 at 1900 LST. The pilot balloon results indicate (Fig. 19) a weak flow from both slopes of the valley towards the lake in a very shallow layer near the ground. There seems to be a return flow, also at both sides of the lake in a much deeper layer reaching to about 1300 ASL. The no lift balloon, released on the western shore traveled very slowly towards the centre of the lake (Fig. 20) staying all the time very close to the surface, until it hit the water (Fig. 21).

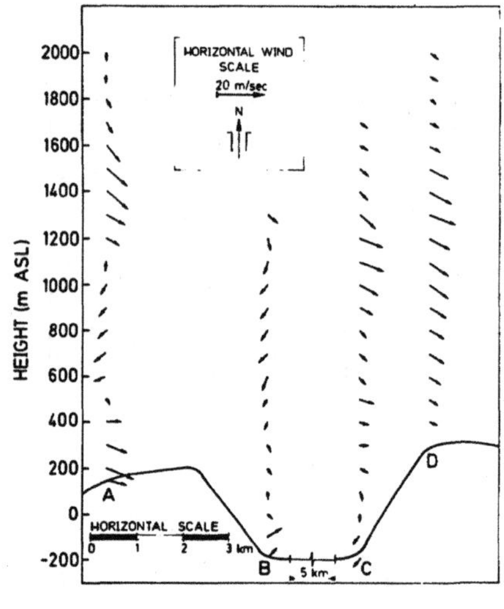

Fig. 19: Horizontal wind as a function of height from pilot balloons at the four surface stations (A,B,C,D).
Experiment 22 on July 12 1900.

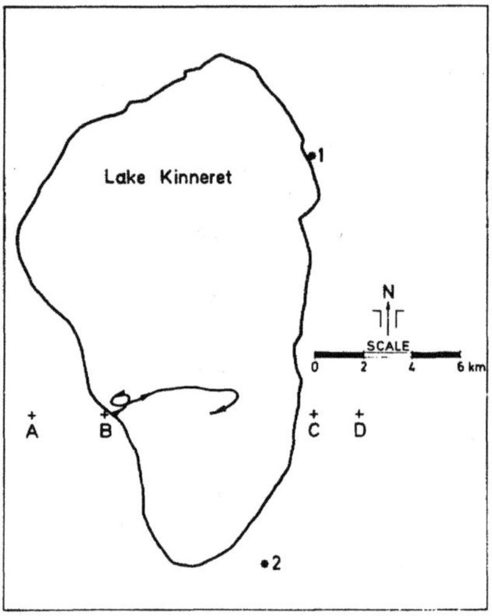

Fig. 20: Horizontal projection of the no lift balloon trajectory. Experiment 22 on July 12 1900.

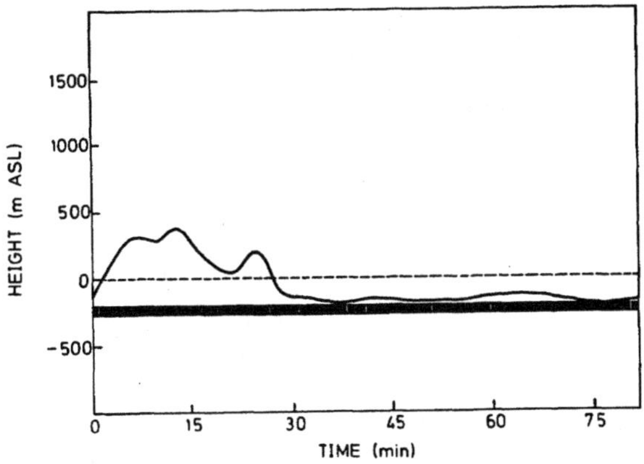

Fig. 21: Height of the no lift balloon and the topography under it as a function of time. Experiment 22 on July 12 1900.

VI. Discussion

As already mentioned in the introduction all numerical simulations in the Lake Kinneret area gave essentially very similar results as regards the main features of the summer circulation. Since we aim to compare our trajectory observations with model results, it is most convenient to use one in which a streamline presentation is given. For that purpose the results presented by Doron and Neumann (1976) were chosen. In that particular model the three main diurnal circulation patterns were each maintained for several hours and we are thus justified in comparing streamlines with trajectories on a time scale of about one hour.

Before trying a comparison between model results and observations it is appropriate to inquire whether the main assumptions made in the construction of the models were actually fulfilled during the experiment. For practical reasons, mainly the availability of computational resources, almost all simulations in the area were made with two dimensional models. The assumption inherent in the use of a two dimensional model designed to simulate an essentially three dimensional phenomenon is that of homogeneity in at least one spatial dimension. It has already been pointed out by Bitan (1981) that this assumption in the direction along the Jordan rift valley (North-South) is not always justified* In addition it is also easily seen (Fig.1) that the Jordan valley is much wider in the lake area than either to the north or south of it.

*Footnote: This is due to the special shape of the lake (its Hebrew name - Kinneret - derives from the Hebrew word Kinor, which nowadays means violin but in medieval and biblical times signified a kind of a lute).

Another assumption made in almost all simulations is that of a constant large scale flow pattern. It should be pointed out that this is not a necessary assumption in two dimensional models. It was usually made because the simulations were not intended to study individual cases but rather to study the average flow patterns. In the particular case of the Doron and Neumann simulation the large scale wind was constant in both space and time with a value of about 4.5 m/sec from the NW. While such a large scale wind prevailed during part of the experiment there were cases in which it came from the SW - e.g. experiments 3 and 4. Because the shape of the lake is not exactly symmetric there is a difference in the effective width of the basin between a northwesterly and a southwesterly flow. Taking these limitations of the models into account we are now in a position to try and compare the experimental results with the model.

Reviewing the results of the present study,it can be stated categorically that the three distinct flow regimes in the diurnal cycle, depicted in the model results,were also observed very prominently in the present experiment. Even though there were significant changes from day to day as regards the time of onset and the strength of the circulation, all three types appeared every day of the experiment.

Let us now examine each of the three regimes separately:

The Morning Regime: The experiments relevant to this period, out of those elaborated on in this paper are 3, 19, and S1. All three are consistent, at least qualitatively, with the streamline presentation in Fig. 2, except for small scale oscillations due to topographic features not included in the model. The convergence zone to the west of the lake valley seems to be responsible for the rapid rise of the balloon observed in experiment 3 (Fig. 7) and the change of direction observed in experiment 19 (Fig. 12). The existence of the closed circulation cell on the western slopes of the lake was demonstrated dramatically in experiment S1. Nevertheless the experiments also bring to light some deficiencies in the models: In the first place it appears that the convergence zone in the model is too shallow. This is apparently a shortcoming of that particular model and is due to the fact that a lid was imposed at about 2000 m. This can easily be remedied by either choosing a different boundary condition or placing the lid much higher. A more serious problem is the size of the closed circulation cell. Whereas, in the model, it extends to almost the entire width of the lake, in nature, it is confined to a very shallow layer near the western slope. It should be pointed out that the horizontal resolution in the model was 2 km so that the minimum horizontal extent of the cell must be 4 km if it is to be depicted at all. None of the above mentioned mesoscale models had a better resolution. Another factor which is certainly important in determining the extent of the cell is the gradient of the topography. In the model there is a drop of 600 meters over a distance of 10 km, whereas in the region where the balloons were released the drop is about 400 m, over a distance of only 1500 m. For such a drop the horizontal and vertical scales of motion became comparable and the hydrostatic assumption made in the model is no longer valid. In order to simulate properly the structure of the flow in that particular region a non-hydrostatic model is needed with a much better horizontal and vertical resolution.

The Afternoon Regime: This type of flow is exemplified in experiments 4 and 21. The main feature of the flow in the model is a strong narrow jet dropping from the top of the mountain in the west and then flowing out of the valley in a much wider channel (Fig. 3). The trajectory presented in Fig. 10 is remarkably close to a streamline in Fig. 2 starting from the top of the mountain to the west at 800 ASL. The sudden drop and the immediate rise are both simulated very well by the model. Similarly the trajectory of Fig. 18 is close to a streamline starting on the west coast of the lake.

At the western shore, changes of the order of one hour in the time of onset of the afternoon regime were observed from day to day. It is still not clear what causes these changes, but presumably they must be connected either with changes in the large scale wind speed and direction or with changes in the low level heat balance. The particular model used here for comparison incorporates a prescribed surface temperature and so cannot simulate the last effect but the problem should not be too difficult to resolve with more sophisticated models.

The Nighttime Regime: An example of the night flow structure is shown in the results of experiment 22. Here most of the information is contained in the pilot balloon observations and the no lift balloon released from the western shore does not yield anything new. The downslope winds on both sides of the lake have already started to develop in a shallow layer. The return flow on the other hand appears to spread in a very deep layer up to about 1100 meters ASL. It also seems that the western cell at least has spilled out of the valley to a distance of about 3 km to the west. This is a phenomenon not observed in the models and must be further investigated, particularly since it was observed only in experiment 22.

VII. Conclusions

The experiment reported in the present work is part of a larger project whose aim is to use mesoscale dynamical models for the prediction of air pollution from local or area sources on that scale. As such, its main aim was to test the adequacy of presently existing models and to point to possible improvements.

The main conclusions of the experiment was that the models are not ready to be used for meaningful air pollution studies in their present form. There is no doubt that numerical models provide a powerful research tool and their use has contributed significantly towards our understanding of the main processes involved in the determination of the mesoscale flow regime. It appears that for practical use the models have to be adapted to the requirements of the specific case. For mesoscale air pollution studies in our region, especially those designed for the prediction of specific episodes, the following improvements are necessary: 1) The model has to be fully three dimensional even in the vicinity of the Jordan rift valley. 2) Much better horizontal resolutions than those used in the models so far are required. A grid spacing of the order of 500 meters seems to be sufficient. 3) With the above mentioned horizontal resolution the model will have to be non-hydrostatic. In the closed cell in Fig. 15 for example the horizontal and vertical scales are about the same.

On the other hand, it turned out that a detailed determination of the synoptic scale flow is not needed in summer. In fact all that is required is the large scale geostrophic wind, which can be obtained from one synoptic upper air observation in the whole region of interest.

The results of the present experiment provide a good starting data base for testing the models with the above mentioned, and other improvements. Nevertheless, the three dimensional thermal structure in the region is an important part of the flow regime. The fact that it was not observed in the present experiment is a serious deficiency which will have to be rectified in subsequent experiments. Temperature and pressure measurements will also be required for a more accurate determination of the vertical coordinate of the no lift balloon.
The no lift balloon technique developed for the purpose of the present experiment proved to be a very reliable and efficient research tool and it appears that it could also be used in specific real time pollution episodes for the determination of the over all danger areas.

References

Alpert, P., Cohen, A., Neumann, J. and Doron, E.: 1982 'A Model Simulation of the Summer Circulation from the Eastern Mediterranean Past Lake Kinneret in the Jordan Valley 'Mon. Wea. Rev. 110,994-1006.

Angell, J.K., Pack, D.H., Machta, L., Dickson, T.R., and Hoecker, W.H.: 1972 'Three Dimensional Air Trajectories Determined from Tetroon Flights in the Planetary Boundary Layer of the Los Angeles Basin' J. App. Meteor. 11, 451-471.

Anthes, R.A. and Warner, T.T.: 1978(I), 'Development of Hydrodynamic Models Suitable for Air Pollution and Other Mesometeorological Studies', Mon. Wea. Rev. 106, 1045-1078.

Anthes, R.A. and Warner, T.T.: 1978(II), 'Simulation of Mesoscale Flows over Israel'Isr.Meteor. Res. Paper. 2, 93-123.

Bitan, A.: 1981, 'Lake Kinneret (Sea of Galilee) and its Exceptional Wind System' Bound. Lay. Meteor. 21, 477-487.

Dickerson, M.H. and Gudiksen, P.J.: 1981 'ASCOT FY-1981 Progress Report', UCID-18878-81, ASCOT 81-2.

Doron, E. and Neumann, J.: 1976 'A Mesometoerological Model with Topography: Initial Results and Problems', Presented at Army Mesometeorological Meeting, Las Cruses, N.M., Dept. of Atmos. Sci., Hebrew Univ. of Jerusalem.

Doron, E., and Neumann, J.: 1977 'Land and Mountain Breezes with Special Attention to Israel's Mediterranean Coastal Plain', Isr. Meteor. Res. Pap. 1, 109-122.

Estoque, M.A.:1961 'A Theoretical Investigation of the Sea Breeze', Quart. J. Royal Met. Soc. 87, 136-146.

Hanna, S.R. and Hoecker, W.H.: 1971 'The Response of Constant Density Balloons Applied to Sinusoidal Variations of Vertical Wind Speeds', J. Appl. Meteor. 10, 601-604.

Hanna, S.R., Nappo, C.J., Hosker, R.P., and Briggs, G.A.:1974, 'Description of the Eastern Tennessee Trajectory Experiment', ATDL cont. 103, Oakridge, Tenn.

Hirsch, J.H. and Booker, D.R.: 1966 'Response of Super Pressure Balloons to Vertical Air Motions', J. App. Meteor. 5, 601-604.

Longhetto, A.: 1971 'Some Improvements in the Balanced Pilot Balloons Technique', Atmos. Environ. 5, 327-331.

Mahrer, Y. and Pielke, R.A.: 1977, 'The Effects of Topography on the Sea and Land Breezes in a Two Dimensional Numerical Model; Mon. Wea. Rev. 105, 1151-1162.

Mahrer, Y. and Segal, M.:1979 'A Numerical Study of the Air Flow over Israel Using a Two Dimensional Mesoscale Model' 4th Conf. on Numerical Weather Prediction. Am. Meteor. Soc., 256-259.

Manins, P.C. and Sawford, B.L.: 1979 'A Model of Katabatic Winds', J. Atmos. Sci. 36, 619-630.

McNider, R.T., Hanna, S.R., and Pielke, R.A.:1981 'A Numerical Simulation of the Transport and Dispersion of Pollutants in Slopes and Mountain-Valley Flows', Proc. of 5th Symp. Turbulence Diffusion and Air Pollution, Atl. Ga.

Nappo, C.J.: 1977 'Mesoscale Flow over Complex Terrain During the Eastern Tennessee Trajectory Experiment (ETTEX)', J. App.Meteor. 16, 1186-1196.

Nappo, C.J. and Snodgrass, H.F.: 1981 'Observations of Nighttime Winds Using Pilot Balloons in Anderson Creek Valley, Geysers, California', J. App. Meteorol. 20, 721-727.
Neumann, Y. and Mahrer, Y: 1971,'A Theoretical Study of the Land and Sea Breeze Circulation', J. Atmos. Sci. 28, 532-542.
Pielke, R.A.:1974 'A Three Dimensional Numerical Model of the Sea Breezes Over South Florida', Mon. Wea. Rev. 102, 115-139.
Rao, K.S. and Snodgrass, H.F.: 1981 'A Nonstationary Nocturnal Drainage Flow Model',Bound. Lay. Meteor. 20, 309-320.
Segal, M., McNider, R.T., Pielke, R.A., and McDougal, D.S.: 1982, 'A Numerical Model Simulation of the Regional Air Pollution Meteorology of the Greater Chesapeake Bay Area - Summer Day Case Study', Atmos. Environ. 16, 1181-1397.
Serruya, S.: 1975 'Wind, Water Temperature and Motions in Lake Kinneret: General Pattern', Verk, Int. Verein Limnol. 19, 73-87.
Terliuc, B., Asculai, E., and Doron, E.: 1983 'The Use of Compensated Aerological No-Lift Balloons to Determine Relatively Long Term Dry Air Parcel Trajectories', J. Clim. and Appl. Meteor. 22,
Thyer, N.H.:1966 'A Theoretical Explanation of Mountain and Valley Winds by a Numerical Method', Arch. Met. Geoph. Bioklim. A15, 318-348.
Wooldridge, G.L. and Orgill, M.M.:1978 'Airflow, Diffusion and Momentum Flux Patterns in a High Mountain Valley', Atmos. Environ. 12, 803-808.

Questions, answers and comments

J.E. CERMAK: Neglect of Coriolis effects in physical modeling of atmospheric flow, introduces some uncertainties in similarity of model and prototypic mean velocity and concentration distribution. What effects of Coriolis accelerations have you found in your measurements of flows over Lake Kinneret and what is their significance?

E. ASCULAI: The model to which the observations were compared includes Coriolis effects as part of its formulation. In this particular experiment it seems that the trajectories are too short so that they either appear to be straight, or other effects dominate the turning. However, a turning and acceleration of the large scale flow as it passes over the mountain range parallel to the coast, has been observed in Israel many times.

J.C.R. HUNT: What kind of upwind temperature profile was assumed in the model calculations?

E. ASCULAI: The initial temperature profile in the model was the following: From the surface up to 930 mb a temperature drop of 6.5°C/1000m, from 930 mb to 900 mb an inversion of 1.5°C/1000m, and above this layer again a drop of 6.5°C/1000m. The inversion layer was introduced in order to simulate the summertime large scale subsidence inversion over the Middle East. The model itself does not include a mechanism for maintaining this inversion over a long period, but in our results, it was maintained for at least three diurnal cycles.

S.R. HANNA: The model predictions represent ensemble averages, while the no-lift balloon provides only a single realization. It would be preferable to generate ensemble averaged observations of balloon motions by averaging five to ten or more balloon trajectories. Do you have any plans to increase the number of balloons followed during a given experiment?

E. ASCULAI: In the present experiment the balloon system is tracked by two radio theodolites. It is possible to track two balloons at the same time, but even this is somewhat difficult. The way to deal with clusters of balloons is by optical tracking and photography, but for short ranges only. For longer ranges, a high resolution radar system may provide the proper solution, if one desires cluster measurements. At present we are planning to continue with our single balloon experiments and to compare them with existing or future models. As the models improve, it may be necessary to utilize cluster measurements to compare not only the general features but the finer details of the flows. In the present experiment, balloons released an hour apart showed good repeatability, having essentially the same flow features.

EXPERIMENTAL AND MODEL TRANSPORT AND DIFFUSION STUDIES IN COMPLEX TERRAIN WITH EMPHASIS ON TRACER STUDIES

Marvin H. Dickerson, Kevin T. Foster and Paul H. Gudiksen
Lawrence Livermore National Laboratory, University of California, Livermore, California 94550

ABSTRACT

The U.S. Department of Energy (DOE) Atmospheric Studies in Complex Terrain (ASCOT) program began in the fall of 1978 as a multiple DOE and other Federal Laboratory program devoted to developing a better physical understanding of atmospheric boundary layer flows in areas of complex terrain. The first technical challenge undertaken by the program was an investigation of atmospheric boundary layer phenomena associated with the development, continuation and breakup of nocturnal drainage wind flows. This paper discusses the general objectives the program has addressed during the past several years and focuses on results from a major field experiment conducted in 1980 in The Geysers area of northern California. Specifically, results from measurements of simultaneous tracer releases are compared to calculations from a mass-consistent wind field model coupled to a particle-in-cell transport and diffusion model. Results of these comparisons show that model calculations agree with measurements within a factor of 5 approximately 50 percent of the time. Part of the difficulty faced by the models in these comparison studies is associated with large variabilities between measurements made by samplers located one or two Δx apart when compared to the resolution of the models. Space and time averaging improves the comparisons considerably, although the design of the field experiment did not allow the determination of optimum spacial and temporal averages.

INTRODUCTION

The Department of Energy is currently sponsoring a program of Atmospheric Studies in Complex Terrain (ASCOT) to improve the technology needed to assess air quality impacts of developing energy resources in areas of complex terrain. The program uses field experiments, theoretical atmospheric physics research, and mathematical models to develop a measurements and modeling methodology that can provide quality assessments in these areas. The ASCOT team is composed of scientists

Boundary-Layer Meteorology 30 (1984) 333–350. 0006-8314/84/0303-0333$02.70

from several DOE-supported research laboratories and university programs.

With the program's initial focus on studying transport and dispersion of materials injected in or near nocturnal drainage flows, three series of field experiments were conducted in The Geysers geothermal area in northern California. The initial experimental series, conducted during July 1979 in the Anderson Creek valley of The Geysers area, were exploratory in nature and of limited scope. Results from this series of experiments were used to plan a second series of experiments during September 1980 in the same valley, to acquire more detailed information about the temporal and spatial characteristics of drainage flows and their interaction with the regional and synoptic scale flows. In contrast to these generic studies of the transport and dispersion of pollutants entrained in drainage flows, a third series of experiments, conducted during August 1981 in the Big Sulfur Creek valley of The Geysers, was designed to (1) acquire data needed to predict the impact of the hydrogen sulfide emissions from future geothermal power plant cooling tower plumes during nocturnal drainage flows, and (2) to perform nocturnal drainage flow studies in a different environmental setting for testing the general applicability of the methodologies developed from the experimental studies in the Anderson Creek valley (Dickerson and Gudiksen, 1984; Gudiksen and Dickerson, 1984). This paper concentrates on the 1980 field studies, particularly on the tracer experiments. The variability between concentration values measured by closely spaced samplers is discussed and the relationship between measured and modeled concentrations is described.

EXPERIMENTAL DESIGN

With the goal of studying the transport and dispersion of materials injected in or near drainage flows, the design of the experiments reflected the following general evaluations:

- the entire nocturnal drainage cycle from initiation to full development and breakdown.
- the temporal and spatial characteristics of the drainage flows within the valley, including the evolution of pooling of drainage flows within the Anderson Creek valley and the subsequent outflow of this air from the valley.
- the influence and the extent of mixing between the external flows and the drainage flows within the valley, and
- the effect on the drainage flows of changing surface roughness due to forest canopies.

The 1980 series of experiments, in the Anderson Creek valley, consisted of five identical experimental plans conducted on different but similiar nights. Each experiment consisted of tracer studies coordinated with a host of meteorological studies. The Anderson Creek valley has the characteristics of a basin. Its topographic features and the layout of the tracer studies are shown in Fig. 1. The valley is bounded by Cobb Mountain to the north, by a ridge to the west and south, and by Boggs Mountain to the east. The Anderson, Gunning, and Putah Creeks, which form the principal drainage areas, merge near Anderson Springs with outflow toward the southeast. The tracer studies included the use of two perfluorocarbons, two heavy methanes, and sulfur hexafluoride gases that were measured using conventional sampling techniques. Also oil fog tracked by lidar and tetroons tracked by radar were used in these experiments. One of the perfluorocarbon tracers (PMCH; C_7F_{14}) was released into the nocturnal drainage flows from an open, but very sheltered area in Anderson Creek; the other perfluorocarbon tracer (PDCH; C_8F_{16}) was released within a forest canpy in Gunning Creek. These sites are located approximately halfway between the floor of the valley and the top of the ridge. Two heavy methane tracers, methane-20 ($^{12}CD_4$) and methane-21 ($^{13}CD_4$) were released within the upper reaches of the Anderson Creek drainage area. The methane-21 was released at the surface directly into the drainage flows; methane-20 was released simultaneously into the transition layer at a height of 60 to 75 m above the surface to investigate the extent of mixing between the transition layer flows and the drainage layer. The sulfur hexafluoride was released in the upper part of the Putah Creek drainage area to evaluate the merging of the flows from Putah Creek with those from the Anderson and Gunning Creek drainage areas. All of the releases were one hour in duration. The downwind surface concentrations were sampled at 30 to 50 locations depending upon the tracer. The sampling periods ranged from 10 minutes

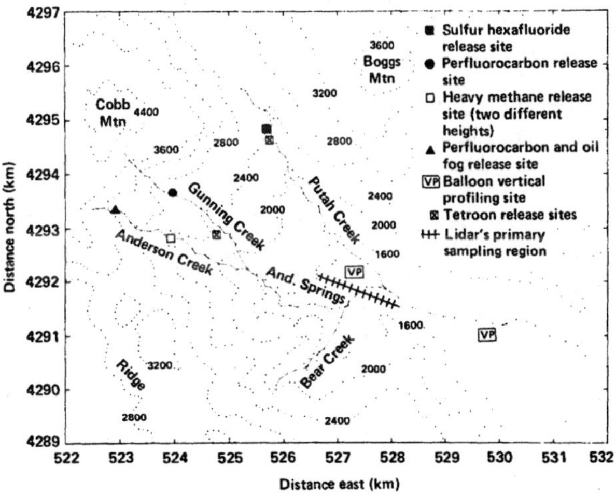

Figure 1. Layout of the ASCOT tracer experiments conducted within the Anderson Springs Valley in 1980.

to 8 hours. In addition, two vertical profiling systems were used to define the temporal variations in the vertical distributions of the tracers within the valley basin and outflow region. These consisted of balloon borne sampling systems. The sampling and analytical techniques for the perfluorocarbon tracers have been reported by Ferber et al. (1981) and by Lovelock and Ferber (1982), while those for the heavy methane tracers have been reported by Cowan et al. (1976) and Fowler (1979).

To acquire more detailed structural information about the three-dimensional evolution of these tracers, oil fog was released at the same site as the PMCH perfluorocarbon tracer and tracked by a lidar. For each release, the lidar, situated near the valley outflow region, performed a series of scans in various vertical planes to observe the evolution of the plume. The region most frequently sampled by the lidar and, hence, the site of the most detailed analyses is shown in Fig. 1. The lidar used in this work has been described by Eberhard (1981). The remaining tracer studies were tetroons tracked by radar within the Anderson and Putah Creek drainage areas. These were released at a height of 100m both individually and in clusters of three from the two sites shown in Fig. 1. Thus, the tetroons were flown in the transition layer overlying the drainage flows within the two valleys, and provided direct measurements of individual air parcel trajectories and dispersion characteristics of these air parcels. A description of the radar and the data analysis techniques is reported by Fosberg and Lanham (1983).

Meteorological measurement systems dedicated to these experiments included an array of nine acoustic sounders, seven tethersondes, eight optical cross-wind anemometers, two rawinsondes, one minisonde, 27 surface meteorological stations with real-time telemetry, as well as ten 10 m and one 60 m meteorological towers. Most of the acoustic sounders, tethersondes, and optical anemometers were dedicated to evaluating characteristics of the drainage flows within the three major drainage areas and for evaluating interactions of these flows with the transition layer flows and the pooling within the valley basin. The surface stations, on the other hand, provided a comprehensive view of the surface flow characteristics over the entire valley. Finally, the rawinsondes and the minisonde, which were located outside the valley, were dedicated to defining the regional scale flow within which the Anderson Creek valley was imbedded. The complete data set for these experiments has been reported by Gudiksen (1983). A more complete description of the experimental plan has been reported by Dickerson (1980).

TRACER AND MODELING STUDIES

The surface distributions of the PMCH and PDCH tracers, released for a one-hour period within the Anderson Creek and Gunning Creek drainage areas, are shown in Fig. 2 for the first two hour period

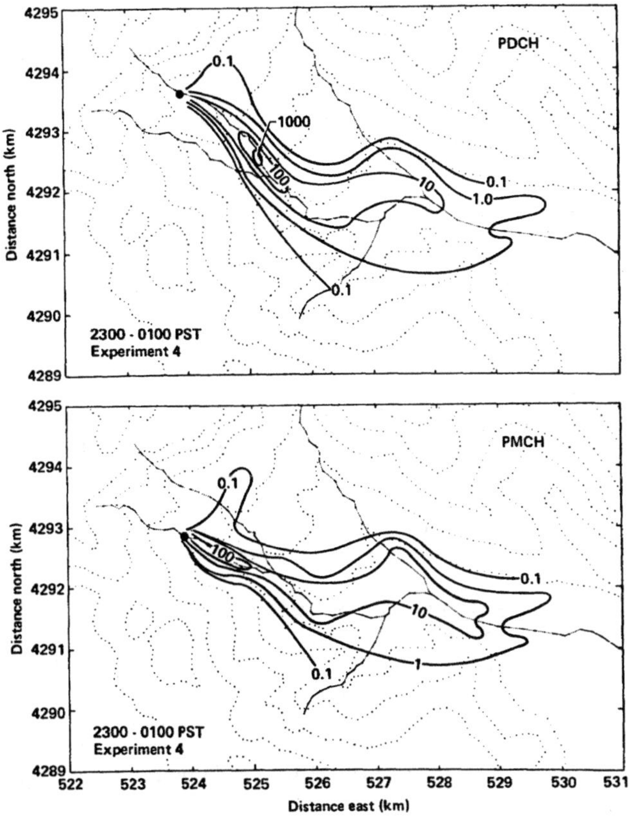

Figure 2. Surface concentration patterns of PDCH (top) and PMCH (bottom) during Experiment 4. Units are ppt and are averaged over the first 2 hours after initiation of the release.

after the initiation of the release. This figure illustrates a typical (smoothed) pattern seen for the perfluorocarbon releases during the 5 experiments. A complete summary of the experiments has been reported by Gudiksen, et al. (1983).

Results from these experiments and the SF_6 and heavy methane releases have been utilized to evaluate a three-dimensional mass-consistent diagnostic wind field model (MATHEW) and a particle-in-cell transport and diffusion model (ADPIC) for simulating pollutant behavior in drainage flows. These models have been described by Sherman (1978) and Lange (1978). MATHEW provides ADPIC with hourly averaged wind fields which are used to compute pollutant transport by the mean winds. ADPIC computes diffusion velocities based on empirically-determined eddy viscosity coefficients and the pollutant concentration gradients. The pollutant is represented by a large number of Lagrangian "marker" particles which are transported by the combined mean and turbulent velocities. The pollutant concentrations are obtained by counting the particles within each grid cell.

To date, the models have been run using the data from Experiments 2 and 4 of the September 1980 series in the Anderson Creek valley. For this study, the calculational domain was 7 by 10 km extending 1100 m above the lowest point in the topography. Individual cells were 250 by 250 by 50 m in the x, y and z directions, respectively. Using the hourly averaged meteorological data from the surface stations, tethersondes, and acoustic sounders, the models computed the concentration distributions of all the gaseous tracers utilized in this experiment. As an example of the results, Fig. 3 shows the computed surface concentration pattern of the PDCH tracer released within the Gunning Creek drainage area. This pattern may be compared directly with the measured distribution shown at the top of Fig. 2. Generally, results of the calculations agree reasonably well with the observed concentration patterns. Ratios of measured to computed concentrations for Experiments 2 and 4 were within a factor of 5 for 40 to 60 percent of the samples as shown in Fig. 4. However, agreement within a factor of two occurred over a range of 12 to 32 percent of the samples.

Previous model comparison studies with measured concentration data are summarized and shown by the curve in the upper left corner of Fig. 4. These studies were conducted in areas with rolling terrain considerably less complex than the Geysers area. For these studies the ratios of measured to computed concentration values by the MATHEW/ADPIC models were within a factor of 2 about 60% and a factor of 5 about 80% of the time. Thus, complexities introduced by the rough

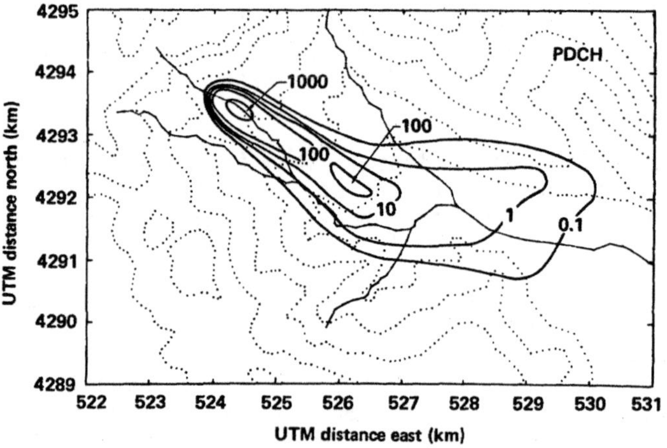

Figure 3. MATHEW/ADPIC calculated concentration pattern of PDCH during Experiment 4.

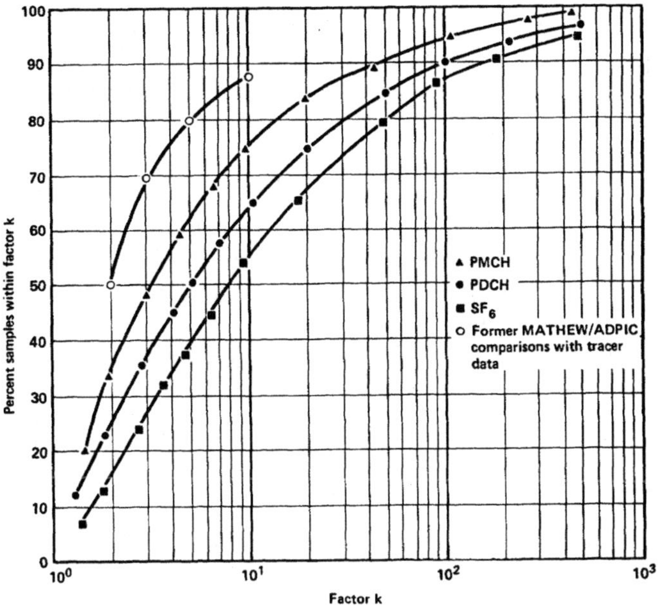

Figure 4. Percent of MATHEW/ADPIC calculations within a factor k of measurements.

terrain coupled with the nightime stable boundary layer tended to noticibly degrade the comparisons.

Variability of Tracer Concentration Measurements

One aspect of transport and dispersion characteristics expected to emerge from a stable boundary layer coupled to complex terrain is the variability in both space and time of measured tracer concentration values, even if the samplers are relatively closely spaced. Although the 1980 Geysers tracer releases were not designed for such a study, three groupings (clusters) of samplers provided an opportunity to investigate the variability of tracer measurements made at samplers located from 0.9 to approximately 6.0 km from the release points. Spacing between these clustered samplers ranged from 100 to 600 m. Fig. 5 shows the relationship of the 3 sampler clusters (C1, C2, C3) chosen for this study to the tracer release points for PMCH, PDCH and SF_6. Table 1 lists average spacing of samplers within each cluster and the average distance from each cluster to the respective tracer release points. Contour intervals shown in the enlargements are 25 m. For purposes of comparison with Gaussian diffusion estimates, σ_y is shown for a stable atmosphere (that includes meander) for distances downwind equal to the distance of the cluster locations from the tracer release points. The maximum difference in elevation between samplers within a cluster is also given in Table 1.

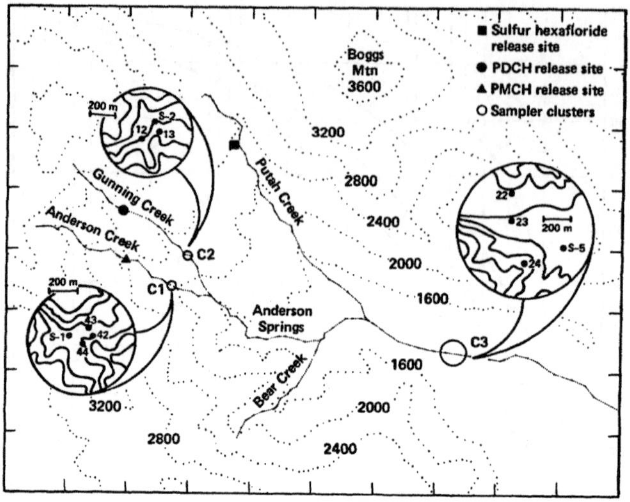

Figure 5. Location of sampler clusters.

Table 1. Relationships of samplers within the clusters and distance of the clusters from the release points.

Cluster #	Average distance from tracer release points (km)			Average distance between samplers (m)	σ_y (m)	Maximum difference in elevation between samplers (m)
	PMCH	PDCH	SF_6			
C1	0.9	--	--	150	120	12
C2	--	1.4	--	150	150	20
C3	5.9	5.9	6.0	440	650	40

For the first four experiments ratios between equivalent two hour average concentration measurements were calculated for every pair of samplers within each cluster. These ratios were expressed as factors for C1 (PMCH), C2 (PDCH) and for the three tracers measured in C3 (PMCH, PDCH, SF_6). Figure 6 shows the results of these comparisons where numbers in parentheses above each column of points indicates the number of comparisons for each group. Results depicted in Fig. 6 show that the variation among sampler values increases with distance from the tracer release points from almost an order of magnitude to about two orders of magnitude (with the exception of one point in C3). Normally a greater variation is expected near the tracer release points since the concentration pattern is not as diffuse and is therefore subject to larger concentration gradients.

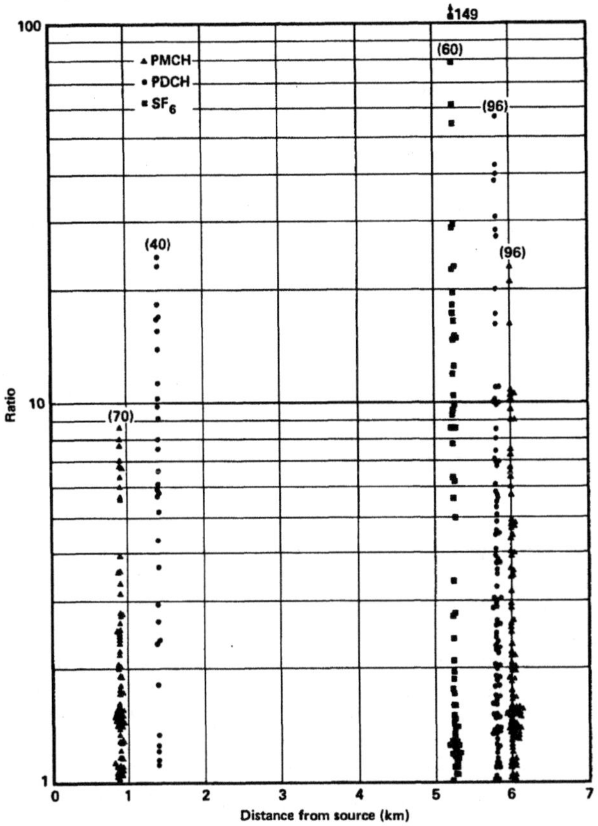

Figure 6. Ratios of 2 hour average concentration measurements within each sampler cluster.

To further illustrate the variation ratios for each 2 hourly concentration measurement to the lowest 2 hourly value measured within a cluster were calculated. Four 2 hourly sets of ratios were produced for each tracer for each experimental. Nightly 8 hour average concentration values and ratios were also calculated, as well as 32 hour averages and ratios for the combined four experiments. Results of these calculations, depicted as the maximum variation between sampler values within a cluster for the 2, 8 and 32 hourly averages are shown in Fig. 7 for C1 and C2. For C1, during the first experiment, variations between the four samplers ranged from about 1.2 to 8 and variations of sampler values in C2 ranged from 17 to 21. With the exception of experiment 3, sampler values in C2 varied by almost an order of magnitude over values in C1. Nightly 8 hour ratios were less than the 2 hour ratios for C1 except for experiment 1. In this case the largest 2 hr ratio occurred during a period that highest concentrations

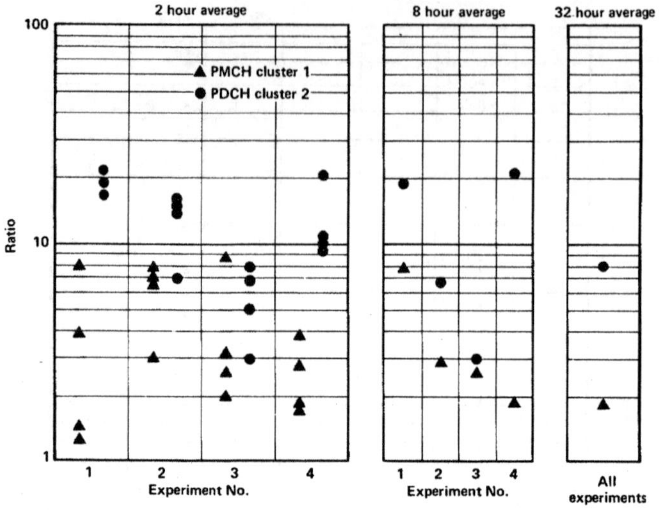

Figure 7. Ratios of sampler measurements in clusters 1 and 2 for 2, 8 and 32 hour averages.

(which could range over several orders of magnitude during the night) were measured. This is the reason that, with the exception of experiment 3, averaging did not significantly improve the ratios in C2. Experiment 3, which was not subject to strong drainage winds is a special case and will be discussed later in this report. Variations between sampler values averaged over the four experiments (32 hours) reduced to less than a factor of 2 for C1 but were a factor of 8 for C2. This points to a systematic bias in the sampler locations chosen to represent C2 although the samplers were relatively closely spaced.

Figure 8 shows comparisons of ratios for concentrations measured in C3 for the three tracers. Again the large variation (generally factors of 10 to 40) between concentration values measured by samplers located relatively close to each other (440m average for C3) is striking, particularly since the cluster is from 5.25 to approximately 6 km from the tracer release points. PMCH, released in Anderson Creek, shows the least variation for 2 hour averages and SF_6 released in Putah Creek shows the greatest variability except for experiment 3. Averaging over each night (8 hours) reduced the maximum ratios by a factor of 5 to 10 with minimal additional reduction gained by averaging over the four experimental nights.

It is difficult to determine specific reasons for the observed variations between sampler values within the clusters chosen for this study; however, it is apparent that both a systematic and a random component exists. In most cases temporal averaging for 8 hours reduced the ratios, with a slight additional reduction gained by averaging

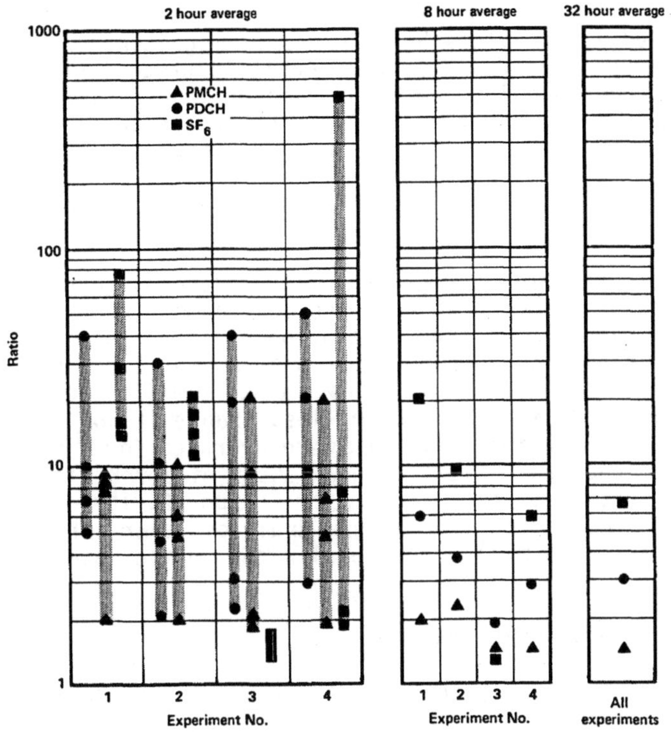

Figure 8. Ratios of sampler measurements in cluster 3 for 2, 8 and 32 hour averages.

over the four experiments. The natural variability associated with the transport and diffusion processes is expected to be a random component and should diminish with averaging. The systematic component (which contributes to the representativeness of the measurements) is attributed to a combination of the complex meteorological processes involved, caused in this case by a stable nighttime boundary layer coupled with underlying complex terrain. Although the samplers in each cluster were closely spaced, small differences in elevation or surrounding terrain features can significantly influence concentration patterns developed under stable atmospheric conditions. As constant density layers develop and flow over one another the lowest layers are influenced and guided by the terrain features. For example, there is evidence from the tracer concentration pattern analysis that the flow near C3 favored the north side of the outflow region as the flow exits the valley. This feature of the flow is observed in the low concentrations consistently measured by Sampler 24. Samplers 22 and S-5 (in the floor of the valley) were the remaining samplers that occassionally reported the lowest 2 hour average concentration measurements.

Although less than 60 m from sampler 13, Sampler 12 in C2 consistently measured lower values by an order of magnitude or greater. In this case sampler 13 and S-2 were probably located in the drainage layer flowing down Gunning Creek and sampler 12 was either shielded from the drainage flow by terrain features or was above the drainage layer. On the other hand all the samplers located in C1 appear to be located in the drainage flow down Anderson Creek since the maximum ratios shown in Figure 7 reduce to factors of 2 or 3 for 8 hour averages.

During experiment 3 the general meteorological conditions were not favorable for strong drainage wind development which resulted in weak drainage winds developing several hours after the tracers were released. Comparisons of 2 hour ratios between samplers shown in Figures 7 and 8 for experiment 3 show values similiar to the other nights for PDCH and PMCH while being less than a factor of 2 for SF_6. When the concentration values were averaged for 8 hours the maximum ratios reduced to less than a factor of 3 for all tracers measured in the three clusters. The major reason attributed to this reduction was the development of a weak drainage layer influenced by relatively strong (∿8 m/s) upper level winds that existed during experiment 3.

Variability of Model Calculations

The large variability that can exist between average concentration values measured at samplers located relatively close to each other poses a difficult problem for comparing calculated model point values to individual sampler point values. One question that immediately surfaces is, can the model produce similar variabilities at these sampler locations? Samplers in cluster 3 were chosen for comparison with model calculations using the first six hours (three 2 hourly averages) for experiments 2 and 4. Results of these calculations and their associated variabilities were compared to equivalent measurements (Fig. 9). The most surprising element shown in this figure is the large variability shown by the model calculations. Resolution of the model was 250 m compared to the average spacing of 440 m for the sampler locations.

Although model calculations and measurements show equivalent variabilities, comparisons between individual 2 hourly average values differed by more than an order of magnitude. Similiar variabilities between model calculations and measurements are also shown for 6 hour and 12 hour averaging times. Elements in the MATHEW/ADPIC model that contribute to the observed variability in the calculated concentration values are the model terrain and hourly changing wind field. When the measurements and calculated values were spacially averaged over the cluster and temporally averaged for 6 hours for each experiment there was good agreement between the model calculations and measurements (Table 2). With the exception of PDCH for Experiment 2 the comparisons were well within a factor of 2.

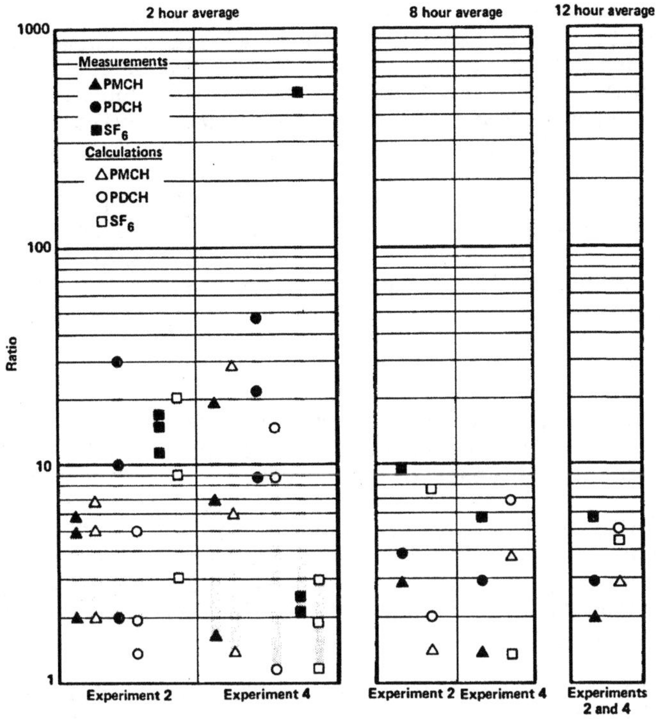

Figure 9. Ratios of sampler measurements compared with ratios of model calculations in cluster 3.

Table 2. Comparison of measured and calculated six hourly concentration values averaged over samplers in cluster 3.

Experiment	Tracer	Measured	Calculated
2	PMCH	4.3	2.5
	PDCH	6.6	0.7
	SF_6	161.0	134.0
4	PMCH	5.1	9.8
	PDCH	8.0	6.6
	SF_6	35.0	38.5

SUMMARY, CONCLUSIONS AND RECOMMENDATIONS

The ASCOT 1980 field experiments in The Geysers area of northern California have been briefly described and the measurements have been used to investigate the spacial variability that exists between concentration measurements made at samplers located relatively close to each other. Examples, using samplers in three clusters, showed variations of more than an order of magnitude for 2 hour average concentrations measured, in some cases, few tens of meters apart and within a kilometer of the tracer release points. These large variations, expressed as ratios to the smallest measured value, persisted and in some cases were larger for samplers located up to 6 km from the tracer release points and spaced a few hundred meters apart. Averaging concentration values for 8 hours tended to reduce the ratios; however, averages over the four experiments produced only a slight additional reduction. For one experiment, when weak drainage winds existed in the lower boundary layer and stratified drainage flows did not fully develop, ratios of 8 hour average concentrations values were less than a factor of 3 for measurements in all three clusters.

A systematic and a random component of the variation between sampler values is suggested to separate the natural variability of the atmosphere from the influence of the measurement environment. Although the ASCOT experiments were not designed to determine these components, there is evidence from this study that the systematic component can be large between relatively closely-spaced samplers. The combination of complex terrain, coupled with a stratified lower boundary layer, appears to contribute significantly to the observed systematic differences between samplers.

Model calculations for two experiments showed the same order of magnitude variability between sampler locations for 2, 6, and 12 hour averages. This model variability was surprising since the model resolution was either less than or slightly greater than the spacing between samplers. The major contributors to the variability in the model calculations were attributed to the hourly changing windfields developed from measurements and the terrain modeled as building blocks. Although comparisons between individual 2 hour measurements and model calculated values varied over an order of magnitude, they agreed for two experiments within less than a factor of 2 when the measurements and calculations were averaged over space and time for each cluster for each experimental night.

Several implications are suggested by this study for making measurements of tracer or pollutant concentrations in complex terrain, particularly under nighttime drainage wind conditions. For example, care should be exercised in the placement of air quality monitoring stations in areas subject to these conditions. Comparisons of ratios measured by samplers in C3 showed that the variations are more a function of sampler placement than of tracer release points. Even 8 and 32 hour average concentration measurements made by closely spaced samplers can vary by an order of magnitude.

Spatially averaging meteorological data by remote sensing instrumentation has been shown to be effective for developing model input data and model testing (Porch and Lange, 1982). Comparison of the temporal spectrum of optical cross-wind sensors over paths ranging from a few hundred meters to several kilometers with tower mounted anemometers shows that averaging times greater than 1-2 hours are required at a point in complex terrain to produce variability similiar to that found in spatially averaged data over a few hundred meters. This increased variability in meteorological flow fields induced by complex terrain is reflected in the tracer variability. Since the spatial averaging feature of optical cross-wind sensors has helped to quantify the variability in the flow fields in complex terrain and provided data more compatible with the spatial scales of numerical models, a possible deployment of double-ended spectrographic systems for spatially averaged tracer concentrations should be explored. Laser or incoherent light detection systems for wavelengths affected and unaffected by tracer material should be developed and applied to quantifying the spatial variability of tracers in complex terrain, as well as providing improved input for numerical model evaluation.

To provide additional insight into the space and time variability expected in measurements due to both the underlying terrain and the associated meteorology an experimental design is suggested that uses clusters of tracer samplers along with the associated meteorological instrumentation. This is a different approach to the standard methods of arcs or random spacings for sampler placement which intercept the tracer material. Optimum spacing among samplers within a cluster and spacing between clusters can be determined by a combination of modeling and statistical studies. In addition, remote sensing instrumentation can be included to determine temporal and spacial relationships between point and space averaged measurements. These experiments can be designed and conducted in complex as well as simple terrain. Examples of objectives for a set of experiments designed in this manner are:

- Determination of systematic variations induced in concentration measurements by complex terrain.
- Determination of optimum spacial and temporal averages of concentration measurements for estimating representative measurements for model evaluation studies.
- Development of guidelines for locating air quality measurement stations in complex terrain.

Although a number of experiments would be required in both the simple and complex terrain areas, an initial set of experiments in each setting would help establish guidelines outlined by the above objectives and provide a framework for developing an understanding of the increased complexities introduced by the atmospheric boundary layer when it comes in contact with complex terrain.

ACKNOWLEDGMENTS

The authors wish to acknowledge Brent Marble, Illinois Institute of Technology who began the analysis for this study during his employment as a summer student at the Lawrence Livermore National Laboratory. Rolf Lange provided the MATHEW/ADPIC model calculations and William Porch included his insights on the use of optical remote sensing for estimating path averaged meteorological tracer concentration data. We also thank Ken Peterson for a critical review of the manuscript and Pam Drumtra for typing the manuscript.

This work was performed under the auspices of the U. S. Department of Energy by the Lawrence Livermore National Laboratory under contract No. W-7405-Eng-48.

REFERENCES

Cowan, G. A., Ott, D. G., Turkevich, A., Machta, L. Ferber, G. J., and Daly, N. R.: 1976, 'Heavy Methanes as Atmospheric Tracers', Science 191, 1048-1050.

Dickerson, M. H. and Gudiksen, P. H.: 1980, 'ASCOT FY-1980 Progress Report', University of California Lawrence Livermore National Laboratory, UCID-18878.

Dickerson, M. H. and Gudiksen, P. H.: 1984, 'Atmospheric Studies in Complex Terrain Technical Progress Report, FY1979 through FY1983', University of California Lawrence Livermore National Laboratory, UCID-19851.

Eberhard, W.: 1981, 'Field Measurements in Support of Dispersion Modeling in Complex Terrain - Annual Report 1980', NOAA Technical Memorandum ERL WPL-78, NOAA Wave Propagation Laboratory, Boulder, CO.

Ferber, G. J., Telegadas, K., Heffter, J. L., Dickson, C. R., Dietz, R. N., and Krey, P. W.: 1981, 'Demonstration of a Long Range Atmospheric Tracer System Using Perfluorocarbons', NOAA Technical Memorandum ERL ARL-101, NOAA Air Resources Laboratories, Silver Spring, MD.

Fosberg, M. A., and Lanham, L. M.: 1983, 'Above-Canopy Dispersion in Nighttime Downslope Flow', Presented at Seventh Conference on Fire and Forest Meteorology, American Meteorological Society, April 25-29, 1983, Ft. Collins, CO.

Fowler, M.: 1979, 'Methane Tracer System Development', Proceedings of the Atmospheric Tracers and Tracer Application Workshop, LA-8144-C, Los Alamos Scientific Laboratory, Los Alamos, New Mexico.

Gudiksen, P. H.: 1983, 'ASCOT Data from 1980 Field Measurement Program in the Anderson Creek Valley', California, University of California Lawrence Livermore National Laboratory, UCID-18874-80, Vol. 1, 2, and 3.

Gudiksen, P. H., Ferber, G. J., Fowler, M. M., Eberhard, W. L., Fosberg, M. A., and Knuth, W. R.: 1983, 'Field Studies of

Transport and Dispersion of Atmospheric Tracers in Nocturnal Drainage Flows', University of California Lawrence Livermore National Laboratory, UCRL-88931. To be published in Atmospheric Environment.
Gudiksen, P. H. and Dickerson, M. H.: 1984, 'Executive Summary: Atmospheric Studies in Complex Terrain Technical Progress Report, FY1979 through FY1983', University of California Lawrence Livermore National Laboratory, UCID-18878-83 Summary.
Lange, R.: 1978, 'ADPIC - A Three-Dimensional Particle-in-Cell Model for the Dispersal of Atmospheric Pollutants and its Comparison to Regional Tracer Studies', J. Appl. Meteor. 17, 320-329.
Lovelock, J. E., and Ferber, G. J.: 1982, 'Exotic Tracers for Atmospheric Studies', Atmospheric Environment 16, 1467-1471.
Porch, W. M. and Lange, R.: "Spatially Averaging Cross-Wind Sensors and Numerical Model Results for Nocturnal Drainage Winds in Complex Terrain', University of California Lawrence Livermore National Laboratory, UCRL-87797.
Sherman, C. A.: 1978, A Mass-Consistent Model for Wind Fields Over Complex Terrain', J. Appl. Meteor. 17, 312-319.

Questions, answers and comments

J.C.R. HUNT: I would be grateful if you could say:
a) At what height were the tracers measured;
b) Did the tracers mix throughout the depth of the drainage winds;
c) What was the relative magnitude of the drainage winds and the incident winds.

M.H. DICKERSON:
a) The surface tracer measurements were made approximately 1 m above ground. These are the measurements I used in this study. In addition, vertical profiles up to approximately 500 m were measured at two locations.
b) The vertical distribution of tracer within the drainage layer varied from one night to another. In general the tracer was not well mixed with, I believe, one exception where the ambient wind direction was opposite to the drainage flow direction.
c) For the five experimental nights the magnitude of the drainage winds was generally 1-2m/s and the ambient wind speed varied from 2-6m/s.

M. GRABER:
a) What type of LIDAR measurements did you make?
b) Did you include this LIDAR data as input to your dispersion model and, if yes, how?

M.H. DICKERSON:
a) LIDAR was used to measure cross-section concentrations of smoke released at the PMCH release point.
b) The LIDAR measurements were not used as input to the dispersion model. We have been testing the inclusion of optical anemometer measurements, in the wind field model input data.

THE WIND REGIME IN COASTAL AREAS WITH SPECIAL REFERENCE TO RESULTS OBTAINED FROM THE SWEDISH WIND ENERGY PROGRAM

Ulf Högström and Ann-Sofi Smedman-Högström
Department of Meteorology, Box 516,
S-751 20 Uppsala, Sweden

ABSTRACT

The paper describes various aspects of the wind regime in coastal areas as obtained from several experimental programs along the Baltic coast of Sweden. The studies include the change with distance inland of mean wind structure as well as the turbulence structure in various conditions. It is found that wind spectra are usually well described by local similarity in its high frequency part, even in complex terrain. The low frequency parts of the spectra show clear 'spectral lag effects'. The effects of a small slope, ca 5 m height change over 300 m travel distance, is clearly seen in some spectra, and it is shown that current 'flow over a hill theory' can be used to account for it. – The change of mean wind speed throughout the entire boundary layer as the wind passes a low but wooded island (Gotland), ca 30 km wide, has been studied in a series of relatively strong wind, near neutral cases. Some unexpected features are found; in particular the wind speed decreases more rapidly with distance inland than predicted by current numerical models. – Two cases with a low level jet are discussed in some detail. Arguments are presented for the phenomenon to be caused by frictional decoupling at the Latvian coast, about 200 km upwind - the jet being thus an analogy in space to the classical Blackadar nocturnal jet frequently observed in continental areas.

1 INTRODUCTION

The most obvious physical factors that are likely to influence the wind regime in coastal areas are: abrupt change in roughness in going from the sea inland or vice versa and a corresponding change in the surface heat flux. The first of these factors is always present (although, of course, the magnitude of the contrast may vary from site to site), the second factor being, however, inherently variable in time. As will be demonstrated later in this paper an additional factor, viz. the distance to an upwind coast is likely to play an important role in some, common

Boundary-Layer Meteorology 30 (1984) 351–373. 0006–8314/84/0304–0333$02.70

meteorological situations.

In situations when the land gets appreciably warmer than the sea surface during the day a classical sea breeze circulation may form, provided the general meteorological situation is favourable. In less ideal conditions, in particular when the synoptic wind is not weak, a complicated wind regime is likely to result. The sea breeze, pure or with superimposed synoptic wind has been the subject of a great many investigations. For an extensive review of this topic the reader is refered to the book by Atkinson (1981).

In this paper we will concentrate on other, sometimes less obvious aspects of the wind regime in coastal areas. The results have been obtained from several field projects that were set up for studying meteorological factors of importance for siting of large wind energy conver-

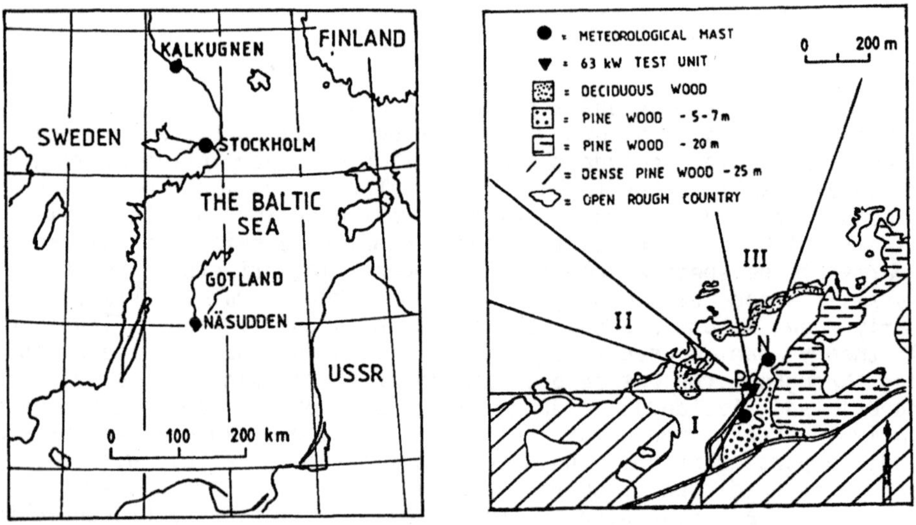

Figure 1(left). Map showing part of the Baltic Sea with adjacent land. The measurements reported in this article were carried out at Kalkugnen on the Swedish coast and at the island Gotland. Figure 2(right). Map of the Kalkugnen test site with the two meteorological masts ('N' and 'S' respectively) and the wind turbine ('P').

sion systems (WECS) in coastal areas in Sweden and were carried out during the years 1977-1982. The measurements were taken in the two areas indicated in Figure 1: the Kalkugnen test site on the Baltic coast, about 180 km north of Stockholm and the southern part of the island of Gotland, which is situated in the middle of the Baltic Sea. As explained in detail in the next Section the land surfaces rises gently from the sea in both these areas, so there are no large scale topography effects present. This means that the coastal effects on the wind regime can be studied more or less in isolation. As will be shown later there are, however, certain topographical effects even at these sites. The term 'wind regime' used losely so far refers to properties of the mean wind field, including its spatial and temporal variation, as well as the corresponding turbulence field.

2. MEASUREMENT SITES AND INSTRUMENTATION

At *Kalkugnen* (cf. Figure 1) a 63 kW test WECS was installed in 1977. Figure 2 is a close up map of the area. The terrain is slowly rising from sea level up to 5 m, and the ground is covered with coniferous forest generally, but the immediate surroundings of the test plant have been cleared.

In late 1977 a meteorological field program was started, and it ran continuously until 1980, when the nacelle was destroyed during a lift operation. The field program was designed to give basic data for load calculations and testing the performance of the power plant. The meteorological measurements comprise continuous wind and temperature profiles on two 42 m towers ('N' and 'S' in Figure 2) together with registrations of radiation and humidity. In addition, turbulence measurements were performed during three campaigns in 1979 and 1980.

As can be seen from Figure 2 the site is far from being "ideal". The roughness of the surface varies a great deal and the terrain is sloping towards the sea. It might be thought that such a complexity in local terrain characteristics would make it impossible to draw any general conclusions concerning the flow regime. As shown below this fear, is, however, ungrounded.

The field experiments at the island of *Gotland* (cf. Figure 1) can conveniently be divided into two groups:

1) An experiment aimed at mapping the boundary layer flow field over the southern half of the island, its direct wind energy application being a search for 'the most windy site on the island'. It took the form of a series of three campaigns in late 1978 with very dense pilot balloon releases, tracked with double theodolites, carried out simultaneously at 5 sites. As shown below this study illustrates in a nice way how the boundary layer flow is affected as it passes over a rather low (less than 80 m high) but relatively densely vegetated, ca. 30 km wide island. Figure 3 is a map of the island, showing the measurement sites and the roughness length characteristics.

2) As a result of the above mentioned field experiment the peninsula Näsudden, situated on the south-west coast of the island, cf. Figure 3, was chosen as the site for a 2.5 MW prototype WECS. Figure 4 is a detailed map of the peninsula, showing the position of the WECS and of the 145 m high meteorological tower, which was erected at the site especially for the wind energy program. The instrumentation in the mast comprises wind and temperature sensors at 7 levels with continuous recording. In addition, turbulence measurements with instruments at up to maximum three levels have been carried out in several campaigns in 1980 and in 1982. The turbulence instrument used is the MIUU instrument

(MIUU = Meteorological Institute, University of Uppsala, described in detail in Högström et al (1980) and in Högström (1982). It is a wind vane based three axial hot wire system with additional sensors for temperature and wet bulb temperature. The instrument has been carefully tested in a big wind tunnel and corrected for an aerodynamical error as described in Högström (1982). This instrument was also used in the Kalkugnen experiment. During the turbulence measurement campaigns at Näsudden there were also carried out double theodolite balloon measurements and radiosoundings.

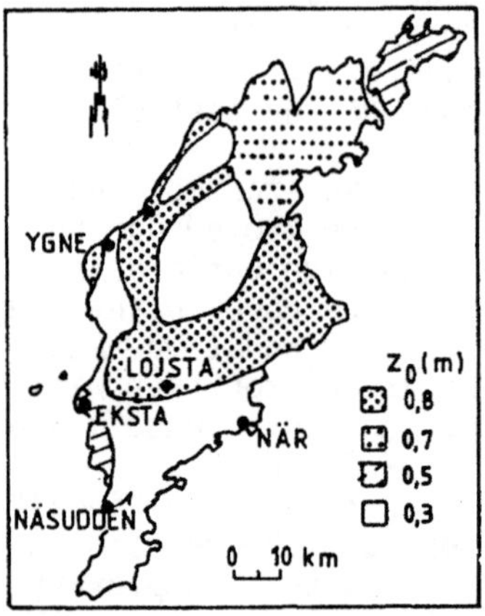

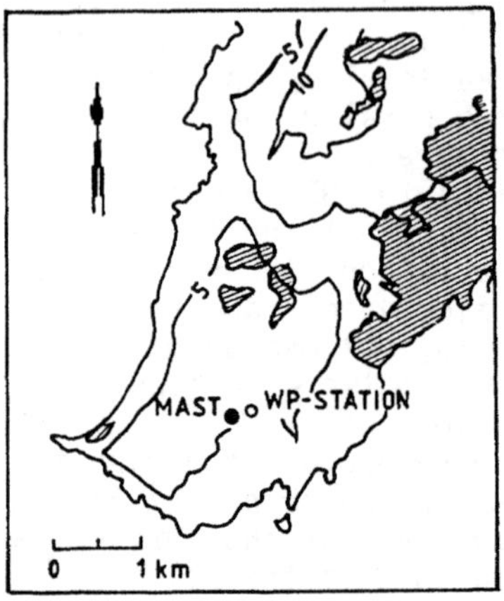

Figure 3(left). Map of Gotland with the PIBAL sites.
Figure 4(right). Map of the Näsudden site. Shaded areas: forest.

As seen from Figure 4 the Näsudden peninsula is generally quite flat, the land rising slowly (for a large sector at a rate of 1:200) to about 5 m above sea level. Most of the area is sheep grazing land, covered with scattered, 1-2 m high juniper bushes. As indicated in Figure 4 there are, however, also some pine groves of very limited extention. To the north-east there is an extended forest area.

3. THEORETICAL FRAMEWORK FOR THE ANALYSIS

When air flows from an area with certain characteristics in terms of roughness length and thermal properties to an area with different characteristics an internal boundary layer (IBL) forms. As is typically the case in the study of turbulent flows, no general theory exists that describes the physical characteristics of this internal boundary layer in terms of the various govering parameters involved. Few field experiments have so far been conducted with the specific aim in mind to study the detailed structure of the boundary layer that evolves during the passage of air over distances of several kilometers downwind of a discontinuity in roughness and/or thermal properties. The IBL field experiment most often quoted is that of Bradley (1968). This experiment, although of very good quality, is of limited use in full scale applications. Also the extensive Risö IBL studies (see Petersen et al, 1976, 1979 and 1980) refer to a very shallow layer and to small horizontal distances. As a contrast, the phenomenon has been simulated widely with the aid of various numerical PBL models. In Section 4 we will compare some predictions with the measurements from the Gotland PIBAL projects.

Schematically, an IBL has the following general structure. Close to the ground an equilibrium layer forms that grows only very slowly with distance from the discontinuity, 1:100 being an often quoted figure for the slope of the upper interface of the equilibrium layer. Above this interface there is a transition layer which is appreciably deeper – the slope of the upper boundary being of the order 1:10, the actual figure varying, however, a lot with stability.

Different properties of the flow vary in different ways with height within the IBL. The equilibrium layer represents the layer in which the turbulent flux of momentum can be considered roughly equal to its surface value, u_*^2. The transition layer on the other hand is characterised by rapid variation with height of the momentum flux. In the text below the local value of the momentum flux at a height z will be denoted by $u_{*\ell}^2(z)$. In contrast to the momentum flux the entire IBL wind profile is often observed to have shape characteristics typical of the 'new' surface, i.e. if the wind profile is being expressed in terms of a power law:

$$\overline{u}(z) = \overline{u}(z_1)\left(\frac{z}{z_1}\right)^{\alpha}, \quad z < h \tag{1}$$

where h is the height of the IBL, α depends (to the first approximation) only on the local roughness length z_0 and on the local stability and not on the distance from the discontinuity. This empirical result was made use of by Smedman-Högström and Högström (1978) to develop a practical method for the estimation of wind profiles from wind measurements at a low level. Some tests of the performance of this method against the Gotland PIBAL measurements are presented in Section 4.

When the ground is inclined to the horizontal, as is the case at a coast, an 'escarpment effect' may be observed. As pointed out by Jensen and Petersen (1978) the result found for flow over low hills and ridges can also be used to quantify the observed features of the wind profiles downwind of an escarpment or over gently sloping terrain. Over sloping terrain the flow can be treated as inviscid flow over an obstacle. An "inner layer" of strong shear can be defined in which variances and stress decrease with height, the thickness of which only depends on the hill length scale (L) and the surface roughness (z_0). Jackson and Hunt (1975) give the following expression for the height of this inner layer ℓ:

$$\ell/z_0 = \frac{1}{8}\left(\frac{L}{z_0}\right)^{09} \tag{2}$$

In the *inner layer* in neutral air the local wind profile is logarithmic and a maximum or an inflection point will occur at a height $\approx \ell$. The fractional speed-up (Δs) depends on the height- and length-scale of the hill but not on the shape of the slope

$$\Delta s = \frac{\Delta u}{u_0} = \frac{2h}{L} \quad (3)$$

u_0 = wind-speed upwind
Δu = the change in wind-speed

Velocity spectra in the inner layer have been treated by Panofsky et al (1982). They analyse Bradley's (1980) data from Black Mountain, Canberra and use the following normalization for neutral stratification

$$nS(n)/(-\overline{u'w'} \cdot u_*)^{\frac{2}{3}} = 0.3f^{-\frac{2}{3}} \quad (4)$$

where $-\overline{u'w'}$ is the local Reynolds stress and u_* is evaluated from the wind profile ($f = nz/\overline{u}$, the normalized frequency).

Over flat homogeneous terrain the nominator becomes u_*^2, which is the usual scaling, first given by Kaimal et al (1972).

In the *outer layer* the variation in shear stress is weak and the flow can be estimated with potential flow theory, with the surface at the top of the inner layer. The turbulence structure may be obtained by rapid-distortion theory (Britter et al, 1981). Up to now no appropriate scaling for velocity spectra has been found for this layer.

As shown in Section 6 certain aspects of the flow characteristics observed at Kalkugnen can be explained in terms of this theoretical framework.

It has been demonstrated by many authors during the last five year period or so that certain turbulence characteristics are in 'local equilibrium'. A closer examination reveals that this equilibrium relates to the high frequency parts of the spectra, where e.g. the *local* 'friction velocity' $u_{*\ell}$ is the relevant scaling velocity, while the low frequency parts of the spectra appear to bear remembrance of the upwind surface. This 'spectral lag effect' was first described and explained independently by Panofsky et al (1982) and Högström et al (1982). In a subsequent paper Höjstrup (1982) explored the spectral lag effect in quantitative terms and demonstrated very good agreement with measurements. As will be shown in Section 6 the spectral lag concept is very useful in explaining observed spectra in coastal environments.

In Section 5 a novel concept will be presented in order to explain a very pronounced low level jet observed at Näsudden in a case when the stratification over the Baltic Sea was stable. The mechanism can be described as an analogy in space to the well-known Blackadar (1957) nocturnal jet, i.e. frictional decoupling at an upwind coast initiates an inertial oscillation. Thus, after a traveltime of the order half a pendulum day a low level jet occurs. The hypothesis is strongly supported by the results of calculations with a two-dimensional higher-order closure model, as shown in Section 5. Although the idea of an analogy in space to the classical Blackadar (1957) nocturnal jet seems not to have been formulated explicitly earlier in the open literature, observations of a coastal low level jet, the origin of which is very likely to be attributable to the same mechanism has been presented by Sethu-Raman (1982). He has also observed the same phenomenon in measurements over the Bay of Bengal (Sethu-Raman, personal communication).

4. MODIFICATION OF THE MEAN AIR FLOW DUE TO INCREASED FRICTION

As mentioned briefly in Section 2 an intensive PIBAL measurement campaign was carried out at Gotland in late 1978. All the measuring data are available in a technical report from MIUU, Faxén (1980). Briefly pilot balloons were released from the 5 sites indicated on the map of Gotland, Figure 3, during three campaigns in the period September 12 to November 23. Each campaign lasted two or three days, and 15-30 balloons were released and tracked by double theodolites at every site on each day during the daylight period. Hourly mean wind profiles were computed for fixed times of the day for every site in order to facilitate comparisons. The method of doing this, which includes the use of a Gaussian weighting function, as well as all other details of the evaluation procedure is reported in Alexandersson and Bergström (1980). As shown in that report the mean wind profiles produced in this way are usually accurate to within ± 0.5 m/s.

The measurements from the last period, 22-23 November 1978 are of particular interest as they represent conditions with minimum thermal effects: the elevation of the sun at this time of the year at the latitude of 57°N is quite low, the period with daylight being only 7 hours; in addition the wind was relatively strong, as shown in the cross section in Figure 5. The figures give the measured wind speed at the sites

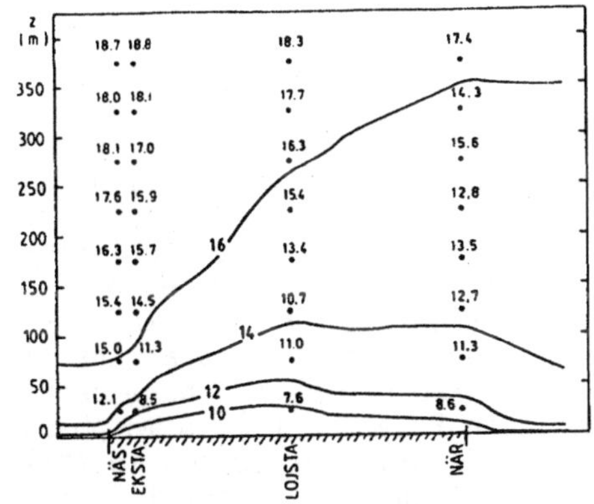

Figure 5. Cross section showing measured wind (the figures) on 23 November 1978 at 14.00 LST at the stations Näsudden, Eksta, Lojsta and När. The isolines are from a model simulation.

Näs, Eksta, Lojsta and När, cf. Figure 3. The lines give the corresponding predicted wind field and is obtained by a two-dimensional higher order closure model, see Enger (1983) for details. This situation has also been used as a test case for the evaluation of several numerical wind field models in an International Energy Agency project, see Pennell (1983). All the models tested give a fair agreement with the measurements, at least in a qualitative sense, i.e. they all show the typical retardation of the flow in the lower layers as the air passes over the island. The wind direction was westerly, so Figure 5 shows a cross section very nearly along the mean flow direction.

A closer look at the model results in relation to the observations do, however, reveal some interesting systematic deficiencies in the predictions. Thus, as can be seen from Figure 5, Enger's model underestimates the retardation at Eksta and Lojsta. This is also evident if one looks at the outcome of a comparison of the predictions with this model and the observations for the entire November test period (three days). For the station Lojsta the model underestimates the retardation in the height interval 75-175 m by more than 50 percent, the observed mean retardation for this height interval between Näs and Lojsta being 4.6 m/s and the corresponding predicted reduction 2.8 m/s. It is interesting to notice that the three-dimensional SIGMET model (see Pennell, 1981, Appendix A, Figure 5.83) gives very nearly the same underestimation for the chosen test time 23 November 1978 at 1300 LST. Thus, the deficiency of Enger's model to predict the retardation correctly is not likely to be due to it's lack of a third dimension. It rather seems as if the effective roughness length of the interior of the island is larger than the chosen values 0.5-0.8 m. This is noteworthy, as the topography is very gentle. In particular it is interesting to observe the very rapid retardation at the station Eksta, which is only 2 km from the coastline. The area at Eksta is quite flat but covered with a relatively dense, but not very high, pine forest.

From a wind energy point of view the predictions with the various numerical models for the Gotland November case are rather bad, considering that the energy output of a wind turbine is proportional to the cube of the wind speed. It is interesting then to see whether other approaches may yield better result. In Section 3 a model by the present authors (Smedman-Högström and Högström, 1978) was described. The model computes the wind profile in the lowest 200 m from measurements of the wind at standard anemometer level and from estimates of the roughness length of areas surrounding the station and an estimate of stability. All the profiles from the pilot balloon measurements at the five stations on Gotland (i.e. from all the three campaigns) have been evaluated for the value of the exponent α in Equation (1). Atmospheric stability was estimated from a Pasquill-Gifford-Turner scheme in three classes: 1, unstable; 2, near neutral; 3, slightly or moderately stable. The result of the analysis is shown in Figure 6, the curves being from the original paper (Smedman-Högström and Högström, 1978). This result strongly supports the use of the model described in that paper for regional wind energy mapping in areas with gentle topography. In Smedman (1980) the model is used for an estimation of the climatological frequency distributions of wind speeds at typical hub height for large WECS.

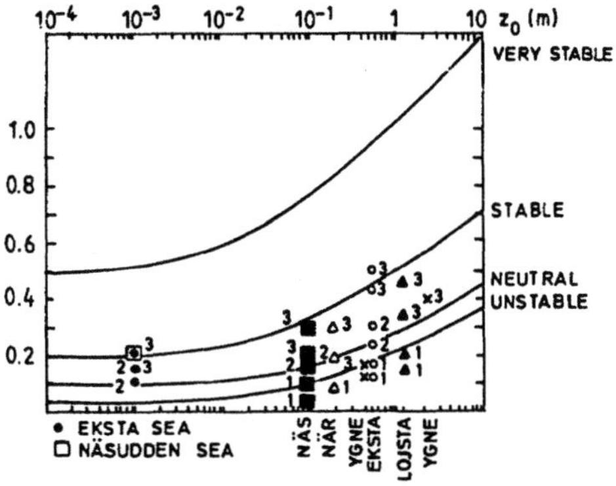

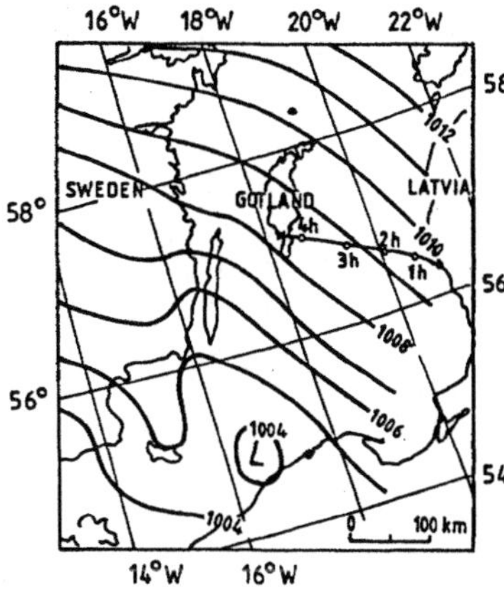

Figure 6(left). The exponent α of Equation (1) as evaluated from the Gotland PIBAL data. 'Eksta sea' and 'Näsudden sea' releate to the upper boundary layer part in the profiles from that sites. The figures relate to stability, according to the scheme: 1, unstable; 2, neutral and 3, slightly or moderately stable. The curves are from Smedman-Högström and Högström (1978). Figure 7(right). Map showing the area of the low level jet case study of section 5.1, with Gotland in the middle of the Baltic Sea. The isobars are for 29 May 1980 0100 h. Also included is a trajectory from the Latvian coast to Näsudden (the cross) at Gotland.

5. LOW LEVEL JETS OBSERVED AT NÄSUDDEN

Two, very different cases, where low level wind maxima were observed will be discussed.

5.1 The May 1980 case

During a period of several days at the end of May 1980 the synoptic situation over the Baltic Sea was characterized by a very steady south-easterly flow (see Figure 7). As the sea surface was appreciably colder than the approach flow - the sea surface temperature was ~ 6 °C and the daytime air temperature near the Latvian coast was above 20 °C - the lowest layers of the atmosphere over the Baltic Sea attained a strongly stable stratification.

This situation brought about a remarkable wind regime at Näsudden, characterized by a strong low level jet present most of the time. As indicated below the observed jet phenomenon is likely to be the result of two independent processes. The entire situation will be the subject of a separate, forthcomming paper. Here we will concentrate on one of the mechanisms, the frictional decoupling effect, introduced in Section 3.

A summary of the observed meteorological characteristics at Näsudden is given in Figures 8 and 9. Further information is contained in Smedman and Högström (1983) (called S-H, 1983, below). As seen from Figure 8 there is a very strong inversion in the lowest few hundred meters during the night. During the day a shallow ~ 90 m deep convective boundary layer is observed at Näsudden (cf. Fig 2 in S-H, 1983), but over the sea the inversion is no doubt ground based all the time. As shown in S-H (1983) the flow at 135 m is only intermittently turbulent, and as can be seen from Figure 2 of S-H (1983) $|\partial \tau / \partial z|$ is always very small at that level. Figure 9 shows the detailed structure of the wind field throughout the lowest 2000 m during the period 28 May 1400 LST to 29 May 0700 LST. As seen from Figure 9 there is always a strong wind gradient in the lowest ~ 200 m, and at times there is a pronounced low level jet present at about that height - see also Figure 8.

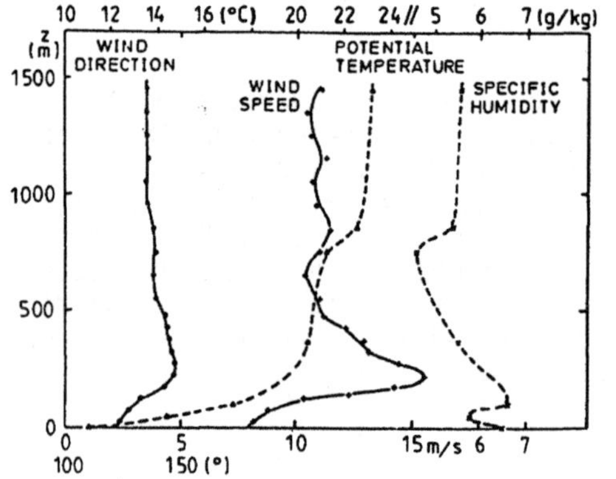

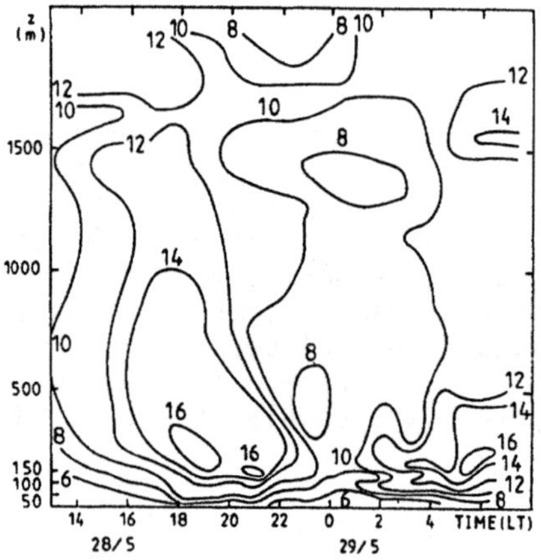

Figure 8(left). Pilot balloon and radiosonde data from Näsudden for 29 May 1980 0500 h, showing wind direction, wind-speed, potential temperature and specific humidity. Figure 9(right). Wind-speed cross section at Näsudden for the period 28 May 1400 h to 30 May 1200 h. Below 145 m wind data are taken from anemometer measurements in a mast. Above that height the wind data have been obtained from a series of double theodolite pilot balloon measurements.

Figure 10 is a wind hodograph for 145 m for the period 28 May 1230 h to 29 May 0600 h. Also shown in the diagram is the mean surface geostrophic wind for the period. The variation of this vector was remarkably small: for the period shown in Figure 10 the direction of the surface geostrophic wind varied between 140° and 150°, and the speed varied between 10-15 m/s, being 13-15 m/s for most of the time.

Figure 10 gives the impression that two independent mechanisms may be active in producing the wind hodograph: one which gives the maximum at 2030 LST and another that produces the maximum at 0500 LST. In fact, there are strong indications that the first, early evening maximum, is of fairly local origin. This is manifested in a local pressure gradient (on the scale of about 20 km) over southern Gotland. In contrast, during the morning hours, i.e. at the time of the second maximum, there is no local pressure gradient on this scale present. Here we introduce the hypothesis of frictional decoupling at the upwind coast, i.e. the Latvian coast, ca. 200 km from Näsudden (cf. Figure 7).

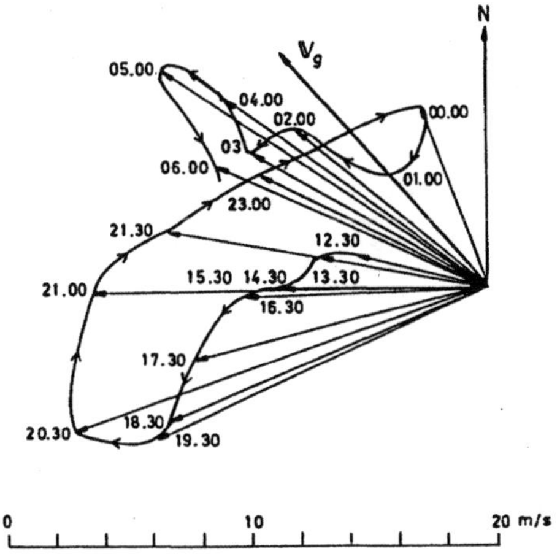

Figure 10. Wind hodograph for the 145 m wind at Näsudden for the period 28 May 1230 h to 29 May 0600 h. Also included is the mean surface geostrophic wind for the period.

In order to support that hypothesis the two-dimensional higher-order closure boundary layer model of Enger (1983) was employed for simulating the case.

The horizontal grid net of the model was placed from the Gotland area, across the eastern parts of the Baltic Sea and inlands over the Baltic states to a distance of 46 km from the coast. The gridpoint distance in the horizontal was 4 km and the total number of gridpoints was 60 in the horizontal and 30 in the vertical direction, ranging from 0.5 m to 2000 m. The roughness lengths used over the sea and over land were 0.00025 m and 0.2 m respectively. The time step was 120 s. The input data to the model is the geostrophic wind and the temperature at screen level. The geostrophic winds were estimated from the analysed pressure fields every third hour and linear interpolation was used in between these values. To include the thermal wind, the geostrophic wind at the 850 mbar level was evaluated. Initially the temperature was set equal to 13°C at the 0.5 m height for every gridpoint in the horizontal and with the stratification stable throughout the boundary layer. The simulation was started at midnight, and after one hour simulated time, allowing for the model to calculate a stable initial field, the geostrophic wind and the screen level temperatures were allowed to vary according to observations. The temperature observation at Liepaja was taken to represent the Baltic states, while the observed sea surface temperature was used over the Baltic Sea.

Comparison of the predicted and measured profiles for 07 LST (Figures 11 and 12) shows remarkable agreement. This is also true for the wind direction.

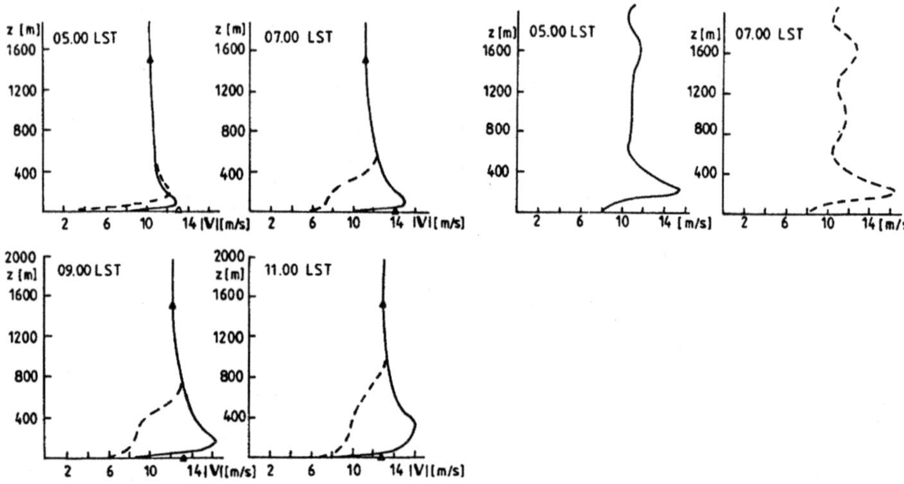

Figure 11(left). Wind profiles simulated with Enger's model. Hatched line: profile over Latvia; full line: profile at Näsudden. Figure 12 (right). Measured wind profiles at Näsudden 29 May at 0500 and 0700 LST.

Figure 11 shows a gradual increase of the wind maximum during the first hours of simulation. This is believed to be an effect of the initialization. From physical reasoning we belive the jet to be permanently present over the sea outside Gotland throughout this fairly stationary meteorological situation. The measurements at Näsudden during the next night, i.e. 29-30 May, indicate that the sequence of events of the previous night repeat themselves.

It is not the place here to go into a detailed discussion of the possible cause of the local effects apparent in the intervening daytime period. Suffice it here to say that in the evening the undisturbed flow over the sea can make itself evident in the measurements at Näsudden.

5.2 The October 1980 case

Another low level jet, albeit not at all so strong as in the May case, was observed during a measuring campaign during the period 8-10 October

1980 at Näsudden. The synoptic-situation of this period was very different from that of the May period discussed above. The surface temperature of the Baltic Sea was not very much different from that over land, and the stratification was fairly close to neutral. Figure 13 shows two, very different, mean wind profiles (presented as relative wind speeds $\overline{u}(z)/u_{120}$). Particularly noteworthy is the wind maximum encountered in the class B profile. This maximum is evident in each of the 14 profiles that make up this data subset. The height of the maximum fluctuates in the interval 75-100 m. The profile for class B differs very much from the class A profile, as seen in Figure 13. Further analysis reveals that the class B profiles are cases with very long over water trajectories (several hundred kilometers), whereas the class A profiles occured during an intervening period of time when the wind turned to a more southerly and even south-easterly direction, so that the air had to pass ca. 10 km over land (the southern tip of Gotland).

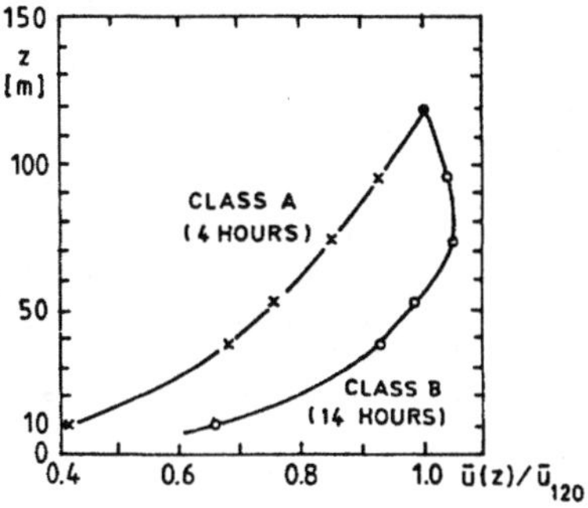

Figure 13. Normalized wind profiles, $\overline{u}(z)/\overline{u}_{120}$), for the October 1980 measurements at Näsudden.

The reason for the appearance of this low level maximum is not clear. A wind maximum at a low level can result when the air flows up a slope. Application of simple 'flow over a hill' theory (Jackson and Hunt, 1975) shows clearly that the insignificant slope encountered at Näsudden (0.5 percent) cannot give a measurable effect. A 'low level jet' can also arise when there are big horizontal temperature contrasts. A study of weather maps for the particular time period shows that the horizontal temperature contrasts were exceedingly small, so that this explanation must also be rejected. This is also substantiated from an evaluation of the geostrophic wind in the area from available surface and 850 mbar maps: although the geostrophic wind was variable in time, there is no significant variation with height, i.e. no appreciable thermal wind in the layer below 1500 m. It is possible that the frictional decoupling effect is responsible for the 'October jet' also, but simulations with Enger's (1983) 2D-model has so far failed to substantiate this. The conclusion is, however, not certain, as the result of the simulation in this case is probably fairly sensitive to assumptions about the upwind conditions. The transport distance is likely to be very large, ca. 400 km, and it cannot be ruled out that there are mesoscale pattern in the sea surface temperature that may have affected the boundary layer flow. It is, however, suggestive that the probable over water transport time is about 7 hours, which happens to be half a pendulum day, which corresponds to the time for the maximum wind speed in an inertial oscillation.

6. TURBULENT CHARACTERISTICS

6.1 Kalkugnen

Measurements with the MIUU turbulence instrument were performed at the Kalkugnen site in three campaigns during the period November 1979 to April 1980. Various alternative measuring heights were used: 14 m, 25 m and 36 m in the N tower (cf. Figure 2) and 25 m in the S tower. In all 10 one hour runs were performed. In six of these two instruments were sampled simultaneously, there being only one instrument in the remaining four runs. Data were sampled at 20 Hz. A 1024-point Fast Fourier Transform (Statistical Analysis System, SAS) was used to compute spectra in two separate, slightly overlapping bands. The stratification was near neutral in six of the runs, unstable in three and stable in one of the runs. In the analysis the data have been divided in three categories, I-III, according to wind direction, as indicated in Figure 2. As seen from that Figure the terrain at Kalkugnen is very heterogeneous. However, when the data were stratified in this way a fairly coherent picture emerges.

a. *Sector I 210° - 270°*

When the wind is blowing from the sector 210°-270° the fetch over the clearing is about 200-400 m (see Figure 2), and farther away there is an extended area with dense pine wood, with tree height 25 m.

Figure 14 shows a typical wind profile in that sector, with the upper part representing the pine wood ($z_0 \approx 1.5$) and the lower part the rough open country over a distance of 200 m ($z_0 \approx 0.4$). The height of the new internal boundary layer has been calculated with the following expression from Elliott (1958):

$$h/z_0 = a(x/z_0)^{0.8}$$
$$a = 0.75 + 0.03 \ln z_0'/z_0 \qquad (5)$$

where z_0 = roughness length upwind
z_0' = roughness length downwind
x = distance to the roughness change

This gives $h \approx 25$ m, in agreement with the wind profile in Figure 14.

Following the suggestions by Panofsky et al (1982) and Högström et al (1982) the velocity spectra have been normalized with the local measured ($-\overline{u'w'}$)-values i.e. with $u_{*\ell}^2$ and plotted against the reduced frequency f: $n \cdot z/\overline{u}$. The measuring heights varied between 14 m, 25 m and 36 m at the north tower and was 25 m at the south tower. Most of

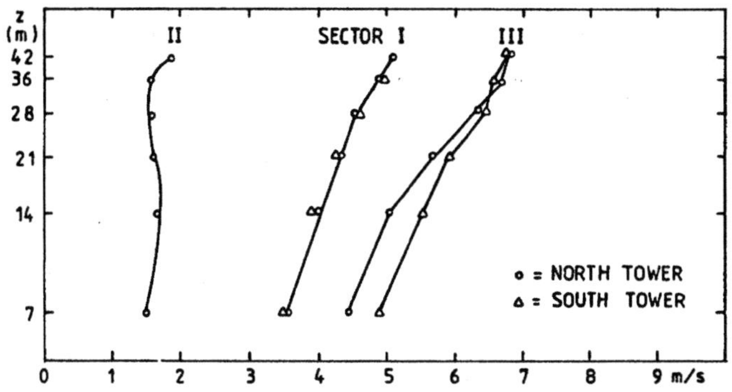

Figure 14. Example of wind profiles at Kalkugnen measured in the N and S tower respectively in the case of winds comming from Sector I - III.

the runs were performed in neutral air, but a few also in slightly unstable stratification. The non-neutral cases have been scaled with $\phi_\varepsilon^{2/3}$ where ϕ_ε = normalized dissipation rate Kaimal et al (1972).

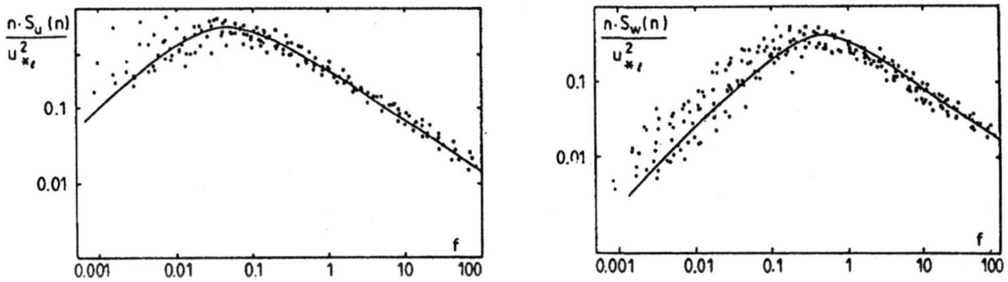

Figures 15 and 16. Normalized spectra for the longitudinal and vertical wind component at Kalkugnen. Wind from sector I (210°-270°).

Figure 15 shows the normalized spectra of the longitudinal component together with the curve for a uniform site (Kansas). The agreement is good over the whole frequency range, which is somewhat surprising. One might expect to find much more energy at lower frequencies, which are likely to bear a memory of the much rougher upwind area (pine wood). The vertical component is given in Figure 16. The measured spectral points scatter around the Kansas curve with no systematic deviation.

The variancies of the three wind components scaled with the local values of $u_{*\ell}^2 = (-\overline{u'w'})$ are slightly higher than the expected values for uniform terrain (Dutton et al, 1979). This could, however, be a stability effect. The result is given in Table 1.

Table 1. Normalized variances measured at Kalkugnen compared to expected values for uniform terrain. (Dutton et al, 1979). Note that $u_{*\ell}$ is the 'local friction velocity'. Measuring height 25 m.

	Kalkugnen Sector I	Uniform terrain	Kalkugnen Sector II
$\sigma_u^2/u_{*\ell}^2$	6.00	5.76	17.56
$\sigma_v^2/u_{*\ell}^2$	5.52	3.84	31.92
$\sigma_w^2/u_{*\ell}^2$	1.93	1.61	7.23

b. *Sector II 290° - 310°*

In this sector the wind is blowing from the sea and the fetch over land is about 250 m. The ground is very rough with bushes and scattered deciduous trees. The height of the ground above sea level at the two towers is 5 m. The wind profile in Figure 14 is, except for the upper part a typical example of a wind profile over sloping terrain. There is an "inner layer" with logarithmic wind profile up to ≈ 14 m and then an upper layer with weak shear. The height of the inner layer can be calculated according to expression (2) with $z_0 = 0.2$ m, which gives ℓ=15 m. The upper part of the wind profile probably represents a tongue of land situated about 1000 m upwind.

Unfortunately the turbulence measurements are taken in the outer layer (25 m). As mentioned in Section 3 there is no appropriate way of normalizing spectra in the outer region. When scaled with their local $-\overline{u'w'}$-values the variances of the wind components have about four times their values for uniform terrain (see Table 1).

c. *Sector III 350° - 20°*

Most of the turbulence runs were carried out with wind in this sector.

The terrain is, as stated above, gently sloping from sea level up to 5 m at the north tower and the distance over the rough clearing is 200-350 m. From the typical wind profile measured at the north tower (see Figure 14) one can distinguish three boundary layers. The upper part of the profile is related to the sea and the lowest part to the new boundary layer which builds up over the rough open surface ($z_0 \approx 0.4$). Between these two layers there is a region in the profile with a slope corresponding to a very high roughness ($z_0' \approx$ 1-2 m) which is difficult to explain.

In the wind profiles from the north tower one cannot find any maximum in wind speed at a theoretical interface between the inner and outer layer. Probably that effect is masked by the variation in roughness and zero displacement. However, at the south tower there is often a weak maximum or an inflection point about a height of 25 m (see Figure 14).

From expression (2) the height (ℓ) of the inner layer can be calculated for the two towers

$$\ell_N \approx 15 \text{ m}$$

$$\ell_S \approx 24 \text{ m}$$

Accordingly the velocity spectra measured at a height of 14 m will be treated as inner layer spectra and normalized in the way proposed by Panofsky et al (1982) and given in expression (4). Figures 17 and 18 show the result for the longitudinal and the vertical wind components respectively. In the inertial subrange the agreement between the curve for uniform terrain and the measurements is surprisingly good. The lower spectral density in the low frequency range for the u-spectra is very likely to have been caused by the spectral lag effect.

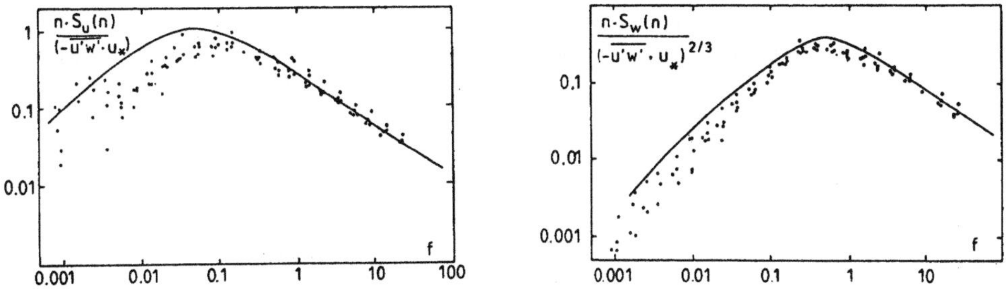

Figures 17 and 18. Normalized spectra of the longitudinal and vertical wind component at Kalkugnen, 14 m, with wind from sector III (350°-20°).

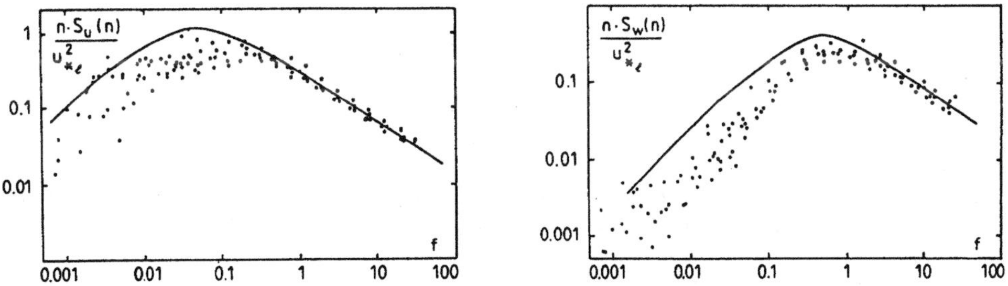

Figures 19 and 20. As Figures 17 and 18, but for 25 and 36 m.

As stated before, there are no appropriate scaling for the velocity spectra in the outer region. However, in this case the big change in roughness around the site seems to dominate over the "sloping terrain effect". Therefore the measured spectra at 25 m and 36 m are normalized with the local $-\overline{u'w'}$-values, as if there was just a change in roughness. Figures 19 and 20 show a fairly good agreement with the "ideal" curve (Kansas) in the inertial subrange, while there is less energy in the low frequency part. Again a spectral lag effect is evident. Here it is worth noting that the profiles measured at Kalkugnen are very variable and do not render themselves suitable for detailed analysis in the same general terms as the spectra do.

6.2 Näsudden, October 1980

The concept of local equilibrium is nicely borne out by Figure 21 which shows the standard deviations of the longitudinal, lateral and vertical wind components respectively, normalized by u_* for the measurements at 11 m at Näsudden, plotted as a function of z/L, where z is height and L is the Monin-Obukhov length. The values for neutral are respectively: $\sigma_u/u_* = 2.40$, $\sigma_v/u_* = 1.60$ and $\sigma_w/u_* = 1.20$, in good, agreement with what have been found at more 'ideal' sites cf. Dutton et al (1979). It is suggested that the stability trend of the data in the near neutral range encountered here ($-0.1 < z/L < 0.2$) is likely to be of general validity.

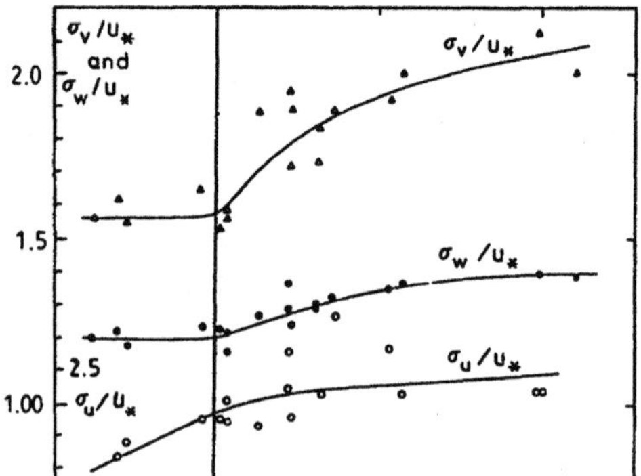

Figure 21. Normalized standard deviations of the wind components as a function of z/L. Measurements at 11 m, Näsudden, October 1980.

Spectra for 11 m and for 77 m have been determined as described under 6.1.

Figures 22-23 give normalized wind spectra for 11 m and Figure 24 the corresponding longitudinal spectra for 77 m.

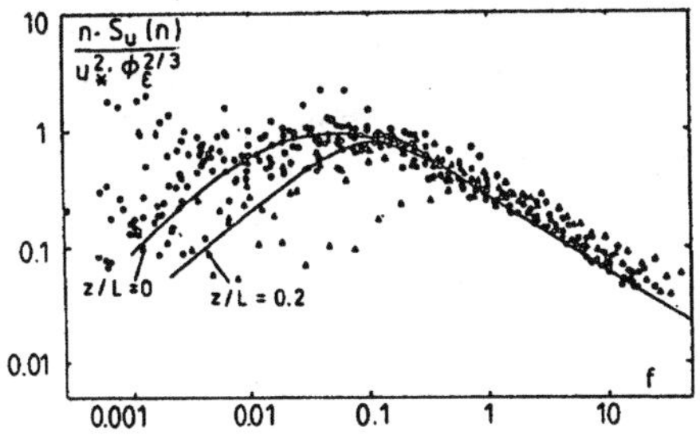

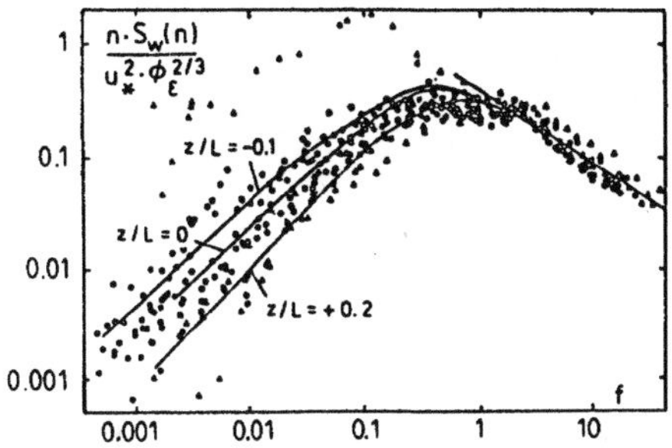

Figures 22 and 23. Normalized longitudinal and vertical spectra for 11 m at Näsudden, October 1980. Legend: rings; wind from 200°-240°; triangles: wind from 150°-180°. Curves from Kaimal et al (1972).

The curves drawn in the various plots are the curves obtained by Kaimal et al (1972) for the very ideal Kansas site. In all the plots there are the curves for neutral stratification (z/L = 0) and in some of the plots also curves for slightly stable conditions, z/L = + 0.2 and very slightly unstable conditions, z/L = - 0.1, which represents the range of stabilities encountered in these runs (cf. Figure 21).

It is seen from the plots that the measurements from 11 m are in general agreement with the Kansas results - in particular this is the case with the w-plot (Figure 23). In the u- and v-spectra there is higher energy in the low frequency range, but this part of the spectrum was suppressed by filtering in the Kansas study. It is clear from Figure 22 that there are energetic motions in the frequency range $f < ca\ 0.004$. This means that meso-scale motions are present, the wavelength of which is typically of the order $\lambda = z/f \approx 10.000$. Figure 24

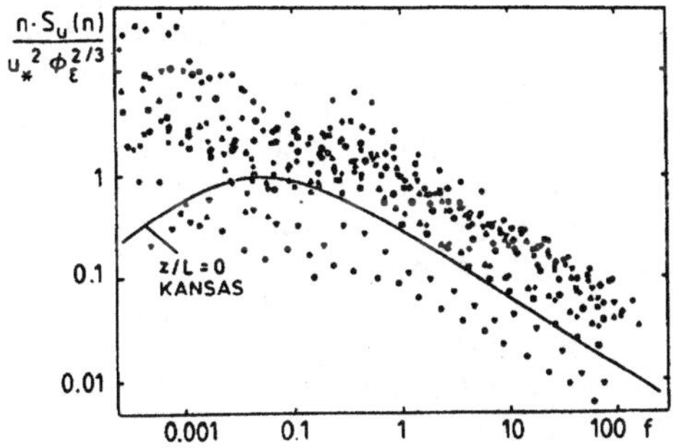

Figure 24. Normalized spectra for the longitudinal wind component, 77 m, October 1980. The symbols represent individual runs; filled symbols: wind direction 150-180°; other symbols: 200-240°. Curve from Kaimal et al (1972).

shows that the spectra for 77 m do not agree with the corresponding curves for Kansas. This is no surprise, considering the shape of the wind profile (Figure 13). It is interesting to notice that meso-scale motions of typical scale 10.000 m is also evident in these spectra (because $f = 0.01$ at 77 m corresponds approximately to $f = 0.001$ at 10 m).

7. CONCLUSIONS AND DISCUSSION

The studies presented in the previous Sections have highlighted some important aspects of the wind regime in coastal areas, whereas other aspects have not been studied at all, in particular sea breeze effects.

The Gotland pilot balloon study clearly illustrates the effects of increased roughness. An important conclusion from comparisons with wind fields obtained by simulation with several 2-, and 3-D numerical models is that the *apparent* roughness length of the wooded areas on Gotland is larger than assumed. This result may be related to the particular processes of momentum exchange over high vegetation (see e.g. Thom et al, 1975 or Garratt, 1980). It is noteworthy that the z_0-value estimated for the rather dense Thetford pine forest from profile measurements is ca. 0.9 m, a value not very different from what was chosen for Lojsta at Gotland. If the failure of the numerical models to simulate the retardation of wind speed over Gotland is indeed correctly attributable to 'the effective z_0-value' for wooded ground, this is an important issue for the application of meso-scale models and, perhaps, also for general circulation models, to real terrain.

Whereas the numerical boundary layer models failed to predict the reduction of wind speed over the island (with the z_0-values employed here) the semi-empirical approach suggested previously by the present authors (Smedman-Högström and Högström, 1978) worked well.

It has been clearly demonstrated that a low level jet of appreciable strength may occur several hundred kilometers downstream of a coastline in situations when the sea surface is colder than the air, i.e. when the marine boundary layer is stably stratified. Over the Baltic Sea this situation typically occurs during about half the year. Thus we expect this kind of low level jet to be a very regular phenomenon in this region and very probably, in other coastal areas as well. It has still not been settled whether this kind of low level jet may also occur in situations when the stratification over the sea is close to neutral, as in the October 1980 case.

The observations of turbulence structure reported from Kalkugnen and Näsudden during conditions relatively close to neutral clearly illustrate the usefulness of the concept of local equilibrium for the high frequency statistics and the spectral lag effect for the low frequency parts of the spectra. The data also show that simple application of the 'flow over a hill' concept applied to a slope, as observed at the Kalkugnen site, is quite effective in describing the observed features.

ACKNOWLEDGEMENTS

The field projects reported in this paper were made possible by the joint effort of the staff at MIUU and Mr Tord Kvick with collegues from SMHI, Norrköping. Dr Leif Enger and Mr Hans Bergström of MIUU have carried out the computations with the 2-D higher order PBL model. The effort of all those people is greatly appreaciated.

The field experiments were sponsored by the National Swedish Board for Energy Source Development (NE Contract No 5061472). The analysis was supported in part by the Swedish Natural Research Council (NFR Contracts No E-EG 3556-100 and E-EG 2684-103).

REFERENCES

Alexandersson, Hans and Bergström, Hans.: 1979, 'Evaluation of Double Theodolite PIBAL Tracking Data'. Reports No 56, Dept. of Meteor., Uppsala. 72 pp.

Atkinson, B.W.: 1981, 'Meso-Scale Atmospheric Circulations'. Academic Press, London. 495 pp.

Blackadar, A.: 1957, 'Boundary Layer Wind Maxima and their Significance for the Growth of Nocturnal Inversions'. Bull. Amer. Meteor. Soc., 38, 283-290.

Bradley, E.F.: 1968, 'A Micrometeorological Study of Velocity Profiles and Surface Drag in the Region Modified by a Change in Roughness'. Quart. J. Roy. Meteor. Soc., 94, 361-379.

Bradley, E.F.: 1980, 'An Experimental Study of the Profiles of Wind Speed, Shearing Stress and Turbulence at a Crest of a Large Hill'. Quart. J. Roy. Meteor. Soc., 106, p. 101-123.

Britter, R.E., Hunt, J.C.R. and Richards, K.J.: 1981, 'Air Flow Over a Two-dimensional Hill: Studies of Velocity Speed-up, Roughness Effects and Turbulence'. Quart. J. Roy. Meteor. Soc., 107, p. 91-110.

Dutton, J.A., Panofsky, H.A., Larko, D., Hampton, N.S., Stone, G. and Vilardo, M.: 1979, 'Statistics of Wind Fluctuations Over Complex Terrain'. Final report. Dept. of Meteorology, Penn. State Univ. University Park, PA 16802.

Elliott, W.O.: 1958, 'The Growth of the Atmospheric Internal Boundary Layer'. Trans. Amer. Geophys. Union. 39, p. 1048-1054.

Enger, Leif.: 1983, 'Numerical Boundary Layer Modelling with Application to Diffusion'. Part I. Report No 70, Dept. of Meteor., Uppsala. 54 pp.

Faxén, Torgny.: 1980, 'The Gotland-Skåne PIBAL Project for WECS Siting'. Part A. Report No 62, Dept. of Meteor., Uppsala. 192 pp.

Garratt, J.R.: 1980, 'Surface Influence Upon Vertical Profiles in the Atmospheric Near-Surface Layer'. Quart. J. Roy. Meteor. Soc., 106, 803-819.

Högström, Ulf.: 1982, 'A Critical Evaluation of the Aerodynamical Error of a Turbulence Instrument'. J. Appl. Meteor., 21, 1838-1844.

Högström, U., Enger, L. and Knudsen, E.: 1980, 'A Complete System for Probing the Detailed Structure of Atmospheric Boundary Layer Flow'. Meteor. Inst. Uppsala. Report No 60, Uppsala. 41 pp.

Högström, U., Bergström, H. and Alexandersson, H.: 1982, 'Turbulence Characteristics in a Near Neutrally Stratified Urban Atmosphere'. Boundary Layer Meteor., 23, 449-472.

Höjstrup, Jörgen.: 1981, 'A Simple Model for the Adjustment of Velocity Spectra in Unstable Conditions Downstream of an Abrupt Change in Roughness and Heat Flux'. Boundary-Layer Meteor., 21, 341-356.

Jackson, P.S. and Hunt, J.C.R.: 1975, 'Turbulent Wind Flow Over a Low Hill'. Quart. J. Roy. Meteor. Soc., 101, 929-955.

Jensen, N.O. and Peterson, E.W.: 1978, 'On the Escarpment Wind Profile'. Quart. J. Roy. Meteor. Soc., 104, 719-728.

Kaimal, J.C., Wyngaard, J.C., Izumi, Y. and Coté, O.R.: 1972, 'Spectral Characteristics of Surface Layer Turbulence'. Quart. J. Roy. Meteor. Soc., 98, 563-589.

Panofsky, H.A., Larko, R., Lipschulz, R., Stone, G.: 1982, 'Spectra of Velocity Components Over Complex Terrain'. Quart. J. Roy. Meteor. Soc., 108, p. 215-230.

Peterson, E.W., Kristensen, L. and Su, C.C.: 1976, 'Some observations and Analysis of Wind Over Non-Uniform Terrain'. Quart. J. Roy. Meteor. Soc., 102, p. 857-869.

Peterson, E.W., Jensen, N.O. and Höjstrup, J.: 1979, 'Observations of Downwind Development of Wind Speed and Variance Profiles at Bognaes and Comparison with Theory'. Quart. J. Roy. Meteor. Soc., 105, p. 521-529.

Peterson, E.W., Taylor, P.A., Höjstrup, J., Jensen, N.O., Kristensen, L. and Petersen, E.L.: 1980, 'Risö 1978: Further Investigations into the Effects of Local Terrain Irregularities on Tower-Measured Wind Profiles'. Boundary-Layer Meteor., 19, 303-313.

Pennell, W.T.: 1983, 'An Evaluation of the Role of Numerical Wind Field Models in Wind Turbine Siting: Appendices. Pacific North West Laboratory PNL-SA-11129 Appendices.

SethuRaman, S.: 1982, 'Observations of the Boundary Layer Wind Structure Near Land-Sea Interface'. In: Proceedings from First International Conference on Meteorology and Air/Sea Interaction of the Coastal Zone'. American Meteorological Society, Boston, USA. p. 4-7.

Smedman, A-S.: 1980, 'The Gotland-Skåne PIBAL Project for WECS Siting'. Part B: Evaluation. Report No 63, Dept. of Meteor., Uppsala. 24 pp.

Smedman-Högström, A-S. and Högström, U.: 1978, 'A Practical Method for Determining Wind Frequency Distributions for the Lowest 200 Meters from Routine Meteorological Data'. J. Appl. Meteor., 7, 942-954.

Smedman, A-S. and Högström, U.: 1983, 'Turbulent Characteristics of a Shallow Convective Internal Boundary Layer'. Boundary Layer Meteor., 25, 271-287.

Thom, A.S., Stewart, J.B., Oliver, H.R. and Gash, J.H.C.: 1975, 'Comparison of Aerodynamic and Energy Budget Estimates of Flux Over a Pine Forest'. Quart. J. Roy. Meteor. Soc., 101, 93-105.

Questions, answers and comments

J.C.R. HUNT: I should like to emphasize the importance of the recent findings that only small scale turbulence adjusts to the local flow in the inner layer over a roughness change or over a hill. Do you use Panofsky's formulation for the rate of change of the spectral components?

U. HÖGSTRÖM: We have not used Panofsky's formulation to calculate the gradual adjustment of the low frequency parts of spectra.

WIND ENERGY ESTIMATES BY USE OF A DIAGNOSTIC MODEL

Roy M. Endlich
SRI International
Menlo Park, California 94025 U.S.A.

Abstract. A diagnostic model is a relatively simple and practical tool for modeling the wind flow of the boundary layer in complex terrain. The model begins with a wind analysis based on available surface wind reports and geostrophic winds (computed from pressure data). The height of the boundary layer top (upper surface of the computational domain) is prescribed to fit local conditions. Using the continuity equation in terrain-following coordinates, the winds at mesh points are adjusted to produce nondivergence while maintaining the original vertical component of vorticity. The method of computing the nondivergent winds uses "direct alterations." This method may be useful for other modeling purposes and will be described. Data for a long period (usually a year) are analyzed to obtain eigenvectors and the associated time series of their coefficients at each observation time. The model is run only for the five or six eigenvectors that explain most of the variance. The wind field at any particular time is reconstructed from the eigenvector solutions and their appropriate coefficients. Comparisons of model results with measured winds at sites representing different types of terrain will be shown. The accuracy and economy of the model make it a useful tool for estimating wind energy and also for giving wind fields for low-level diffusion models.

1. INTRODUCTION

The concept of modeling the wind flow in the planetary boundary layer using the continuity equation was introduced by Sherman (1978). The basic procedure is to make an initial wind analysis using whatever data are available for the region of interest, and to adjust the winds within the boundary layer to be nondivergent. This causes the winds to speed up over higher terrain or to pass around obstacles through lower pathways. Sherman used the variational method of Sasaki (1970) to compute nondivergent winds. Bhumralkar et al. (1980) used terrain-following coordinates to facilitate application of the method. (Terrain heights, variable from point-to-point, form the base of the computational mesh.) Ludwig and Byrd (1980) introduced the use of the eigenvectors of a series

Boundary-Layer Meteorology 30 (1984) 375–386. 0006–8314/84/0304–0375$01.80

of wind observations at several stations, and the eigenvector coefficients at each hour, to represent a history of winds. Since winds above the anemometer level are measured only at radiosonde stations (often far from areas of interest), Endlich et al. (1982) used geostrophic winds (computed from sea-level pressure reports at stations nearest to the site) to represent winds at the boundary-layer top. Also, an algorithm was developed for specifying the height of the boundary layer top (variable from point-to-point) based on parameters for the average boundary thickness in the region and the slope of the top relative to the terrain. Model runs are made only for the five or six eigenvectors that explain most of the variance. The wind field at each particular time is reconstructed from the nondivergent eigenvector solutions and the appropriate coefficients.

The reasons for using a diagnostic model, rather than a more complete and complicated primitive equation model, are the following. The diagnostic model can operate with generally available surface data, and does not require complete initial sets of hydrodynamic variables. Diagnostic models do not require time-dependent boundary conditions as generally used in primitive equation model integrations. Finally, the computational requirements are much less for diagnostic models.

2. MODEL FEATURES

2.1. Setting the Boundary Layer Top (BLT)

The thickness of the boundary layer and the shape of the top (whether curved or flat) are important factors in producing the characteristics of the nondivergent flow. In many instances, radiosonde data that describe the boundary layer top (BLT) are not available. In such cases, average thickness over the grid (taken from climatological values) and a slope factor are used to specify the height of the BLT at each grid point. The height of the top at a particular point of the grid is denoted by H; the average thickness of the boundary layer in the area of interest (surrounding a wind turbine site) is denoted by H_a; h is the terrain height at the point of interest; h_s is the terrain height at the site; and k is the slope factor. Then the basic equation is

$$H = H_a + kh + (1 - k)h_s \tag{1}$$

If k = 0, the boundary layer top is flat; if k = 1, the top parallels the terrain. Values between zero and one give intermediate slopes, and values greater than one give slopes steeper than the terrain slope. Negative values give slopes opposite to the terrain slopes. Figure 1 shows typical boundary layer tops for a nighttime case (H_a = 500 m, k = 0.2) and a daytime case (H_a = 1500 m, k = 0.8). The parameters H_a and k can be treated as functions of time of day and season. An additional feature is that in the vicinity of coastlines the BLT usually has a general slope from ocean to higher values over land in the daytime, with the reverse situation at night. To account for this, two additional

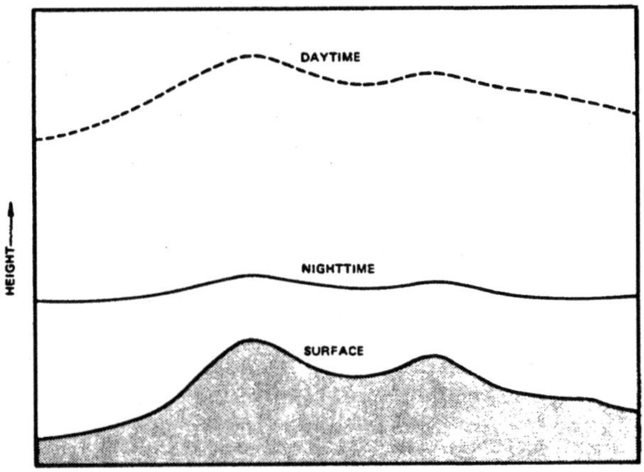

Fig. 1. Typical configurations of the boundary-layer top for daytime and nighttime.

parameters are used to specify a tilt to the BLT. One parameter specifies the west-to-east difference in the thickness of the boundary layer across the grid, and the other parameter specifies the south-to-north difference. There is also a parameter for the minimum thickness of the boundary layer over isolated mountain peaks. These five parameters are used in an iterative solution for the height of the BLT (Endlich and Lee, 1983).

Han et al. (1982) performed numerical model experiments that show the height and shape of the BLT in complex terrain depend upon the wind direction. This effect has not been included in the current diagnostic model.

2.2. Application of the Geostrophic Wind

Winds at anemometer height are analyzed at grid points from standard surface reports. The geostrophic wind vector (usually a single value for the area) is applied at the BLT unless the boundary layer is unusually deep. In the latter case, the geostrophic wind is applied at a height between 500 m and 900 m above ground. The lower value is used over smooth terrain, and the higher value is used over rough terrain. Between the anemometer height and the top of the boundary layer, winds are interpolated by applying a logarithmic wind profile. This technique produces smooth changes of both wind speed and direction with height, with the largest shear near the ground.

2.3. Solution for Nondivergent Winds

A three-dimensional relaxation with appropriate boundary conditions has been used for the solution for nondivergent winds by Sherman (1978),

Bhumralkar et al. (1980), and others. The convergence of the method is slow unless certain constraints are imposed. The principal constraint causes adjustments to the vertical components of the flow to be small compared to adjustments to horizontal components. (This is realistic for most cases of interest.) This constraint is tantamount to the assumption that the flow is parallel to the terrain-following coordinates (sigma surfaces), with no flow through them. For parallel flow, a more efficient method of solution than relaxation can be used, as described below.

A volume element of the grid mesh is shown in Figure 2. The upper and lower surfaces of the elements are sigma surfaces; the vertical separation between them is known over the mesh. As mentioned earlier, we assume there is no flow through the sigma surfaces. Thus, the flow into or out of a volume element takes place through the vertical faces; i.e., it is due to the horizontal wind components. The flux through a face is equal to the u component (or v component as appropriate) multiplied by the vertical separation Δz between sigma surfaces. (In general, this vertical separation varies from point-to-point.) Let $u^* = u\Delta z$ and $v^* = v\Delta z$ so that the winds are weighted by the vertical separation. Then, for a volume element the divergence is

$$D = \partial u^*/\partial x + \partial v^*/\partial y \quad . \tag{2}$$

Similarly, the vertical component of vorticity is

$$\xi = \partial v^*/\partial x - \partial u^*/\partial y \quad . \tag{3}$$

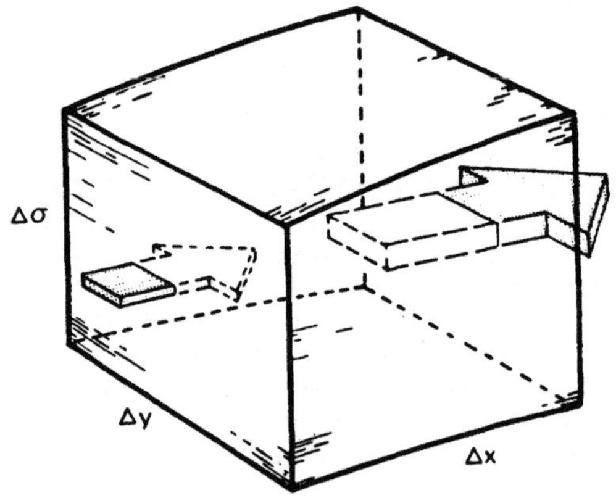

Fig. 2. Volume elements of the grid mesh. The wind components in the x-direction (arrows) indicate net outflow. For nondivergence this must be compensated by net inflow in the y-direction (not shown).

The procedure for calculating nondivergent winds is based on altering the wind components directly, without use of a streamfunction or potential function (Endlich, 1967). The method has been applied by Endlich (1968), Liu and Goodin (1976), Chen (1980), Goodin et al. (1980) and others. The basic concept is to make local alterations to u and v components to eliminate divergence. For example, if there is net convergence into a volume element owing to greater inflow than outflow, this can be corrected by decreasing the wind components on inflow faces and increasing those on outflow faces. The changes to wind components made to eliminate divergence are usually quite small; nevertheless they disturb the vorticity and divergence at surrounding grid points. Therefore, additional changes are made to neighboring wind components to restore the original vorticity. Volume elements are treated in sequence, and 15 to 20 passes through the grid are required to reduce divergence (whether positive or negative) to small magnitudes while retaining the original vorticity field. Formulas for computing the wind alterations are given in the papers cited above. Lateral boundary conditions are not used; instead the final nondivergent winds are adjusted by a constant vector (usually quite small) that causes the final winds averaged over the grid to equal the average of the original winds. With this procedure, the flow in any layer can be made nondivergent independently of the flow in neighboring layers. Thus, the layers of the grid can be treated in sequence; this makes the computations simpler than three-dimensional relaxation.

The computational stencil used is shown in Figure 3. The original wind analysis is made at the points indicated by +s; however, for computational purposes in Equations 2 and 3, the u and v components are computed as averages of values to the north and south, or east and west, as appropriate. In the present model, these averaged data are used in the correction formulas. In general, this method can be used to alter

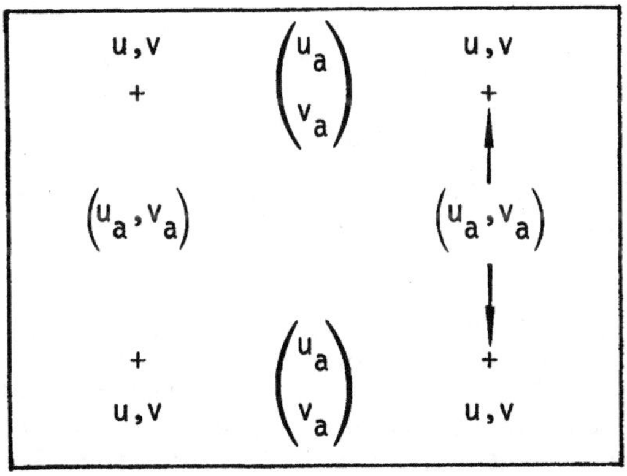

Fig. 3. Original wind analyses are at points denoted by +s. Components (u_a, v_a) averaged for two grid points are used in computing divergence and vorticity.

an initial wind field that is only an approximation to a desired field, when the latter is specified by values of vorticity and divergence at each grid point. For some applications the desired divergence is zero, but nonzero divergence values can be used if appropriate values can be deduced from dynamical reasoning or from data such as precipitation amounts. Also, the desired vorticity values can be different from those associated with the original wind analysis.

The mathematical properties of computational methods of this type have been investigated by Dinar (1979). He described multi-grid methods (alternate use of coarse and fine grids) that drastically reduce the amount of computation required. The writer plans to use such methods in the future.

In comparison to results obtained for a given initial wind field using relaxation, the wind alteration technique gives slightly different results. The differences are caused mainly by the boundary conditions of the relaxation, which hold the wind components on the uppermost sigma surface (the BLT) fixed. This is not true of the direct alteration technique, which allows readjustments to occur at the BLT.

3. TYPICAL RESULTS

An example of results is shown for Goodnoe Hills, Washington. The location of the wind measuring site (which is now occupied by three Boeing Mod 2 wind generators) is at an altitude of 805 m on the crest of hills north of the Columbia River, approximately 50 km northeast of Dallesport, Oregon. The weather data for 1964 were chosen for the model simulations because reports from Dallesport and Yakima (both reasonably close to the site) were available for that time.

The geostrophic winds for this site were computed from reports of sea-level pressure at Yakima, Dallesport, and Pendleton. These three stations form a regular triangle with the site located fairly close to its center; thus, the stations are well distributed. The geostrophic vectors are generally quite strong and are in good agreement with the observed surface winds at the three stations. The mean geostrophic wind for the year 1964 is from 297° at 11.6 m s^{-1}. This relatively strong wind is a major factor in producing high wind speeds in the simulation for the site.

The nondivergent calculation changes initial u or v components up to 2.0 m s^{-1}. The results are shown in Figure 4. The simulated annual average wind speed is 8.3 m s^{-1}. The simulated diurnal curves for the four seasons are shown in Figure 5. Marked differences from winter to summer are indicated. The wind speed distribution (Figure 6) shows that 34 percent of the winds are higher than 10 m s^{-1}; thus, the simulation indicates that this site is quite windy. The strongest winds of the simulation are predominantly from the west to northwest. There are no site measurements available for 1964; however, measurements were made.

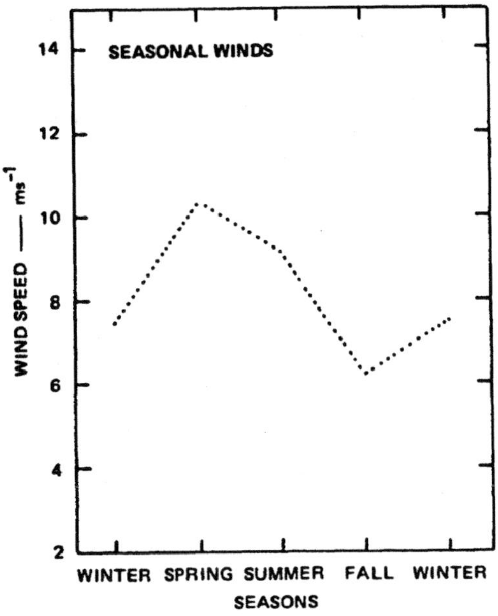

Fig. 4. Seasonal wind speed curves for Goodnoe Hills, Washington, as simulated by the wind flow model.

there by Pacific Northwest Laboratory from January through September 1981 (see Wendell et al., 1981). These data show an average speed of 7.9 m s^{-1} at a height of 61 m above ground. The simulations were performed before these data were received.

Results for other locations are shown in Table 1. Simulations for some sites are much better than for others, probably depending on the degree to which modeling assumptions are met. At Stratton Mountain, Vermont (a 1200 m peak), measured wind speeds are too few to provide reliable statistics. The worst results are for Romero Overlook at Pacheco Pass, California. This pass through low mountains east of Gilroy, California, joins the coastal valleys (dominated by cool marine air) and the interior San Joaquin Valley (very hot and dry in summer). The arrangement of the temperature pattern is such that the thermal wind consistently opposes the surface geostrophic wind. Prior to being used in the model, the very strong surface geostrophic winds should have been corrected for the thermal wind effect. This simple procedure would have produced more reasonable results for this site.

4. DISCUSSION

The method described above has proved to be practical and of useful accuracy in estimating wind statistics for a variety of locations in the

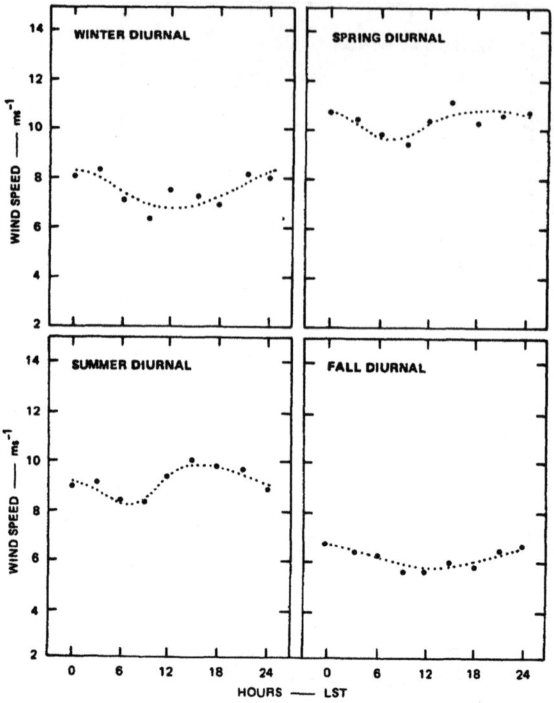

Fig. 5. Diurnal wind speed curves for Goodnoe Hills, Washington, as simulated by the wind flow model.

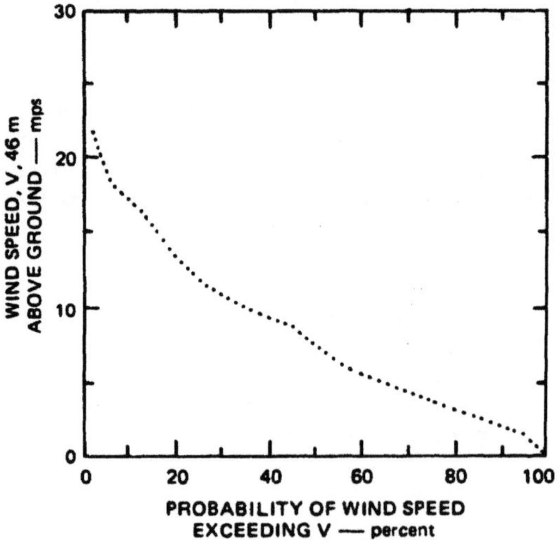

Fig. 6. Simulated frequency distribution of wind speed for Goodnoe Hills, Washington.

Table 1

Comparison of measured and simulated wind speeds

Site	Terrain Characteristics	Annual Average Wind Speed (m s^{-1}) Measured	Simulated
Clayton, New Mexico	Flat, mountains to west	7.6	6.2
San Gorgonio Pass, California	Mountains and desert	8.0	7.7
Block Island, Rhode Island	Offshore island	7.5	7.5
Meade, Kansas	Flat	8.0	7.9
Stratton Mountain Vermont	Mountain Peak	Unknown	12.2
Pacheco Pass, California	Coastal mountains	6.1	9.9

United States. Where radiosonde data are available to describe the boundary layer winds, the geostrophic wind can be replaced with actual wind measurements; this is expected to lead to improved accuracy in the initial wind analyses and in the nondivergent flow. Current studies are using nondivergent boundary layer winds in relation to air pollution transport and in analyses of processes related to acid rain.

ACKNOWLEDGMENT

This work was supported by Pacific Northwest Laboratory, Richland, Washington, under subcontract B-D5789-A-E to the U.S. Department of Energy. I thank Dr. David S. Renné of PNL for advice and assistance.

REFERENCES

Bhumralkar, C.M., Mancuso, R.L., Ludwig, F.L., and Renné, D.S.: 1980, 'A Practical and Economic Method for Estimating Wind Characteristics and Potential Wind Energy Conversion Sites,' Solar Energy 25, 55-65.

Chen, T.-C.: 1980, 'On the Energy Exchange Between the Divergent and Rotational Components of Atmospheric Flow over the Topics and Subtropics at 200 mb During Two Northern Summers,' Mon. Wea. Review 108, 896-912.

Dinar, N.: 1979, 'Fast Methods for the Numerical Solution of Boundary Value Problems,' PhD Thesis, Weizmann Institute of Science, Rehovot, Israel.

Endlich, R.M.: 1967, 'An Iterative Method for Altering the Kinematic Properties of Wind Fields,' J. Appl. Meteorol. 6, 837-844.

Endlich, R.M.: 1968, 'Direct Computation of Geostrophic Winds from Observed Winds using the Balance Equation,' J. Appl. Meteorol. 7, 994-1003.

Endlich, R.M., Ludwig, F.L., Bhumralkar, C.M., and Estoque, M.A.: 1982, 'A Diagnostic Model for Estimating Winds at Potential Sites for Wind Turbines,' J. Appl. Meteorol. 21, 1441-1454.

Endlich, R.M. and Lee, J.D.: 'An Improved Diagnostic Model for Estimating Wind Energy,' Final Report to Pacific Northwest Laboratory, Subcontract B-D5789-A-E, SRI International, Menlo Park, CA 94025.

Goodin, W.R., McRae, C.J., and Seinfeld, J.H.: 1980, 'An Objective Analysis Technique for Constructing Three-Dimensional Urban-Scale Wind Fields,' J. Appl. Meteorol. 19, 98-108.

Han, Y.-J., Ueyoshi, K., and Deardorff, J.W.: 1982, 'Numerical Study of Terrain-Induced Mesoscale Motions in a Mixed Layer,' J. Atmos. Sci. 39, 2464-2476.

Liu, C.Y. and Goodin, W.R.: 1976, 'An Iterative Algorithm for Objective Wind Field Analysis,' Mon. Wea. Review 104, 784-792.

Ludwig, F.L. and Byrd, G.: 1980, 'An Efficient Method for Deriving Mass-Consistent Flow Fields from Wind Observations in Rough Terrain,' Atmos. Environ. 14, 585-587.

Sasaki, Y.: 1970, 'Some Basic Formalisms in Numerical Variational Analysis,' Mon. Wea. Review 98, 875-885.

Sherman, C.A.: 1978, 'A Mass-Consistent Model for Wind Fields over Complex Terrain,' J. Appl. Meteorol. 17, 312-319.

Wendell, L.L., Barchet, W.R., Connell, J.R., Miller, A.H., Pennell, W.T., and Renné, D.S.: 1981, Annual Report of the Wind Characteristics Program Element for the Period October 1979 Through September 1980, Report PNL-3876, Pacific Northwest Laboratory, Richland, Washington.

Questions, answers and comments

L. BERKOFSKY: You have published a paper in which the wind is separated into a rotational part and a divergent part. Have you applied this idea in your present work?

R.M. ENDLICH: Yes, the basic method that you referred to is used in the present work. The rotational part of the flow is retained and the divergent part is discarded.

S.P.S. ARYA: Since your model is based on, or accounts for, the surface observations of wind, it is most likely to be accurate near the surface. How about the predicted winds in the upper part of the boundary layer? How good is your assumed distribution of the wind, which is based on the classical Ekman spiral, over a complex terrain, when we know that the observed wind profiles often deviate considerably from the Ekman profile due to the effects of stability, baroclinity and other factors, even over a flat and homogenous terrain?

R.M. ENDLICH: We do not have measured data to verify the model winds above approximately 100m. It may be true that accuracy is lost with height particularly in the most complex terrain. In part we can compensate for baroclinicity by using a thermal wind correction to the geostropic wind based on surface temperature measurements.

J.C.R. HUNT: Does the model make any allowance for changes in the flow regime over obstacles depending on thermal stability?

R.M. ENDLICH: The thickness of the boundary layer and the slope of the boundary layer top are changed from day to night. This gives an effect analogous to changing stability. Otherwise stability is not explicitly considered.

N. DINAR: Eigenvector expansion is a linear process. Your process of solution includes non-linear elements (like interpolation). What is the justification to this eigenvector expansion?

R.M. ENDLICH: The use of the eigenvectors provides a great saving of computations. The non-linearities introduced are relatively minor and in my opinion do not disturb the orthogonality to an appreciable degree. This point can be checked empirically.

P. ALPERT: Could you please explain the physical basis of your assumption that the two-dimensional horizontal divergence should vanish?

R.M. ENDLICH: The assumption is made that the flow is constrained by the boundary layer top and it is everywhere parallel to the terrain-following coordinate surfaces. This allows for vertical motions in cartesian coordinates. The solution for non-divergent flow on these surfaces is two dimensional.

PART IV

DISPERSION OF POLLUTANTS IN THE ATMOSPHERE AND ITS DEPOSITION ON THE SURFACE

OVERWATER DISPERSION IN COASTAL REGIONS

Steven R. Hanna, Robert J. Paine, and Lloyd L. Schulman
Environmental Research & Technology, Inc.
696 Virginia Road
Concord, MA 01742 USA

ABSTRACT

Air pollution sources such as oil platforms and ships are sometimes located near coastlines where the plumes can impact population centers. Models of overwater dispersion must account for the stability of the overwater boundary layer. An overwater dispersion model based on the standard Gaussian formula is described which uses measurements of the air-sea temperature difference, the wind speed, and the mixing depth to predict concentration patterns at the coastline. Internal boundary layers and complex terrain at the coastline are accounted for. This new model is evaluated using the results of three tracer experiments in United States coastal zones.

1. INTRODUCTION

There is concern over air quality at sea and in coastal regions due to increased shipping traffic, construction and operation of offshore oil and gas platforms, and burning of wastes at sea. Some naval operations are also dependent on overwater diffusion. Most diffusion models are equally valid over land and water, as long as the surface boundary layer is properly parameterized. It is important for the modeler to recognize that there are vast differences between the overwater and overland boundary layers, which led Holland (1981) to write:

> "For mid-latitude continental meteorologists, accustomed to thinking of temperature as the driving agent for convection and surface heat flux, the encounter with the atmosphere overlying the tropical ocean was like a visit to another planet."

Latent heat fluxes are observed to be ten times the magnitude of sensible heat fluxes over water, and the difference between air and water temperature is frequently positive during the day and negative

Boundary-Layer Meteorology 30 (1984) 389–411. 0006-8314/84/0304-0389$03.45

during the night (the opposite of typical differences over land). This difference can be seen in the observed diurnal variations of air and water temperatures in the North Sea (Nieuwstadt 1977) plotted in Figure 1 for all months of the year. Under these conditions the Pasquill-Gifford-Turner stability classification scheme is clearly no longer valid, since it is keyed to overland heat fluxes. A clear annual cycle is also evident on the figure, with stable conditions in summer when the air is relatively warm, and unstable conditions in the winter when the air is relatively cold. This behavior is also the opposite of that observed over land. However, Lenschow et al. (1980) show that overwater wind, temperature, and turbulence profiles do agree with convective scaling laws derived over land if the latent heat flux is combined with the sensible heat flux in defining the Monin-Obukhov length L and the convective velocity scale w_*.

Near coastal regions the boundary layer can be complicated by the presence of surface inhomogeneities. Stable onshore flow during the daytime is eroded from below by convection as it passes inland, leading to an upward sloping Thermal Internal Boundary Layer (TIBL) that can intercept an elevated plume from an offshore source and fumigate it to the ground (Lyons 1975). Deardorff and Willis (1982) developed a model that simulates this phenomenon using convective scaling concepts. Horizontal gradients of sea surface temperature can lead to advection inversions (over colder waters) or sea smoke (over warmer waters). In general, maximum daily mixing depths over water are lower than over land because of the reduced intensity of boundary layer convection (Hsu 1979). Conditions with low mixing depths and nearly constant wind directions can persist for days due to the absence of diurnal variations in surface temperature.

The dispersion model to be described in this paper accounts for the overwater surface boundary layer and the changes at the coastline. It was developed for application to determining onshore air quality from releases at offshore oil and gas platforms, and is sufficiently simple that it can be run using hourly data for a period of one year. The basic model is straight-line Gaussian. The model was evaluated using the results of tracer experiments at three offshore sites, showing that maximum concentrations can be expected during advection inversion conditions.

2. DIFFUSION MODEL FRAMEWORK

The core of the model is the standard Gaussian diffusion formula (Gifford 1968):

$$C = \frac{Q}{2\pi u \sigma_y \sigma_z} \exp(-y^2/2\sigma_y^2)(\exp(-(z-h_e)^2/2\sigma_z^2) + \exp(-(z+h_e)^2/2\sigma_z^2)) \qquad (1)$$

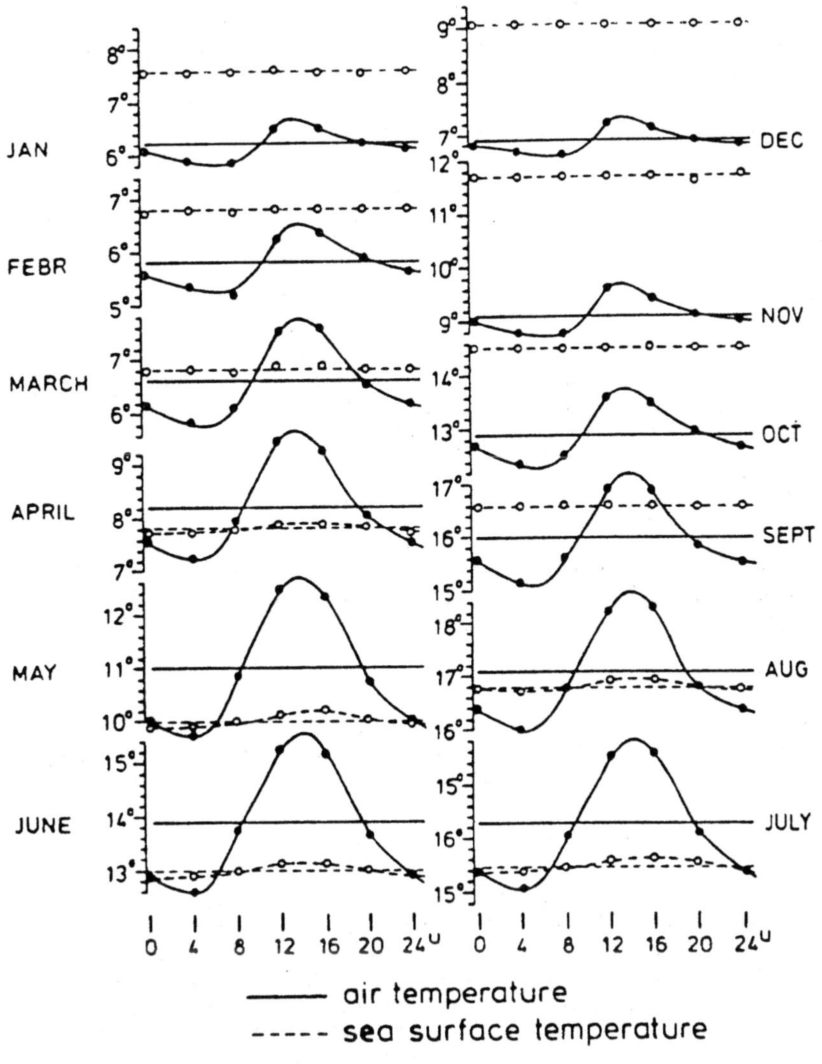

Figure 1 Average Daily Cycle of Air and Water Temperature for Each Month as Measured at the Light-Vessel Noordhinder (Nieuwstadt 1977) in the North Sea

where y is crosswind distance from the mean plume axis and z is elevation of the receptor above sea level. The parameters σ_y and σ_z are the lateral and vertical standard deviations of the concentration distribution, u is mean wind speed, h_e is effective plume release height, Q is hourly average source strength, and C is mean concentration. There are many detailed formulas for the model components described by Hanna et al. (1983) which would take up too much space to repeat in this overview. Here we will describe the qualitative features of the model, emphasizing the boundary layer parameterizations. The model has the following components:

2.1 Building Effects

The effects of aerodynamic downwash and mixing of the plume in the wake of the structure are include since most offshore sources have short stacks. Techniques listed in the BLP model users' guide (Schulman and Scire 1980) are used.

2.2 Plume Rise

Standard formulas by Briggs (1969) are used so that the model will be consistent with EPA regulatory models. An option for partial penetration of elevated inversions was developed using Briggs' (1975) suggestions.

2.3 Wind Speed

The model prefers to use an observed wind speed at stack top, but if this is not available it can calculate the wind speed from boundary layer theory (Businger 1973) to be described in Section 3.

2.4 Dispersion Coefficients: σ_y and σ_z

The model assumes that $\sigma_y = i_y \times S_y(x)$ and $\sigma_z = i_z \times S_z(x)$, where i_y and i_z are turbulence intensities (σ_v/u and σ_w/u) and S_y and S_z are dimensionless universal functions that equal 1.0 near the source and equal about 0.5 at downwind distances x of about 10 km (Pasquill 1976). The model prefers to use turbulence intensities observed at stack top, but can generate these values from overwater boundary layer theory, if needed. For example, given estimates of the air-water temperature difference, the relative humidity, and the wind speed, the model can predict i_y and i_z. For very stable conditions ($d\theta/dz > 5°/100$ m) a new formula is used in which σ_z is a function of $d\theta/dz$.

2.5 Mixing Depth: z_i

The mixing depth over the water is usually much lower than over the land. It enters the model in the same way that it enters EPA models and is used to determine the degree of penetration of the plume into the stable air aloft and to calculate reflection of the plume from the top of the mixed layer. If the mixing depth is very low, and the plume does not penetrate the stable air, the plume becomes well-mixed in the vertical a short distance from the source.

2.6 Changes at the Shoreline

- At the sea-land interface, the rate of growth of σ_y and σ_z will change to reflect the stability of the overland boundary layer. Multiple changes of underlying surface (sea or land) can be treated. The model handles receptors on elevated terrain using features taken from the RTDM model developed by ERT; impaction can occur in stable conditions and the "half height correction" is used in neutral and unstable conditions. Overland diffusion rates are identical to those in EPA models.
- Fumigation of elevated plumes that are initially located in stable overwater flow but subsequently encounter a convective layer at the shoreline is modeled using the Deardorff-Willis (1982) approach. This new fumigation model has been shown to be more realistic than the EPA fumigation model suggested by Turner (1969).

2.7 Data Input

It is clear from the discussion above that the model can accept more extensive data input than the standard EPA models. For example, observations of overwater turbulence intensities can be used to greatly improve the model predictions. Because the overwater boundary layer does not have the same diurnal variation of stability as the overland boundary layer, it is important that a measure of stability be obtained. The model prefers observations of the sea-air temperature difference and the wind speed as input; if these are not available, climatological defaults can be used. Also, because overwater mixing depths are much lower than overland mixing depths, the model predictions are improved if mixing depth is measured at the source.

3. BOUNDARY LAYER PARAMETERIZATIONS OVER WATER

The proposed model is intended for application to a variety of air quality problems, and it is not reasonable to expect that detailed observations of winds, temperature, and turbulence to elevations of several hundred meters will be available at any given site. In fact, in most cases we will feel lucky to have access to buoy or platform data in the vicinity of the source. Consequently the model has been designed to accept minimal data and use boundary layer profile equations to predict the required meteorological parameters at the level of the plume. The minimum required observations are:

- wind speed u at a single level, preferably 10 m;

- sea temperature T_s;
- air temperature T_a at a single level, preferable 10 m;
- relative humidity RH.

We stress that a measurement of the mixing depth z_i would be very useful, even though the model can operate without it.

The roughness length, z_0, is calculated from a formula suggested by Hosker (1974):

$$z_0(m) = 2.0 \times 10^{-6}\, u_{10}^{2.5}(m/s) \tag{2}$$

where u_{10} is the wind speed observed at an elevation of 10 m.

3.1 Mean Profiles

Following Businger (1973), it is assumed that the wind and virtual temperature profiles obey similarity theory:

$$\frac{.4z}{u_*}\frac{du}{dz} = \phi_u\left(\frac{z}{L_v}\right) \tag{3}$$

$$\frac{.4z}{\theta_{v*}}\frac{d\theta_v}{dz} = \phi_T\left(\frac{z}{L_v}\right) \tag{4}$$

where θ_v is the virtual potential temperature:

$$\theta_v = \theta(1+0.61W) \tag{5}$$

and W is the mixing ratio of water vapor, which is a function of temperature and relative humidity. The friction velocity u_* and the scaling temperature θ_{v*} are defined from the momentum and heat fluxes:

$$u_*^2 = -\overline{u'w'} \tag{6}$$

$$u_*\,\theta_{v*} = \overline{w'\theta_v'} \tag{7}$$

This heat flux definition includes both sensible and latent heat fluxes. The Monin-Obukhov length, L_v, is also defined so that it includes the effects of moisture:

$$L_v = \theta_v u_*^2/0.4g\theta_{v*} \tag{8}$$

Standard formulas for ϕ_u and ϕ_T given by Businger (1973) are assumed:

$$\phi_u(z/L_v) = (1-15z/L_v)^{-1/4} \qquad z/L_v < 0 \qquad (9)$$

$$= 1 + 4.7z/L_v \qquad z/L_v > 0 \qquad (10)$$

$$\phi_T(z/L_v) = (1-9\ z/L_v)^{-1/2} \qquad z/L_v < 0 \qquad (11)$$

$$= 1 + 6.5z/L_v \qquad z/L_v > 0 \qquad (12)$$

Businger gives the formulas for u(z) and θ(z) that result when equations (3) and (4) are integrated from z_o to z.

$$u = \frac{u_*}{.4}\left(\ln\frac{z}{z_o} - \Psi_u\left(\frac{z}{L_v}\right)\right) \qquad (13)$$

$$\theta_v - \theta_{vs} = \frac{\theta_{v*}}{.4}\left(\ln\frac{z}{z_o} - \Psi_\theta\left(\frac{z}{L_v}\right)\right) \qquad (14)$$

where

$$\Psi_u\left(\frac{z}{L_v}<0\right) = 2\ln\left(\frac{1+\phi_u^{-1}}{2}\right) + \ln\left(\frac{1+\phi_u^{-2}}{2}\right) - 2\tan^{-1}(\phi_u^{-1}) + \frac{\pi}{2} \qquad (15)$$

$$\Psi_u\left(\frac{z}{L_v}\geq 0\right) = -4.7z/L_v \qquad (16)$$

$$\Psi_T\left(\frac{z}{L_v}<0\right) = 2\ln\left(\frac{1+\phi_T^{-1}}{2}\right) \qquad (17)$$

$$\Psi_T\left(\frac{z}{L_v}\geq 0\right) = -6.5z/L_v \qquad (18)$$

Alternatively, the parameters u_*, θ_{v*}, and L_v could be estimated using bulk aerodynamic and heat transfer coefficients C_u and C_T defined by Schacher et al. (1982):

$$u_*^2 = C_u u^2(10\ m) \quad (19)$$

$$u_*\theta_{v*} = C_u^{1/2} C_T^{1/2}\ u(10\ m)(\theta_{vs} - \theta_v(10\ m)) \quad (20)$$

$$L_v = (C_{uN}\ u^2(10)/0.4)/(gC_{TN}^{1/2}(\theta_v(10\ m) - \theta_{vs})/\theta_v) \quad (21)$$

Subscript N refers to neutral conditions, for which $C_{TN} = 1.3 \times 10^{-3}$. The coefficient C_{uN} is observed by Kondo (1975) to be a slight function of wind speed:

10 m wind speed (m/s)	C_{uN}
0.3 - 2.2	$1.08\ u^{-0.15} \times 10^{-3}$
2.2 - 5.0	$(0.77 + 0.86\ u) \times 10^{-3}$
5 - 8	$(0.87 + 0.57\ u) \times 10^{-3}$
8 - 25	$(1.2 + 0.025\ u) \times 10^{-3}$

3.2 Turbulence Intensity Calculations

Our method of calculating σ_y and σ_z assumes that the turbulent component of these parameters is proportional to the lateral and vertical turbulence intensities, i_y and i_z, respectively:

$$i_y = \sigma_v/u = (u_*/u)F_y(z_i/L_v) \quad (22)$$

$$i_z = \sigma_w/u = (u_*/u)\ F_z(z/L_v) \quad (23)$$

where the following formulas are valid for F_y and F_z at heights less than about 100 m (Panofsky et al. 1977):

$$F_y = 1.7 \qquad z_i/L_v > 0 \quad (24)$$

$$= (4.9 - 0.5z_i/L_v)^{1/3} \qquad z_i/L_v < 0 \quad (25)$$

$$F_z = 1.3 \qquad z/L_v > 0 \quad (26)$$

$$= 1.3(1-3z/L_v)^{1/3} \qquad z/L_v < 0 \quad (27)$$

The leading constant (4.9) in Equation (24) is different from that (12) recommended by Panofsky et al. (1977), but is used so that F_y

is continuous at $z_i/L_v = 0$. Equation (13) can be used to estimate (u_*/u) in the equations for i_y and i_z. If the effective source height is greater than about 100 m, then i_y and i_z can be corrected for height using empirical methods outlined by Caughey (1982).

It is possible that i_y may be measured directly at an offshore buoy or tower, but i_z is difficult to measure properly even over a land surface. In general, we have more confidence in Equation (23) than in measurements of i_z over water.

4. DISPERSION PARAMETERS σ_y AND σ_z

We assume that the dispersion coefficients σ_y and σ_z have several independent components that can be added quadratically.

$$\sigma_y^2 = \sigma_{yt}^2 + \sigma_{yb}^2 + \sigma_{ys}^2 + \sigma_{yo}^2 \tag{28}$$

$$\sigma_z^2 = \sigma_{zt}^2 + \sigma_{zb}^2 + \sigma_{zo}^2 ,$$

where subscripts t, b, s and o refer to turbulent, buoyancy enhancement, shear, and building effect components, respectively.

To estimate the turbulence components, σ_{yt} and σ_{zt}, the model follows the recommendations of the AMS Workshop on Stability Classification Schemes and Sigma Curves (Hanna et al. 1977):

$$\sigma_{yt} = i_y \times S_y(x) \tag{29}$$

$$\sigma_{zt} = i_z \times S_z(x) \tag{30}$$

We use Briggs' (1973) recommendation for the functions $S_y(x)$ and $S_z(x)$.

$$S_y(x) = (1+0.0001x)^{-1/2} \quad , \quad x \leq 10^4 \text{ m} \tag{31}$$

$$\left.\begin{aligned} S_z(x) &= 1 && \text{Class A and B} , && x \leq 10^4 \text{ m} \\ &= (1+0.0002x)^{-1/2} && \text{Class C} , && x \leq 10^4 \text{ m} \\ &= (1+0.0015x)^{-1/2} && \text{Class D} , && x \leq 10^4 \text{ m} \\ &= (1+0.0003x)^{-1} && \text{Class E and F} , && x \leq 10^4 \text{ m} \end{aligned}\right\} \text{ for } \frac{d\theta_v}{dz} < \frac{5°\text{C}}{100\text{m}} \tag{32}$$

An additional class is added for very stable conditions, based on suggestions by Strimaitis et al. (1983):

$$S_z(x) = (1+s^{1/2}x/0.32u)^{-1/2} \qquad \frac{d\theta_v}{dz} \geq \frac{5°C}{100\ m} \tag{33}$$

where s equals $(g/\theta_v)(d\theta_v/dz)$. In order to make the model consistent with observations of σ_y (Heffter 1965) at mesoscale distances, S_y evaluated at 10 km is used in the model for x greater than 10 km. Effective Pasquill stability class is estimated using Golder's (1972) method, assuming a roughness length, z_o, in the range from 10^{-4} to 10^{-3} m:

L_v Range	Stability Class
$-10\ m \leq L_v < -5\ m$	B
$-25\ m \leq L_v < -10\ m$	C
$\lvert L_v \rvert > 25\ m$	D
$10\ m < L_v \leq 25\ m$	E
$5 < L_v \leq 10\ m$	F

The initial components, σ_{yo} and σ_{zo}, due to building downwash are assumed to be given by formulas in the BLP model (Schulman and Scire 1980). We use formulas recommended by Pasquill (1976) to calculate the buoyant plume enhancements, σ_{yb} and σ_{zb}, and the shear enhancement, σ_{ys}:

$$\sigma_{yb}^2 = \sigma_{zb}^2 = (\Delta h)^2/10 \tag{34}$$

$$\sigma_{ys}^2 = 0.03(\Delta WD)^2 x^2 \tag{35}$$

where Δh is local plume rise above stack top and ΔWD is wind direction shear (in radians) over the depth of the plume, which is assumed to be equal to the plume rise in the plume rise stage and $4.3\sigma_z$ further downwind.

At the land/sea interface, where stabilities and turbulence intensities may change, the new dispersion rates are accounted for by means of a virtual source. The change does not take place until the plume intercepts the rising internal boundary layer, h_T, defined by

Neutral: $h_T = 0.1x$ (Jackson 1976) (36)

Stable: $h_T = 0.1x$, with maximum $h_T = 0.4(u_* L_v/f)$ Zilitinkevich (1975) (37)

Unstable: $h_T = 0.19x^{0.47}$ $u < 5$ m/s (38)
$= 0.20x^{0.29}$ $5 < u < 7$ m/s

$$= 0.25x^{0.14} \quad u \geq 7 \text{ m/s}$$
$$(h_T \text{ and } x \text{ in km, Raynor et al. 1979}).$$

The parameter f in equation (37) is the Coriolis parameter. The equations for unstable conditions are empirical, based on observations taken over Long Island by Raynor et al. (1979). When the plume passes through h_T, the rate of diffusion appropriate for land is used.

When overwater conditions are stable and overland conditions are unstable, the Deardorff-Willis (1982) fumigation model may be used, with the slow entrainment hypothesis. This model is too lengthy to repeat here, but is thoroughly described in the reference. The fumigation model accounts for entrainment of the plume into the rising layer, and subsequent transport of plume material to the ground with a convective layer, and subsequent transport of plume material to the ground with a downward velocity of about $0.5w_*$. Because the model is relevant only for thin elevated plumes, it is often not applicable to oil platform and ship plumes, whose effective height is not above 100 m.

4. MODEL EVALUATION

The model was evaluated using data from three tracer experiments in Ventura and Pismo Beach, California, and Cameron, Louisiana. The first two experiments were sponsored by the U.S. Department of the Interior, and the last experiment was sponsored by the American Petroleum Institute. In each case nonbuoyant SF_6 tracer gas was released at a height of 10 to 20 m from a location about 5 to 10 km offshore. Figures 2, 3 and 4 are maps of the experimental sites, showing the source positions and the locations of meteorological and aerometric instruments. Table 1 gives some details concerning each experiment.

TABLE 1
SUMMARY OF DIFFUSION EXPERIMENTS

Site and Time	Source	Number of Monitors	Number of Hours
Ventura, CA Sept. 1980 Jan. 1981	5 km offshore	41	37
Pismo Beach, CA Dec. 1981 June 1982	6 km offshore	65	62
Cameron, LA July 1981 Jan. 1982	4 to 8 km offshore	70	53

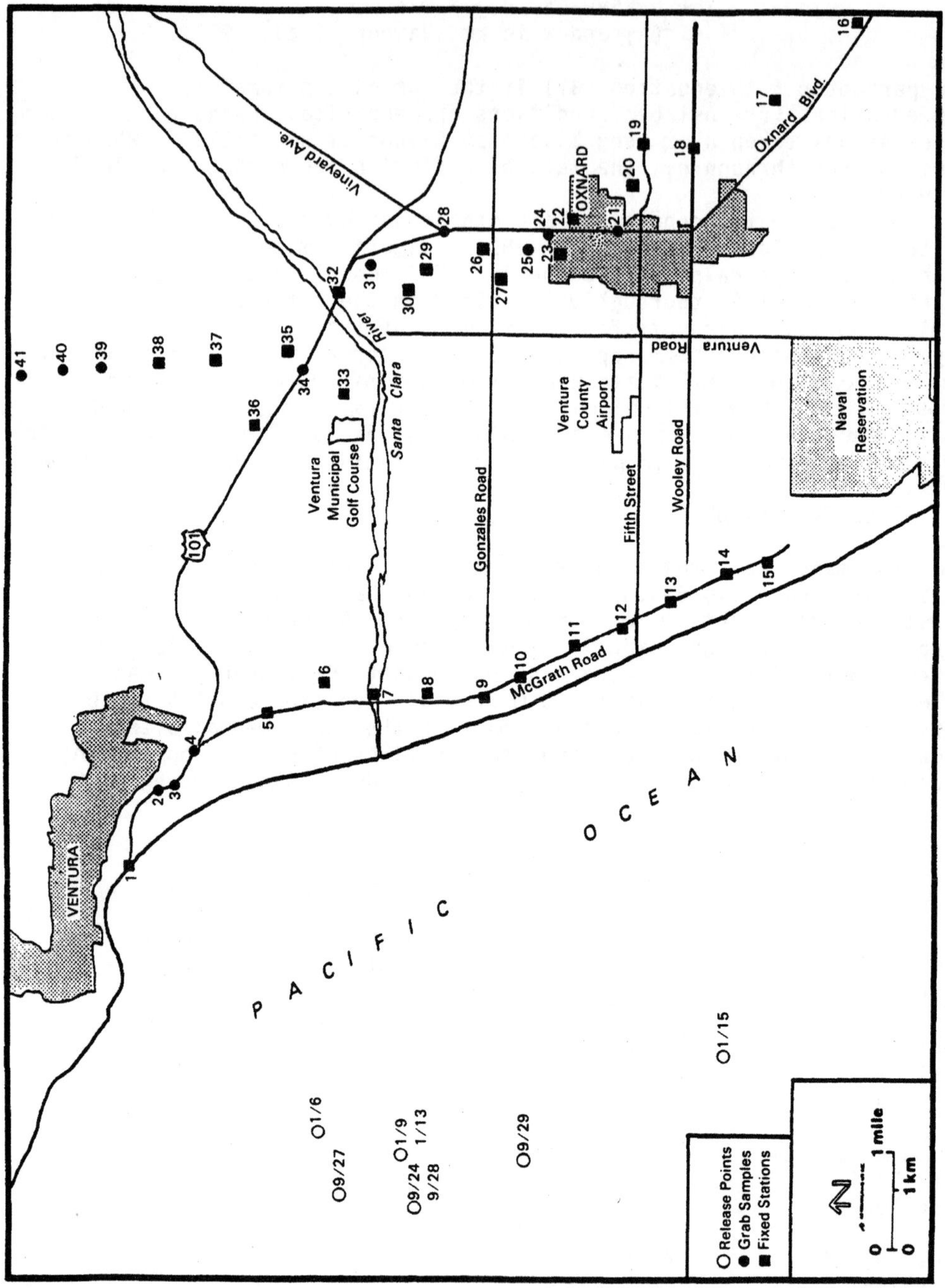

Figure 2 Tracer Sampling Sites Used in the Ventura Area Experiment

8308050

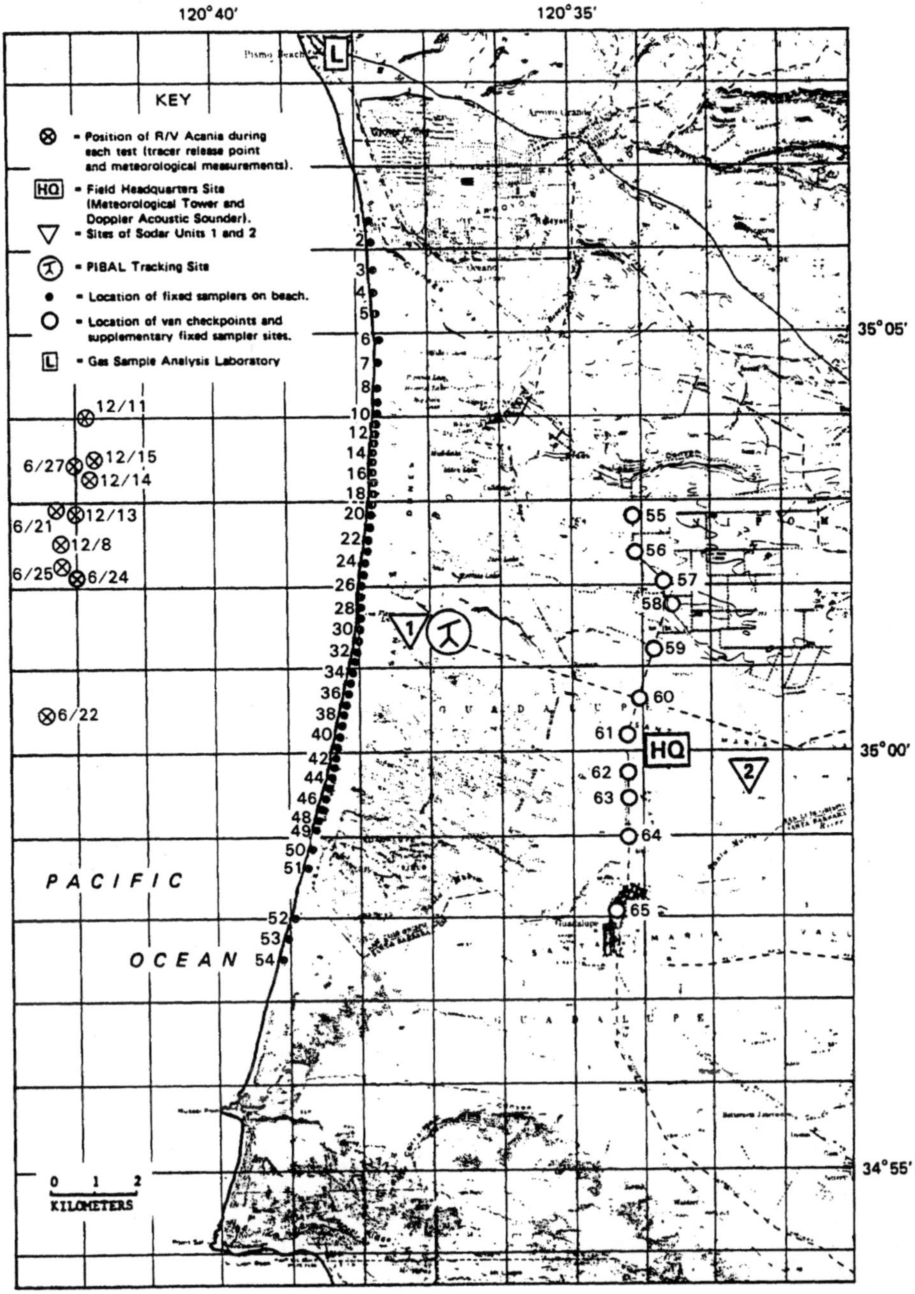

Figure 3 Tracer Sampling Sites Used in the Pismo Beach Experiment

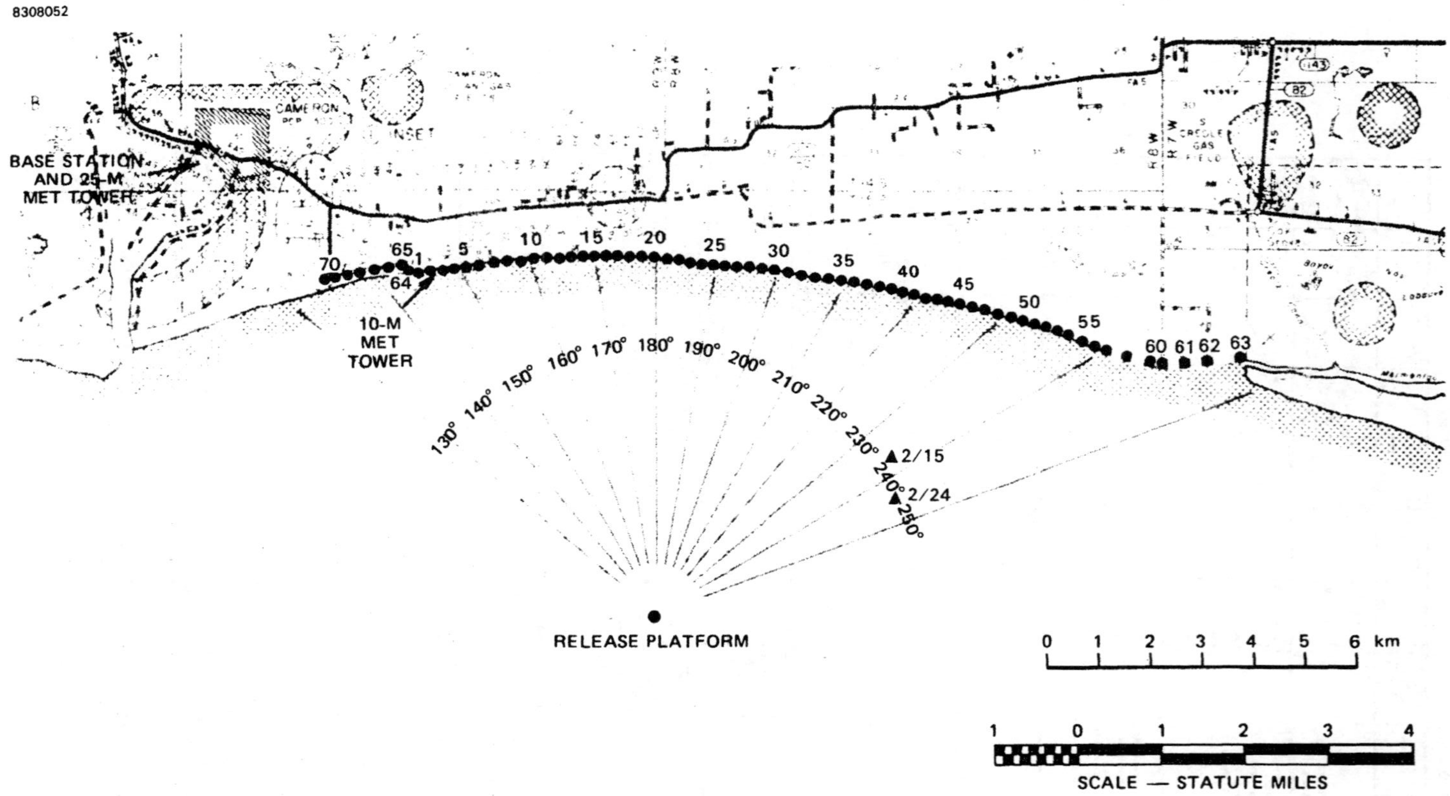

Figure 4 Tracer Sampling Sites Used in the Cameron, LA Experiment
The Diamonds give the Release Locations for 2/15 and 2/24

Meteorological data were quite extensive, including offshore turbulence measurements and vertical profiles from acoustic sounders, radiosondes, and aircraft. The data were randomly divided into developmental and test data sets. The model was refined using the developmental data set, which led to the inclusion of a formula for very stable σ_z based on the analysis of Cameron winter data. This analysis also showed that low mixing heights, z_i, frequently limited vertical mixing, for z_i was less than about 100 m in more than half of our data hours. Following this analysis, this model and the EPA MPTER model were evaluated using the independent test data base.

The model performance measures that were used are based on a simplification of the American Meteorological Society recommendations (Fox 1981) as reported by Hanna (1983). The basis of this simplified procedure is three standard statistical measures related to the differences in means (t), variances (F), and correlations (R).

Measure 1. $$t = \frac{\overline{x - y}}{s_{x-y}} \sqrt{n} \tag{39}$$

Measure 2. $$F = \max\left(s_x^2/s_y^2 \, , \; s_y^2/s_x^2 \right) \tag{40}$$

Measure 3. $$R = (z(r_1) - z(r_2)) \sqrt{n-3} \tag{41}$$

where $$z(r) = 0.5 \ln((1 + r)/(1-r)) \tag{42}$$

$$\bar{x} = \sum_{i=1}^{n} x_i/n \qquad \bar{y} = \sum_{i=1}^{n} y_i/n \tag{43}$$

$$s_x^2 = \sum_{i=1}^{n} (x_i - \bar{x})^2/n \qquad s_y^2 = \sum_{i=1}^{n} (y_i - \bar{y})^2/n \tag{44}$$

$$s_{x-y}^2 = \sum_{i=1}^{n} (x_i - y_i)^2/(n-1) - (\bar{x} - \bar{y})^2 \frac{n}{n-1}$$

$$r_j = \sum_{i=1}^{n} (C_{pji} - \overline{C_{pj}})(C_{oi} - \overline{C_o})/n s_{pj} s_o \qquad j = 1,2 \tag{45}$$

The parameters x and y are means, s_x^2 and s_y^2 are variances, and r is the correlation between the predictions of a model and observations. Correlation is transformed to the variable z by Equation (42) because z is more likely to be normally distributed.

For the top 25 concentrations unpaired in time and space, the t variable is modified by means of the Wilcoxen matched-pair, summed-rank test (Panofsky and Brier 1963). In this test, the top 25 predictions and observations are ranked from 1 to 50 (lowest to highest). Let T be the sum of the ranks of the predictions. Then,

$$t = \frac{T - 637.5}{51.54} \tag{46}$$

where 637.5 is the expected value of T (the sum of the ranks) if the two data sets have the same mean, and 51.54 is the expected standard deviation (for n=25).

Because these tests are normally used to assess whether a given statistic is significant at some confidence level, they are not immediately suitable for an extensive model comparison exercise in which a final "score" is estimated for each model. Therefore a weighting scheme was devised in which points are continuous in t, F, and R, as listed in Table 2. Note also that the scoring scheme is broken into two parts: model versus data and model 1 versus model 2.

The model was evaluated using all the monitors in Figures 2 through 4. The comparisons or pairings of observed and predicted concentrations were done in several ways, reflecting the need to choose pairings that are important for regulatory decisions as well as pairings that are important for scientific reasons:

Data Set 1: High 25 observed and predicted concentrations unpaired in space and time.

Data Set 2: Maximum concentration observed and predicted during a given hour at any receptor on a sampler arc (data paired in time but not in space).

Data Set 3: Maximum concentration observed and predicted at a given receptor at any hour (data paired in space but not in time).

Data Set 4: All data paired in time but with predicted plume axis lined up with direction to observed maximum concentration during that hour.

Data Set 5: All data paired in space and time.

The order of the five data sets is in decreasing priority of regulatory significance. The details of the weighting scheme are not given here but are described in the report by Hanna, Schulman, and Paine (1984). When the points were totaled, the new model's score was 50% higher than the MPTER model's score. The highest observed and predicted concentrations are listed in Table 3, showing agreement within a factor of two for the new model. The results for February 15 and 24 at Cameron are questionable because the tracer source was moved several kilometers from its usual location on an oil platform on those days. Consequently the meteorological data taken on the platform may have been unrepresentative of the source location. By chance, these two days happened to record the highest C/Q of any of the three experiments, due to the presence of a strong advection inversion.

TABLE 2

SUMMARY OF SIMPLIFIED SCORING SCHEME

	A Model vs. Data	B Model 1 vs. Model 2
Test 1	$x = C_o$, $y = C_p$ Points = 1 if $\lvert t\rvert \le 1$ $= \lvert t\rvert^{-1}$ if $\lvert t\rvert \ge 1$	$x = C_o - C_{p1}$, $y = C_o - C_{p2}$ All points to better model (smaller $C_o - C_p$) Points = 1 if $\lvert t\rvert \ge 3$ $= \lvert t\rvert/3$ if $t < 3$
Test 2	$x = C_o$, $y = C_p$ Points = F^{-1}	$x = C_o - C_{p1}$, $y = C_o - C_{p2}$ All points to better model (smaller s^2) Points = 1 if $F \ge 3$ $= (F-1)/2$ if $F < 3$
Test 3	r_1 is correlation between C_o and C_p r_2 is zero Points = 0 if $r_1 \le 0$ $= R/4$ if $r_1 > 0$ $= 1$ if $R > 4$	r_1 is correlation between C_o and C_{p1} r_2 is correlation between C_o and C_{p2} All points to model with higher r (assuming one r is greater than 0) Points = 0 if both $r \le 0$ Points = $R/4$ if one $r > 0$ Points = 1 if $R > 4$

TABLE 3

HIGHEST CONCENTRATIONS C/Q

Percent Over or Underprediction is Given in Parentheses

	Highest Hourly C/Q Observed	Highest Hourly C/Q New Model	Highest Hourly C/Q MPTER Model
Ventura	3010	2520 (16% under)	3410 (13% over)
Pismo Beach	9210	15490 (68% over)	24020 (161% over)
Cameron (with Feb. 15 and 24)	37010	20330 (45% under)	5240 (86% under)
Cameron (without Feb. 15 and 24)	9680	13590 (40% over)	4180 (57% under)

5. DISCUSSION

The model evaluation exercise illustrates that model predictions can be greatly improved if overwater observations of meteorological parameters can be obtained. High concentrations may be produced on shore if the mixing depth is low, as it was in over half of our data hours ($z_i \leq 100$ m), or if a strong inversion exists. In the winter, water temperatures near the shore are usually less than those offshore, leading to a surface-based advection inversion when onshore flow is present. Unfortunately, mixing depths and vertical temperature gradients are usually not available from offshore meteorological systems on buoys or platforms, and there is a great need for the development of observing systems for these parameters. Care must be taken in locating observing systems, since the shoreline region is by definition a transition zone, where conditions can vary rapidly over distances of 10 km or less.

It is helpful to the model if turbulence intensities i_y and i_z are observed on-site. Experience with these three data sets suggests that i_y can be measured fairly well by current instruments, but i_z is unreliable due to the difficulty in precise orientation of the instrument. Some problems were also found with surface water temperature observations, T_s, if they were not taken at the same location as the air temperature observations. In about half of our test runs, the observations of i_z and T_s-T_a become irrelevant because the plume at the shoreline was well mixed vertically between the surface and height z_i. This phenomenon is caused by the combination of low mixing heights ($z_i \leq 100$ m), low-level neutrally-buoyant releases, and plume travel distances of more than 5 km. In these cases, the only effective difference between models is in their σ_y formulation, where the new model uses i_y to predict σ_y,and the MPTER model uses the Pasquill-Gifford curves. This difference can lead to overpredictions by the MPTER model (for example, the highest concentration at Pismo Beach in Table 3) during stable, light wind conditions, when observed i_y is relatively high.

These tracer studies did not use buoyant plumes, and therefore the question never came up of penetration of the inversion capping the relatively shallow mixed layer. We expect that inversion penetration will be a real problem for buoyant plumes from oil platforms and ships, due to the high frequency of mixing depths of 100 m or less over water. In these situations the partial penetration capabilities of the new model should be quite realistic. The EPA models treat penetration as an all or none situation, and if the plume rise is calculated to be only one meter above z_i, no ground level impact of the plume is predicted. In contrast, our model would predict that about one-half of the plume would remain in the mixed layer in this situation.

Acknowledgements

This research was supported by the U.S. Department of the Interior, the Exxon Production Research Company, and the Western Oil and Gas Association.

REFERENCES

Briggs, G.A.: 1969, Plume Rise, USAEC Critical Review Series, TID-25075, Clearinghouse for Federal Scientific and Technical Information.

Briggs, G.A.: 1973, 'Diffusion Estimates for Small Emissions', ATDL Cont. 79, ATDL, P.O. Box E, Oak Ridge, TN.

Briggs, G.A.: 1975, 'Plume Rise Predictions', Lectures on Air Pollution and Environmental Impact Analyses, American Meteorological Society, Boston, MA, pp 59-111.

Businger, J.A.: 1973, 'Turbulence Transfer in the Atmospheric Surface Layer', Workshop in Micrometeorology, Am. Meteorol. Soc., 45 Beacon St., Boston, MA 67-100.

Caughey, S.J.: 1982, 'Observed Characteristics of the Atmospheric Boundary Layer'. Atmospheric Turbulence and Air Pollution Modeling, F.T.M. Nieuwstadt and H. Van Dop (eds.) Reidel Dordecht, Holland.

Deardorff, J.W. and G.E. Willis: 1982, 'Ground Level Concentrations Due to Fumigation into an Entraining Mixing Layer; Atmos. Environ., 16, 1159-1170.

Fox, D.G.: 1981, 'Judging Air Quality Model Performance: A Summary of the AMS Workshop on Dispersion Model Performance; Bull. Am. Meteorol. Soc., 62, 599-609.

Gifford, F.A.: 1968, 'An Outline of Theories of Diffusion in the Lower Layers of the Atmosphere; In: Meteorology and Atomic Energy - 1968. D.H. Slade (ed.). USAEC Report TID-24190, pp 66-116.

Golder, D: 1972, 'Relations Among Stability Parameters in the Surface Layer', Bound. Lay. Meteorol, 3, 47-58.

Hanna, S.R.: 1983, 'A Simplified Scoring Scheme for Air Quality Models; Paper 83-36.6 at Annual Air Pollution Control Association Meeting, available from APCA, Pittsburgh.

Hanna, S.R., G.A. Briggs, J. Deardorff, B.A. Egan, F.A. Gifford, and F. Pasquill: 1977, 'AMS Workshop on Stability Classification Schemes and Sigma Curves', Bull. Am. Meteorol. Soc., 58, 1305-1309.

Hanna, S.R., R.J. Paine and L.L. Schulman: 1984, 'Users Guide to the Offshore and Coastal Dispersion (OCD) Model; prepared for the U.S. Department of the Interior by ERT, Inc., 696 Virginia Rd., Concord, MA.

Heffter, J.L.: 1965, 'The Variation of Horizontal Diffusion Parameters with Time for Travel Periods of One Hour or Longer; J. Appl. Meteorol., 4, 153-156.

Holland, J.Z.: 1981, 'Contributions of GARP to our Knowledge of Tropical Atmospheric Phenomena', AAAS Annual Meeting, Toronto, Canada, 8 January.

Hosker, R.P.: 1974, 'A Comparison of Estimation Procedures for Overwater Plume Dispersion', Proceedings of the Symposium on Atmospheric Diffusion and Air Pollution, American Meteorological Society, 281-288.

Hsu, S.A.: 1979, 'An Operational Forecasting Model for the Variation of Mean Maximum Mixing Heights across the Coastal Zone,' Bound. Lay. Meteorol., 16, 93-98.

Jackson, N.A.: 1976, 'The Propagation of Modified Flow Downstream of a Change in Roughness,' Q.J. Roy. Met. Soc., 102, 924.

Kondo, J.: 1975, 'Air-Sea Bulk Transfer Conditions in Diabatic Conditions,' Bound. Lay. Meteorol., 9, 91-112.

Lenschow, D.H., J.C. Wyngaard, and W.T. Pennell: 1980, 'Mean Field and Second-Moment Budgets in a Baroclinic, Convective Boundary Layer,' J. Atmos. Sci. 37: 1313-1326.

Lyons, W.A.: 1975, 'Turbulent Diffusion and Pollutant Transport in Shoreline Environments,' Lectures on Air Poll. and Environ. Impact Analyses, American Meteorological Society, 136-208.

Nieuwstadt, F.T.M.: 1977, 'The Dispersion of Pollutants over a Water Surface," Eight Int. Tech. Meeting on Air Poll. Modeling and its Applic. NATO/CCMS Doc. No. 80, 337-359.

Panofsky, H.A. and G.W. Brier: 1963, 'Some Applications of Statistics to Meteorology', The Pennsylvania State University Press, University Park, PA, 224 pp.

Panofsky, H.A., H. Tennekes, D.H. Lenschow and J.C. Wyngaard: 1977, 'The Characteristics of Turbulent Velocity Components in the Surface Layer under Convective Conditions; Bound. Lay. Meteorol., 11, 355-361.

Raynor et al.: 1979, 'Formation and Characteristics of Coastal Internal Boundary Layers During Onshore Flow,' Bound. Lay. Meteorol., 16, 487-514.

Schacher, G.E., K.L. Davidson and C.W. Fairall: 1982, 'Atmospheric Marine Boundary Layer Convective Mixing Velocities in the California Coastal Region,' Atmos. Environ., 16, 1183-1192.

Schulman, L.L. and J.S. Scire: 1980, 'Development of an Air Quality Dispersion Model for Aluminum Reduction Plants', Document P07304A. Environmental Research & Technology, Inc., Concord, MA.

Strimaitis, D.G.., A. Venkatram and T.F. Lavery: 1983, 'A Model to Estimate Concentrations during Plume Impingement,' Preprints, Sixth Symposium on Turbulence and Diffusion, Am. Meteorol. Soc., 28-31.

Turner, D. 1969, Workbook of Atmospheric Dispersion Estimates. Office of Air Programs Publication No. AP-26. U.S. EPA, Research Triangle Park, NC.

Zilitinkevich, S.S.: 1975, 'Resistance Laws and Prediction Equations for the Depth of the Planetary Boundary Layer,' J. Atmos. Sci., 32, 741-752.

Questions, answers and comments

J. TADMOR: Referring to the fumigation case which you described for a continuous release from a stack located at the sea-shore, would you expect that such an "explosive vertical mixing" might occur following an accidental, instanteneous puff release, e.g. from a nuclear reactor at the same site?

S.R. HANNA: Yes, the same boundary layer physics will apply to the instanteneous puff as well as to the continuous plume. We would expect the prediction of ground level concentration for the puff to be highly uncertain though, because a given puff may become entrained in a particular updraft or downdraft. The average concentration over a great number of puffs would correspond to the model prediction.

T.W. HORST: With regard to wind-shear direction, is there a default scheme for lack of input data?

S.R. HANNA: In this case we use climatological measurements for the input data.

F.B. SMITH: Since the source was elevated above the sea surface, were there any occasions when, due to the stability over the sea, the concentrations were found to be higher at the more distant inland samplers than at the nearer coastal samplers?

S.R. HANNA: The source height for all three tracer experiments was about 10 to 15m. Since the sources were about 5km from the coastal samplers, there was very little effect of source height on the observed concentration patterns, and the maximum concentrations were consistently observed at the row of samplers closer to the source.

R. BERKOWICZ: How do you estimate stability classes over water? Are the measurements of σ_w accurate enough for determination of σ_z?

S.R. HANNA: Stability classes were estimated using the Golder technique, based on a knowledge of the roughness length z_0 and the Monin-Obukhov length L. Overwater measurements of σ_w are difficult and hence are usually unreliable for input to the model, in most cases. I would trust only σ_w data obtained by experienced boundary layer research groups.

THE MODIFICATION OF PLUME MODELS TO ACCOUNT FOR DRY DEPOSITION

Thomas W. Horst
Pacific Northwest Laboratory
Richland, WA 99352, U.S.A.

ABSTRACT. The assumptions and predictions of four diffusion-deposition models are compared, and two simple plume depletion models are recommended. One model applies an analytical, constant eddy-diffusivity solution of the advection-diffusion equation as a deposition correction to the general Gaussian plume model. Predictions of this model compare moderately well with those of the surface depletion model, an exact treatment of plume depletion, and it is particularly useful for estimating the transport and deposition of settling particles. The second model is a correction to the simple source depletion model that also accounts for the change in the vertical concentration profile caused by deposition. The computational requirements of this model are similar to those of the unmodified source depletion model, while its predictions near the surface are very close to those of the surface depletion model.

1. INTRODUCTION

Models for the estimation of airborne pollutant concentrations commonly focus on atmospheric transport and diffusion processes, and ideally these models are evaluated with data from experiments using conservative, nondepositing, nonreacting tracers. However, real atmospheric pollutants (and, often, convenient atmospheric tracers) usually deposit at some rate on the earth's surface. As a consequence, atmospheric diffusion models must be modified to account for the loss of airborne pollutants by dry deposition, as well as to estimate the resulting surface contamination.

Methods for modifying atmospheric transport and diffusion models to account for deposition depend strongly on the type of model being considered. For this reason, it is useful to divide diffusion models into two broad classes. The first type estimates airborne pollutant concentrations by directly solving the advection-diffusion equation for all points in the spatial domain of interest, e.g., by using a gradient-transfer assumption. Since deposition is naturally expressed as a flux at the lower boundary of the atmosphere, it is readily incorporated

Boundary-Layer Meteorology 30 (1984) 413–430. 0006–8314/84/0304–0413$02.70

into a solution of the advection-diffusion equation as one of the required boundary conditions. The other type of diffusion model, e.g., Gaussian plume, similarity, and statistical models, infers the shape and dimensions of the pollutant concentration distribution from the mean and turbulent structure of the atmosphere and/or from the behavior of atmospheric tracers. Since this type does not explicitly treat the vertical transport process, it is difficult to modify for deposition in a way that is physically realistic. Modification of this second type of diffusion model, which I will call plume models, is the main subject of this review.

Depositing pollutants can also be divided into two classes, depending on the relative contributions of gravitational settling and turbulent diffusion to vertical atmospheric transport. For gases and for particles smaller than approximately 10 μm radius, settling may be neglected compared to turbulent diffusion. These nonsettling pollutants will be the chief concern of this review since they include respirable particles and since they are transported to greater distances from the source than settling particles. For nonsettling pollutants, modification of the diffusion model is principally confined to a layer near the surface that is depleted by deposition. For settling particles, the additional vertical transport caused by gravitational settling must be included throughout the plume.

We begin with a short discussion of the inclusion of dry deposition in direct solutions of the advection-diffusion equation, followed by a general discussion of plume models, and then by the main topic of this review, a comparison of four plume depletion models for non-settling pollutants. These four models include an exact, surface depletion model, two source depletion models, and a constant eddy-diffusivity solution of the advection-diffusion equation.

2. DIRECT SOLUTION OF THE DIFFUSION EQUATION

Direct solution of the differential, advection-diffusion equation is a very fundamental approach to estimating concentrations of airborne pollutants, and the inclusion of deposition is straightforward. With gravitational settling and deposition and, e.g., the gradient-transfer assumption, the required boundary condition at the surface is (Calder, 1961),

$$-K(z)\frac{\partial C}{\partial z} - v_s C = F = -v_d C(z_d) \quad , \qquad (1)$$

where $K(z)$ is the vertical eddy diffusivity, C is mean airborne pollutant concentration, and v_s is the pollutant settling velocity. The vertical pollutant flux F at the surface is a downward deposition flux, parameterized as a deposition velocity v_d multiplied by the airborne concentration at some near-surface height z_d.

Solution of the advection-diffusion equation with the gradient-transfer assumption requires specification of the spatial distribution of wind and eddy diffusivity, and realistic distributions of these variables generally require solution by a numerical (e.g., finite-difference) method. Since computer costs increase rapidly with the number of grid points, it is desirable to make the spacing of grid points as large as possible. This requires some caution in the application of Eq. (1). The first point to be noted is that v_d is a function of the reference height z_d. Assuming that the deposition flux near the surface is independent of height, integration of Eq. (1) shows that the concentration C increases with height and v_d decreases

$$C(z) = C(z_d)[1 + v_d(z_d)R(z,z_d)] \tag{2}$$

$$v_d(z) = v_d(z_d)/[1 + v_d(z_d)R(z,z_d)] \quad , \tag{3}$$

where the atmospheric resistance R is

$$R(z,z_d) = \int_{z_d}^{z} \frac{dz'}{K(z')} \quad , \tag{4}$$

and for clarity we have neglected gravitational settling. Thus for a finite vertical grid spacing, the deposition velocity must be reduced to account for the atmospheric resistance of the lowest grid interval, and the inverse of the atmospheric resistance provides an upper limit to the deposition velocity that can be used in the model (Doran and Horst, 1977).

The second point is that the lowest grid interval should be small enough to neglect divergence of the vertical flux below the height where the deposition boundary condition is applied. This divergence is a necessary consequence of horizontal divergence or changes with time of the near-surface concentration (Slinn, 1984). Considering the horizontal divergence resulting from diffusion downwind of a point source, the requirement is roughly $s/x \ll v_d/u$, where s is the height of the lowest grid interval, x is distance downwind of the source, and u is the mean wind speed. It should be noted that a smaller deposition velocity requires a smaller value of the lowest grid interval.

Since realistic solutions of the advection-diffusion equation generally require numerical methods, they are expensive to implement. In addition, unless the eddy diffusivities are artificially made a function of distance from the source, the gradient-transfer assumption is strictly applicable only to vertical diffusion from ground-level sources. It does not properly describe lateral plume diffusion, vertical plume diffusion in the vicinity of an elevated source, or vertical diffusion in daytime convective conditions. Although the latter difficulties may be overcome with more sophisticated techniques, e.g., second-order closure or large eddy simulations, these are also likely to increase the computational complexity and cost.

3. PLUME MODELS OF ATMOSPHERIC DIFFUSION

Because solution of the advection-diffusion equation is, in general, computationally complex and the assumptions necessary to obtain a solution are in many cases theoretically unsatisfactory, several methods have been developed to infer the shape and dimensions of a plume from the mean and turbulent structure of the atmosphere and/or from the behavior of atmospheric tracers. Often these models are analytical, and thus they are easier to apply and their predictions are easier to understand than is the case for a numerical solution of the diffusion equation. However, their analytical form and their empirical basis also makes them more difficult to modify to account for dry deposition.

The most apparent example of a plume model is the Gaussian plume model. The binormal spatial distribution of the Gaussian model follows from the constant eddy-diffusivity solution of the advection-diffusion equation, but it is principally based on statistical arguments and direct observations of plumes from point sources of atmospheric tracers. The Gaussian model is easy to understand because the binormal distribution is an analytical function that is amendable to simple mathematical analysis. The input to the model is also quite simple: wind speed, wind direction and the dimensions of the plume as a function of distance from the source. Since estimates of the plume dimensions are based on observations, the model predictions can be fairly realistic in some situations.

Despite the utility of the Gaussian plume model, it has been found in many cases to be only an approximate description of the vertical distribution, even for a nondepositing contaminant. Some examples are the vertical diffusion from a ground-level source (Neiuwstadt and van Ulden, 1978) and vertical diffusion during daytime convective conditions (Lamb, 1982). Because these non-Gaussian distributions are also empirical, it is quite important to note that many of the plume depletion techniques are adaptable to any plume model.

Since the atmospheric motions contributing to the horizontal, crosswind diffusion of a plume are usually of a different nature from those effecting the vertical diffusion, virtually all plume models treat these two processes separately. For example, the binormal Gaussian distribution is easily divided into its horizontal and vertical components, and its horizontal and vertical dimensions are commonly estimated from different empirical curves or from quite different theoretical analyses of the diffusion process. With the assumption that the vertical and horizontal distributions are independent of one another (neglecting, for one thing, the effects of vertical wind shear), it follows that deposition does not alter the horizontal distribution, and our discussion can deal solely with the crosswind-integrated distribution,

$$C(x,z) = \int_{-\infty}^{\infty} C(x,y,z)\, dy = Q(x)D(x,z) \quad . \tag{5}$$

The preceeding equation also introduces the horizontal, downwind pollutant flux Q(x),

$$Q(x) = \int_{z_d}^{\infty} u(z)C(x,z)\, dz = Q_o - \int_o^x v_d C(x',z_d)\, dx' \quad . \tag{6}$$

and the vertical diffusion function D, defined as the crosswind-integrated concentration C(x,z) normalized by the horizontal flux,

$$D(x,z) = C(x,z)/\int_{z_d}^{\infty} u(z)C(x,z)\, dz \quad , \tag{7}$$

For a nondepositing pollutant, we define $D=D_o$ and $Q(x)=Q_o$.

Q(x) and D contain the two effects of dry deposition on the plume. The suspension ratio, $Q(x)/Q_o$, describes the total loss of material from the plume by deposition, and D/D_o (defined later as a profile correction P) describes the effect of deposition on the vertical distribution. Q(x) is related to $D(x,z_d)$ by the mass balance, Eq. (6), and hence

$$Q(x) = Q_o \exp[-\int_o^x v_d D(x',z_d,h)\, dx'] \quad . \tag{8}$$

Thus the principal task of a plume depletion model is to describe the depleted diffusion function D.

4. SURFACE DEPLETION MODEL

The surface depletion model (Horst, 1977) is an exact modification of nondepositing plume models to account for dry deposition. It is a special case of the surface flux model (Horst, 1978), which estimates the airborne contamination resulting from a distributed, surface source of depositing material. If the source per unit area is S(x,y), then the net flux at the surface is

$$F(x,y) = S(x,y) - v_d C(x,y,z_d) \tag{9}$$

and the crosswind-integrated, airborne contamination is

$$C(x,z) = \int_o^x F(x')\, D_o(x-x',z,h=o)\, dx' \quad . \tag{10}$$

A formal discussion of this equation can be found in Horst and Slinn (1984), but it is based simply on the linearity of the advection-diffusion equation and the principle of superposition.

The surface depletion model for a plume follows by adding the conventional point source at height h and neglecting the surface source S,

$$C(x,z) = Q_o D_o(x,z,h) - \int_o^x v_d C(x',z_d)\, D_o(x-x',z,o)\, dx' \quad . \qquad (11)$$

The loss of airborne material by deposition is effected by a sink at the surface, whose influence is propagated vertically by the diffusion function for a ground-level source. Note that the depleted plume $C(x,z)$ is determined directly from the diffusion function D_o for a non-depositing pollutant. In general, the depletion integral in Eq. (11) must be evaluated numerically, but in contrast to Eq. (8) this task is exacerbated by the fact that the integrand of the depletion integral is a function of the receptor coordinate x. Yamartino (1981) has derived a formal solution of Eq. (11) for particular forms of $\sigma_z(x)$.

The Lagrangian similarity diffusion model (van Ulden, 1978; Horst, 1979) currently provides one of the best descriptions of diffusion from a ground-level source, and thus is a leading candidate for specifying the diffusion function $D_o(x,z,o)$ in the integrand of Eq. (11). However,

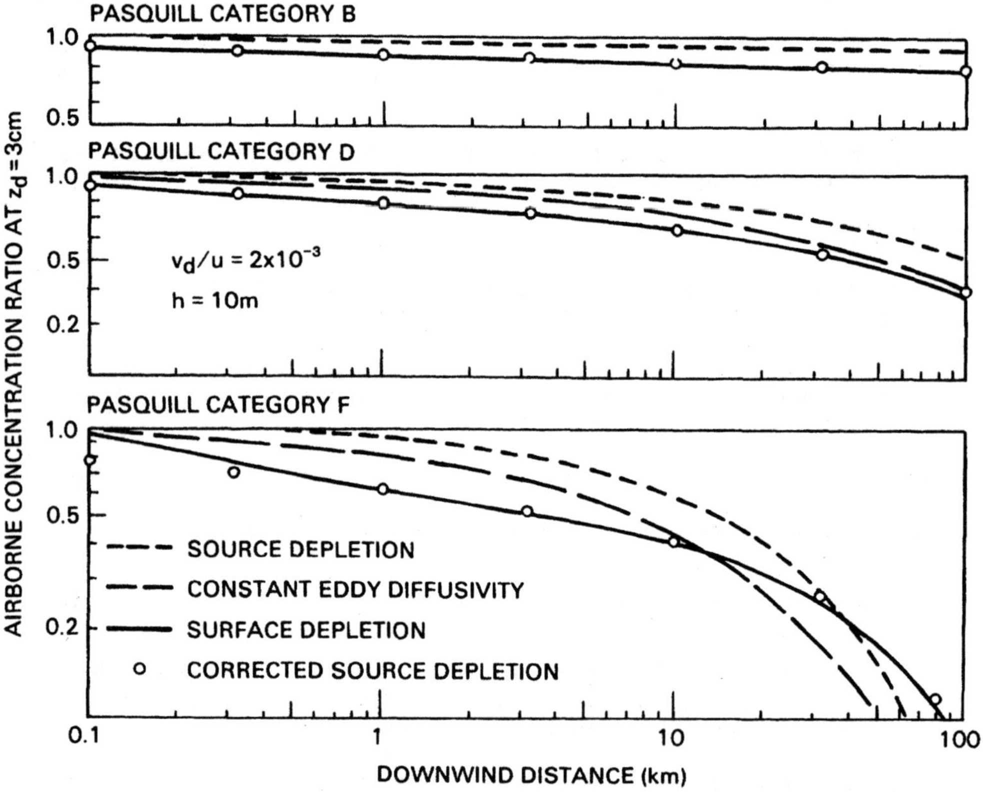

Figure 1. The ratio of near-surface (z_d=3cm) airborne concentration with deposition to that without deposition, as a function of downwind distance and stability, for v_d/=0.002 and h=10m.

to compare the plume depletion models discussed below, the Gaussian plume diffusion model will be used with the surface depletion model. The Gaussian diffusion function for a point source of nondepositing pollutant at height h is

$$D_o(x,z,h) = \frac{1}{\sqrt{2\pi}\,u\sigma_z}\left\{\exp[-(h-z)^2/2\sigma_z^2] + \exp[-(h+z)^2/2\sigma_z^2]\right\} . \quad (12)$$

where σ_z is the vertical dimension of the concentration distribution. The Gaussian model is used here because of its simplicity, generality and widespread application. The identical diffusion model is used with each plume depletion model in order to examine the differences in the depletion models, isolated from differences in deposition caused by differences in the diffusion models.

Figures 1-4 compare the predictions of three simpler plume depletion models to those of the surface depletion model. The discussions of these simpler models relate the differences in the model predictions

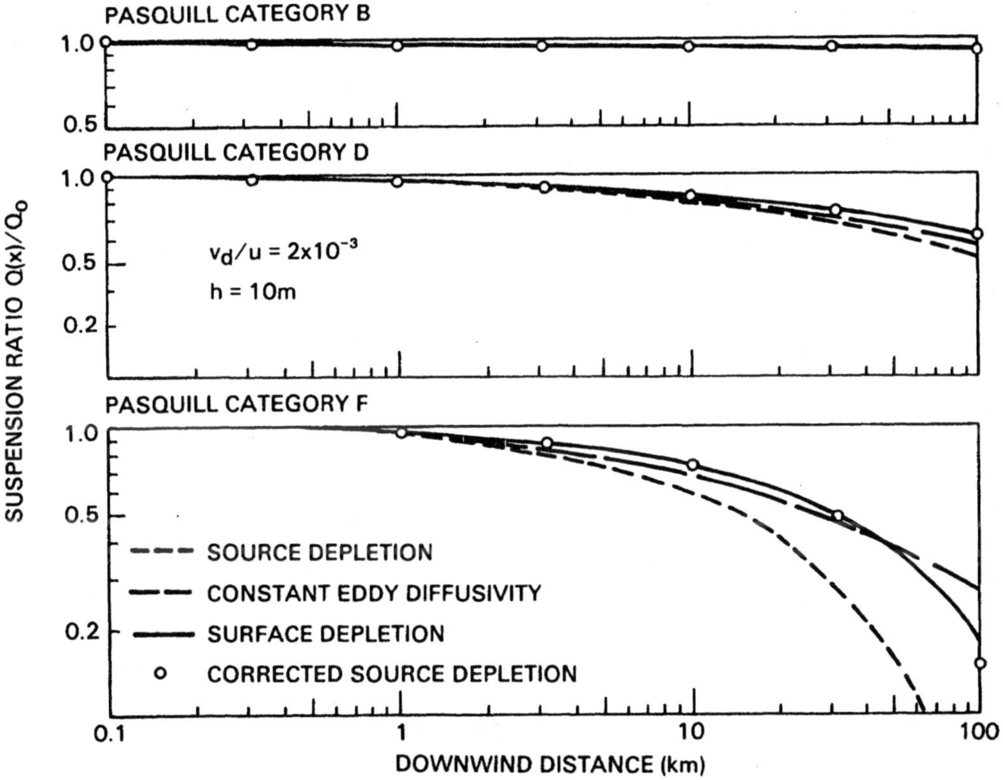

Figure 2. The suspension ratio, as a function of downwind distance and stability, for $v_d/u=0.002$ and h=10m.

to the assumptions on which the models are based. The comparison in Fig. 1 and 2 is done for a source height of 10 m, $v_d/u=0.002$, $z_d=3$ cm and Pasquill diffusion categories B, D and F. With a wind speed of 5 m/s, the ratio $v_d/u=0.002$ corresponds to a deposition velocity of 1 cm/s. Values for σ_z were obtained from Briggs' formulas for rural dispersion (Gifford, 1976). Figure 1 shows the ratio of the near-surface pollutant concentration with deposition to that without deposition, $C(z_d)/C_o(z_d)$, as a function of downwind distance, and Fig. 2 shows the suspension ratio. The dependence of plume depletion on deposition velocity and source height can be inferred from Fig. 3 and 4, which show similar model comparisons for Pasquill category D, $v_d/u=0.01$, and h=10 m and 100 m.

5. SOURCE DEPLETION MODEL

One of the earliest, and probably the most used, of the plume depletion models makes the simple assumption that the vertical distribution is unaffected by deposition; i.e., $D=D_o$ (Chamberlain, 1953). It is called a source depletion model because it is described completely

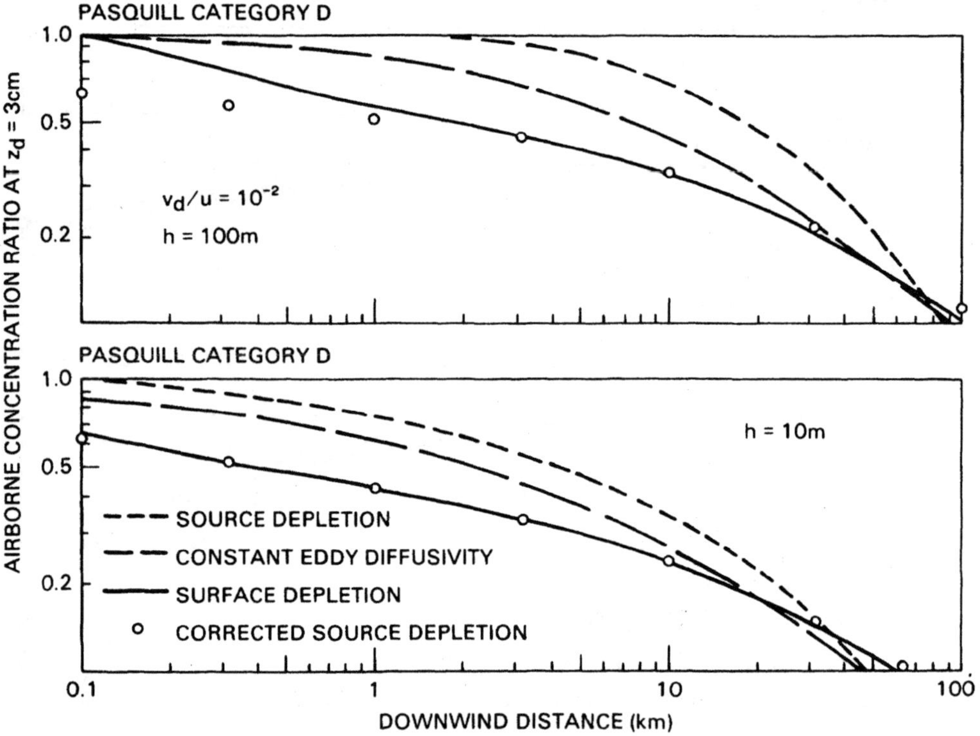

Figure 3. The ratio of near-surface (z_d=3cm) airborne concentration with deposition to that without deposition, as a function of downwind distance, for Pasquill stability category D, $v_d/u=0.01$, h=10m and h=100m.

by Eq. (8), which accounts for the deposition losses by appropriately reducing the source strength with downwind distance. For comparison with the surface depletion model, the source depletion model can be written in the form,

$$C(x,z) = Q_o D_o(x,z,h) - D_o(x,z,h) \int_0^x v_d C(x',z_d)\, dx' \quad . \qquad (13)$$

Comparison of the diffusion functions in the depletion terms of Eq. (11) and Eq. (13) emphasizes the assumption of the source depletion model that deposition is a loss at the source rather than locally, at the surface.

The source depletion model instantaneously distributes the effect of deposition throughout the vertical extent of the plume, whereas in fact the plume is preferentially depleted at ground level. As a consequence, it overestimates the ground-level pollutant concentration, overestimates the deposition, and underestimates the suspension ratio. This is evident in the comparison with surface depletion, Fig. 1-4. However plume depletion is a self-limiting process that reduces errors

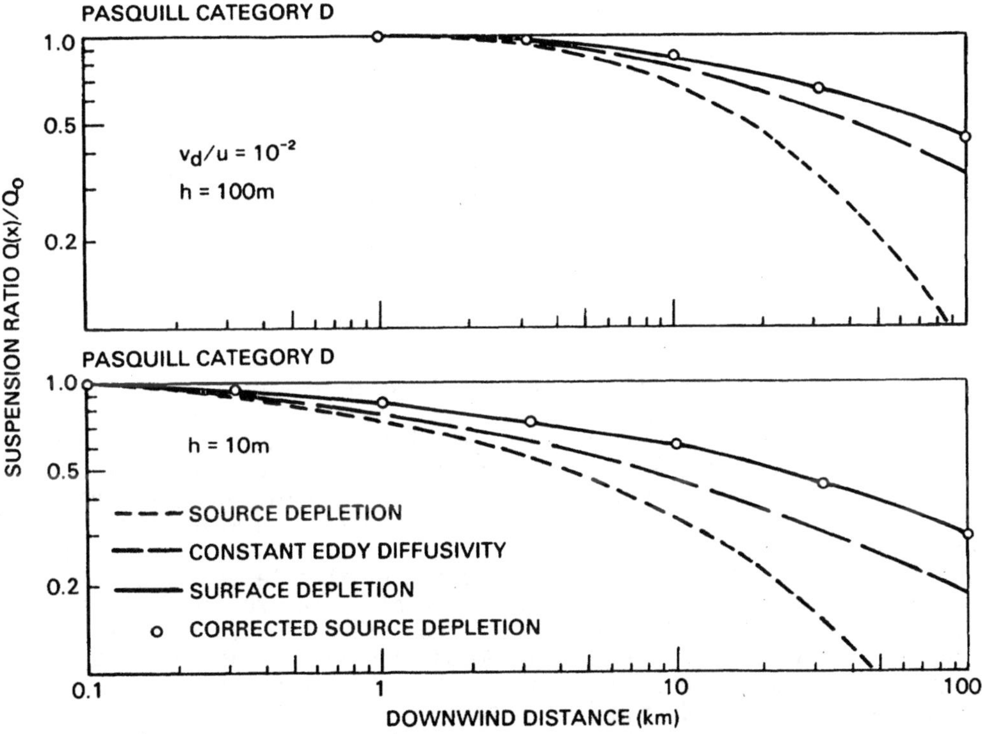

Figure 4. The suspension ratio, as a function of downwind distance, for Pasquill stability category D, $v_d/u=0.01$, h=10m and h=100m.

in the model: underestimation of the suspension ratio, caused by overestimation of the deposition, reduces overestimation of the ground-level concentration. It may be seen in Fig. 1 and 3 that, as a consequence, the ground-level concentration is overestimated near the source and is underestimated far from the source.

From Fig. 1-4 it is also evident that the errors of the source depletion model are small except for situations with limited atmospheric mixing (Pasquill F) or high deposition ($v_d/u=0.01$). Source depletion is an acceptable approximation when the rate of deposition is small compared to the vertical growth rate of the plume, $v_d \ll d\sigma_z/dt$ (Pasquill, 1974). This criterion is consistent with the comparison of source and surface depletion illustrated in Fig. 1-4. Berkowicz and Prahm (1978) and Corbett (1981) have derived similar criteria from gradient-transfer solutions of the advection-diffusion equation. They compared solutions that use the source depletion assumption to account for deposition to solutions that directly use the deposition flux boundary condition, Eq. (1). Corbett found that for low-level releases the source depletion model estimates the suspension ratio within a factor of two when $v_d/u<\sigma_z/x$.

6. A CORRECTION TO THE SOURCE DEPLETION MODEL

6.1 The Profile Correction

Horst (1980, 1983) has developed a modified source depletion model that gives results in close agreement with the surface depletion model, while the computational requirements are comparable to those of the source depletion model. This model calculates the suspension ratio using Eq. (8), but modifies the nondepositing diffusion function D_o by a profile correction $P(x,z)$ to account for the change in the vertical distribution caused by deposition,

$$C(x,z) = Q(x)D(x,z,h) = Q(x)D_o(x,z,h)P(x,z) \quad . \tag{14}$$

To estimate the profile correction P, we note that it will play its most important role near the surface: the change in the vertical distribution caused by deposition is greatest near the surface, and a correct value for the near-surface concentration is necessary for determining the deposition flux and the suspension ratio. When there is no deposition, there is a layer near the surface in which the pollutant concentration is independent of height. For a Gaussian plume at a downwind distance such that $\sigma_z>h$, the depth of this layer is roughly $0.2\sigma_z$. With deposition, the vertical concentration profile in this layer will be determined solely by the deposition flux to the surface. Assuming that within this layer the deposition flux is independent of height, we can substitute Eq. (14) into Eq. (2) and use the near-surface approximation $D_o(x,z)=D_o(x,z_d)$ to obtain

$$P(x,z) = P(x,z_d)[1 + v_d R(z,z_d)] \quad . \tag{15a}$$

Conserving mass by substituting Eq. (14) and (15a) into the first half of Eq. (6), we get

$$P(x,z_d) = 1/[1 + v_d \int_{z_d}^{\infty} uD_o(x,z,0)R(z,z_d)\, dz] \quad . \tag{15b}$$

Here the integrand has been simplified with the approximation $D_o(x,z,h) \simeq D_o(x,z,0)$. This is justified by the observation that almost all deposition occurs beyond the downwind distance where $\sigma_z = h$. The diffusion function D_o has its maximum value at that distance and goes rapidly to zero at smaller downwind distances, e.g., $D_o(\sigma_z = h/4) = D_o(\sigma_z = h)/400$. Thus Eq. (15b) need only be accurate for $\sigma_z > h$, and $h \simeq 0$ is a reasonable approximation in that region.

6.2 Atmospheric Resistance

The atmospheric resistance to vertical transport, Eq. (4), is a familiar concept in deposition modeling (Wesley and Hicks, 1977). For a passive contaminant, the eddy diffusivity near the surface is

$$K(z) = u_* kz/\phi \tag{16}$$

where u_* is the eddy friction velocity, k is von Karman's constant, and ϕ is a function of atmospheric stability often obtained from observations of the turbulent transport of sensible heat. For neutral stability ϕ_H is a constant α and thus

$$R(z,z_d) = \frac{\alpha}{u_* k} \ln(z/z_d) \tag{17}$$

and

$$P(x,z_d) \simeq 1/\left\{1 + \frac{v_d \alpha}{u_* k}\left[\ln(\sigma_z/z_d) - 1\right]\right\} \quad . \tag{18}$$

Corresponding results for non-neutral atmospheres can be derived from the known dependence of ϕ_H on stability (Paulson, 1970).

For the corrected source depletion model to be physically realistic, it is important that the atmospheric structure used to calculate the resistance be consistent with that used to calculate the diffusion. This is not straightforward for the usual Gaussian diffusion model where the atmospheric structure is implied by the rate of growth of the plume. However, the integrand of Eq. (4) can be heuristically transformed from a function of K to a function of σ_z by considering the solution of the advection-diffusion equation for a constant eddy diffusivity. The solution for a point source is a Gaussian plume with

$$\sigma_z(d\sigma_z/dx) = K/u \quad . \tag{19}$$

Further, the first and second moments of a Gaussian plume from a ground-level source are related by

$$\bar{z} = \sqrt{2/\pi}\,\sigma_z \quad . \tag{20}$$

Substitution of Eq. (19) and (20) into Eq. (4) gives

$$R(z,z_d) = \sqrt{\frac{2}{\pi}}\,\frac{1}{u}\int_{z_d}^{z}\frac{dz'}{z'(d\sigma_z/dx)} \quad . \tag{21}$$

As an example, for $\sigma_z = ax$:

$$R(z,z_d) = \sqrt{\frac{2}{\pi}}\,\frac{1}{au}\,\ln(z/z_d) \tag{22a}$$

$$P(x,z) \simeq 1/\left\{1 + \frac{v_d}{u}\sqrt{\frac{2}{\pi}}\,\frac{1}{a}\left[\ln(\sigma_z/z_d) - 1\right]\right\} \quad . \tag{22b}$$

Examples for other analytical forms of σ_z are given by Horst (1983).

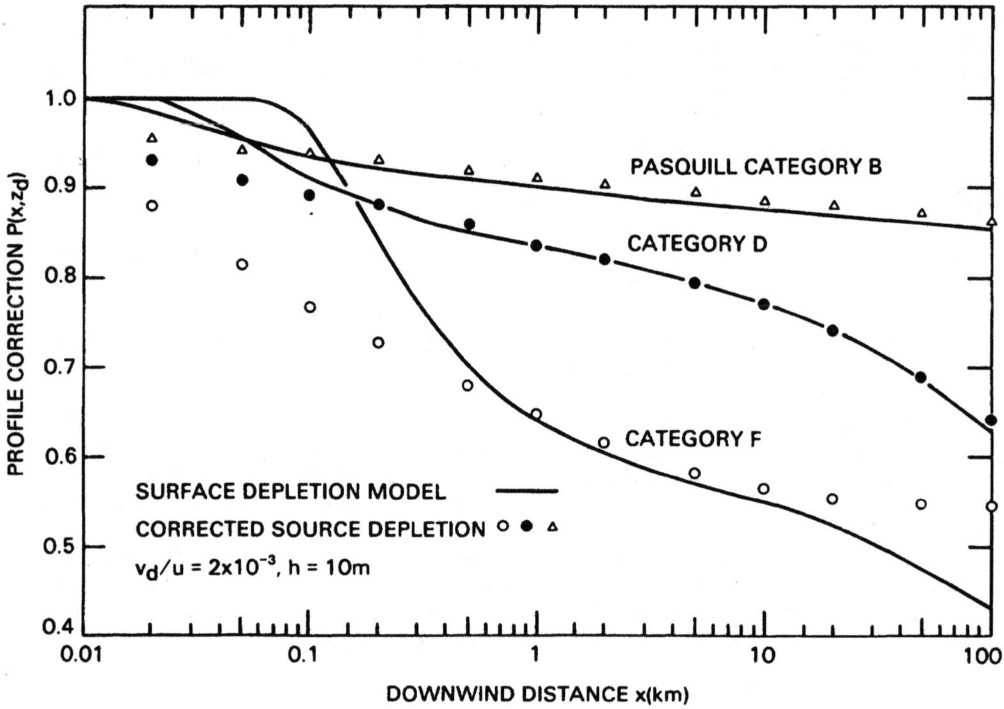

Figure 5. The near-surface (z_d=3cm) profile correction as a function of downwind distance and Pasquill stability category.

6.3 Comparison to Surface Depletion

The assumptions and approximations used to derive Eq. (15) for the profile correction and Eq. (21) for the atmospheric resistance can be tested by comparing the predictions of the corrected source depletion model with those of the surface depletion model. Figure 5 shows the near-surface profile correction $P(x,z_d)$ as a function of downwind distance and atmospheric stability, for the conditions of Fig. 1 and 2. The unmodified source depletion model assumes P is equal to one, corresponding to instantaneous vertical mixing of the losses due to deposition. Since atmospheric resistance is low for Pasquill category B and high for category F, the profile correction is close to unity for category B and furthest from unity for category F.

The profile function predicted by the corrected source depletion model is in good agreement with that calculated from the surface depletion model, particularly near and beyond the distance where σ_z=h (80 m for Pasquill B, 200 m for D and 800 m for F). This is consist-

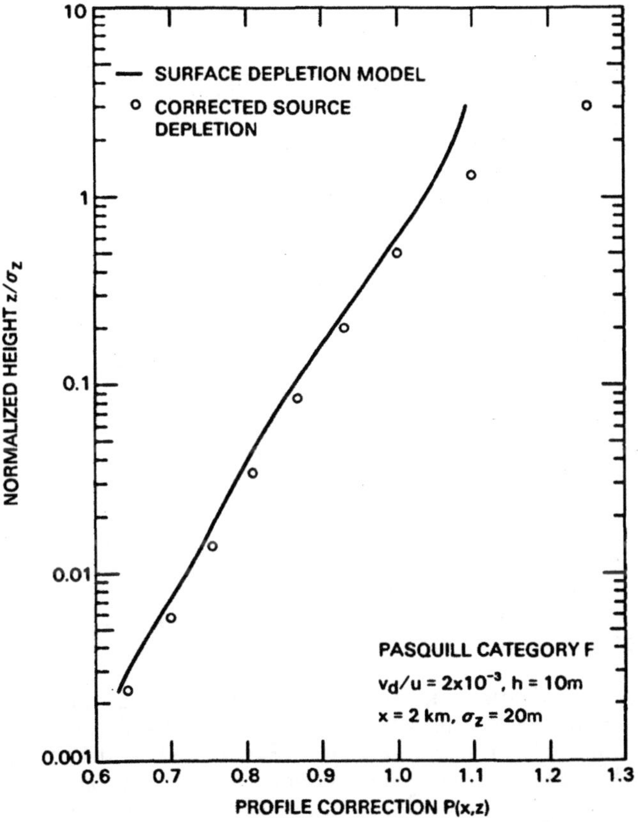

Figure 6. The profile correction as a function of height for a case with σ_z/h=2.

ent with the assumptions made in deriving Eq. (15b). The errors in $P(z_d)$ upwind of that point are of little consequence because the elevated plume has not yet reached the surface and hence both the near-surface concentration and the deposition are negligible. The differences beyond 10 km in $P(z_d)$ for Pasquill F are principally due to approximations used in deriving analytical expressions for the integrals in Eq. (21) and (15b) (Horst, 1983).

Figure 6 shows the profile correction $P(x,z)$ as a function of z/σ_z for Pasquill category F and a distance of 2 km, which is downwind of the distance where $\sigma_z=h$. The corrected source depletion model agrees quite well with the surface depletion model, and the agreement extends well above the height $z=0.2\sigma_z$ to which Eq. (15a) is expected to be valid. Note that the slope $\partial P/\partial z$ is predicted especially well. Since $\partial P/\partial z=v_d\partial R/\partial z$, the good agreement of the two models below $z=\sigma_z$ supports Eq. (21), the expression for the atmospheric resistance as a function of σ_z.

The near-surface concentration and suspension ratios predicted by the corrected source depletion model Eq. (8), (14) and (15) are compared to the predictions of the surface depletion model in Fig. 1-4. Since these estimates are for a Gaussian plume, Eq. (21) has also been used to calculate the atmospheric resistance. The corrected source depletion model compares very well with the surface depletion model, particularly for $\sigma_z/h>1$, the region downwind of the source where most of the deposition occurs. Note in particular that the error in $P(z_d)$ beyond x=10 km does not produce comparable errors in $C(z_d)$. This is again a consequence of the self-limiting nature of plume depletion.

7. THE CONSTANT EDDY-DIFFUSIVITY SOLUTION

There has been a continual interest in constant eddy-diffusivity solutions to the advection-diffusion equation with the deposition flux boundary condition (e.g., Smith, 1962; Ermak, 1977; Corbett, 1981; Llewelyn, 1983). This interest is sustained by the existence of solutions in an analytical form and by the similarity of the non-depositing solution to the Gaussian plume model. Ermak (1977) and Rao (1981) have suggested that these solutions provide a framework for modifying the Gaussian plume model for deposition that is preferable to the source depletion model. Despite its theoretical shortcomings, this approach has some appeal because of the simplicity of the analytical solution and because the solution also includes a plausible treatment of gravitational settling.

Rao (1981) transforms the solution of the advection-diffusion equation from a dependence on the eddy diffusivity K to a dependence on σ_z with the substitution $\sigma_z^2=2Kx/u$. Since the solution assumes a constant eddy diffusivity, this procedure is correct only for the case where σ_z varies as $x^{1/2}$. The sole justification in the case of a general σ_z is the recovery of the Gaussian plume when there is no deposition. Nevertheless, the result is,

$$C(x,z) = \frac{Q_o}{\sqrt{2\pi}\, u\sigma_z} \exp\left[-\frac{v_s}{u}\,\frac{(z-h)x}{\sigma_z^2} - \frac{v_s^2 x^2}{u^2 2\sigma_z^2}\right]\left\{\exp\left[-\frac{(z-h)^2}{2\sigma_z^2}\right] + \exp\left[-\frac{(z+h)^2}{2\sigma_z^2}\right]\left[1 - 2\sqrt{2\pi}\,\frac{Vx}{u\sigma_z}\exp\xi^2 \operatorname{erfc}\xi\right]\right\} \quad (23a)$$

$$V = v_d - v_s/2 \quad (23b)$$

$$\xi = \frac{(z+h)}{\sqrt{2}\,\sigma_z} + \sqrt{2}\,\frac{Vx}{u\sigma_z} \quad (23c)$$

Rao makes the interesting observation that for nonsettling particles, Eq. (23) accounts for deposition by reducing only the conventional image-source term as suggested by Csanady (1955) and Overcamp (1976). For settling particles, the leading exponential function that depends on v_s tilts the plume downward at an angle whose tangent is v_s/u.

The constant eddy-diffusivity solution, Eq. (23), does not conserve mass with a general functional form for $\sigma_z(x)$. This can be corrected by dividing Eq. (23) by its vertical integral, Eq. (7), to get the normalized concentration D and then multiplying D by the actual horizontal flux Q(x) determined from the second half of Eq. (6). Rao (1981) provides analytical expressions for the vertical integral of Eq. (23).

The constant eddy-diffusivity model, corrected for mass conservation, is compared to the preceeding models in Fig. 1-4. In all cases it is at least as good as the source depletion model, and in most cases the difference between the constant eddy-diffusivity model and the surface depletion model is roughly one-half that of the source depletion model. The remaining differences between the constant eddy-diffusivity model and the surface depletion model are caused by the assumption of the constant eddy-diffusivity model that σ_z varies as $x^{1/2}$. The constant eddy-diffusivity model specifies the correct value of σ_z at the receptor location where the model is applied, but since σ_z is generally a linear function of x near the source, it overestimates σ_z upwind of the receptor. Thus, just as in the source depletion model, the deposition losses that occurred upwind of the receptor are mixed through a vertical depth that is too large, the near-surface airborne concentrations are overestimated, and the suspension ratio is underestimated.

8. CONCLUSIONS

Four plume depletion models have been compared: a surface depletion model, two source depletion models, and an adaptation of the constant eddy-diffusivity solution of the advection-diffusion equation. The surface depletion model is an exact modification of nondepositing plume models to account for the effect of dry deposition. However it is applicable only to nonsettling contaminants and is computationally complex.

The source depletion models are applicable to both settling and nonsettling contaminants, and they are easier to use than the surface depletion model. Chamberlain's (1953) source depletion model is biased in its estimates of deposition, but these errors are small for all situations except those with limited atmospheric mixing and high deposition. With the addition of a profile function to correct for alteration of the vertical distribution caused by deposition, the predictions of the source depletion model are in close agreement with those of the surface depletion model. The corrected source depletion model predictions are best for the near-surface airborne concentrations, particularly for $\sigma_z/h>1$. As a consequence it also gives a good prediction of the deposition and of the suspension ratio.

The constant eddy-diffusivity plume-depletion model is also easier to use than the surface depletion model. It directly includes gravitational settling, but this aspect of the model has not been tested here. With a correction to assure that mass is conserved, this model is at least as good as the uncorrected source depletion model, and in most cases its predictions are closer to those of the surface depletion model by a factor of roughly 1/2. In contrast to the other plume depletion models, however, the constant eddy-diffusivity model is applicable only to diffusion that can be adequately described by the Gaussian plume model.

ACKNOWLEDGEMENTS

The author is grateful for helpful discussions of these models with K. S. Rao and J. C. Doran. This research was supported by the U. S. Department of Energy, Office of Health and Environmental Research, and by the U. S. Environmental Protection Agency, under contract DE-AC06-76RLO 1830.

REFERENCES

Berkowicz, R. and Prahm, L. P.: 1978, 'Pseudospectral Simulation of Dry Deposition from a Point Source', Atmos. Environ. 12, 379-387.

Calder, K. L.: 1961, 'Atmospheric Diffusion of Particulate Material Considered as a Boundary Value Problem', J. Met. 18, 413-316.

Chamberlain, A. C.: 1953, Aspects of Travel and Deposition of Aerosol and Vapour Clouds, UKAEA Report No. AERE-HP/R-1261, Harwell, Berkshire, England.

Corbett, J. O.: 1981, 'The Validity of Source-Depletion and Alternative Approximation Methods for a Gaussian Plume Subject to Dry Deposition', Atmos. Environ. 15, 1207-1214.

Doran, J. C. and Horst, T. W.: 1977, Comments on 'Long-Range Travel of Airborne Material Subjected to Dry Deposition', Atmos. Environ. 11, 1246-1247.

Csanady, G. T.: 1955, 'Dispersal of Dust Particles from Elevated Sources', Aust. J. Phys. 8, 545-550.
Ermak, D. L.: 1977, 'An Analytical Model for Air Pollutant Transport and Deposition from a Point Source', Atmos. Environ. 11, 231-237.
Gifford, F. A.: 1976, 'Turbulent Diffusion Typing Schemes: A Review', Nucl. Saf. 17, 68-86.
Horst, T. W.: 1977, 'A Surface Depletion Model for Deposition from a Gaussian Plume', Atmos. Environ. 11, 41-46.
Horst, T. W.: 1978, 'Estimation of Air Concentrations Due to the Suspension of Surface Contamination', Atmos. Environ. 12, 797-802.
Horst, T. W.: 1979, 'Lagrangian Similarity Modeling of Vertical Diffusion from a Ground-Level Source', J. Appl. Meteor. 18, 733-740.
Horst, T. W.: 1980, 'A Review of Gaussian Diffusion-Deposition Models', Atmospheric Sulfur Deposition, D. S. Shriner, C. R. Richmond, and S. E. Lindberg (ed.), Ann Arbor Science, Ann Arbor, MI, 275-283.
Horst, T. W.: 1983, 'A Correction to the Gaussian Source Depletion Model', Precipitation Scavenging, Dry Deposition and Resuspension, H. R. Pruppacher, R. G. Semonin and W.G.N. Slinn, Elsevier North Holland, Amsterdam, The Netherlands, 1205-1218.
Horst, T. W. and Slinn, W.G.N.: 1984, 'Estimates for Pollution Profiles Above Finite Area-Sources', to appear in Atmos. Environ.
Lamb, R. G.: 1982, 'Diffusion in the Convective Boundary Layer', Atmospheric Turbulence and Diffusion Modeling, F.T.M. Nieuwstadt and H. van Dop (ed.), D. Reidel, Dordrecht, Holland, 159-230.
Llewelyn, R. P.: 1983, 'An Analytical Model for the Transport, Dispersion and Elimination of Air Pollutants Emitted from a Point Source', Atmos. Environ. 17, 249-256.
Nieuwstadt, F.T.M. and van Ulden, A. P.: 1978, 'A Numerical Study on the Vertical Dispersion of Passive Contaminants from a Continuous Source in the Atmospheric Surface Layer', Atmos. Environ. 12, 2119-2124.
Overcamp, T. J.: 1976, 'A General Gaussian Diffusion-Deposition Model for Elevated Point Sources', J. Appl. Meteor. 15, 1167-1171.
Pasquill, F.: 1974, Atmospheric Diffusion, 2nd Ed., John Wiley & Sons, New York.
Paulson, C. A.: 1970, 'The Mathematical Representation of Wind Speed and Temperature Profiles in the Unstable Atmospheric Surface Layer', J. Appl. Meteor. 9, 857-861.
Rao, K. S.: 1981, Analytical Solutions of a Gradient-Transfer Model for Plume Deposition and Sedimentation, NOAA Tech. Mem. ERL ARL-109, Air Resources Laboratories, Silver Spring, MD.
Slinn, W.G.N.: 1984, 'A Potpourri of Deposition and Resuspension Questions', Precipitation Scavenging, Dry Deposition and Resuspension, H. R. Pruppacher, R. G. Semonin and W.G.N. Slinn, Elsevier North Holland, Amsterdam, The Netherlands, 1361-1416.
Smith, F. B.: 1962, 'The Problem of Deposition in Atmospheric Diffusion of Particulate Matter', J. Atmos. Sci. 19, 429-434.
van Ulden, A. P.: 1978, 'Simple Estimates for Vertical Diffusion from Sources Near the Ground', Atmos. Environ. 12, 2125-2129.

Wesely, M. L. and Hicks, B. B.: 1977, 'Some Factors that Affect the Deposition Rates of Sulfur Dioxide and Similar Gases on Vegetation', J. Air Poll. Control Assoc. 27, 1110-1116.

Yamartino, R. J.: 1981, 'Solutions to the Equation for Surface Depletion of a Gaussian Plume', 12th NATO/CCMS International Technical Meeting on Air Pollution Modeling and Its Application, Palo Alto, CA.

WIND TUNNEL SIMULATION OF DIFFUSION IN A CONVECTIVE BOUNDARY LAYER

M. Poreh, Technion, Haifa, Israel and J. E. Cermak,
Professor-in-Charge, Fluid Mechanics and Wind Engineering
Program, Colorado State University, Fort Collins, CO 80523,
USA.

A study of turbulent dispersion of passive tracers in unstable boundary layers, conducted in the Meteorological Wind Tunnel of the Fluid Dynamics and Diffusion Laboratory at Colorado State University, is described. The measured mean and turbulent velocities are found to be similar to those measured in atmospheric convective boundary layers. The diffusion pattern, from ground-level and elevated sources over both a smooth floor and a rough floor, is found to be the same as that measured in the water-tank experiments of Deardorff and Willis (1975) and in numerical models. The measurements show an initial rapid descent of plumes from elevated sources and a subsequent plume rise at $t^* > 0.5\ h/w^*$. Ground-level concentrations from elevated sources are found to be larger, at certain distances from the source, than those from a ground-level source of equal strength. The measurements of the cross-wind spread σ_y are in agreement with the Prairie Grass measurements and confirm earlier predictions that the initial cross-wind spread for ground-level sources is larger than that for elevated sources.

1. INTRODUCTION

When the wind speed is very weak and the temperature of the ground is higher than that of the air, the upward flux of heat becomes one of the important physical parameters that determines the turbulent structure of the unstable atmospheric boundary layer (ABL) and the mechanics of transport processes within this layer. Usually such an unstable layer is capped by an inversion at height h above the ground, which acts as a lid for upward fluxes. Such a capped unstable ABL is also termed a convective boundary layer (CBL).

The characteristics of the CBL are well known and documented. A convenient reference is the book edited by F. T. M. Nieuwstadt and H. van Dop (1982) which contains several lectures related to this subject, given in 1981 at The Hague by several noted researchers.

The CBL is usually idealized as composed of several layers (Caughey, 1981): the <u>surface layer</u>, where the wind shear plays a

Boundary-Layer Meteorology 30 (1984) 431–455. 0006-8314/84/0304-0431$03.75

dominant role; the free convection layer extending to approximately 0.1 h, where the shear is no longer important, but height z is a significant parameter; the mixed layer, where the turbulence structure is independent of both height and shear; and the entrainment interfacial layer, which extends roughly from 0.8h to 1.2h and where a negative heat flux might appear (see Caughey, Figure 4.12).

The characteristic velocity scale of the CBL is usually defined as

$$w^* = (h\ g\ \overline{w'\theta'}/\theta)^{1/3} \tag{1}$$

where $\overline{w'\theta'}$ is the heat flux and θ is an average potential temperature. Note that both θ and the velocity U are approximately constant in the major part of the CBL.

The Monin-Obukhov length L, which is a measure of the thickness of the surface layer, is defined as

$$L = -(v^*)^3\ \theta/k\ g\ \overline{w'\theta'}\ . \tag{2}$$

where v^* is the shear velocity.

Until the early seventies it was assumed that diffusion in the CBL could be described by the conventional Gaussian model or by the conventional Eulerian type mass transport equation with appropriate values for turbulent diffusivities. The pioneering numerical work of Deardorff (1970, 1972), and the experiments of Deardorff and Willis (1975) on diffusion from a simulated ground-level cross-wind line source in a water-tank model of the convective mixed layer (with zero mean flow) and later, more advanced numerical modeling (see Lamb, 1981), showed that the diffusion of clouds and plumes in a CBL has unique and surprising features which cannot be described by the above models.

Some of the surprising features are shown in Figures 1 and 2 from Lamb (1979). The lower portion of Figure 1 shows that for a travel time on the order of 0.5 h/w^*, the centerline of a ground-level plume begins to rise rather rapidly. The abscissa in this figure is the dimensionless distance

$$X = x/(h\bar{U}/w^*) \tag{3}$$

where $\bar{U}$ is a representative horizontal velocity. Further downstream, at $X \sim 0.9$, one finds that the maximum concentration, due to a source located at $z^s = 0.025h$, is not at ground level, as in neutral boundary layers, but above the ground. On the other hand, plumes from elevated sources appear to descend quite rapidly but subsequently rise at larger values of X. Figure 2 shows the calculated nondimensional ground-level concentration $C(x,y,0)h^2\bar{U}/S$ (where S is the source strength) for point sources at different elevations, z^s. The most surprising result of this and similar simulations, which is seen in this figure, is that at certain distances ground-level concentrations from the elevated sources are larger than those from the ground-level source.

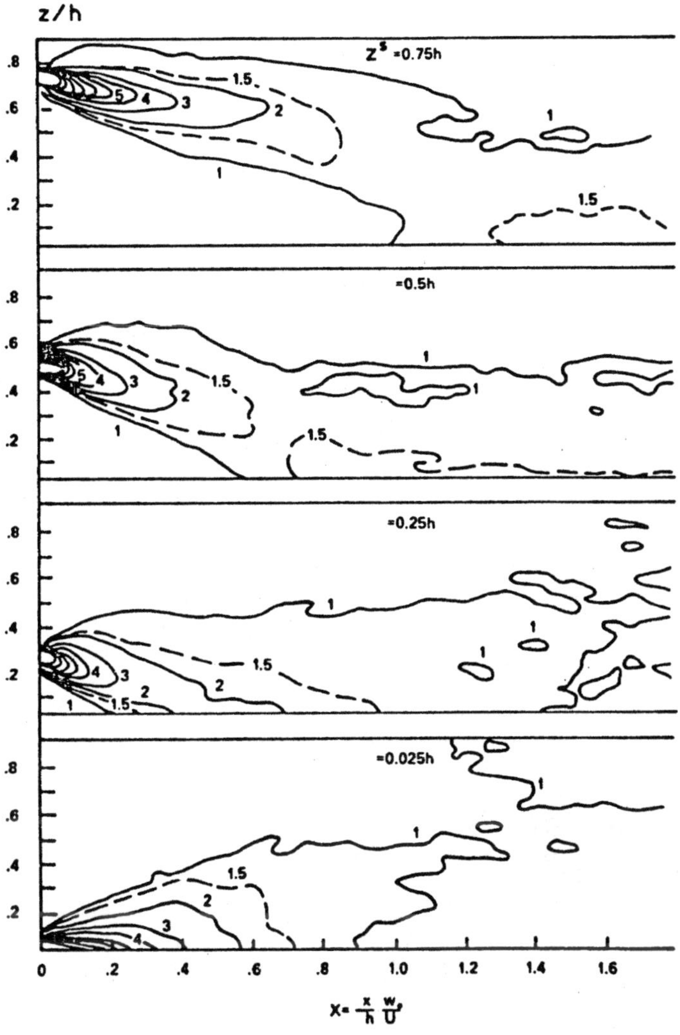

Figure 1. Calculated non-dimensional cross-wind integrated concentration $C^y(x,z)hU/S$ for point sources of heights z^s = 0.75 h, 0.5 h, 0.25 h and 0.025 h (Lamb, 1979).

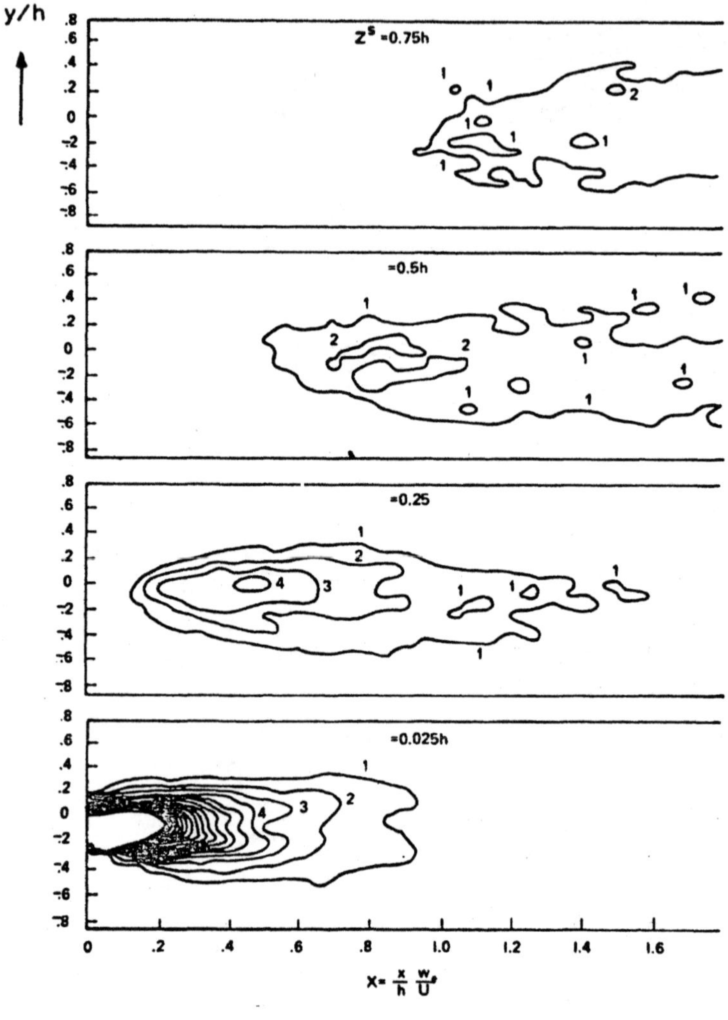

Figure 2. Calculated non-dimensional ground-level concentration $C(x,y,0)h^2U/S$ for point sources of heights z^s = 0.75 h, 0.5 h, 0.25 h and 0.025 h (Lamb, 1979).

Another surprising feature is the effect of the source height on the lateral spread of the plume, σ_y. In neutral flows σ_y increases with height, whereas the numerical simulations of Lamb indicate that at short distances from the source σ_y for $z_s = 0.25$ is approximately 70 percent larger than for $z_s = 0.025$ (see Lamb, 1981, Figure 5.3).

As Lamb describes it, these findings were so grossly out of line with the conventional models that it was speculated by several scientists that they are inapplicable to atmospheric diffusion, except perhaps under very low wind-speed conditions. According to Lamb (1981) when $-h/L > 10$ or $\bar{U} < 6w^*$).

Although the predicted turbulent structure of the CBL has been supported by field measurements, the predicted pattern of diffusion in the CBL has not been directly observed in either the atmosphere (probably because of difficulties to perform extensive field studies), or in wind-tunnel simulations with finite mean velocity. The purpose of the present experimental investigation has been to simulate diffusion in an unstable ABL and to examine whether such a physical simulation would reveal the foregoing phenomena.

2. EXPERIMENTAL APPRATUS AND PROCEDURES

The experiments were conducted in the meteorological wind tunnel (MWT) in the Fluid Dynamics and Diffusion Laboratory, Colorado State University. The design and operation of the 2 m x 2 m x 27 m MWT are described in detail by Cermak (1981).

Four 91 cm high spires were installed at the entrance to the test section, see Figure 3. The 12.3 m long, electrically heated aluminum plate, heat source started at a distance of 12 m from the spires. The 12 m section of the wind tunnel between the spires and the smooth aluminum plate had a relatively high surface roughness when compared with that of the aluminum plate. As will be seen later this rough-to-smooth change of roughness had some effect on the velocity field over the aluminum plate.

The temperature of the plate was maintained at approximately T = 157°C for the smooth-floor configuration and T = 146°C for the rough-floor configuration. The measurements of the heat flux, indicated that the specific heat flux per unit area was approximately $\overline{w'\theta'}$ = 0.63°K-m/sec for the smooth floor configuration and $w'\theta'$ = 0.67°K-m/sec for the rough-floor configuration.

The rough-floor configuration was created by 1.6 cm twisted-link steel chains spaced at 21.6 cm intervals along the entire test section, as shown in Figure 3.

The velocities were measured with a constant temperature cross-film anemometer sensor, type TSI 1241-10, with a resistance wire temperature sensor mounted in close proximity. The sensors were calibrated in a temperature and velocity controlled calibrator. An empirical equation relating anemometer voltage, velocity, and temperature was developed and incorporated into the on-line computer data collection system, permitting instantaneous readings of the velocities and the temperature. Calibration checks were performed

Figure 3. Upstream view of the MWT with the rough-floor configuration.

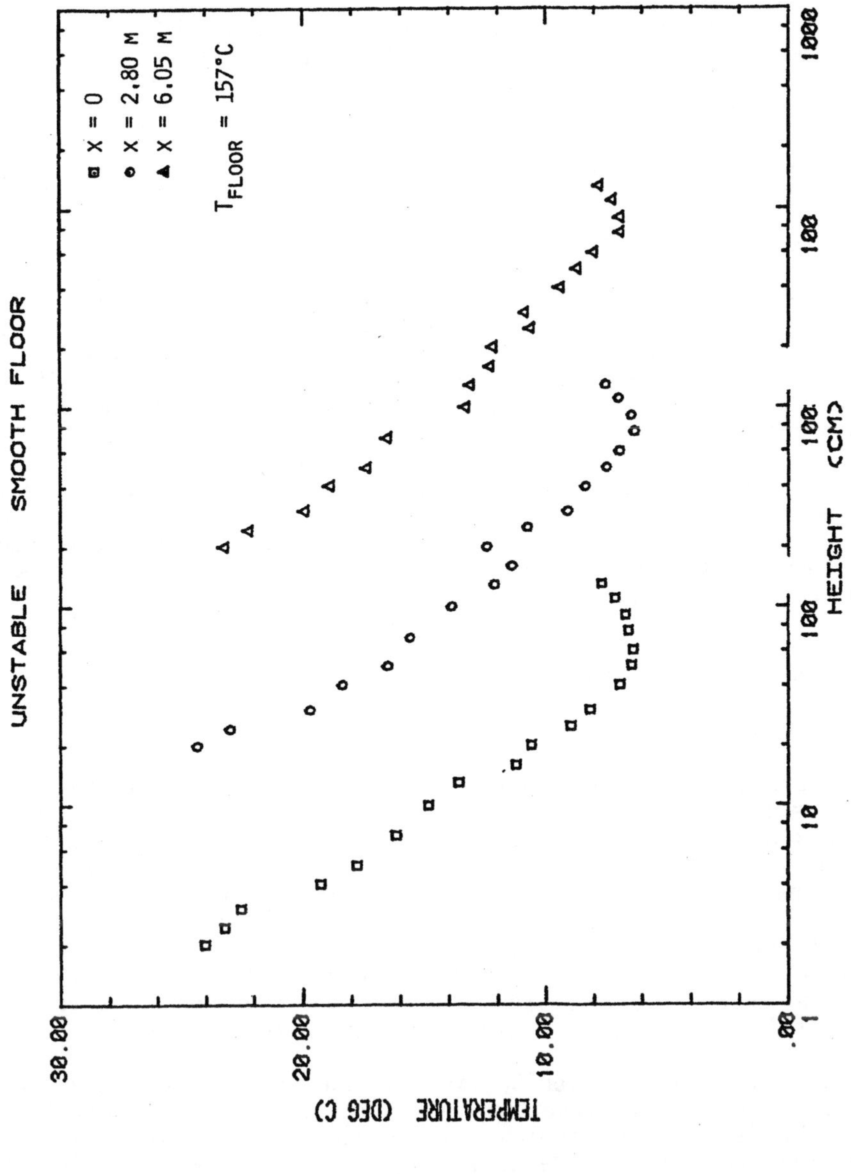

Figure 4. Temperature profiles, smooth floor.

prior to each continuous series of data collection. The mean and cross-correlations were calculated by the computer from 1000 instantaneous readings taken over a period of 150 seconds.

A neutrally buoyant hydrocarbon tracer was released from horizontal 0.63 cm brass tubes located at z_s^s = 0, 10 and 20 cm above the floor. The effective height of the z^s = 0 source was about 3 mm above the floor, but will be referred to as a ground-level source. The average horizontal exit velocity of the tracer gas was 0.9 m/sec. The temperature of the tracer gas was maintained at the average temperature at z_s^s by passing the gas through 90 cm long, thin walled, 3.1 mm diameter, brass tubings located at the source height.

The sources were located 3.88 m from the leading edge of the aluminum plate. The position of the source was designated as x = 0.

Samples of the air-tracer mixture downwind from the source were withdrawn through 1.59 mm I.D. tubes at an average velocity of 1.80 m/sec. Five-minute samples were collected simultaneously from different locations. The concentration of the tracer was analyzed using a gas chromatograph.

Flow visualization with TiO_4 smoke was used to observe the behavior of the plumes and to examine whether large rolling eddies were present over the heated plate. No evidence of such eddies was found. Concentration measurements revealed a very slight continuous lateral shift of the plume (on the order of 20 cm) at 6.05 m downstream from the source.

3. THE EXPERIMENTAL RESULTS

3.1 The Velocity Field over the Smooth Floor

Measurements of the temperature profiles at the position of the source (x=0) and at x = 2.80 m and x = 6.05 m, are shown in Figure 4.

The measurements reveal an inversion of the temperature gradient at a height h above the ground. The height h is slightly increasing in the direction of the flow, from h = 0.55 m at x = 0 to h = 0.85 m at x = 6.05 m. The value h = 0.75 m, at x = 2.80 m appears to be a representative height of the unstable layer for the smooth floor configuration. Using this value, a specific heat flux $\overline{w'\theta'}$ = 0.63°K-m/sec and an average temperature θ = 283°K, the representative value of the convective velocity is estimated to be approximately w* = 0.25 m/sec.

The measured mean velocity distributions u(z) at x = 0, x = 2.80 m and x = 6.05 m are presented in Figure 5. A mild horizontal nonhomogeneity is observed indicating, as shown by Antonia and Luxton (1971), that the boundary layer as a whole recovers very slowly after a rough-to-smooth change in the surface roughness. Nevertheless, both the horizontal and the vertical nonhomogeneities in the convective layer are mild and one may calculate the dimensionless distance $X = xw^*/(h\bar{U})$ in this layer with sufficient accuracy using the average value $\bar{U}$ = 1.90 m. According to Antonia and Luxton the shear stress, after a rough-to-smooth change in the surface roughness, increases throughout the central region of a boundary layer instead of monotonically decreasing. On the other hand, the surface shear stress is drastically reduced, then increases at a lower rate with x. Indeed the shear stress measurements at x = 0, 2.80 m and 6.05 m presented in Figure 6 show a similar behavior.

The average value of the shear velocity in this region is estimated from Figure 6 to be $v^* = 0.085$ m/sec, which implies that the Monin-Obukhov length for this flow was on the order of 0.07 m and $-h/L \simeq 10.7$. As explained earlier, the Monin-Obukhov length is a measure of the thickness of the surface layer where the shear stress plays a dominant role. Since in this experiment the shear stresses in the central region of the boundary layer were slightly larger than the shear stress at the wall, it is plausible that the surface layer extends beyond the above estimated height, or in other words, the effective value of -h/L is smaller.

The measured normalized vertical velocity fluctuations $\overline{w'^2}/(w^*)^2$ are plotted in Figure 7. The values for the flow over the smooth floor vary between 0.1 to 0.3 and are of the same order of magnitude as the field measurements in the Ashchurch and Minnesota experiments (see Lamb, 1981, Figure 4.18).

Thus, it may be concluded that although the unstable boundary layer developed in the wind tunnel over the smooth floor is not a perfect model of an ideal CBL, its major characteristics are typical of many real convective atmospheric boundary layers.

3.2 The Velocity Field over the Rough Floor

The temperature distributions for this case are shown in Figure 8. The values of h at the different stations are slightly higher for this case and the representative value of h is estimated to be h = 0.90 m.

The mean velocity profiles, shown in Figure 9, indicate that the horizontal nonhomogeneity of the velocity field in this case is smaller, apparently because the relative change in the surface roughness was smaller. Distribution of shear stresses at the different stations, shown in Figure 10, are also more similar. Using these data, the average horizontal velocity for this flow is estimated to be $\bar{U} = 1.70$ m/sec and the average shear velocity is estimated to be $v^* = 0.115$ m/sec. Thus, the convective velocity scale and the Monin-Obukhov scale for the flow over the rough floor are of the order of $w^* = 0.28$ m/sec and L = 0.16 m. One may conclude that the added roughness increased the thickness of the surface layer and reduced the convective nature of the flow. The ratio -h/L for this case is smaller than six and the vertical velocity fluctuations have also decreased, as seen from Figure 7.

3.3 The Measured Concentration Field

More than 40 concentration profiles were measured in the study. Representative dimensionless vertical concentration profiles, $C(X,z,0)\bar{U}h^2/Q$ plotted versus z/h, are shown in Figures 11a-11ℓ. The

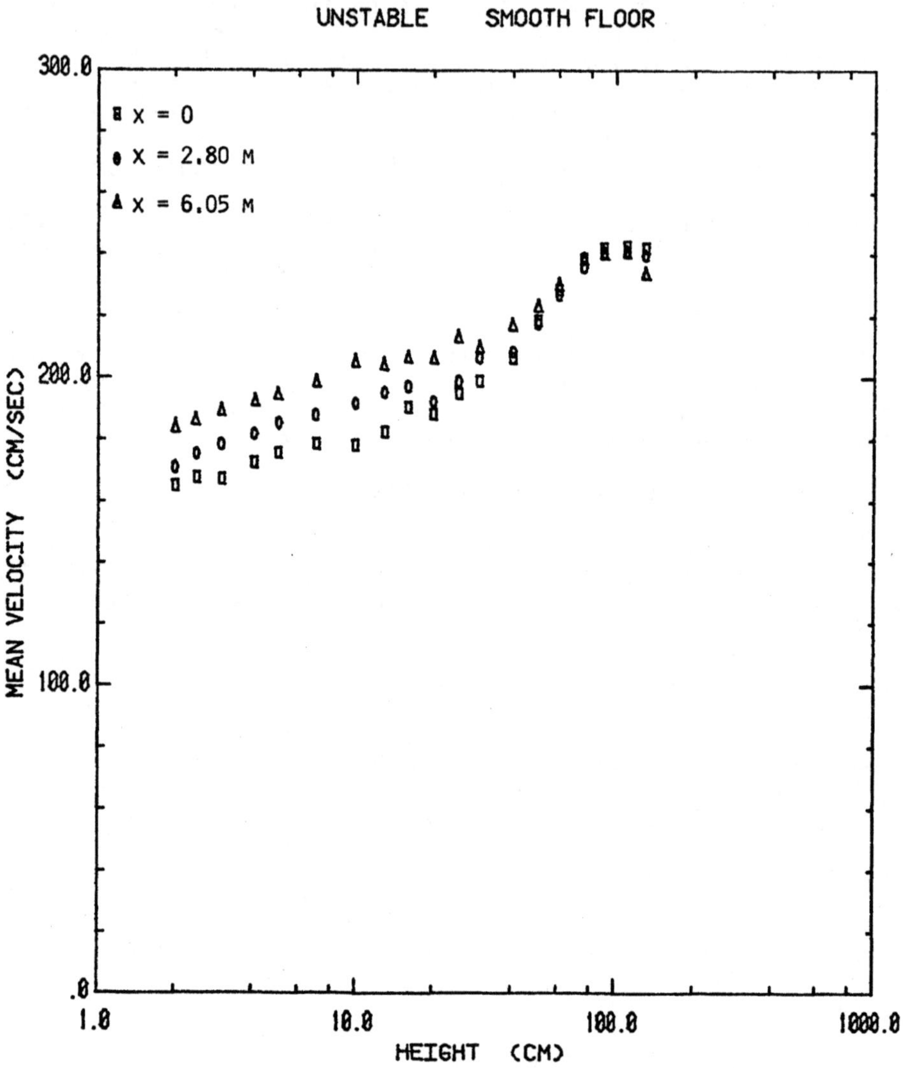

Figure 5. Mean velocity profiles, smooth floor.

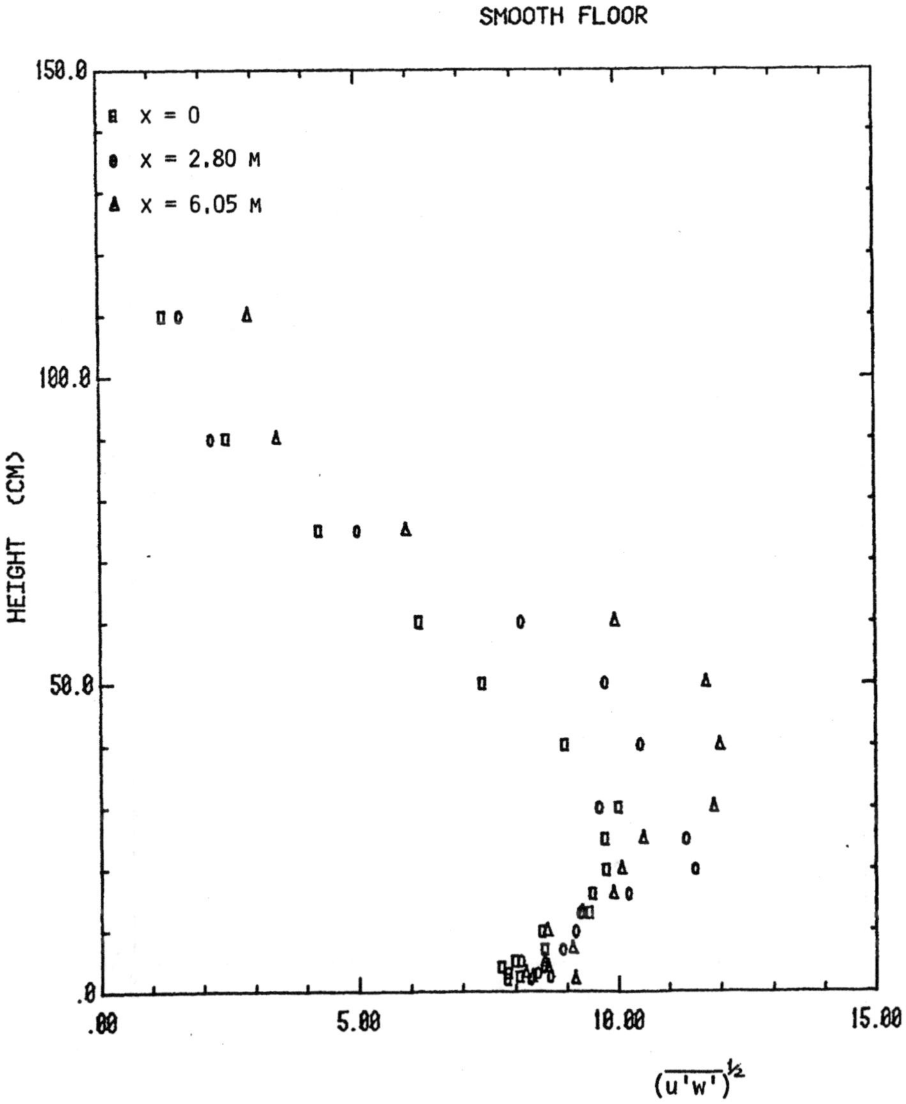

Figure 6. Vertical distribution of $(\overline{u'w'})^{\frac{1}{2}}$, smooth floor.

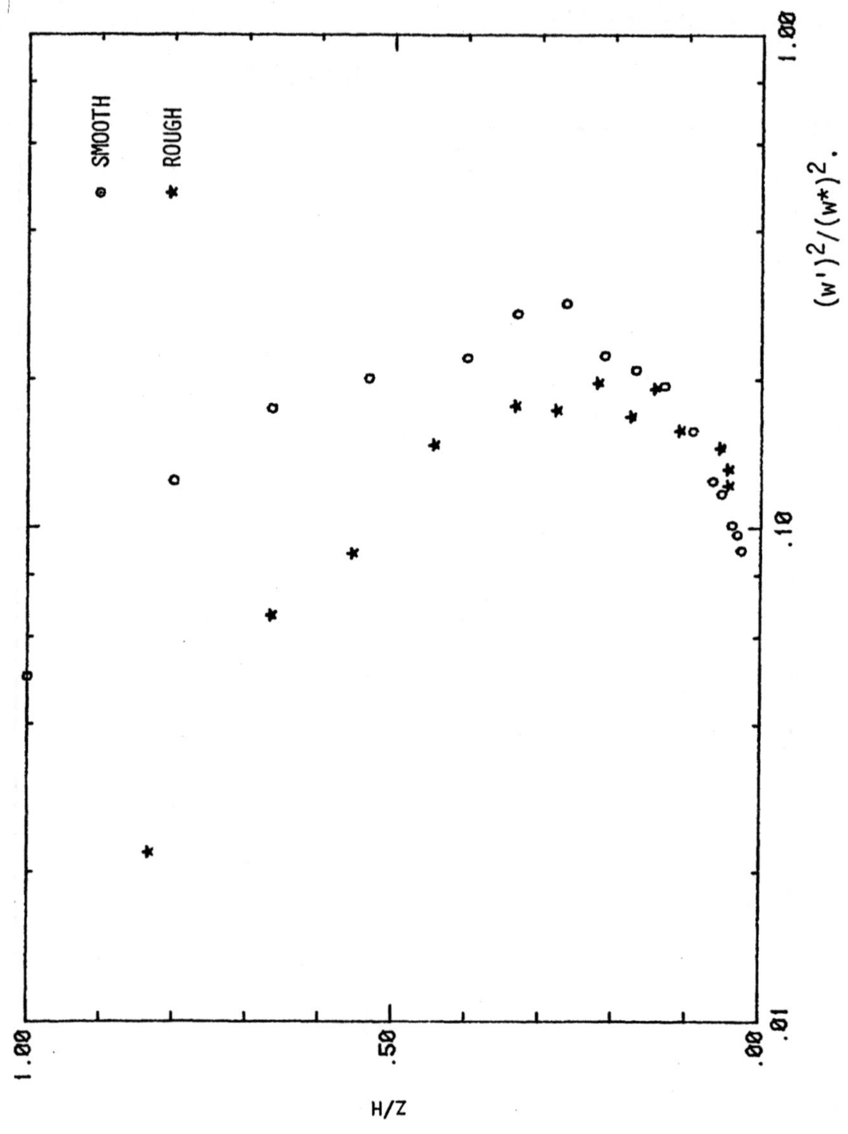

Figure 7. Distribution of the vertical velocity fluctuations $(w')^2/(w^*)^2$.

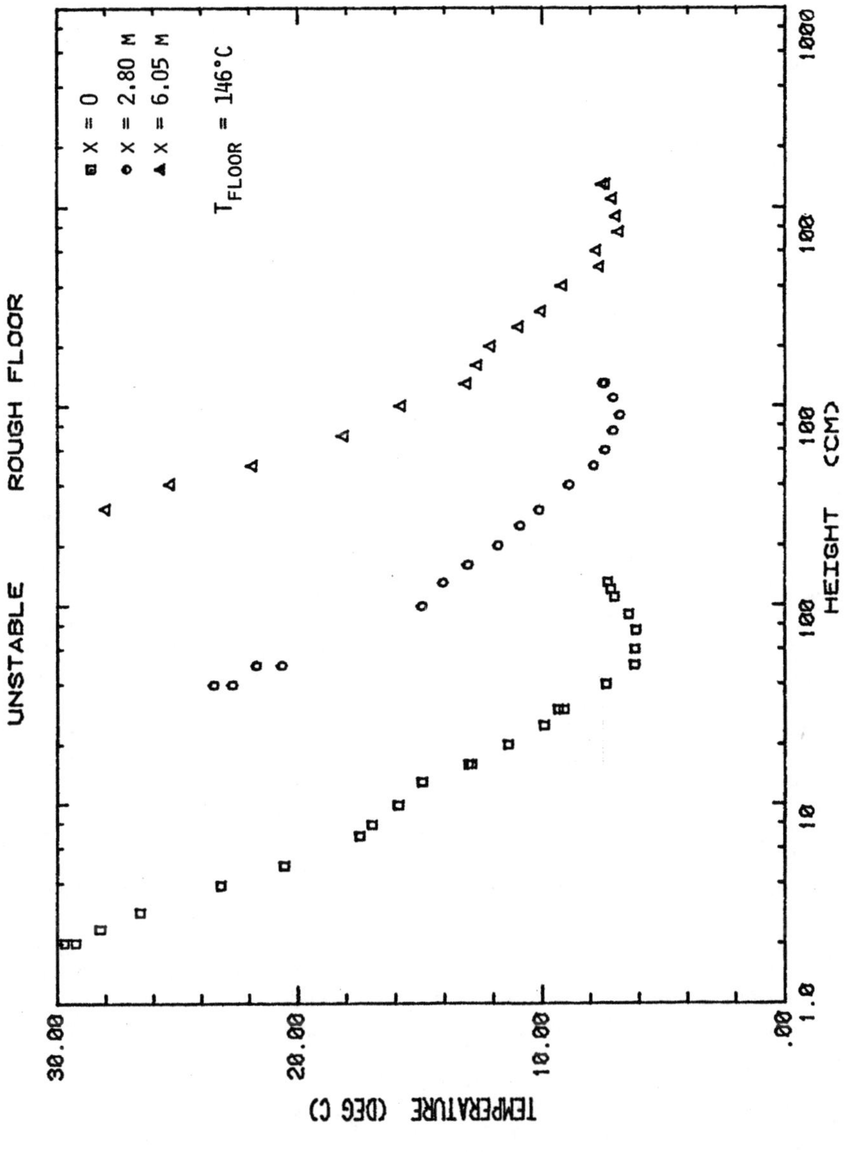

Figure 8. Temperature profiles, rough floor.

concentration profiles measured immediately downwind of the ground-level source Figure 11a,b,c have an approximate Gaussian shape. The profiles are, however, better described by the function

$$C = C(0) \exp [-\beta(z/\lambda_z)^s] \tag{4}$$

where $\beta = \ln(2)$, λ_z depends on X, and s is a power smaller than 2. The calculated values of s are 1.21, 1.31 and 1.41 and 1.39 at x = 0.45, 0.85, 0.135 and 1.95 m from the source for a smooth boundary. Figures 11d and 11f indicate that the maximum concentrations at x = 2.8 m, $X = 0.49$, x = 3.9 m, $X = 0.68$, and at x = 6.05 m, $X = 1.06$, are no longer at ground level.

The heights of the maximum concentration of the vertical profiles downwind from the elevated sources, Figure 11g, are initially at the height of the source, but rapidly approach the ground. Figure 11h shows for example, that it is impossible to tell whether the concentration profile at $X = 0.49$ and $z^s/h = 0.133$ is due to an elevated or a ground-level source. This pattern is also observed in neutral flow, however, the vertical profile at $X = 1.06$, Figure 11i, shows that the position of the maximum concentration is beginning to rise above the ground.

The same pattern is observed for the unstable flow above the rough boundary.

The variation of the measured maximum ground-level concentration $C(X,0,0)Uh^2/Q$ is plotted in Figure 12 for the smooth boundary and in Figure 13 for the rough boundary. As seen from these figures, the ground-level concentrations from an elevated source exceed after a certain distance the ground-level concentrations from lower-level sources. Comparison with Figure 2 shows a very close similarity between the calculated and the measured diffusion pattern, and the dimensionless concentration values in the two figures are of the same order of magnitude, although not identical. Comparison of the rough and smooth floors data shows that the increased shear has slightly decreased the maximum ground-level concentration for the source at $z^s/h = 0.133$ and has slightly modified the values of the dimensionless concentrations at different points. However, the basic pattern of diffusion has not changed.

Figures 14 and 15 show the cross-wind spread of the plume and its dependence on the height of the source. One sees that the σ_y is larger for ground-level sources, as predicted by Lamb (1979). The measured values of σ_y/h are, however, larger than the calculated values (see Lamb, 1981, Figure 5.3).

Figure 16 compares the measured values of the cross-wind spread over the smooth floor with the Prairie Grass data and the results of Deardorff and Willis (1975) and Lamb (1979). It appears that the wind-tunnel measurements at small values of $xw^*/(hU)$, give larger values of σ_y/h, however, they are consistent with some Prairie Grass measurements. At larger distances the wind-tunnel data is in better agreement with the few available field measurements.

The difference between the values of σ_y/h for short distances could be due to the effect of the surface layer present in the wind-tunnel experiment.

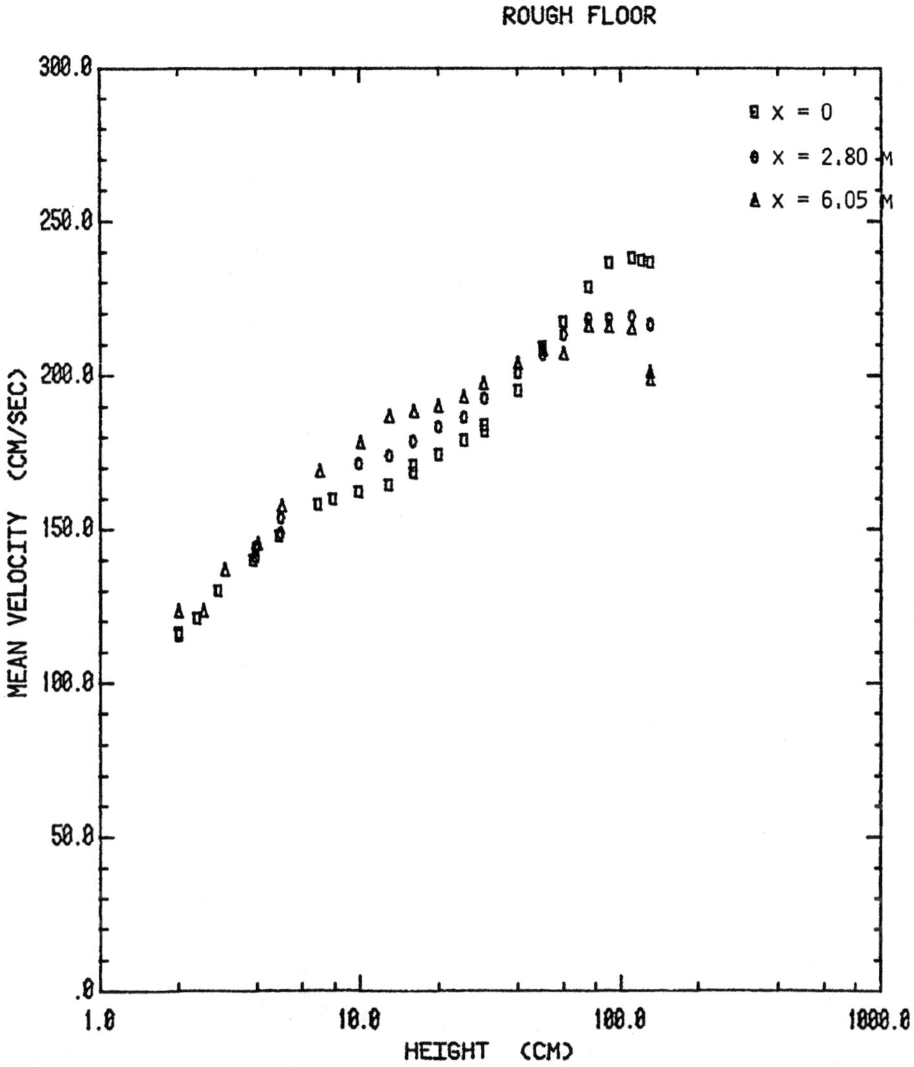

Figure 9. Mean velocity profiles, rough floor.

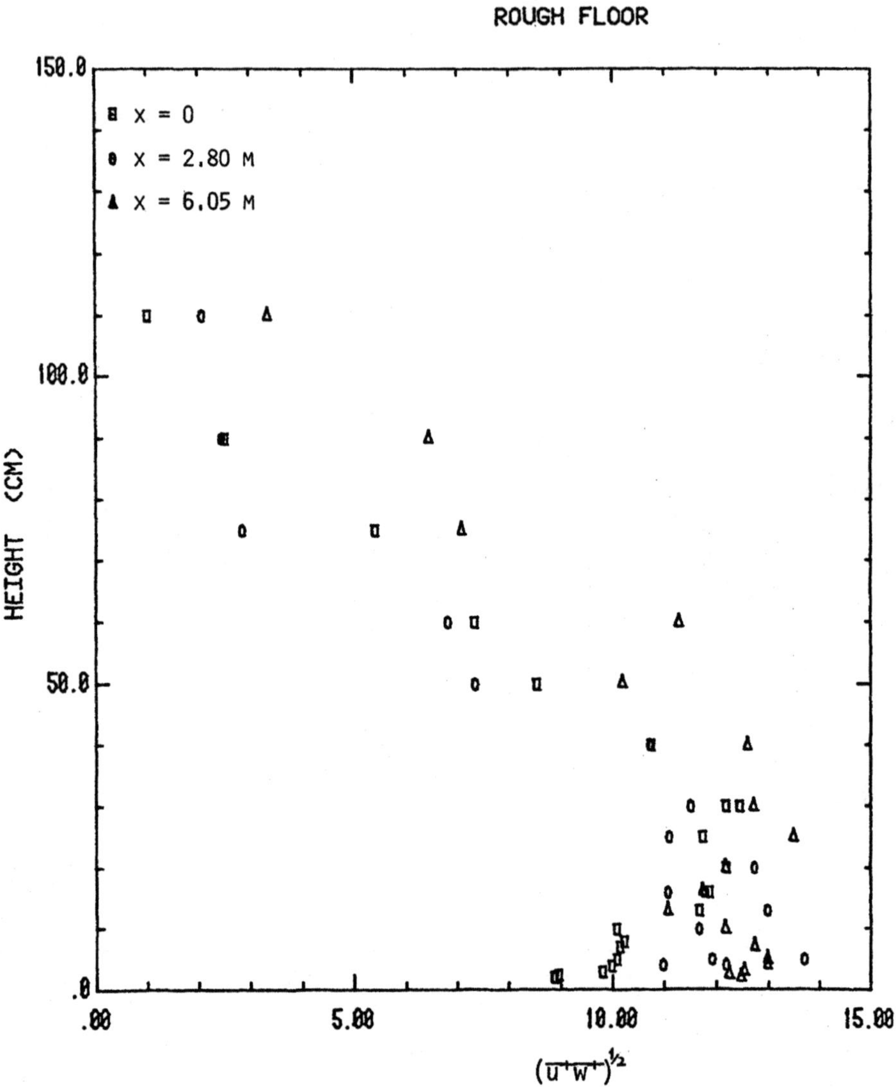

Figure 10. Vertical distribution of $(\overline{u'w'})^{\frac{1}{2}}$, rough floor.

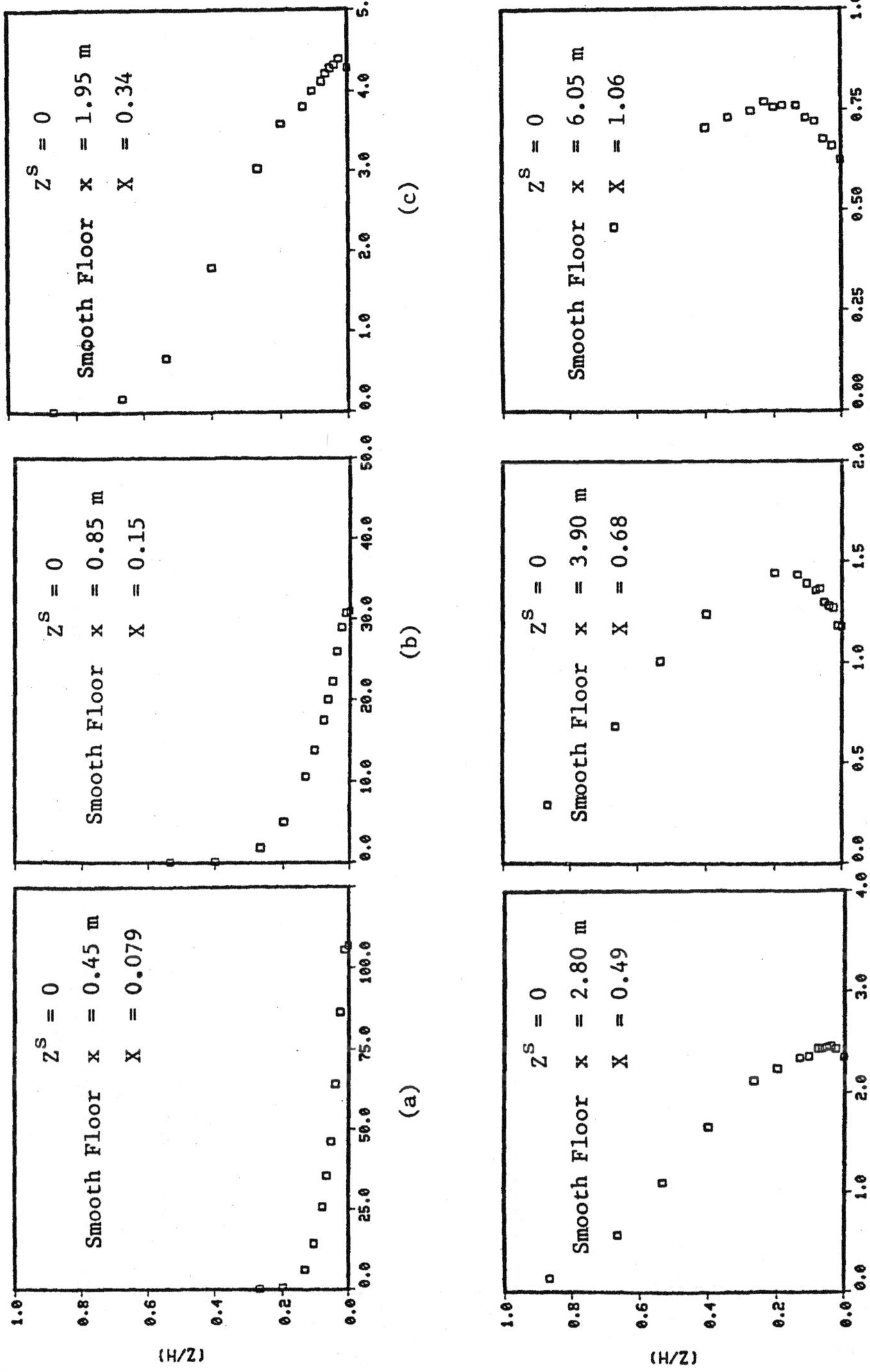

Figure 11. Selected dimensionless concentration profiles $C(x,0,z)\ Uh^2/Q$, for smooth floor $z^S = 0$.

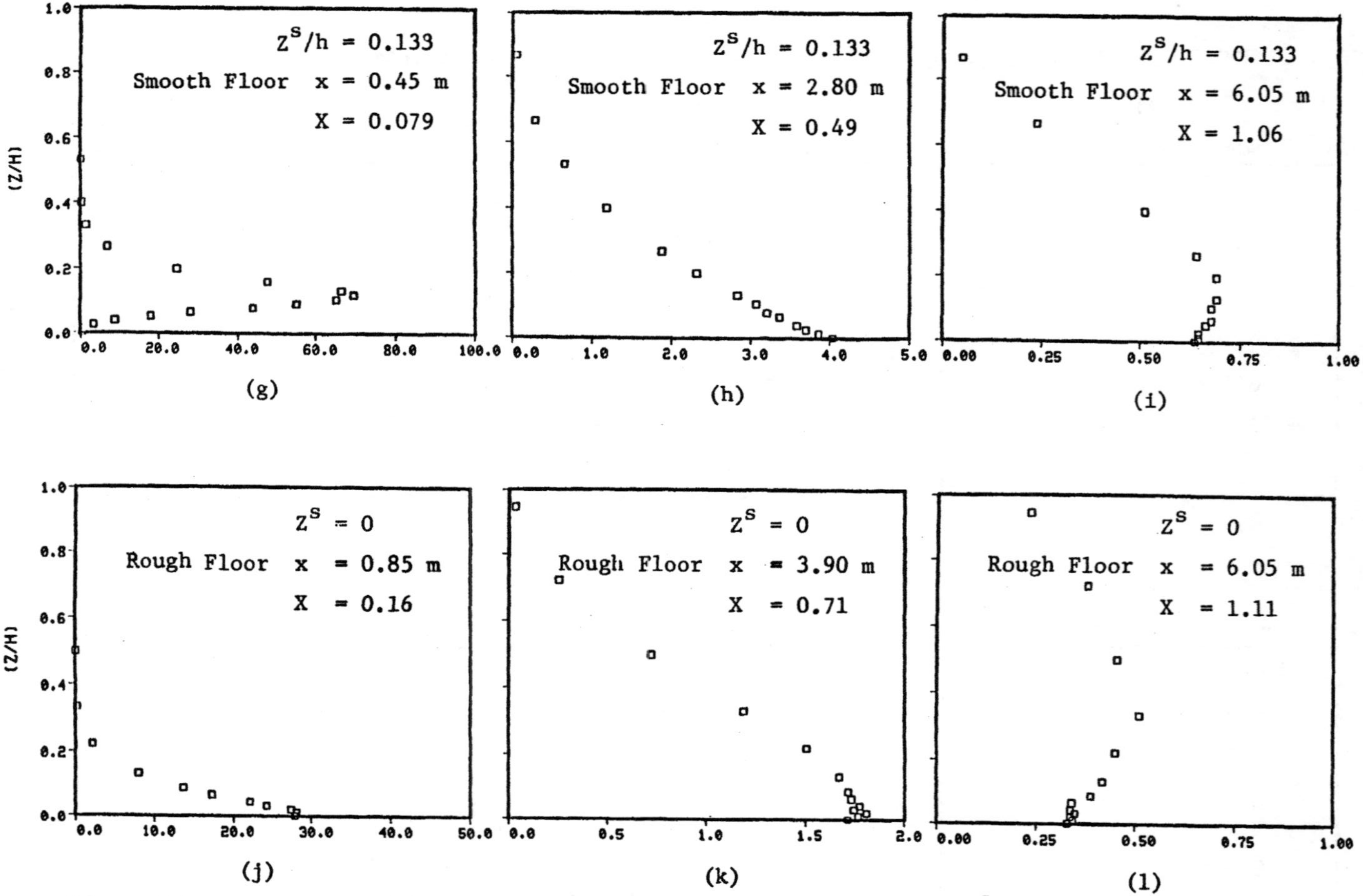

Figure 11. Selected dimensionless concentration profiles $C(x,0,z)\ Uh^2/Q$, for smooth floor (SR) $z^S = 0$ and rough floor (RF) $z^S = 0$.

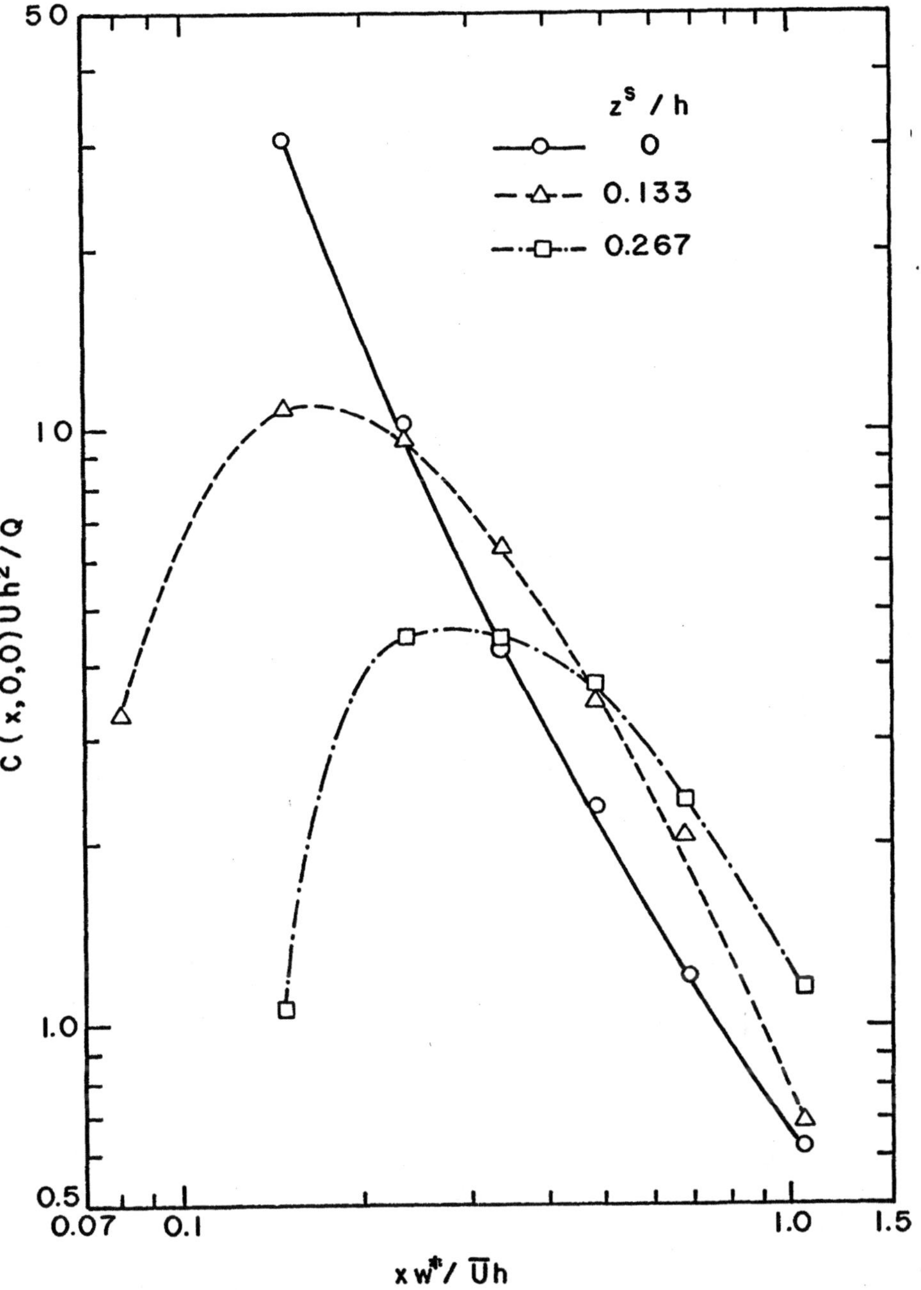

Figure 12. Dimensionless maximum ground-level concentrations smooth floor.

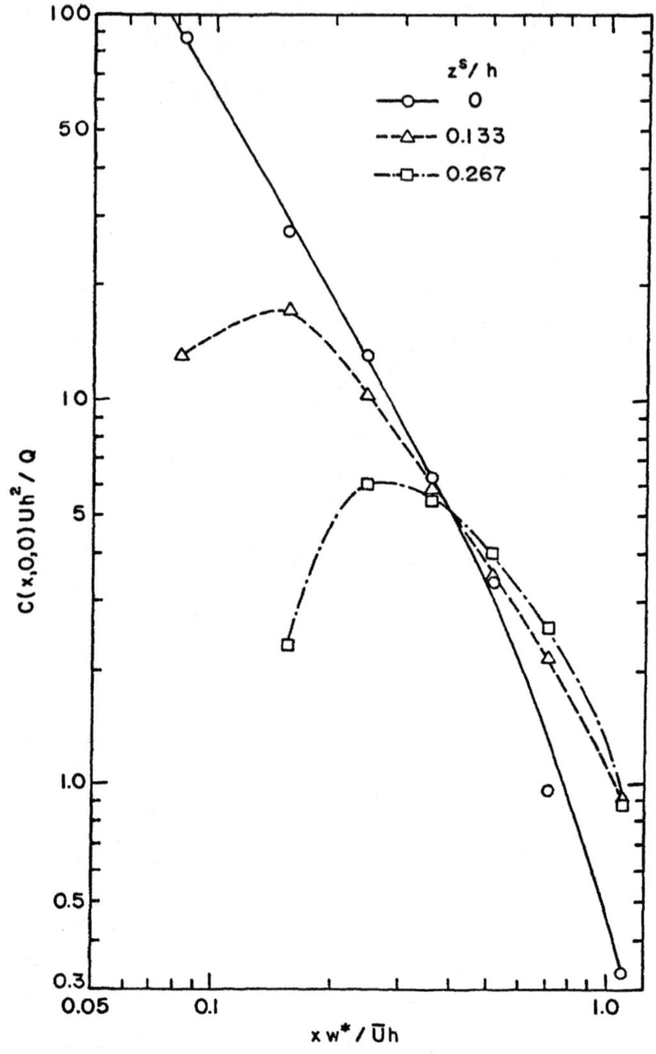

Figure 13. Dimensionless ground-level concentrations, rough floor.

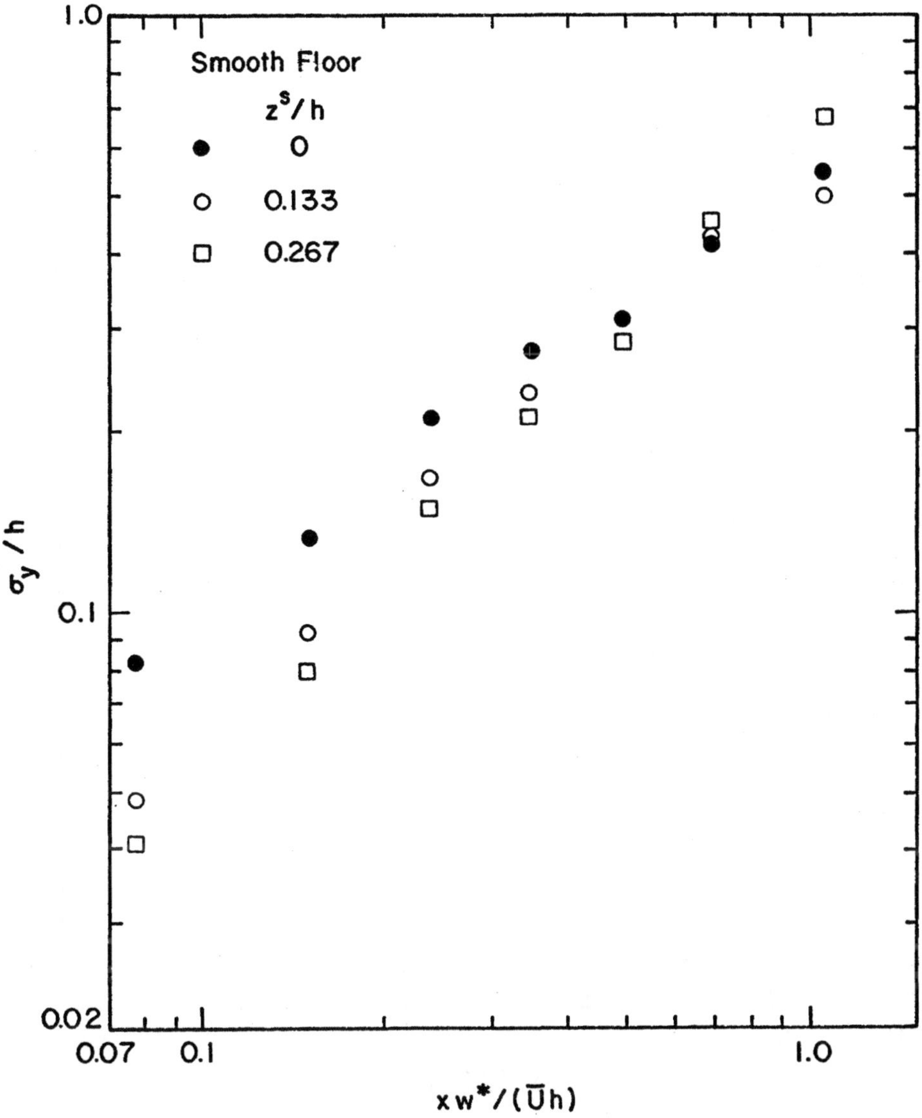

Figure 14. Cross-wind spread, smooth floor.

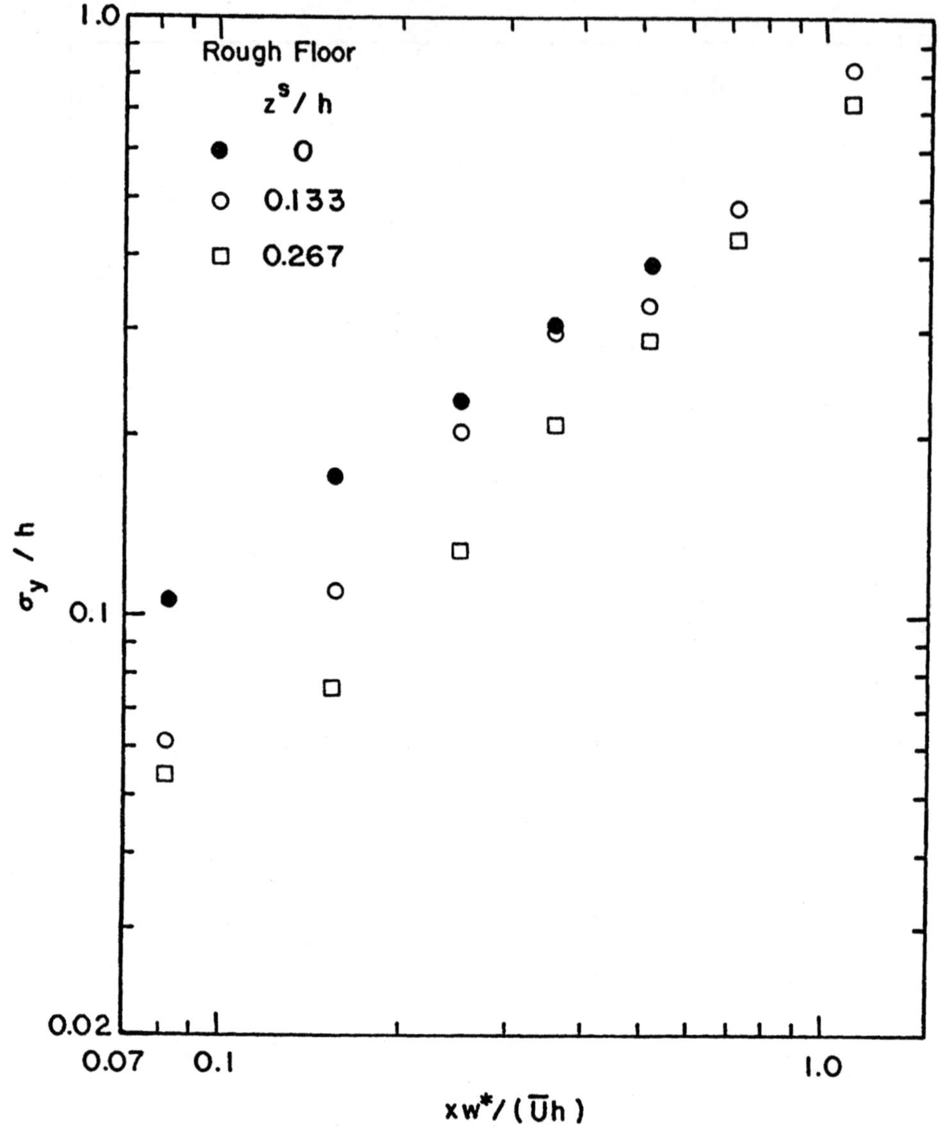

Figure 15. Cross-wind spread, rough floor.

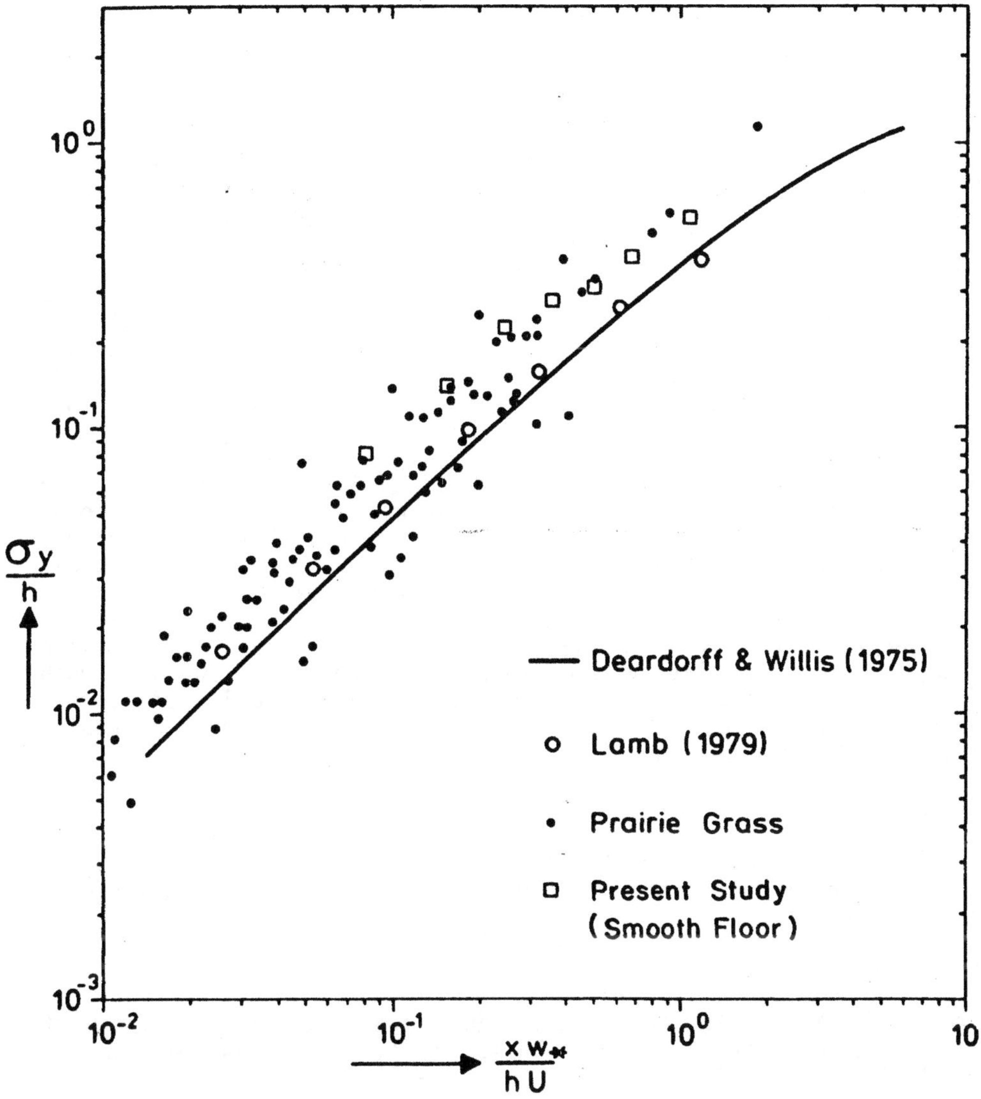

Figure 16. Cross-wind spread of plumes from ground-level sources.

3.4 Conclusions

Measurements of mean and turbulent velocities and temperatures in unstable boundary layers over smooth and rough floors in the meteorological wind tunnel at Colorado State University were found to be similar to those measured in the atmospheric CBL. The wind-tunnel CBL flows had a height to Monin-Obukhov length-scale ratios of approximately -11 and -6.

The patterns of diffusion from ground-level and elevated sources, for both the smooth floor and the rough floor, are similar to that found in the water-tank experiment of Deardorff and Willis (1975) and in numerical models. The measurements confirm the predicted rise of plumes from ground-level sources after a travel time on the order of 0.5 h/w^* and an initial rapid decrease of plumes from elevated sources. They also show that after a certain distance from the source, ground-level concentrations from an elevated source are higher than ground-level concentrations from sources at lower elevations.

The measured cross-wind spread, σ_y, was found to be in agreement with the Prairie Grass measurements. The measurements confirm the numerical predictions that σ_y is larger for ground-level sources, than for elevated sources.

Thus, it may be concluded that the wind-tunnel measurements confirm the predicted unique features of diffusion in a CBL, which cannot be described by a simple Gaussian model or by a conventional Eulerian type mass balance differential equation. They also indicate the CBL diffusion pattern can be found for values of $-h/L$ on the order of 6 which are often encountered in the atmosphere. The effect of surface roughness on the diffusion pattern is found to be small.

The work also exhibits the value of physical simulations in wind tunnels, which can also be used to model convective flows with much more complicated boundary conditions as well as diffusion of buoyant and negatively buoyant flows in CBL. Such problems cannot be easily simulated numerically--Lamb (1981).

Acknowledgements

The contribution of J. A. Beatty, J. C. Maxton and P. Shrivastava in the experimental work and in the data analysis is gratefully acknowledged.

References

Antonia, R. A. and Luxton, R. E.: 1971, The Response of a Turbulent Boundary Layer to a Step Change in Surface Roughness, J. Fl. Mech. 48, p. 721.

Caughey, S. J.: 1981, in Niewstadt and van Dop (1982).

Cermak, J. E.: 1981, Wind Tunnel Design for Modelling of Atmospheric Boundary Layer, J. of Eng. Mech. Div., ASCE 107, No. EM3, 623-642.

Deardorff, J. W.: 1970, Preliminary Results from Numerical Integration of the Unstable Boundary Layer, J. Atm. Sci. 27, 1205-1211.

Deardorff, J. W.: 1972, Numerical Investigation of Neutral and Unstable Planetary Boundary Layers, J. Atm. Sci. 29, 91-115.

Deardorff, J. W. and Willis, G. E.: 1975, A Parameterization of Diffusion into the Mixed Layer, J. Appl. Meteo. 14, 1451-1458.

Lamb, R. G.: 1979, A Numerical Simulation of Dispersion from an Elevated Point Source in the Convective Planetary Boundary Layer. Atm. Environ. 12, 1297-1304.

Lamb, R. G.: 1979, The Effects of Release Height on Material Dispersion in the Convective Planetary Boundary Layer. Preprint vol. AMS Fourth Symposium on Turbulence, Diffusion and Air Pollution, Reno, Nevada.

Lamb, R. G.: 1981, Diffusion in the CBL. in Nieuwstadt and van Dop (1982).

Nieuwstadt, F. T. M. and van Dop, H.: 1982, Editors, Atmospheric Turbulence and Air Pollution Modeling - A course held in The Hague, September 1981. D. Reidel Publishing Co., Boston, Massachusetts, U.S.A.

PART V

ROUND TABLE DISCUSSION

ROUND-TABLE DISCUSSION

HANNA

The purpose of this portion of the program is to allow everyone a chance to voice their opinion on how the papers given at this conference may be used in specific applications. For example, some of the applications that have been brought up are the wind flow over mountains, sea coasts, and urban areas, and the environmental effects of smoke stacks in complex terrain. The panel consists of six distinguished members of our group, and each member of the panel will first present a brief statement. After the completion of this introductory portion of the program, the floor will be open to all questions and comments.

I would like to begin the discussion with some points concerning the predictability of models in general. In any application it is desirable to choose the model which gives the optimum predictability and therefore it is necessary to recognize the various components of model error: errors in input data, errors caused by stochastic fluctuations in the predicted field, and errors in model formulation. Errors in input data can be very significant if the instruments are poorly sited and/or poorly calibrated. Errors caused by stochastic variability in the predictive fields can also be important, especially close to a pollution source or near the edge of a plume. Every relevant parameter has a turbulent component, including pollutant concentrations, pressures, and velocities, and it is not possible to predict the second-by-second variability of these parameters. The final component in total model error is the error in the model formulation itself. For example, a highly simplified model would have much error due to imperfections in the physics in the model. But there are also disadvantages with a very complex model, because it requires so much input data that the errors in the data themselves swamp the prediction. In many applications there is an optimum level of complexity where the total error is reduced to a minimum. Below this level, the total error increases due to errors in model physics, and above this level of model complexity the total error increases due to the input data. An example of this is the simple ATDL urban diffusion model which can predict concentrations just as well as more complex three-dimensional numerical models. The explanation for this apparent paradox is that there is so much error in the input to the more complex models that the improvements in the physics do not reduce the total model errors.

CERMAK

I surpose my presence on the panel derives primarily from my interest in wind tunnel modelling, and I will try to categorize some of the areas which I consider to be of interest in this particular subject area. The general relationships are outlined in Figure 1. One of the

Boundary-Layer Meteorology 30 (1984) 459–470. 0006-8314/84/0304-0459$01.80

major areas studied in the laboratory is characterization of winds, in an effort to identify the structure of turbulence and mean velocity distributions with particular boundary conditions. One of the prime purposes of the wind tunnel is as an analog computer to simulate the appropriate wind field for a site that is being considered for power plants, industrial development, or other applications. Once this has been done with sufficient accuracy,that wind field can be used to study a great variety of different problems.

One of the most highly developed areas of research involves wind forces on structures. In the U.S.A. it is now common practice to do a wind tunnel study on buildings which exceed about 100 m in height. The resulting data are used to improve the design of the building. We also get information on the dynamic forces acting on the structure, so that the building will be comfortable to work in, and so that sufficiently strong materials can be used to ensure that the building will not deteriorate.

A central interest of this conference is environmental quality and safety. One of the prime applications of wind tunnels is the design of proper stack heights for particular power plants or industrial sites. First, the model must be designed with appropriate scales so that the model does not block more than about 10% of the cross section of the wind tunnel. Then the surrounding geometry and topography must be realistically included. In this kind of study, one is interested in the ground level concentration field and the distribution of mean concentrations with height for different wind directions for a particular site. One can systematically vary the stack height, the exit velocity of the stack gases, and the gas temperatures, so that one can systematically obtain all the information needed to decide the height of the particular stack.

Another application relates to the siting of complexes of laboratory or hospital buildings. Currently, we are looking at a university hospital situation in which there are about 100 buildings in the complex, with about 200 different sources and receptors. A major problem is to avoid a design that will give cross-contamination.

Many aspects of the siting of LNG and toxic chemical storage facilities can be studied systematically in a wind tunnel. We looked at about 20 different potential plant sites in California by systematically studying the diffusion characteristics. Assessment of transport and dispersion over complex terrain is of importance where there is an accidental spill of toxic materials and there is a need to assess the impact on people. Dust control problems can also be studied systematically, such as the protection of working areas by putting up screens.

EGAN

I was very interested and pleased to see that the concept of dynamic emission control is applied in Israel. About 1977 the Clean Air Amendments of the U.S. were passed, requiring constant emission controls rather than permitting the continuation of the dynamic emission control systems used at several plants. The latter systems required reduced emissions during time periods when adverse meteorological conditions are present or are anticipated, so that ambient concentrations are minimized. Figure 2 is a schematic of this approach. During periods of good dispersion, and industry is allowed to burn fuel of high sulfur content, for example. The reasons for dropping this system in the U.S. relate to acid rain and total sulfur volume in the atmosphere. Nevertheless for several years up until 1977 my company worked extensively with these systems. Much of the new technology of the past several years gives us even better reason to think that dynamic emission control would be technologically very successful.

The components of the system center around the air quality prediction scheme, which can be a dispersion model of an empirical trend predictor of ambient air quality from a real time monitoring system. Forecasts and air quality observations are used by the system to predict ambient air quality levels and then determine the required control technology or strategies. For example, we are gaining much more information on modeling dispersion in coastal areas, based on extensive field measurements in Canada and fluid model studies by Deardorff. With temperatue profiles and turbulence measurements, (σ_V, σ_W), major improvements can be made in predictive capabilities. Also, by incorporating trend information in the system, there is a tendency to counter some of the wide swings that you often get with changes of wind direction or wind speed in ordinary model predictions.

GIFFORD

My own work in the last few years has been concerned with large scale and longer range diffusion. The larger scales have played a role in the consideration of most of the models that have been discussed, particularly those that are aimed at applications. I should mention that it is my own view, now having looked at the longer range problem, that properties of the scales of motion that lie just above the boundary layer scales are more and more going to be influencing the modellers. These mesoscale eddies in fact control the boundary layer scales. They do not dominate it particularly in the active buoyant case, but they are part of the whole scheme of motions in the atmosphere whose sole effect, according to my own view, is to set up the boundary layer properties that we experience, in order to provide the proper dissipation of the energy which is created at large scales. Some of the future applications will show more and more that it is necessary to consider the larger scales of motions as dominating some of the properties to be moddelled.

I would like to make a general comment on modelling. Sitting through this meeting, I was thinking that in some ways the development of modelling, particularly as it aims at applications, is like many other aspects of life. If you compare physics to music and art, the classical period is when the investigator is creating fundamental new concepts. Occasionally he will dash out to his laboratory or wind tunnel to get a number that he needs and then come back to continue with his developemnt of concepts. And then of course, when other people want to apply these theories, and find that they do not quite apply, the classical period becomes a Baroque one. In this new period you convolute the concepts a little bit and add a little curve here and there. The problem really is to avoid sliding into the Rococo period where you are adding the curvatures just for their own sake. This is the situation that modellers, I feel, have occasionally fallen into. It is very healthy when applied people and modellers talk to one another. When you are in the Baroque period you have a strong eye on the classical roots of your subject. The problem comes in when you slide into the Rococo period and there is nobody to talk to but other modellers.

HUNT

At this conference we are considering a wide range of predictions, beginning with the local problem where we are interested in real time predictions. For example, these could be problems concerned with air pollution as a function of the local changes in meteorology, or possibly situations when we want to know how the wind energy is going to change in a matter of hours. On the other hand, we have the climatological prediction problems in which statistics are evaluated over many years of, for example, wind energy or dispersion probabilities. To some extent, similar models are used for both of these situations. But equally there are models that are specifically appropriate to the real-time or local predictions, and I think we are looking forward to a new era when every major power station or wind energy site will have significant remote sensing capabilities. We may be able to have as, say, French nuclear stations do, acoustic sounders and lidar so that we will be able to tell what the wind is doing for a radius of 20 km around. Also, we may soon be employing receivers from satellites to determine the wind field.

This brings us to the question of how to use all this information. If you have a set of primitive (differential) equations, you need to set up some initial conditions, which may not be consistent with the range of data that is coming in. In the presentations this week, a number of papers have discussed interpolative models, such as mass consistent models. I feel strongly that the use of this approach is to be welcomed but that the mass of data can be used more effectively if it is combined with an increase in understanding of the mechanics of the flow. This may come from simple physics, analytical models, wind tunnel studies, or possibly from second order modelling. I feel that this new area of interpolation using on-line data, such as the local wind statistics or the local wind field, should be given priority in the next few years.

LUMLEY

I hope that our Cornell model is not from the Rococo period. The Cornell second order model is the only model that handles buoyant effects correctly. None of the K-theory and Gaussian plume models can handle the effect of buoyant dispersion. Our model is primarily applicable to a regional scale, 20 kn or so, where many details are needed, and it is not very good for localized sources (see below). It is excellent for studying distributed sources of pollution or water vapor, terrain effects, changes in boundary conditions, military communication problems related to gradients and fluctuations of the index of refraction, and a wide range of other problems related to ground fog, photochemical smog, and on-shore breezes from both stable and unstable situations.

It recently came to light that all the second order models do very poorly close to small or isolated sources. This is because there is an assumption subtly built into their structure that all the time scales are very close to each other. If you are studying a situation in which, say, a young energetic field of turbulence comes out of the stack in the middle of a well-developed field of boundary layer turbulence, time scales are very different and the predictions will be very bad. The time scales will of course approach each other very rapidly. After they come to some sort of equilibrium, the second order model can provide good predictions.

I realize that this is an atmospheric meeting and I shouldn't say anything about other applications, but it perhaps will give you confidence to know that these same models are being used elsewhere. For oceanographic problems, we have used second order models to describe the surface mixed layer in the ocean, wave/turbulence interactions, and the structure of layers of alternating thermal stability. We are using the same models in gas turbine combustor design and the study of turbine blade cooling, including swirl and particle transport, density fluctuations, and diffusion flames with fast chemistry.

NIEUWSTADT

I would like to concentrate on the things that have been said in the session "Boundary Layer Structure". There seems to be a considerable gap between the papers given in that session concerning details on turbulent models and practical goals and applications. It is possible to greatly simplify when predicting some parameters, such as the mean structure of turbulence or the entrainment rate at the top of the mixed layer, as shown by Tennekes. Perhaps this is just a return to first principles. Hopefully, we have returned to the Renaissance but not to the Middle Ages.Nevertheless, I would like to point out that considerable achievements have been made in turbulence in recent times and a good example is the convective boundary layer. Our knowledge of that layer has certainly improved during the last ten years. It seems that the Lagrangian time scale of the diffusion of ideas is about 10 years. It takes time for ideas to diffuse and get accepted. And on that subject, I would like to agree with Dr. Gifford that scientists like to do science because it is fun. But practical applications should also be undertaken and that is usually hard and not so rewarding. Several groups are presently trying to incorporate new knowledge of the convective boundary layer in practical models. I certainly think that this is a good step forward, but it takes time for these new models to be accepted and put into practice. In other boundary layers, such as the stable boundary layer or the cloudy boundary layer, a theoretical basis for practical application will not become available for a few years.

QUESTIONS AND COMMENTS FROM THE AUDIENCE

POREH

Many of you have been to Haifa and have noticed the inhomogenous terrain and its interaction with coastal areas. We are faced with the practical problem of evaluating the impact of large power plants in that area, which certainly cannot be handled by conventional models. Particular problems include plume impact and interaction with the mountains in this complex site. In this connection, I feel that we need to be cautious when we extend the scales of the physical models. Prof. Cermak pointed out that he had a rough theoretical justification for using a 1:10000 scale for modelling, so that supposedly the length scale divided by the kinematic viscosity for the real atmosphere and the laminar viscosity for the model would match. I wonder if he could give some examples of cases where this is a good model assumption. I would like to know from his experience and theoretical background whether there is further justification for such an approach, and its utility in simulating the basic motion of the plume, such as its meandering in complex terrain and its ability to cross over a mountain in stable flows. Later we can apply other models to correct for dispersion, if possible. Secondly, I have a question related to the covective boundary layer. We often think about turbulent flow as random, but in the convective boundary layer the location of hot plumes may not be so random. For example, if you heat a pan in the kitchen, you will find that imperfections will cause boiling to occur at certain points. Because of nonhomogeneous surface conditions on the ground, could I generate a situation where hot plumes always are formed at certain positions? These positions could be designed in such a way around a power plant that the dirty air will always be taken up by thermals and ground-level concentrations significantly reduced. Could this be modeled theoretically using higher order closure?

EGAN

In regard to the impaction and high concentrations on high terrain around Haifa, one needs to have data in order to construct a model. I don't know the extent of your present ambient measurement program,but if you could very quickly identify the conditions under which the highest concentrations occur,I could suggest a few approaches. One approach would be to use a stratified tunnel to study night time conditions, or an unstratified tunnel for day time. This would give you an idea of where you should take some more measurements in the region of interest. As a first numerical modelling approach, I would suggest using the concept of critical streamline height, as discussed in my review paper.

HUNT

You may have a big model that includes all possible effects but I think that in order for your mind to operate properly, you are bound to prepare a list. For example, you would systematically consider the effect of a sea breeze, the effect of a roughness change, and then the effect of topography. You then combine these effects and wonder what happens if a sea breeze is added to a hill, or if a roughness change is added to a sea breeze. The mind thinks linearly but doubtless your models will do more than that. The problem of shoreline transport and diffusion is very difficult, obviously, to simulate in a wind tunnel. You have to do special experiments to simulate, for example, the impaction of a sea breeze on a hill, which may have an extraordinary effect of a splashing upwards. There are many things that may or may not be relevant, but still must be considered in any scientific study of shoreline diffusion in complex terrain.

CERMAK

As an additional comment, I think the physical model will answer a lot of questions related to shore breezes. In Haifa, the situation is complicated a bit by the addition of topography. When we were doing a lake breeze study, we found that the weakest link was getting adequate information for the field conditions that existed at that particular site. This is generally true in physical modelling, since it is always necessary to integrate the model results with the actual meteorology at the site. It is a challenge to meteorologists and to the plant owners to provide information that can be used in a physical model.

With respect to Dr.Poreh's second comment, we are more justified in extending model scales when the dispersion is created by advective transport due to extremely complex geometry. Under conditions where turbulent dispersion is a secondary part of the mixing process, it is not necessary to require equality of the two Reynolds numbers (one based on eddy viscosity and one based on molecular viscosity). When modelling dispersion in very stable conditions over extremely complex topography, the advective transport controls the dispersion process, and the analogy is an excellent one.

LUMLEY

My colleague Sidney Leibovich and I, in collaboration with Julian Hunt, are trying to combine our second order model with some of Julian's ideas about flow over terrain to develop a model that will be capable of predicting flow and dispersion in complex terrain. The situation described by Dr. Poreh is certainly inherent in the type of model I presented, which has the ability to represent nonuniformities in albedo, wetness, or roughness of the surface. In order to assure that pollutants are carried as far aloft as possible, I would surround each power plant by a large asphalt parking lot.

NIEUWSTADT

I could add a very practical point. Any glider pilot will tell you that there are certain locations where thermals occur. The power plants also influence the weather patterns in their regions very much, for it is frequently observed that power plants make cumulus clouds. I do not think that the only criterion for siting a power plant should be the thermal patterns of a region, but you could include that in your requirements.

BERKOFSKY

I would like to add to Dr. Hunt's comment in regard to the initial value problem. What he referred to was the problem in weather prediction known as the initialization problem, which is being pursued very actively. In regard to a comment Dr. Hunt made about stochastic dynamic modelling, this is now being done with reference to sea surface temperature, where the daily fluctuations of temperature are considered to be stochastic forcing functions on the longer period of ocean temperatures. I think that this approach will probably have some merit in treating the long period long range transport of pollutants.

HUNT

I think that the problem of weather forecasting on a much larger scale and the problem of initialization of turbulent flows at the scale of boundary layers are quite different, and this is one of the advantages of the sort of field experiments now going on in Boulder, Colorado. In these field experiments dispersion measurements are being made at the same time as very detailed surveys of the turbulent structure of the boundary layer. At the moment this is rather novel, but I am quite sure that in 10 or 15 years these routine surveys of turbulent structure will be made frequently. But in order to make use of this information, we have got to understand how the observed turbulent structure relates to diffusion.

ALPERT

First, I want to comment on modelling the complex problem of the flow near Haifa. I think there are models today which give quite realistic descriptions of the evolution of the sea breaze near topography in the area of northern Israel, including Haifa. Of course, we cannot go to a scale yet which is smaller than about 3-4 km but we can estimate the average behavior of the sea breeze when it encounters the shoreline and the heights of Haifa.

I have a question for Dr. Lumley: You stated that the second-order closure model is good for regional modelling and you mentioned 20 km as the size of the region. Can you state the expected validity of second-order closure models for sea breeze studies, such as the Mahrer-Pielke model, or our model of the Kinneret area? Is the second order model really useful for an hourly average wind over a complex area with sea breezes on a scale of about 5 km?

LUMLEY

Yes it is. It is hard to justify. The sea breeze is essentially driven by buoyancy. I think that whenever buoyancy is the dominant physical mechanism, you require a model that includes buoyancy in a dynamically correct way. In many respects the buoyant effect on the sea breeze is a mean effect rather than an effect on the turbulent transport itself. And in that sense other models do a reasonable job, so it may not be essential to use a fancy model. There are situations in fluid mechanics, such as the prediction of flows in machinery, or say, the prediction of environmental pollution from power plants, where the primary effect is due to inertia or mean buoyancy. Then simple models do quite adequately, because the turbulence is not being modified.

HUNT

The Mahrer-Pielke model is a hydrostatic model, so that the vertical pressure gradient is dominated by the density gradient. But you cannot get a sea breeze front (a gravity current front), without having quite significant vertical accelerations. This may not be important for the mean flow, but it might be important for dispersion, if you want to get the details of the flow near the front. Equally, as pointed out in my paper, for air flow over hills, there are significant differences between the hydrostatic solution and the solution of the inviscid equations of stratified flow. This may or may not be important, but that has nothing to do with the second order turbulence models.

ALPERT

If you are interested in gradients, the sea breeze front has significant gradients over distances of tens of meters.

ARYA

In the convective boundary layer, the very concept of eddy diffusivity is in question because we have the same source at the ground as near the top of the boundary layer, but the effective eddy diffusivity at a given downstream position is different in the two cases. If the eddy diffusivities are not going to be dependent on the flow but rather on the source characteristics, then how are they going to be specified?

HASSID

There are models in which the eddy diffusivities are proportional to the time of travel close to the source, becoming dependent on the turbulence characteristics of the flow later. This concept is more or less consistent with the development of the Lagrangian time scale, as given in Taylor's original paper.

ARYA

My point was that these recent results indicate that the original concept has to be reexamined, because the problem is not whether the eddy diffusivities are varying as a function of travel time (you can probably prescribe that effect) but whether the eddy diffusivities are going to be dependent on the source characteristics. It is hard to visualize why eddy diffusivity, if it is the correct concept, should depend on the source characteristics.

LAKRETZ

In the burning of a fuel oil product, a lot of H_2O vapor is emitted and this affects plume rise. When you burn coal and collect the ash by electrostatic precipitator, you get a different plume rise. What is the effect on plume rise of these processes?

HANNA

There is an interesting review by Schatzmann and Policastro of this subject in the March 1984 Bulletin of the American Meteorol. Soc., discussing effects of moisture additions. There are a few models available for moist plume rise, most of which are based on earlier cooling tower plume models. Scrubbed plumes usually do not have as high a plume rise as other plumes because their total buoyancy flux is less.

LAKRETZ

On the other hand, if you burn fuel oil and you don't use an electrostatic precipitator, what is the difference?

EGAN

You have to do a computation of the difference of the buoyancy flux that you would get with the scrubbed plume versus the one that would be prescribed without scrubber. I would say that it is unlikely that this would make more than a factor of two difference in the buoyancy flux. Using the standard plume-rise formula, where plume rise is proportional to the buoyancy flux raised to the one-third power, there would probably be not more than 20-30% difference in the total plume rise.

DINAR

My question is to Dr. Hanna. How are you going to implement the concept of model predictability? I would like to hear more general ideas about the strategy you are trying to use to use for this purpose?

HANNA

It is possible to use existing theories to estimate the variance of the stochastic component of the concentration fluctuations. This can be compared with the variance of the differences between the model predictions and observations in any given application. If the variance of the differences is much less than the stochastic variance, then the model is good, and if this variance is much greater than the stochastic variance, then the model is bad. The larger the stochastic variance, the less likely it is that you will decide that you have a good model for a given variance of the difference between model predictions and observations.

SMITH

It seems to me that there is a class of practical problems which we haven't really addressed in any detail during these last few days. For example, there is the question of the accidental release of inflammable gases and the possibility of flames travelling along the plume. There is the question of camouflage, using smoke, to hide tanks or whatever in warfare. Another practical problem involves agricultural pests which may be homing in on hosts, and have to follow certain scents that are emitted into the atmosphere. All these problems really depend upon there being some continuity in concentration within plumes at certain specified threshold levels. I wonder whether the panel thinks that there is any hope in tackling these rather difficult problems?

HUNT

One of the ways in which you can explore these problems is by stochastic modelling. In fact, Dr. Smith has the only paper on this subject in the conference. This is a good opportunity to comment on Dr. Lumley's claim that his method of modelling was the only one likely to work for diffusion in highly convective conditions. I think that many people feel that stochastic modelling is quite good for estimating the mean concentrations in this very non-Gaussian inhomogenous situation, and these models are certainly at the moment proving to be very useful in terms of predicting concentration fluctuations. There is certainly a controversy about the subject of concentration fluctuations, because, rather like the Richardson number at the top of the stable boundary layer, some people believe the ratio of the standard deviation of the concentration fluctuations to the mean goes to zero, some people believe it goes to infinity, and some people believe it goes to one. So the subject is in a very interesting stage of development. But if we can get beyond the fluctuation variance to these practical questions that Dr. Smith is talking about, the study of individual or pairs of particles won't tell us any more than the variance, and will not give us much guidance on whether the contaminant is rather lumpy or whether it is a continuous sheet of material. You should release many more particles in your model and simulate their movement in detail. At the moment, a number of groups are studying an economical way of finding out the instantaneous shape of contaminant surfaces.

HANNA

For some of the problems that Dr. Smith mentioned, the mean concentration is almost irrelevant. You could predict that the probability is less than 50% that a gas plume is not going to blow up, but I wouldn't like to light a match in it. Current models and observation of concentration fluctuations show that you could have pockets of concentrations 2 to 10 times higher than the mean. Chatwin has suggested that you need to predict the probability density function, and we now have models for concentration fluctuations that can predict these density functions with fair success.

CERMAK

This subject is being studied in the laboratory using a source of buoyant particles and measuring the integrated transmissibility of light beams through the plume. These measurements give the fluctuations in concentration in the total integrated path that this beam is traversing.

SUBJECT INDEX

ANNOUNCEMENT

Boundary Layer Structure. Modeling and Application to Air Pollution and Wind Energy.

Editors: H. Kaplan and N. Dinar.

Please note that a hardbound edition of these special issues of Boundary-Layer Meteorology, Vol. 30 Nos. 1-4 (September - December 1984), is available from the publishers.

ISBN 90-277-1877-6.

Prices: Dfl. 190,- / $69.50 / £48.50.

GPSR Compliance

The European Union's (EU) General Product Safety Regulation (GPSR) is a set of rules that requires consumer products to be safe and our obligations to ensure this.

If you have any concerns about our products, you can contact us on ProductSafety@springernature.com

In case Publisher is established outside the EU, the EU authorized representative is:

Springer Nature Customer Service Center GmbH
Europaplatz 3
69115 Heidelberg, Germany

Batch number: 10815974

Printed by Printforce, the Netherlands